HABITAT, POPULATION DYNAMICS, AND METAL LEVELS IN COLONIAL WATERBIRDS

A FOOD CHAIN APPROACH

Marine Science Series

The CRC Marine Science Series is dedicated to providing state-of-the-art coverage of important topics in marine biology, marine chemistry, marine geology, and physical oceanography. The series includes volumes that focus on the synthesis of recent advances in marine science.

CRC MARINE SCIENCE SERIES

SERIES EDITOR

Michael J. Kennish, Ph.D.

PUBLISHED TITLES

HABITAT, POPULATION DYNAMICS, AND METAL LEVELS IN COLONIAL WATERBIRDS

A FOOD CHAIN APPROACH

JOANNA BURGER
MICHAEL GOCHFELD

CRC Press
Taylor & Francis Group
Boca Raton London New York

CRC Press is an imprint of the
Taylor & Francis Group, an **informa** business

CRC Press
Taylor & Francis Group
6000 Broken Sound Parkway NW, Suite 300
Boca Raton, FL 33487-2742

First issued in paperback 2020

© 2016 by Taylor & Francis Group, LLC
CRC Press is an imprint of Taylor & Francis Group, an Informa business

No claim to original U.S. Government works

ISBN 13: 978-0-367-57476-5 (pbk)
ISBN 13: 978-1-4822-5112-8 (hbk)

Library of Congress Cataloging-in-Publication Data

Names: Burger, Joanna. | Gochfeld, Michael.
Title: Habitat, population dynamics, and metal levels in colonial waterbirds
: a food chain approach / Joanna Burger and Michael Gochfeld.
Description: Boca Raton : Taylor & Francis, 2016. | Series: CRC marine
science series ; 36 | Includes bibliographical references and index.
Identifiers: LCCN 2015042672 | ISBN 9781482251128 (alk. paper)
Subjects: LCSH: Water birds--Ecology--Northeastern States. | Colonial
birds--Ecology--Northeastern States. | Water birds--Effect of metals
on--Northeastern States. | Colonial birds--Effect of metals
on--Northeastern States. | Estuarine ecology--Northeastern States.
Classification: LCC QL683.N73 B87 2016 | DDC 598.40974--dc23
LC record available at https://lccn.loc.gov/2015042672

Visit the Taylor & Francis Web site at
http://www.taylorandfrancis.com

and the CRC Press Web site at
http://www.crcpress.com

Dedication

We dedicate this book to Fred Lesser, who was completely devoted to helping us study the birds of Barnegat Bay for 40 years, to all our students who provide hope for the future study and conservation of colonial waterbirds, and to the international team of shorebird biologists who migrate to Delaware Bay each year to help unravel the biology of shorebirds.

Fred Lesser in the field with an Egret Chick.

And to the students who are carrying on the research and conservation work with colonial birds. From left to right in top row: Brian Palestis, Joanna Burger, Steve Garber. Front row: Taryn Pittfield, Nellie Tsipoura, Susan Elbin, Carl Safina, Sheila Shukla, Christian Jeitner.

The international team of shorebird biologists.

Contents

Part I
Introduction to Barnegat Bay and Northeast Estuaries

Part II
Habitat and Populations Dynamics

Part IV
Implications, Conclusions, and the Future

Foreword

Although I have worked on every continent and in every ocean, many of my most precious memories are in the coastal bays of my youth, including Barnegat Bay. It was on Barnegat's shores and marshes that I watched some of the East Coast's last Ospreys survive through the DDT era and raise their young (they are now abundant), worked as a college student to reintroduce some of the first captive-raised Peregrine Falcons (they had disappeared, but are again common), heard my first Whip-poor-wills (they are vanishing), weathered awe-inspiring, sometimes terrifying summer lightning storms, caught crabs, gathered clams, fished, and encountered some of the most memorable and momentous people in my life. I soon found myself in a boat with Fred Lesser, Joanna Burger, and Michael Gochfeld, counting tern and skimmer nests on salt marsh islands under the bay's wide-open sky. For those coming-of-age experiences, Barnegat Bay was a vast, magical place. I feel lucky to have had them and been there. I am thankful to all those who afforded me the opportunities to contribute what little skill I had. Having been repaid with such precious memories, it is clear that I got by far the better deal. But I hope I have since, in the wider world, lived up to the investment that Barnegat Bay and some special people made in a hopeful and impressionable young person who so fervently, so simply, wanted to be out in the beauty, somehow contributing something positive to a glorious place and its wondrous creatures. The fact that the only painting I own is framed and hanging in my home is an amateurish watercolor that I painted on Barnegat Bay's Sedge Island in 1976 of an Osprey nest with two chicks—the first I had ever seen—is the best testament to how much I value my time there.

The world has twice as many people now as when I was born. But it is not twice as good a place as it was. This implies something of the scale of the challenges, and their urgency. The once vast bays that you could get lost in are far more crowded. The once clean fish and crabs are far more contaminated. The bird colonies that seemed wild and eternal now suffer more frequent inundation in a world of more frequent storm surges atop ever-rising sea levels.

Reading this book will make you an expert of sorts on Barnegat Bay and the other Northeast Bays. That might seem an ambitious goal for the authors as well as the reader. But it really is not the goal. It is merely the starting point. The bays need advocates and defenders. And advocates and defenders need experts. That is where you will come in.

The authors have spent their careers putting this information together. This work, inspired by the beauty of mornings and urged onward by the cries of terns, compiled boat ride by boat ride, step by muddy step, mosquito bite by mosquito bite, through long hours under a hot sun and longer under desk lamps, is the work of their lifetime. The authors have bequeathed to the next generation not the culmination of all things knowable about Barnegat Bay—though this book may make it seem that way. Rather, their gift is to make it that much easier to get started on a new journey, our own.

But be warned. As this pathfinding book indicates, the future is not what it used to be. But therein is the next generation's opportunity. Turn the page and take your first muddy step.

Carl Safina
Director
The Safina Center at Stony Brook University
Stony Brook, New York
www.safinacenter.org

Preface

Huddled deep in my down jacket, grasping my arms around my waist to keep in the heat, I faced into the wind. Beside me stood Fred Lesser, stoically steering the boat through the chilly fog of an early summer morning. I could hardly see the front of the 18 ft whaler, yet he was focused on a distant island, its position marked on the map in his head. "We need an early start," he had said. What in the world was I doing skimming across Barnegat Bay, in the Ocean County Mosquito Commission boat? Fred, an ecologist, was the director of the Commission. He was unusual for his time and place. He had a holistic view of mosquito control, and the importance of maintaining the marsh ecosystem with as little management as possible, while still controlling mosquitos. Coastal ecosystems and mosquito control were important to the New Jersey State economy, even then, in the early 1970s. People living in beach communities, walking the beaches of New Jersey, fishing in the surf, or playing with their children in the sand did not want to be harassed by mosquitos. Mosquito control was big business, but mosquitos were part of a complex estuarine ecosystem that we were studying.

In the early dawn, Fred had picked me up from my hunting hut on Clam Island, where I was studying the social interactions among nesting gulls. We were bound for the Lavallette islands in the north end of the bay. We would census the birds there, check their breeding status, then start south stopping systematically at all the suitable islands to count the number of nesting gulls, terns, and skimmers on each salt marsh island. I had a coastal chart in my pack, but Fred was working off his mental map, honed since he was a youngster. He grew up on the bay, and his love for it was infectious. Channels are for boaters who want to avoid running aground, but we had to reach islands, and Fred knew the way into each, at almost any tide.

The fog burned off before we reached the northern end of the bay. The blue sky peeked through the disappearing clouds, and the sun grew warmer. He cut the engine, and as we drifted noiselessly up to the first island, a few Common Terns flew overhead, hurrying away to plunge-dive for fish and bring back their courtship feeding offering to a waiting mate. When we bumped into the island, a cloud of terns arose from the *Spartina* mats, circling overhead and calling loudly. I sat and watched, and was hooked on the swirling mass of delicate terns, some with a faint blush of pink on their breasts. More practically, Fred hooked the anchor into the marsh.

We walked through the nesting colony that day, counting adults, counting nests, counting the number of eggs in each nest, and recording any eggs washed out of nests or that were cut open by predators. "No tern chicks yet," we noted, "another week at least." We counted the numbers of other species nesting on each of the islands—mainly, Herring Gulls on the islands in the northern part of the bay—they nested earlier and were already feeding their downy chicks, still cute at this stage. In the middle of the bay, down by Barnegat Inlet, some of the islands had small Cherry and Poison Ivy bushes that held the nests of Great and Snowy Egrets, Black-crowned Night-Herons, and other species. This morning we only counted from afar, and tried not to flush the incubating birds. Farther south, the Common Tern colonies were next to Laughing Gull colonies; and still farther south, there were Black Skimmers nesting on sandy patches or on the wrack, the dark brown mats of dead vegetation washed up on the salt marsh islands by winter storms.

It was enchanting to see the subtle differences in habitats on these salt marsh islands, and all the different combinations of colonial nesting birds. I not only got to see the mosaic of different vegetation types, but to map their distribution as part of our research on habitat suitability and availability. Did the terns know something about habitat quality not yet evident to our eyes? Hidden in the *Spartina* grass, there were nests of Salt Marsh and Seaside Sparrows, Red-winged Blackbirds, Marsh Wrens, and the occasional Clapper Rail or Willet. Pairs of Oystercatchers would circle around us calling loudly, a distinctive piping note, luring us away from their nests or chicks. It was amazing to see this avian diversity residing on small salt marsh islands. By the time we passed under the Manahawkin Bridge, there was a steady stream of beach-bound traffic. In midsummer, the barrier beaches were teeming with thousands of tourists, the roads often crammed with cars unable to move faster than a Diamond-back Terrapin crossing the sand to reach its nesting habitat. I fell in love with the ambiance and have been enthralled for more than 40 years by the birds of Barnegat Bay, the subtle hues of the vegetation that change through the season, and the islands themselves that have changed over the years. I have always felt incredibly privileged to be out on the bay, and that this research was my job as a university professor. "They pay you to do that," my father had once said when I explained my typical week.

Like the gulls, terns, and herons that form dense colonies to breed, people who study these species flock together at professional meetings, seeking out others to compare observations or collaborate. How do the habitats differ? How do the colonies differ? How do the predators differ? and How do people influence the success or failure of these colonies? The nearest Common Tern colonies were on Long Island, which led me to Mike Gochfeld, who had also been studying the terns. Mike's interest in toxicology rubbed off on me, and my interest in behavior rubbed off on him, and we began collaborating.

We visited each other's colonies, and over the years, focused our attention on Barnegat Bay. In part, this shift was due to the abandonment of many of the Long Island colonies because of encroaching predators and people, and in part the shift reflected the protected nature of the Barnegat Bay salt marsh islands. Being surrounded by shallow water, still inhabited by healthy numbers of Salt Marsh Mosquitos and Greenhead flies, and devoid of any sandy beaches, they were inhospitable to boaters. The islands were left alone by people if not by the environment. After a preliminary bay-wide study mapping all of the islands in the mid-1970s, we began to notice that islands had begun to disappear—trimmed around the edges by erosion, washed over by severe winter tides, and broken apart by ice and wind. We soon realized that we were witness to a drama unfolding on a global scale.

As the years went by, Mike, Fred, and I continued to visit the colonies day after day, week after week, and year after year. Every spring, when I heard the first Killdeer call over our Somerset home, I could hear in my mind the calls of the gulls and terns, and we were soon bound for Barnegat Bay and another field season—40 plus at last count. Our studies widened to include the population dynamics and heavy metal levels in the gulls and terns, herons and egrets, and even to the fish they consumed. Although the bird colonies are on islands, they are not isolated from other parts of the ecosystem, and our interests include fish, crabs, Horseshoe Crabs, and Diamond-back Terrapin, as well as the interactions between people and the birds. Although Barnegat Bay was our focus, we also worked in the New York–New Jersey Harbor Estuary and in Delaware Bay.

This book is a result of our more than 40 years of study on the behavior, populations, and heavy metals in the colonial birds nesting in Barnegat Bay and the nearby estuaries and bays in the Northeastern United States. Some data sets are more complete than others; some questions required only a few years of data to answer, and some required collaboration with other colleagues. Many remain unanswered for the new generation of researchers who have studied with us. Just as the flavor of nesting colonies of birds requires understanding the habitat, avian coinhabitants, predators, and competitors, the evolving research approaches and new technologies require many collaborations with a wide range of scientists and disciplines. And it requires integration of our findings with those of others in disciplines including toxicology, geology, and climate change science. We are eternally grateful to the birds and their ecosystems, and to the many people who have been our collaborators and friends over the years.

Joanna Burger

Preface

Even in medical school, I tried never to miss a fall weekend to visit Jones Beach. It is a famous bird watching spot, particularly on fall migration. Exhausted migrants literally fall out of the sky into the short beach vegetation where they are easily seen (or eaten), exhausted after a long night flight. If they overshoot and crash at sea, gulls are only too willing to lift them from their watery doom and swallow them. Migrating Merlins will often pause to make a pass at a flock of sparrows feeding on an open lawn. There, in the loop of the beach access road, I found a number of dead birds—fledgling Common Terns that had not made it. They were banded—with metal rings on one skeletal leg. I dutifully turned in the band numbers, dozens of them. There began my field studies of the colonial waterbirds of the Jones Beach strip, part of the New York–New Jersey Estuary.

The following spring, I found a thousand pairs of Common Terns and a hundred pairs of Black Skimmers, and dozens of Roseate Terns, nesting just inside the bend of the highway. Nearby, Least Terns and Piping Plover nested on the beaches, just waiting to be trapped, banded, and studied. Other banders up and down the coast were doing the same thing, and we would soon learn about migration, death on the wintering ground, and declining populations.

My studies were not very systematic, but when I met Joanna at a party after a Christmas bird count, she was showing around a scrapbook illustrating how she was studying Franklin's Gulls in a prairie marsh in northwestern Minnesota. Our colonies offered stark contrasts—not just beach versus marsh or beautiful tern versus beautiful gull. I could drive right up to my colony; park close to the nesting birds, and walk to a hot dog stand when the need arose. Joanna reached her study area by boat through Moose-infested, seemingly endless and trackless marshes, then wading chest deep in icy water to mark her nests, and sleeping night after night in a wooden blind to make her observations—"collecting data," she called it. I made a mental note to marry her some time.

Several years later, When Joanna was studying the three-dimensional pattern of nesting in heron and egret colonies, she asked me to show her some of the recently formed colonies on Long Island. To reach Seganus Thatch, a small island near Captree, with a bustling heronry, we had to wade. By wade, I mean chest deep over unfamiliar substrate, gear held high overhead, for a hundred yards or so, and then clamber up onto a rather shaky muddy island covered with Bayberry and Poison Ivy. An explosion of Great and Snowy Egrets, Black-crowned Night-Herons, Little Blues, and Glossy Ibis flushed from their nests, offering a spectacular vision as they circled away from the colony. We had chosen a cloudy morning so their eggs would not be overheated in their absence, as Joanna planned to measure nest height and nearest neighbor distance, and then take fish-eye photos upward to assess neighbor visibility. We allotted ourselves an hour. Initially thwarted by the dense tangle of vegetation, Joanna hurled her body physically against the bushes and clambered over the Poison Ivy vines to reach each nest. I followed suit. "Watch the Poison Ivy," I cautioned in vain since it was everywhere. "You don't get Poison Ivy?" I questioned. "Very badly," she answered, as she crashed forward to reach the first nest, a Night-Heron with four pale blue eggs. This was much too interesting.

We both felt Seganus was a great success. Lots of accessible nests and good data. Later in the car, I asked Joanna, almost casually, how she would go about studying the Common Tern nesting ecology. I had been laboriously measuring vegetation around each nest and at matched points one and two meters away. Joanna started, unhesitatingly, to talk about random points and sample size, vegetation characteristics, substrate quality and different habitats, and offered to use her fish eye to see if the Common Terns care as much about the proximity of neighbors as did Franklins Gulls and egrets. I would later learn that this was her stock in trade, being able to grasp a whole problem at once, recognize its context and visualize it as individual components amenable to collecting, analyzing, and publishing data. This book reflects that approach, integrating disparate disciplines into a meaningful story. It was lunchtime. I bought her a hot dog at the Captree Fishing Station. It was courtship feeding.

Michael Gochfeld

Note: Unless otherwise noted, all photographs are by the authors.

Acknowledgments

We especially thank Fred Lesser for his invaluable help over the years in monitoring the bird colonies in Barnegat Bay, and in providing thousands of hours of friendship and invaluable information about the bay. Christian Jeitner, Taryn Pittfield, and Brian Palestis went above and beyond in helping us finish this book (Figure A.1). We are grateful to a number of Joanna's graduate students who have gone on to have their own distinguished careers but continue to work with us, and although we mention some in the relevant chapters, we thank them now, including Michael Allen, Bill Boarman, John Brzorad, Chris Davis, Amanda Dey, Susan Elbin, Jeremy Feinberg, Tom Fikslin, Steve Garber, Amy Greene, Caldwall Hahn, Chris Jeitner, Larry Niles, Brian Palestis, Kathy Parsons, Taryn Pittfield, Carl Safina, Jorge Saliva, Dave Shealer, Nellie Tsipoura, Laura Wagner, and Wade Wander. Several people have taken part in surveys of bird populations, collection of eggs and feathers, and metal analysis, and we thank them now, including Brian Palestis, Chris Jeitner, Taryn Pittfield, Mark Donio, Sheila Shukla, Tara Shukla, Tom Benson, and Jim Jones. We thank Dave Jenkins, Larry Niles, Mandy Dey, Chris Davis, Kathy Clark, and many others at the Endangered and Nongame Program (NJDEP) for data and support over the years. In addition to the above, we have had many valuable discussions with others about population dynamics, colonial birds, habitats, heavy metals, or Barnegat Bay, and they have shaped our thinking, including Ken Able, Jim Applegate, Peter Becker, Colin Beer, Paul and Francine Buckley, Keith Cooper, John Coulson, Liz Craig, Tom and Chris Custer, Emile DeVito, Mike Erwin, Peter Frederick, Michael Fry, Michael Gallo, Bernard Goldstein, Steven Handel, Helen Hays, Joe Jehl, Mike Kennish, David Kosson, Jim Kushlan, Charlie and Mary Leck, James Shissias, Kenneth Strait, Paul Lioy, Brooke Maslo, Clive Minton, Dan and Aileen Morse, Bert and Patti Murray, Ian Nisbet, David Peakall, Todd Plover, Chuck Powers, Nick Ralston, Barnett Rattner, Robert Risebrough, B. A. Schreiber, Ellen Silbergeld, Humphrey Sitters, Marilyn Spalding, Jeff Spendelow, Alan Stern, Ted Stiles, Niko Tinbergen, HB Tordoff, Dick Veitch, Dwain Warner, Judy and Pete Weis, Chip Weseloh, Chris and Paula Williams, Bob Zappalorti, Ed Zillioux, and many others in the Waterbird Society. The list of friends and colleagues who have helped over the years is endless, and we hope you know who you are.

We especially want to note and thank the next generation of researchers who have made such wonderful contributions to the literature and conservation of these species, and will take on and continue long-term studies that we believe are so important for science, adaptive management, conservation, and the long-term stewardship of the resources we love and care for. Among numerous and relatively new investigators are Carolyn Mostello for Massachusetts bays and estuaries; Susan Elbin, Nellie Tsipoura, and Liz Craig for the New York–New Jersey Harbor Estuary; Brian Palestis for Barnegat Bay; Larry Niles, Mandy Dey, Nellie Tsipoura, and David Mizrahi for Delaware Bay; and Brian Watts for the Chesapeake. We are grateful to our editors, John Sulzycki, Jill Jurgensen, and Linda Leggio at CRC Press/Taylor & Francis, who provided invaluable aid and advice throughout this project.

Finally, we thank our parents (Melvin and Janette Burger, Anne and Alex Gochfeld), our brothers and sisters and their spouses (Melvin Burger Jr., Christina and Fritz Wiser, John and Linda Burger, Barbara Kamm, Roy and Anne Burger; Bob and Elizabeth Gochfeld) for providing support and encouragement over the years, our children and spouses (Deborah Gochfeld, Marc Slattery, Julia Schafhauser, and David Gochfeld) for their wonderful contributions to our lives and work, and the next generation (Edward Burger, Kathy, Greg and Caroline Drapeau, Michael, David, and Daniel and Charlie Wiser, Jacob and Lisa Burger, Andy Burger, Ben Kamm, Eric, Beth, Emily, Allison, Alexis and Amanda Burger; Jennifer Wolfson and Douglas Gochfeld) as our best hope for a sustainable world that enriches their lives, engages their imagination, and protects the natural world we so love.

We especially thank Fred Lesser (top left), Christian Jeitner (top right), Taryn Pittfield (bottom left), and Brian Palestis (bottom right), for making this book happen.

Funding for this project over the years has been provided by the New Jersey Department of Environmental Protection (Endangered and Nongame Species Program, Science and Research), U.S. Fish & Wildlife Foundation, the U.S. Department of the Interior (USFWS), the U.S. Environmental Protection Agency, the National Science Foundation, National Institute of Mental Health (NIMH) National Institute of Environmental Health Sciences (NIEHS), the Trust for Public Lands, the New Jersey Audubon Society, the Consortium for Risk Evaluation with Stakeholder Participation (CRESP, DE-FC01-06EW07053), Rutgers University, the Environmental and Occupational Health Sciences Institute, and the Tiko Fund.

Authors

Joanna Burger, PhD, is a distinguished Professor of biology in the Departments of Ecology Evolution and Natural Resources, and Cell Biology and Neuroscience, Rutgers University, Piscataway (New Jersey). She is a member of the Environmental and Occupational Health Sciences Institute and the Rutgers School of Public Health. Her main scientific interests include the social behavior of vertebrates, ecological risk evaluations, ecotoxicology, and the intersections between ecological and human health. Dr. Burger has conducted research on all classes of vertebrates, and many invertebrates on all continents. Her specialty is combining pure science with human dimension to find solutions so that both ecological and human communities can coexist in an increasingly complex and overpopulated world. Her research has included the interactions between Horseshoe Crabs and shorebirds, birds and their competitors and predators, development and Pine Snake ecology, and contaminants in fish, fish consumption and risk in humans and other animals. Dr. Burger has worked with the Consortium for Risk Evaluation with Stakeholder Participation (CRESP), which was funded by the Department of Energy (DOE) for more than 20 years, examining ecological risks at DOE sites and integrating human and ecological risks. She has served on many committees for the National Research Council, Environmental Protection Agency, U.S. Fish & Wildlife Service, Scientific Committee on Problems of the Environment (SCOPE), and several other state, federal, and nongovernmental organizations. She has published more than 700 refereed papers and more than 20 books. She is a Fellow of the American Association for the Advancement of Science (AAAS), the American Ornithologists Union, the International Union of Pure and Applied Chemistry, and the International Ornithological Society; in addition, she has received the Brewster Medal from the American Ornithologists' Union (AOU), the Lifetime Achievement Award of the Society for Risk Analysis, and an honorary PhD from the University of Alaska.

Michael Gochfeld, MD, PhD, is Professor Emeritus at Rutgers Robert Wood Johnson Medical School (New Brunswick, New Jersey). He is an occupational physician and environmental toxicologist at the Environmental and Occupational Health Sciences Institute of Rutgers University. His main research interest has encompassed ecotoxicologic studies, primarily of birds. His biomedical interest focuses on heavy metal exposure and risk assessment for humans from consumption of fish, balancing the benefits against the toxicity of methylmercury. He also studies exposure to arsenic, cadmium, and lead. Dr. Gochfeld chaired the international working group on cadmium for the Scientific Committee on Problems of the Environment. He has developed protocols for screening communities for adverse effects of exposure to toxic chemicals at Superfund Sites. He also teaches evidence-based medicine, including epidemiology and biostatistics, to medical students, and lectures in courses in toxicology and global health. Dr. Gochfeld has coauthored or coedited eight books on protecting hazardous waste workers, on avian reproductive ecology, and on New Jersey's biodiversity, as well as a textbook, *Environmental Medicine*. He was elected to the Collegium Ramazzini, an international college devoted to the prevention of disease from hazardous exposures in the home, community, or workplace environment.

*****

Together, the authors have been studying colonial nesting birds and other social behavior of vertebrates for 40 years in several countries and all of the continents. They have worked together on the ecological and human health risk issues at various DOE sites and have coauthored numerous papers and books. They live with their 64-year-old parrot, Tiko, in Somerset, New Jersey.

Introduction to Barnegat Bay and Northeast Estuaries

CHAPTER 1

Introduction

Seacoasts are the transition zone between land and ocean, and between freshwater coming from streams and rivers and saltwater coming from the sea. Estuaries receive nutrients from land and sea, and are revitalized with each new tide. Estuaries and coastal environments are some of the most productive areas in the world in terms of biomass produced each year. Tropical rain forests may have the highest density of trees, shrubs and other vegetation, and a high standing crop (biomass), but the new production each year does not compare with that of salt marshes (Odum 1959). Estuaries serve as nurseries for fish and shellfish, receiving schools of breeding adults, and later sending juveniles back to the oceans, and yet estuaries are among the places people congregate for living, commerce, and recreation. Wilson and Fischetti (2010) noted that: (1) although the country's saltwater edges account for only 8% of the nation's counties, they contain 29% of its population; (2) coastal edges contain 5 of the 10 most populous U.S. counties; and (3) growth in coastal counties outstripping that of noncoastal counties—the coastal population grew from 47 million in 1960 to 87 million in 2008. However, these spaces are filling up!

People work, live, and vacation along coasts, which provide us with a wide range of goods, services, and ecocultural values (Burger et al. 2008a, 2012a; Weis and Butler 2009; Duke et al. 2013; Lockwood and Maslo 2014). Throughout the world, coastal areas are critical environments because more than one-half of the people live within 100 mi (160 km) of coastlines, and there is an increasing trend for people to move there (Crosset et al. 2013). Beaches, bays, and estuaries are iconic for many countries and states, attracting tourists and residents alike, and they account for a significant portion of regional economies. People are interested in their own health and well-being, as well as that of their homes, communities, and places where they work and play. This makes coastal environments important to all communities and countries, including landlocked countries and states that depend on ports in adjacent countries.

Katrina, *Sandy*, and other storms have taught us the hard way about the importance of coastal dunes and marshes as buffers against severe storms and the surges. It is difficult to monetize the value of healthy beaches, dunes, and salt marshes, and the spiritual, social, and cultural value of beaches can transcend economic values (UNEP 2006). To some extent, it is a "commons issue" (Hardin 1968; Burger 2001a,b). We all own the sea, tides, and beaches, making it difficult to assign costs, benefits, and responsibilities.

The state of New Jersey, although small in size, has a relatively long coastline, and the "Jersey Shore" is the state's main recognizable geographic feature. For many New Jerseyans (Philadelphians and New Yorkers, too), the Jersey Shore was part of their growing up. They treasure those memories and want their children to experience the same pristine shore. Fifty years ago, urban planner Sanford Farness warned that this was unrealistic for the region (Farness 1966). Yet, this sentiment is no doubt true for people living along seacoasts everywhere. Even people who do not visit the shore appreciate undisturbed beaches, sand dunes, and back bays—it has existence value. Just knowing the beach is there is comforting—aesthetic values are important.

Direct consumptive uses of coasts include fishing and seafood harvesting, salt hay production, grazing, salt production, wood extraction, fish/shellfish farming, and sand mining (Figure 1.1) (Burger 2002a; Burger and Gochfeld 2011a). Nonconsumptive activities include swimming, surfing, walking, jogging, sunbathing, sailing, boating, bird-watching, photography, or just relaxing along the shore (Piehler 2000; Burger 2002b, 2003a,b; EPA 2007a–d; Burger et al. 2010a; Frank 2010). Conflicts occur with new recreational sports, including parasailing and jet skiing (Burger 1998a,b;

(a)

(b)

Figure 1.1   Bays and estuaries are vital parts of ports. (a) Here, a container ship waits to unload near Liberty State Park, and (b) a fisherman and sunbathers use an exposed sandbar that would otherwise be used by foraging egrets and gulls.

Burger et al. 2010a). An array of businesses and infrastructure support these activities, including marinas, bait shops, boutiques, bars, restaurants, supermarkets, fast food shops, and motels.

Bays and estuaries are used as ports for importing and exporting goods and services, including oil. Major seaports, such as Boston, the New York–New Jersey Harbor, Philadelphia, and Baltimore, require an enormous infrastructure in loading docks, railway and truck routes, holding docks, container storage areas, and warehouses, as well as a large workforce to sustain the port (Figure 1.1). These, in turn, produce a vast array of wastes and pollutants, including diesel fumes, oil spills, polyaromatic hydrocarbons (PAHs), metals, and noise.

The positive benefits that people derive from coastal areas, however, are in themselves stressors on these systems. People walking close to or entering waterbird colonies cause disturbances to nesting birds (Parsons and Burger 1982; Safina and Burger 1983; Burger and Gochfeld 1983a). Fishing can lead to overexploitation of fish stocks, with declines in fish takes and the mean size of the catch, and eventual collapse of the fishery. Worldwide, many marine food webs are being overfished, with resultant declines in the populations of some fish (Pauley et al. 1998, 2002; Safina 1998), and population changes of other marine species (Niles et al. 2014). Fish required by the birds may be depleted by fisheries, or released when their fish predators are removed. Boating can result in disturbance to nesting birds and turtles, and injuries to sea grass beds, coral reefs, and fish (Burger 1998a, 2002b, 2003c; Lester et al. 2013). Pollution of coastal waters with urban runoff, fuel oil, and ship ballast water is widespread. Shrimp production destroys mangrove swamps, which provide nesting and feeding sites for herons and other species. Salt production removes mudflats as habitat for shellfish and fish, and as foraging areas for birds. Coastal ecosystems are vital breeding and nonbreeding areas for birds.

Roosting or foraging birds can be disturbed by people driving or walking along beaches (Goss-Custard et al. 2006; Stillman et al. 2007; Burger and Niles 2013a,b, 2014). Not only do organisms suffer from direct loss of habitat, but loss of prey items as well. Commercial and recreational fishing can reduce prey fish for birds, both locally and regionally (Atkinson et al. 2003; Stillman 2008). Every human activity has both direct and cascading effects on the natural coastal ecosystem (Maslo and Lockwood 2014). For example, tourists walking along a beach may step on bird eggs or collect eggs for "fun" or food, but they also lead dogs and other predators to nests (Burger 1991a, 1994a; Burger et al. 2007a). Recreation can negatively impact ground nesting birds because of increased levels of disturbance and introduction of predators (Burger 1991b). Growing human populations and migration to coasts increases these stresses.

Stressors play a key role in the health of any ecosystem and population, and stressors can be physical, chemical, or biological. Physical stressors include weather in the short term, and climate change and sea level rise in the long term. Chemical stressors include metals, pesticides, and petroleum. Biological stressors include invasive species. Stressors affect coastal and estuarine environments, thereby requiring ecosystems to have mechanisms for recovery and resiliency (Pratt and Cairns 1996; Burger 2015b). Recovery is the process of an ecosystem returning to its previous state, and resiliency is the intrinsic ability of ecosystems to deal with stressors without undue disruption, and to recover to a functional (but not necessarily original) state. Recovery requires various periods of stressor-free time to allow populations of plants, invertebrates, fish, birds, and mammals to recover and establish functional ecosystems.

All three types of stressors (physical, chemical, biological) can be either natural or anthropogenic, or a combination of the two. For example, human activities, particularly emissions of carbon dioxide, contribute to a warming climate trend, resulting in melting glaciers and ice caps, leading to sea level rise (IPCC 2007, 2014). With increasing warmth, seawater undergoes expansion, which also contributes to sea level rise. The type, severity, and effects all need to be assessed and monitored as a precursor to management, conservation, stewardship, and the development of public policy (Burger et al. 2013a).

A familiar example of pollution of marine, coastal, and estuarine waters is oil spills. Massive oil spills often receive considerable media attention, incurring large cleanup costs, causing massive injuries to wildlife, and causing disruptions to human and ecological systems (Burger 1994b; Gundlach

2006; Montevecchi et al. 2012; Bodkin et al. 2014). Several oil spills from ships and well blowouts have shed light on this problem, including the *Exxon Valdez* (Bodkin et al. 2014; Ballachey et al. 2015) and the *Deepwater Horizon* (McNutt et al. 2012; Michel et al. 2013; Fulford et al. 2015). Just navigating ports is difficult for large oil tankers, and all major U.S. ports now have specially trained and experienced harbor captains who pilot oil tankers to safe locations. More widespread, however, is chronic oil pollution from natural seeps or recurrent pollution from ships and oil wells. The cumulative impact of chronic releases may have a greater effect on ecosystems and populations because the species and ecosystems never have time to recover fully before the next small oil spill (Burger 1997a).

It is the patterns of tidal flow, water circulation, sedimentation, geomorphology, and energetics that make estuaries some of the most productive areas in the world (Kennish 2002). Estuaries have brackish water, with a salinity that varies as a function of the inputs from the land, sea, and sky (rainfall/ snowfall). It is both the complexity of the estuary and the apparent simplicity of beaches and seashores that attract people—they seem simple because they essentially encompass the ocean, waves, sand, beaches, and dunes in an endless strip of horizontal habitat. Coastlines are complex because they are the interface between the sea and the land, serve as nurseries for many fish, shellfish, and other invertebrates, and buffer the land from daily wave action and storm surges. Habitats typical of coasts are shallow open water, mudflats, and salt marshes in the Northeastern Atlantic coast, as well as river deltas and mangrove swamps farther south, and rocky shores and intertidal pools farther north (Davis and Fitzgerald 2004). Our understanding of the dynamics within and the interplay among these habitats grows continually. In the salt marsh, for example, ecologists continue to discover complexities and unique features in their quest for unifying principles (Weinstein and Kreeger 2002).

Barrier islands and high dunes stabilized with native vegetation are particularly important in preventing overwash and dampening the effects of storm surges (Nordstrom 2008; Pries et al. 2008). This is particularly true given global warming, sea level rise, and increases in the intensity and severity of coastal storms and hurricanes, such as Hurricanes *Katrina* and *Sandy* (IPCC 2007, 2014; Kharin et al. 2007; Russo and Sterl 2012). This will be a recurrent theme in the book. Storms and hurricanes are predicted to increase along the coasts (Lane et al. 2013; NPCC2 2013), where more than half of the U.S. population lives (NOAA 2012a; Crosset et al. 2013). Sea level rise in the Northeast is 3–4 times greater than the global average (Sallenger et al. 2012), and could be as much as 1 m over the next 50 years (NPCC2 2013). Increases in the frequency and severity of storms and hurricanes will increase damages to coastal communities, and further increase their vulnerability and exposure to coastal flooding (Shear et al. 2011; Genovese and Przyluski 2013).

For birds, the ecological consequences of sea level rise and increases in storm severity and frequency are also high. Severe storms and sea level rise have the potential to render previously available nesting habitats no longer suitable for nesting birds and turtles (Burger 2015b). In the long run, "normal" high tides will flood low-lying islands, causing erosion and killing less salt-tolerant vegetation, and eventually covering them. In the short run, flood tides coupled with heavy rainstorms during the nesting season can directly wash out eggs or kill chicks, whereas frequent floods can result in vegetation changes (e.g., bushes used for nesting by herons and egrets die from salt exposure). Sea level rise also reduces the available nesting space along barrier beaches, as well as foraging space for coastal migrants, such as shorebirds (Galbraith et al. 2002, 2014; Maclean et al. 2008). The connections between sea level rise, global temperature change, and the ecology and behavior of birds must be explored to protect human and ecological health and well-being (Caro and Eadie 2005; Burger et al. 2013b,c). Ornithologists, ecologists, behaviorists, and ecotoxicologists need to become active in conservation activities and public policy (Stillman and Goss-Custard 2010; Caro and Sherman 2013). Global environmental stewardship will be difficult and controversial, but effecting environmental change is not a spectator sport.

Birds are an important component of coastal habitats, nesting singly or in groups called colonies of a few to thousands of pairs, and they are of interest to the public because they are large, diurnal, and iconic of beaches and bays. Bird colonies are usually located in places that are inaccessible to predators, such as islands or in trees (Figure 1.2). These habitats continue to face threats from development (see Chapters 2 and 4). Habitat loss, contaminants, introduced predators, competitors, and invasive species, as well as direct and indirect human activities, global warming, and sea level rise impose stresses on breeding birds. Species such as gulls extending their range can become nest site competitors and predators (Burger 1978a, 1979a). Habitat loss is the biggest threat birds face along coasts, and habitat use and selection are discussed further in Chapter 4. In the 40+ years of our studies, we have seen former nesting islands disappear completely because of rising sea level, coupled with subsidence.

Contaminants, even at low levels, can pose a threat to the behavior and reproductive success of waterbirds, particularly top-level predators. Some classic examples of severe effects of metal and

(a)

(b)

Figure 1.2 (a) An egret colony and Night-Heron colony in Barnegat Bay and (b) a cormorant colony in Delaware Bay, New Jersey.

organic contaminants include: (1) lead poisoning of California Condors (Finkelstein et al. 2012, 2014), albatrosses (Sileo and Fefer 1987; Burger and Gochfeld 2000c), and waterfowl (Bellrose 1959; Franson and Pain 2011); (2) selenium effects on birds at Kesterson Reservoir in California (Ohlendorf et al. 1989, 1990; Ohlendorf 2011); (3) mercury poisoning of seed-eating birds and their predators in Europe; (4) mercury effects in fish-eating birds and their prey (Frederick et al. 1999, 2002, 2004; Frederick and Jayasena 2010); (5) dichlorodiphenyltrichloroethane (DDT) causing serious declines of fish-eating pelicans and raptors in the 1960s (Anderson et al. 1969; Jehl 1973; Anderson and Hickey 1970; Shields 2002); and (6) Great Lakes embryo mortality, edema, and deformity syndrome in several species attributed to polychlorinated biphenyls (PCBs) and related compounds (Gilbertson et al. 1991).

With declines in environmental levels of many contaminants because of laws, regulations, and social change (Burger 2013a; Burger et al. 2015a), some of these effects have been reduced. Tracking levels of old and new contaminants in environmental media and biota is an important environmental investment because such data provide early warning of any problems before population levels are adversely affected. Marine and estuarine birds exhibit a range of trophic levels, foraging methods, and foraging habitats, and thus are excellent bioindicators of ecological health and well-being (Fox et al. 1991; Burger 1993; Custer 2000; Erwin and Custer 2000; Goutner et al. 2001). Contaminant levels in birds are the subject of Part III of this book.

Although birds are only one component of marine and coastal ecosystems, they are diurnal, conspicuous, numerous, and easy to see, making them an iconic feature of coastal environments. They are indicators of the foods they eat, as well as indicators for the species that eat them. Shorebirds following the surging tide, terns swirling over fish schools, and gulls converging over children throwing them bread, are an important part of the shore experience. People care about them.

OBJECTIVES OF THIS BOOK

Our overall objective is to examine habitat use, population dynamics, and heavy metal levels in colonial waterbirds of Barnegat Bay and other bays and estuaries in the Northeast. Our primary study area is Barnegat Bay, New Jersey, and our secondary coverage is from Boston and Buzzards Bay to the Chesapeake Bay (Figure 1.3). Although the latter is not strictly in the Northeast, we use the term for convenience. We also describe temporal and spatial patterns, colony sites, and metals in prey items and in the birds themselves. The birds we study are considered "colonial" because several to hundreds or even thousands of pairs nest in close proximity, usually in direct visual contact. A colony can contain a single species or several, and most of those we study in Barnegat Bay have several species. The colonies of gulls and terns are two dimensional; they spread horizontally over marshes or beaches, whereas the colonies of egrets, herons, and ibis are three-dimensional, and typically occur in bushes, reeds, or trees with some birds nesting several meters above the ground. We use colonial waterbirds as the focal point for an ecosystem approach that ranges from invertebrates through prey fish, to humans, and we do so in a context of estuaries from Maine to Florida. We have had to draw boundaries in this book, thereby excluding the sources and pathways of metals in the sediment, water, and the marsh vegetation itself.

The data and analyses presented here are mainly new or extend the timeline to 2015 for previously published data. Data previously published will be so noted, but this is the first synthesis of information on habitat use, population dynamics, and metal levels in waterbirds within a context of other species groups (invertebrates, fish), and several estuaries along the East Coast. For some species, we can describe temporal trends in population levels and contaminants in samples that run almost annually from the early 1970s to the present. For others species, we have a shorter temporal period and fewer samples. In the following chapters, we describe the bays and our methods, habitat use, population dynamics of colonial birds in Barnegat Bay and other Northeast bays, and global

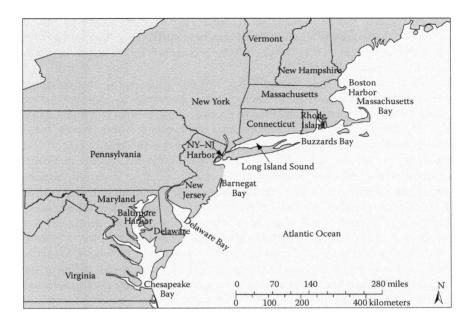

Figure 1.3   Map of coastal areas from Boston Harbor to the Chesapeake Bay.

change and sea level rise, followed by data on temporal and spatial patterns of metals in birds from Barnegat Bay and elsewhere, and we discuss the future for colonial birds in the Northeast.

We examine the following hypotheses: (1) there are temporal differences in overall population trends in colonial birds for estuaries with sufficient data; (2) there are temporal changes in suitable habitats in Barnegat Bay for colonially nesting species; (3) there are temporal differences in population levels in indicator species in Barnegat Bay (and in other bays); (4) there are temporal differences in metal levels in several indicator species in Barnegat Bay over the past 40 years; and (5) the temporal and spatial pattern of metal levels in colonial birds among different trophic levels may differ. Testing these hypotheses is very ambitious, and we do not present data for all indicator species for all hypotheses. Examining all these for each of our key indicator species (gulls, terns, Skimmers, herons, egrets, Night-Herons) would be nearly impossible. However, we examine some of these hypotheses for several species.

The main contaminants we discuss are metals (lead, mercury, cadmium, manganese, and chromium) and metalloids (arsenic and selenium). We will call them "metals" in the rest of the book. Other important studies have examined nonmetal contaminants in the same species, sometimes in the same bays, but we have not discussed them in detail (e.g., Shaw-Allen 2005; Kim and Koo 2007; Grasman et al. 2013). The role of contaminants affecting bird populations is complex and controversial (Nisbet 1994). There are few examples as dramatic or a widespread as the devastating impact of the chlorinated organic pesticides (e.g., DDT) on high trophic level birds of prey (Bald Eagles [*Haliaeetus leucocephalus*], Ospreys [*Pandion haliaetus*], Peregrines [*Falco peregrinus*], or Brown Pelicans [*Pelecanus occidentalis*]). Over the course of barely one generation, widespread pesticide use extirpated Peregrines and Pelicans from the eastern United States and substantially reduced Osprey and Eagle populations as well. Population recovery followed the banning of use of these persistent pesticides. We will mention this historic episode again in several contexts.

Infectious diseases can be a major cause of avian mortality killing thousands to tens of thousands of individuals. These include various strains of avian influenza, avian cholera, mycotoxicosis, Newcastle Disease and many other epidemics scattered around North America since 1970. We have encountered several such outbreaks—dead and dying chicks and adults, unable to walk, their lower abdomen stained with green diarrhea. Botulism C has been particularly devastating (Friend 2006). Metals such as lead contribute to these epidemics, by impairing immune response, predisposing to high mortality (Vallverdú-Coll et al. 2015).

In this chapter, we briefly describe and define concepts discussed in the remaining chapters of the book, including biomonitoring and bioindicators, waterbirds as indicators, contaminants, and the human dimension. A brief description of ecotoxicology in birds is presented in Chapter 8, and exposure and effects of the metals are presented in Chapter 9. These provide the context for Chapters 10–13. Some terms used throughout this book are given in Table 1.1.

One pair of terms requires species mention: *assessment endpoint* and *measurement endpoint*. Usually, only the term "endpoint" is used and the reader is left to determine which is meant. Assessment endpoint is the overarching characteristic that is important to policy makers, managers, and the public (Burger et al. 2007b,c). Measurement endpoint is the characteristic that can be measured and is an indicator of the assessment endpoint. For example, people are interested in whether Piping Plover (*Charadrius melodus*) populations are sustainable, which is an assessment endpoint. The health of a population, however, may be measured by the number of breeding pairs or the number of young produced (measurement endpoints). Furthermore, different stakeholders may be interested in different endpoints. Photographers may want to take photographs of displaying pairs, conservationists may want to see improvement in the number of Piping Plovers, and beach-goers may simply want to see small shorebirds running with the waves along the surf. All three stakeholder groups are interested in the stability of populations of an endangered species. The key question is—"Is the population of Piping Plovers stable along the Atlantic coast?" If not, what needs to be done to restore and sustain it? The number of breeding pairs is the measurement endpoint (it can be measured or counted using a comparable method in different regions). This number is used to address the question—"Is the population stable?" It is never possible, however, to know the exact population size, and stability requires comparing the number of breeding pairs from year to year.

BIOMONITORING AND BIOINDICATORS

The regular, periodic assessment of an environmental or ecological variable (diversity of species, numbers of a species, contaminant levels) is called monitoring. Monitoring, including surveying and censusing, is essential to understand and document the health and well-being of ecosystems and their component parts. The centerpiece of ecological assessment is monitoring, using standard indicators that can be compared within and among ecosystems, over time (Peakall 1992; Cairns and Niederlehner 1996; EPA 1997a; Burger and Gochfeld 2001a; Krabbenhoft et al. 2007; Li et al. 2010; Burger et al. 2013a–d), including coastal landscapes (Lomba et al. 2008). Biomonitoring is the regular, periodic evaluation of a species (population levels), or trait (tumor types, hatching rate, growth rates), or process (food webs and energy flow) to assess the health and well-being of ecosystems. The evaluation can be qualitative (presence of tumors or abnormalities) or quantitative (32 fish with tumors out of 1000 fish examined). Biological monitoring data can be obtained from many sources, and at many biological scales, from tissues and organs, to whole organisms (disease, injuries and mortality), to large landscapes, seascapes, and the Earth.

Although for humans each individual is the unit of interest, for most biological systems, it is populations that are of interest (except for endangered/threatened species, where individuals do count). Considerable time and energy are spent on managing for the recovery of threatened and endangered species by the U.S. Fish and Wildlife Service, state agencies, and nongovernmental organizations (NGOs). In addition to officially listed "endangered" and "threatened" status, which requires formal rule-making, states may designate species of special concern (Table 3.2). Other groups, including Native American Tribes, may monitor species of interest to them. Among NGOs, the American

Table 1.1   Definitions of Key Concepts Used throughout This Book

Term	Definition	Example
Assessment endpoint	The characteristic of interest to the public and policy makers.	Are salmon populations stable (Burger et al. 2015b,c)? Are Great Egret (*Ardea alba*) populations stable? Are Red Knots (*Calidris canutus*) declining?
Bioamplification Biomagnification	Increase in concentrations of certain chemicals in the organisms at each higher trophic level of an ecosystem.	Larger birds, such as Great Black-backed Gulls (*Larus marinus*), have higher levels of mercury in their tissues than the fish they eat, which in turn have higher levels than the smaller fish they consume, which in turn have higher levels than their plankton prey.
Bioconcentration	Accumulation and concentration of a chemical in tissues, resulting from excreting less than is consumed.	The concentration of methylmercury in a large predatory fish may be 100,000 times higher than the concentration in the water column.
Biodiversity	Number of species present in a location or ecosystem.	There are 15 species of waterbirds nesting in New Jersey. This can also be computed using a formula (see Terborgh 2009).
Bioindicator	Species of biota used to provide information on an environmental quality or condition.	Common Terns (*Sterna hirundo*) and Bluefish (*Pomatomus saltatrix*) are collected and analyzed for methylmercury.
Bioindicator species	Biota developed to provide data on environmental quality, conditions, population trends, or productivity.	Reproductive success of Common Terns, Herring Gulls (*Larus argentatus*), and Great Egrets.
Colonial	Individuals that nest together in close proximity.	Great Egrets nest close together in a clump of *Phragmites* and Poison Ivy (*Toxicodendron radicans*) on an island.
Colony	Group of birds nesting together in one place, within communication distance.	250 Common Terns nesting together on a small salt marsh island.
Colony site	Place where a colony of birds nest.	Mike's Island in Barnegat Bay, where 150–350 pairs of Common Terns nest.
Community	A group of potentially interacting species living in an ecosystem.	Common Terns, Black Skimmers, and Fiddler Crabs living on a salt marsh island.
Competition	When two or more individuals use the same resource that is in short supply.	Common Terns and Forster's Terns (*Sterna forsterii*) both plunge-dive for small fish in shallow bays.
Concentration	The amount of any nutrient or contaminant in water, sediment, prey, bird tissues, eggs, or feathers.	Mercury concentration in feathers (nanograms of mercury per gram of feathers or parts per billion). Blood lead concentration (in μg/dL).
Conceptual model	Graphic visualization of sources, pathways, and exposure to receptors.	A graph of radionuclides moving into water, being taken up by many different organisms, and accumulating up the food chain in the tissues of top predators (Burger et al. 2006).
Dose	Amount of a contaminant received by an organism or target tissue.	0.05 mg of lead per kilogram of body weight fed to gull chicks. 0.1 mg/g body weight injected in a gull chick.
Ecosystem	A spatially defined unit, containing abiotic and biotic components, primary producers, herbivores, carnivores, and decomposers.	The Barnegat Bay ecosystem of soil, water, vegetation, and animals.
Endpoint	An attribute of an ecosystem, or species or individual that can be measured to indicate an effect.	Biodiversity, reproductive success, population size of Black Skimmers (*Rynchops niger*), level of mercury in eggs of Common Terns. Reproductive impairment, altered behavior, abnormal gait, death.

(Continued)

Table 1.1 (Continued)   Definitions of Key Concepts Used throughout This Book

Term	Definition	Example
Environmental assessment	Determining the relative health or status of an ecosystem or species; also for physical measurements.	The status and trends of populations of Snowy Egrets (*Egretta garzetta*) in Barnegat Bay. The health of Black Skimmer populations. The amount of rainfall.
Environmental justice	Disproportionate exposure of some groups of people to chemicals or other hazards.	Large Striped Bass (*Morone saxatilis*) caught by recreational fishermen in tournaments are donated to homeless shelters (large Bass bioaccumulate mercury; Gochfeld et al. 2012).
Environmental monitoring	Periodic sampling of environmental media or ecological endpoints.	Counting breeding birds. Twenty-year trends in mercury levels in feathers of Great Egrets from Barnegat Bay (Burger 2013a).
Exposure	Contact between an organism and chemical or physical stressor, allowing the chemical to enter the body.	Birds and people eating fish are generally exposed to mercury.
Exposure pathway	Route of a chemical from a source in an environmental medium (air, water, sediment, food) to a receptor through ingestion, inhalation, or direct absorption.	Mercury in sediment, derived from atmospheric mercury from power plants, is converted to methylmercury by bacteria, taken up by plants, which are eaten by small fish that are consumed by larger fish that are in turn eaten by Great Black-backed Gulls.
Global warming	Gradual increase of temperature worldwide affecting ice cap melting, expansion of seawater, and sea level rise.	Temperatures are expected to rise by an average of 2°C over the next century (IPCC 2007, 2014).
Habitat	Physical and biological environment in which organisms live.	*Spartina alterniflora* (cord grass) salt marsh in the middle of Barnegat Bay.
Hazard	A substance with the potential to cause harm.	Lead exposure can cause cognitive deficits in Herring Gulls (Burger and Gochfeld 1994a, 1995a,b, 2000a).
Host factors	Factors within an organism that affect responses to an agent in the environment.	Species, age, sex, diet, and other exposures (Burger et al. 2003a).
Indicator	Species or endpoint used to evaluate the health of a system or species.	Mercury levels in soil, lead levels in blood of terns, selenium levels in bird eggs.
Landscape ecology	Broader ecosystem patterns (e.g., edges, fragmentation, and corridors).	The Pine Barrens of New Jersey is part of the coastal ecosystem (Forman and Gordon 1986; Forman 1996).
Measurement endpoint	A trait or attribute that can be measured, often as an index for an assessment endpoint.	Number of salmon that pass a given fish ladder/year as an index of salmon population stability; number of Red Knots counted at Delaware Bay in May.
Monitoring	Periodic sampling of endpoints.	Yearly census of the number of Laughing Gulls (*Leucophaeus atricilla*) nesting at the end of the runway at John F. Kennedy (JFK) International Airport.
Perception	How people view the environment, an issue or a risk.	People may believe fish are safe from mercury because they do not look "spoiled."
Population	Number of organisms of a species in a defined geographical area.	The population of Common Terns (*Sterna hirundo*) in Bird Island is 1200 pairs.
Receptor	Organism, tissue, or cell exposed to a contaminant.	Bluefish, Forster's Terns, Great Egret testis, human kidney.
Recovery	Ability of an ecosystem or species to return to its previous state after a stressor event.	Populations of Great Egrets recovered after Superstorm *Sandy* because their nesting shrubs were not totally destroyed.

(Continued)

Table 1.1 (Continued)   Definitions of Key Concepts Used throughout This Book

Term	Definition	Example
Resiliency	The ability of an ecosystem or species to withstand stressors, or perturbation.	Salt marshes in New York–New Jersey Harbor recovered after Superstorm *Sandy* because the water flowed over the top of the marsh.
Risk	The potential for harm owing to exposure to a hazard.	Risk can be physical, biological, or chemical/radiological.
Risk assessment	Estimating the probability of an adverse outcome to an exposure (NRC 1983, 1993): hazard identification, dose–response, exposure, risk characterization.	The risk from exposure is often given as one in a million excess cancer deaths or as a Hazard Quotient = Exposure divided by a reference dose.
Sea level rise	Increase in the level of the oceans of the world.	Sea level rise along the New Jersey–New York Atlantic coast has been 2.5 mm/year from 2003 to 2008 (Titus 1990; Ben Horton, personal communication).
Sentinel	Organism monitored as an early warning to detect potential harm to other organisms in the same environment.	Dogs living in urban areas can be used as sentinels of human exposure to airborne chemicals. Largemouth Bass serve as sentinels of ambient water quality for mercury.
Subsidence	Sinking of the earth's surface.	Islands in Barnegat Bay have less elevation partly because they are slowly sinking.
Subsistence	People who need to catch their own fish or other wildlife for protein.	Some low income people living in the New York–New Jersey Harbor catch fish to provide protein for their families.
Sustainability	The ability of the environment to meet the needs of the present without compromising its ability to meet needs of future generations.	Managing exploitable resources so that the annual yield will remain constant in perpetuity. Managing the take of Bluefish so that future generations can have the same take, of the same size fish, with the same fishing effort.
Synchrony	Performing a particular activity together.	Laughing Gulls lay their eggs in a 2-week period, rather than over a longer period (Gochfeld 1980a).
Toxicity	Harm that results from a given dose of a chemical.	Levels of lethality from different chemicals; toxicity differs by organism.
Toxicosis	Toxic response in an animal.	Condors (*Gymnogyps californianus*) died from excess lead (lead poisoning).
Trophic level	Steps in the food chain including (1) primary producers, (2) herbivores, (3) predators, (4) top-level predators, and (5) decomposers.	Great Egrets eat large fish and are therefore top-level predators, with the greatest exposure to contaminants that have bioaccumulated in the food chain.

Sources: Developed partly from Peakall, D., *Animal Biomarkers as Pollution Indicators*, Chapman & Hall, London, UK, 1992; Bartell, S.M. et al., *Ecological Risk Estimation*, Lewis Publishing, Boca Raton, FL, 1992; NRC, *Issues in Risk Assessment*, National Academy Press, Washington, DC, 1993; Burger, J., *Environ. Bioindicators*, 1, 136–144, 2006a; Burger, J., *Environ. Bioindicators*, 1, 22–39, 2006b; Burger, J. et al., *Int. J. Environ. Sci. Eng. Res.*, 4, 31–51, 2013a; Burger, J. et al., *Nat. Sci.*, 5, 50–62, 2013b; Burger, J. et al., *J. Environ. Protect.*, 4, 87–95, 2013c.

Bird Conservancy has a Watch List, which includes Piping Plover, Least Tern (*Sternula antillarum*), Henslow's Sparrow (*Ammodramus henslowii*), Saltmarsh Sparrow, Seaside Sparrow (*Ammodramus maritimus*), Clapper Rail, Red Knot, Roseate Tern (*Sterna dougallii*), and Black Skimmer (ABC 2007). Fisheries agencies monitor finfish, shellfish, or recreational species. For example, Striped Bass along the Atlantic coast are monitored by state and federal agencies because of their importance to recreational and commercial fisheries, and stakeholders are often involved in research and monitoring (Burger 2009a; Burger and Gochfeld 2005a, 2011b; Burger et al. 2013d,e). Salmon are monitored in the Northwest because of fisheries interests, and they hold special fisheries and ecocultural value to several Tribes (Butler and O'Connor 2004; Bohnee et al. 2011; CRITFC 2013).

Figure 1.4   Several species can serve as bioindicators, including Herring Gulls (a) and Osprey (b). The graph shows that Ospreys are increasing in Barnegat Bay, New Jersey (c).

In a hypothetical healthy ecosystem, populations of component species are more or less balanced and sustainable over time. If a species declines greatly or disappears from that ecosystem, it may upset the balance, leading to a major change and loss of some species and their services or overgrowth of others. Biomonitoring can provide early warning of any changes that could result in significant risk to individual species (including humans), populations, or ecosystems. Biomonitoring (or biological monitoring) is part of a larger set of potential monitoring tools (Figure 1.4, Table 1.2). To truly assess the condition of a given ecosystem, it is also essential to understand changes in the physical environment, ecological systems, human dimensions, and interactions between these three. Both economic and cultural considerations have a major effect on the physical and biological environment, especially in coastal systems (Burger 1988a; Niles et al. 2012; Maslo and Lockwood 2014; Alford et al. 2015).

Table 1.2 Types of Monitoring with Examples Relevant to Species in Coastal Aquatic Environments

Type of Monitoring	Examples of Types of Indicators
Physical monitoring	Seismic activity; number and intensity of storms, number of tornados or hurricanes; rainfall amount and periodicity, soil type, oxygen levels, water pH, water depth; mercury levels in sediment; sediment toxicity
Ecological monitoring	Species present and biodiversity. Population levels; number of breeding pairs, number of fish in a school; age structure; energy flow; nutrient flow; productivity; predators; competitors
Ecological monitoring within a framework of human use	Fish landings; mean size of fish catch; recreational rates; number of fishing or hunting licenses sold
Ecotoxicological monitoring	Levels of contaminants in species and tissues; disease rates; abnormality rates associated with reproductive rates; number of tumors; mercury levels in sediment
Human health assessment and monitoring	Risk from eating contaminated fish and ducks; using data from animals as indications of human health risk; recreational rates
Social/economic assessment and monitoring	Changes in income from fishing licenses, recreational activities or sand mining; land use of different types of habitats; number of marinas, number of fishing clubs
Sustainability monitoring	Ability of ecosystems to remain the same, without further inputs; ability to continue to provide goods and services required by people, including ecocultural values
Ecological monitoring, ecotoxicological monitoring, and regional sustainability monitoring	Trends in population levels, reproductive success and overall health of animal populations in relationship to human growth indices, and the ability for the system to provide the same level of goods and services to humans; ability of the system to sustain ecocultural values

Sources: Modified from Burger, J. et al., *Int. J. Environ. Sci. Eng. Res.*, 4, 31–51, 2013a; Burger, J. et al., *Nat. Sci.*, 5, 50–62, 2013b; Burger, J. et al., *J. Environ. Protect.*, 4, 87–95, 2013c.

Monitoring requires having indicators or bioindicators that can be used to assess changes in some aspect of biological systems. Because the multitude of species in an ecosystem cannot all be monitored, it is essential to develop a suite of bioindicators that can be used to assess status and trends within that ecosystem, as well as ecosystem integrity (Peakall 1992; Heink and Kowarik 2010). There needs to be improved linkage between ecological monitoring and ecotoxicology. In 2000, the National Research Council (NRC) published a valuable book, *Ecological Indicators for the Nation* (NRC 2000), but did not mention contaminants in their index. The NRC basically noted indicators for ecosystems (land cover, land use), ecological capitol (species diversity, native species, nutrient runoff, soil organic matter), and ecological function indicators (carbon storage, production capacity, net primary productivity, lake trophic status, stream oxygen). Although these are useful for examining overall ecosystem health and well-being, they do not address species and community health, or contaminant issues. An advantage to combining ecological and ecotoxicological indicators is improved understanding of how human and ecosystem health are intertwined (Beyer et al. 1996; Dell'Omo 2002; Harris et al. 2006; Conti 2008). The rise of the "One Health" movement linking human and animal health embodies this principle (Rabinowitz and Conti 2013).

Bioindicator selection is the most important part of the assessment process, because to be most useful, they should be applicable for long periods (NRC 2000; Carignan and Villard 2001), and for assessing the ecosystems at wider scales (Markert et al. 1999). Bioindicators are thus the measuring tool used in biomonitoring plans. Some indicators are part of several different monitoring plans. For example, mercury level in sediment could be part of a physical monitoring plan, as well as an ecotoxicology monitoring plan. The specific indicators listed in Table 1.2 are only examples; site-specific ones are required to address particular problems.

Bioindicators can be used to assess current state, past environmental injuries, restoration or remediation options or success, and temporal and spatial trends in the quality of the environment (EEA

2003; Bartell 2006; Burger 2006a,b, 2008a; Heink and Kowarik 2010; Hirata et al. 2010). Bioindicators are developed for the condition of individual species or populations, community dynamics, ecosystem structure and function, and landscapes. Indicators can be developed that assess both exposure and effects, and are most useful when they apply to a particular spatially defined area that can be tracked over time, and when they provide information about both human and ecological health (Burger and Gochfeld 2001a; Burger 2006a,b; Bartell 2006; Heink and Kowarik 2010). The qualities to be considered when developing indicators include: biological relevance, methodological relevance, legal and societal relevance, cultural relevance (Burger et al. 2013a,d; Table 1.3) as well as long-term availability, logistical feasibility, and cost. Sustaining investment in monitoring continues to be challenging.

Usually, bioindicators are a species, such as Bluefish, Blue Crabs (*Callinectes sapidus*), Bald Eagles, or the Common Loon (*Gavia immer*). A wide range of species can be used as bioindicators (Golden and Rattner 2003)—selection depends on the question being asked and the place being studied. The attributes that are monitored are often called "endpoints" or "metrics." Bioindicators and endpoints can be used for a variety of assessment issues, such as contamination, reproductive success, or population levels (EPA 1997a). For example, the widespread Common Terns may be developed as a bioindicator, but the endpoints measured could be mercury levels in feathers, lead levels in eggs, number of tern nests or adults present in the colony, or number of young fledged per nest. These endpoints can be sampled year after year. From a population ecology perspective, the recruitment of new breeding adults to the population is critical for sustainability. However, it is a complex multiyear assessment that is impractical for monitoring. It is essential to develop endpoints and bioindicators with a particular question or problem in mind. In this book, for example, we use the number of adults or pairs of birds in a breeding colony as an endpoint to examine temporal and spatial trends of populations, and we measure levels of metals in eggs and feathers as an indication of temporal trends in avian exposure and environmental contamination.

Ensuring protection of human health and the environment requires measurement and assessment of the condition of ecosystems and their component parts. For example, one can measure mercury levels in Cod (*Gadus morhua*) or number of nesting Green Sea Turtles (*Chelonia mydas*) on a

Table 1.3   Criteria for Selection of Indicators

Relevance	Criterion
Biological	Provides early warning
	Exhibits changes in response to stressors, soon enough to act (sensitivity)
	Changes are real and attributable to a particular stressor (specificity)
	Changes are important and matter to people as well as to the ecosystem
Methodological	Easy to use in the field
	Easy to analyze, interpret, and present to the public
	Cost effective (money, time, schedule)
	Easily repeatable
	Useful over a broad geographical or temporal scale
	Can be conducted by minimally trained personnel
Societal	Of interest to the public (people value and care about the species)
	Transparent to the public
	Can be used to evaluate human exposure and effects as well as ecological
	Contributes to improved human well-being
	Can be used to test public policy or adaptive management
Cultural	Important to specific cultural groups
	Of interest to the general public including recreational and aesthetic interest
	Takes into account environmental justice
Economic	Viable in terms of schedule, time, cost, and degree of training required
Legal	Relates to environmental laws and regulations, and to regulators

Sources: Developed from Burger, J., *Environ. Bioindicators*, 1, 136–144, 2006a; Burger, J., *Environ. Bioindicators*, 1, 22–39, 2006b; Burger, J. et al., *Int. J. Environ. Sci. Eng. Res.*, 4, 31–51, 2013a; Burger, J. et al., *Nat. Sci.*, 5, 50–62, 2013b; Burger, J. et al., *J. Environ. Protect.*, 4, 87–95, 2013c; Unpublished data.

beach, or the energy that is available in a corn crop. Fish and fishery harvests are frequently used to monitor aquatic systems, both because they are consumed by people and to set future harvest quotas (Corsi et al. 2003). Assessing biological health for species, populations, communities, and ecosystems involves evaluating the current condition or status, and monitoring changes over space and time, either retrospectively or prospectively.

Bioindicators monitored to assess status or trends are also being used to assess sustainability of populations or environmental conditions (McCool and Stankey 2004; Marcarenhas et al. 2010; Hirata et al. 2010). A suite of indicators that can be monitored for many years is most useful for assessment of the environment (Niemeijer and deGroot 2008). For example, the South Florida Water Management District uses a set of indicators to evaluate the condition of the Everglades, including biomass of submerged aquatic vegetation, numbers of fish, and pairs of colonial water-birds (SFWMD 2014). Even with a set of agreed-upon bioindicators, it is still sometimes difficult to arrive at consensus among stakeholders who have different perspectives and conflicting interests (MacCoun 1998; Small et al. 2014). Data may help achieve common interests, and an absence of sufficient data may fuel worsening disagreements (Apostolakis and Pickett 1998; Burger et al. 2001a, 2005, 2007b, 2013e; Sarewitz 2004).

Finally, managers, conservationists, and public policy makers, among many others, want to have indicator tools that can be used to determine if a given system is sustainable, or requires adjustments (Hart 1999; Leitao and Ahern 2002) or aggressive management, such as habitat restoration or invasive species control. Often local indicators are essential for assessing regional sustainability (Marcarenhas et al. 2010). Long-term biomonitoring is necessary to assess the efficacy of management, restoration, and Natural Resource Damage Assessment, as well as evaluating stewardship (Burger 2008a,b; Burger et al. 2009a; Peterson et al. 2011). In some cases, types of perceptions can be indicators of public opinion, which in turn affects management and public policy (NRC 2000, 2008; Burger 2008b, 2009a,b, 2012), especially for environmental justice communities (Bullard 1990; Mikula and Wenzel 2000; Chess et al. 2005; EPA 2009; Burger et al. 2010b; Burger and Gochfeld 2011a; Gochfeld and Burger 2011).

USING COLONIAL WATERBIRDS AS BIOINDICATORS AND SENTINELS

Colonial birds have served as bioindicators for ecological health, as sentinels for human health, and as indicators of temporal and spatial changes in contaminant exposures (Walsh 1990; Burger 1993; Burger et al. 2013a,d; Fox 1991, 2001; Custer 2000; Erwin and Custer 2000; Becker 2003; Becker and Dittmann 2009; Dittmann et al. 2011, 2012). The National Research Council (1991) published *Animals as Sentinels of Environmental Health Hazards*, in which they identified four attributes for a "sentinel": (1) should have a measurable response to the agent of concern (sensitivity); (2) should live in the area where monitoring is conducted; (3) should be easily counted and captured; and (4) should have a sufficient population size to permit enumeration. Many waterbird species, such as terns, gulls, and herons, fulfill these criteria. Although migratory species are not optimal (NRC 1991), the hatching and growing young birds are raised on food from the area of concern. These birds are top-level predators that feed on fish, and are thus exposed to bioamplification of a variety of metals and organic pollutants. In addition, by virtue of numbers, colonial bird species provide the opportunity to obtain adequate sample sizes for trend analysis.

Waterbirds exhibit different foraging strategies, from specialists to generalists, and consume a range of foods. Because gulls, terns, and herons usually lay three or more eggs, removing one egg per nest for chemical analysis has negligible impact on populations. Likewise, preflying, feathered young can be captured, and body feathers can be sampled without jeopardizing their survival (Palestis and Stanton 2013). Collecting feathers from prefledging young is noninvasive, easy to do with little disturbance, and has the advantage of reflecting local exposure, because parents collect food within flying range of colonies (Burger 1993; Kim and Oh 2014; Brzorad et al. 2015). Breast feathers grow back in only 2–3 weeks (Burger and Gochfeld 1992a; Burger et al. 1992a; Nisbet 2002). Information about migration and molting chronology, and the ecology and biology of the birds is necessary for interpretation of metals in feathers (Bortolotti 2010).

Colonial nesting birds (Figure 1.5) may be sentinels because they spend part of their daily cycle in the same environments, and some eat the same foods that we do. In many parts of the world, and sometimes even in New Jersey (though not legally), waterbird eggs are collected for food, and subsistence hunting of waterbirds is conducted in many countries. This book takes a food web, ecosystem approach to contaminants, examining metal levels in the context of habitat selection, population dynamics, and inputs to bays and estuaries. It includes the human dimension, and what metal levels in birds tell us about human exposure, as well as stakeholder involvement in these issues.

Some species, such as Glossy Ibis (*Plegadis falcinellus*), feed primarily on invertebrates, and they are generally considered to be at a low trophic level, whereas Great Black-backed Gulls, Herring Gulls, and herons feed on fairly large fish, especially when they eat dead fish washed up on shore. Similarly, Great Egrets feed on larger fish than do Snowy Egrets, mainly because the former have longer and larger bills and longer legs (e.g., they can feed in deeper water). Great Egrets can feed everywhere Snowy Egrets can, but they can also forage in additional places that are too deep for the Snowy Egrets. Not only do the two egrets forage on different prey, often of different sizes, but they sometimes forage in different places and in different ways (Lantz et al. 2010, 2011). Similarly, Forster's Terns forage mainly in salt marsh pools and inland ponds, whereas Common Terns forage mainly in bays and nearshore (Safina and Burger 1988a; Safina et al. 1988; see Chapter 4). Although

Figure 1.5　Colony view of Black Skimmers nesting on a salt marsh island in New Jersey, within sight of barrier island communities.

they nest together in the same colonies, they usually forage in different places on different fish. Even within a colony there may be individual differences; not all terns forage on the same prey items (Nisbet 2002).

Large, fish-eating colonial birds have been used as bioindicators for the health and well-being of themselves, the prey they eat, and the predators that eat them (Custer 2000). They have been used extensively as indicators of pollution (Wolfe et al. 2007). Indeed, the world was alerted to the toxic effects of DDT by rapid declines in the populations of pelican and other fish-eating bird populations that ultimately resulted in DDT being banned (Anderson and Hickey 1970; Shields 2002). DDT interfered with the deposition of calcium in the eggshell; eggs broke under the weight of the incubating adult, and their population crashed—our first indication of endocrine disruption (Colborn and Clement 1992). When these effects were recognized in the 1960s (Gochfeld 1975a), there were very few long-term population monitoring or contaminant monitoring programs.

In a sense, careful selection of a bioindicator, or a suite of bioindicators, can provide information about several nodes of the food chain. In the example shown in Figure 1.6, Common Terns are the indicator, and mercury levels are measured in eggs and feathers. Common Terns capture fish by plunge-diving for fish, usually in bays, inlets, or out in the open ocean (Safina and Burger 1988a). They forage close to the colony, and bring back the fish, mainly one at a time, to feed their chicks. Thus, metal levels in the feathers of fledgling Common Tern chicks represent local exposure. Striped Bass, Bluefish, and other predatory fish feed on small fish, and in fact, Bluefish force schools of small fish to the surface where they are accessible to the terns flying above (Safina and Burger 1988a; Safina et al. 1988). Otherwise, dense schools of prey fish would be too far below the water surface for the terns to capture by plunge-diving.

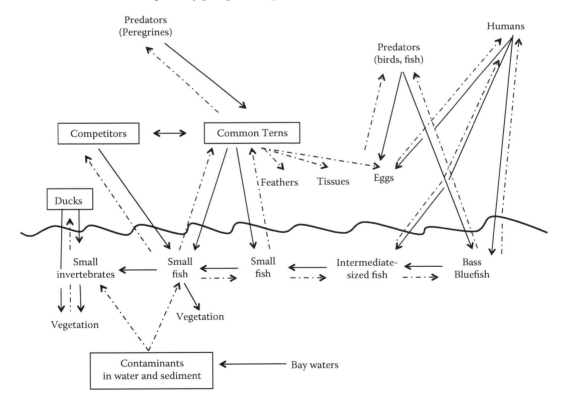

Figure 1.6 Food chain of Common Terns, showing the direction of consumption (solid arrows), and the direction of contamination bioaccumulation (dotted line).

In Figure 1.6, the solid lines indicate the food chain, from small invertebrates and vegetation to small prey fish, to intermediate-sized fish, to large fish, and finally to the birds (or people) who eat them. Contaminants increase up the food chain (biomagnification). The dotted line indicates the pathway of contaminant exposure, which increases with each step on the food chain. Other stressors, such as predators and competitors, affect populations of Common Terns (the bioindicator). The endpoints can be ecological, such as number of nesting pairs, number of eggs/nest, number of chicks fledged/nest, or ecotoxicological, such as levels of contaminants (e.g., lead, mercury, PCBs) in tissues of any node on the food chain.

Figure 1.7 illustrates the metrics related to Common Terns, showing some of the endpoints that can be developed ranging from contaminant levels in eggs, feathers, and other tissues (pollutant endpoints, shown as ovals), to longevity and reproductive success (individual endpoints, shown in rectangles). Metrics related to populations are shown in thicker black rectangles. Those for individual birds are shown in thin-lined rectangles.

Endpoints for a species, such as the Common Tern (Figure 1.7), provide information on that species. However, it is also possible to examine the full suite of endpoints that can be used as indicators

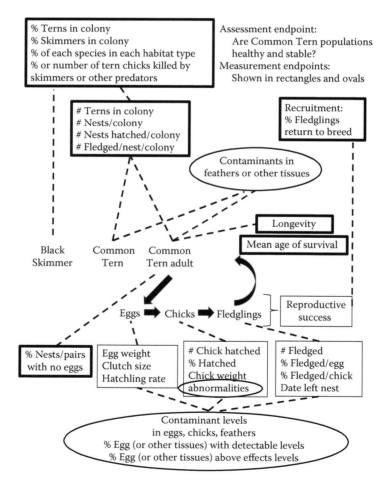

Figure 1.7 Endpoints possible with Common Terns can include an assessment of contaminants (ovals), individual biological endpoints (thin-lined rectangles), and species or population endpoints (thick-lined rectangles).

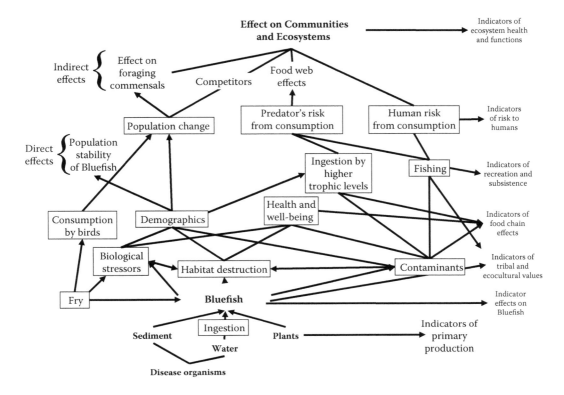

Figure 1.8 Endpoints developed for Bluefish, showing the different types of indicators, including direct and indirect effects.

for a wide range of effects (direct and indirect), and for a number of levels of biological organization, including risk to humans. Figure 1.8 shows the types of endpoints that can be developed for Bluefish, including those for human and ecological health.

HABITAT DIVERSITY AND CHANGES

The barrier beaches were summer homes for the Lenapé Indian tribes before the arrival of Europeans. The first recorded description of our coast is attributed to Robert Juet, first mate on Henry Hudson's *Half Moon*, who in 1609 described the coast near Barnegat Inlet as "all broken islands," with a "great lake of water behind." And "this is a very good land to fall in with, and a pleasant land to see" (Lavallette 2016). We share this sentiment.

Our coastal birds nesting on salt marsh islands and sandy barrier beaches face the everyday cycle of tides, and the longer monthly occurrence of higher tides and flooding threats. Unpredictable summer rainstorms coinciding with the chick-raising period can be devastating, and ultimately the inexorable rise of sea level erodes or submerges habitats (Figure 1.9). The usual conditions have effects on vegetation, wrack available on the marshes, and sand available on islands, thereby influencing habitat suitability and availability and ultimately bird distribution (discussed in Chapter 4) (Burger 1979b).

Perhaps the greatest threat to stable avian populations is habitat loss through natural and anthropogenic forces, as well as interactions between the two. Over geologic time and the millions of years that modern bird species have existed, the pulsing of glaciers, changes in sea level, volcanic activity, and changes in productivity, have had profound, largely undocumented effects on bird populations

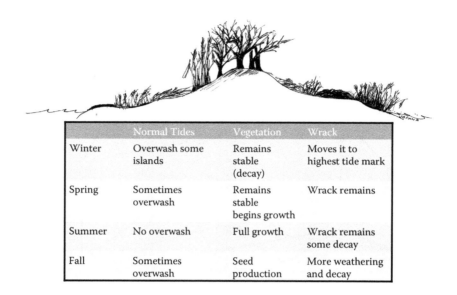

	Normal Tides	Vegetation	Wrack
Winter	Overwash some islands	Remains stable (decay)	Moves it to highest tide mark
Spring	Sometimes overwash	Remains stable begins growth	Wrack remains
Summer	No overwash	Full growth	Wrack remains some decay
Fall	Sometimes overwash	Seed production	More weathering and decay

Figure 1.9 Seasonal changes in tides, vegetation, and wrack that affect the salt marsh habitat for coastal-nesting species.

and distributions. We have no record of historic ecosystems comparable to the La Brea Tar Pits (Los Angeles, California), which preserved fauna from 50,000 to 10,000 years ago. However, as most of that fauna comprised modern extant species such as egrets (La Brea 2015), it is probably safe to infer that our study species were present along our coasts at the same time. But what was our coast? At the peak of the Ice Ages, more than 10,000 years ago, water was locked up in the ice sheet. Sea level was hundreds of feet (100 m+) lower than today, and the seacoast was more than a hundred miles (160 km) further east (Uchupi et al. 2001).

We do know that over the past several hundred years, bird populations have been affected by hunting, egg collecting, and disturbances to nesting birds (Nettleship et al. 1994). Most of the colonial waterbird species we discuss in this book were heavily exploited in the late 1800s until about 1910, and were virtually eliminated from any accessible nesting areas of the Atlantic coast. Eggs were collected commercially, and bird feathers, particularly egret plumes, fed the millinery trade. Also, terns were mounted whole on hats. Frank M. Chapman (1899) described the slaughter and virtual elimination of terns and egrets up and down the Atlantic coast for the munificent sum of 10 cents each, with milliners claiming that demand exceeded supply. A few terns in remote areas of New England or Florida survived to repopulate the coast once the hunting exploitation was terminated in the early twentieth century (McCrimmon et al. 2011).

Habitat loss, however, continues to threaten waterbird populations, particularly as more coastal habitat and beaches are usurped by human development. In 1922 to 1954, there was a 6.5% loss of coastal wetlands in the whole coterminus United States, and although the rate of loss has decreased in the past few decades, it has still amounted to more than 1% each decade (Valiela et al. 2009). More than 50% of salt marshes were claimed mainly for agriculture or drained for vector control (Weinstein and Kreeger 2002). Birds nesting in colonies on islands face stressors that act over a short period of days to months, intermediate periods of years to decades, and longer-term effects (IPCC 2007, 2014; Figure 1.10). These different scales of stressors also interact. Changes may be continuous or episodic. They are also inexorable or subject to long periods of stability.

For almost all of the twentieth century, coastal areas, estuaries, and salt marshes were treated as wastelands, "reclaimed" for farmland, shipping ports, oil terminals, airports, and garbage dumps, while eliminating mosquitos. Ports were located near large bays and estuaries, such as Boston

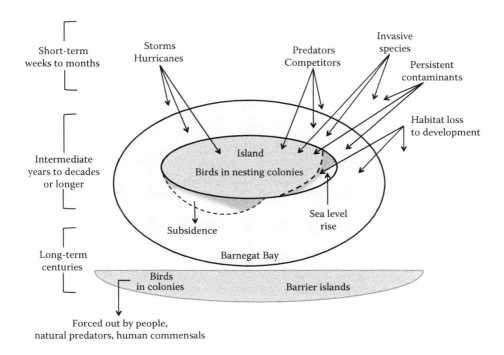

Figure 1.10 Stressors that affect birds nesting on island colonies over the short, intermediate, and long term. Subsidence results in islands sinking, whereas sea level rise results in the water rising, and they combine to result in loss of habitat for birds. Human commensals include dogs and cats that prey directly on the birds.

Harbor, the New York–New Jersey Harbor, Delaware Bay, and Baltimore Harbor. Barnegat Bay was spared this indignity because the bay is shallow, and there are no major industrial cities adjacent to it. There were no shipping lanes, land trade routes, or warehouses to receive goods from overseas. In the early 1900s, the mosquito-ridden Jersey Shore was considered worthless. Our dear friend and collaborator, Fred Lesser, told us how his parents received their house lot as an incentive to subscribe to the *Philadelphia Inquirer*.

However, with the advent of mosquito control by ditching in the 1930s and by pesticides in the 1940s, after the Second World War, and the ubiquity of the automobile (Steers 1966), coastal areas once considered too infested became attractive for tourists and residents alike (Burger and Shisler 1978a, 1979). Residential communities sprang up quickly, drawing businesses such as marinas, restaurants, gas stations, hardware stores, and fish markets, and eventually the full array of fast food restaurants, tourist boutiques, and motels (Hanowski et al. 1997; Niemi et al. 1999; Knight et al. 2003). In the 1950s, coastal development accelerated. People built summer homes and bungalows along the coasts. More and more land was developed, decreasing markedly the amount of wild habitat, especially on the barrier islands and on the edge of the mainland. Coastal wetlands disappeared as they were considered wastelands to be exploited (Lee et al. 2006). Auto traffic increased, boats increased, and some species of beach nesting birds were "evicted."

Least Terns and Piping Plover remained nesting on the beaches of barrier islands, faring poorly as people encroached, heedless of the vulnerability of the eggs and chicks. Estuarine feeding birds also faced shrinking habitat. Birds were forced into smaller and smaller areas. In New Jersey, in particular, the public, managers, planners, and public policy makers ignored the important role and dynamic nature of coastal beaches, dunes, and salt marshes (Nordstrom and Lotstein 1989; Nordstrom and Mitteager 2001; Psuty and O'Fiara 2002; USGS 2010, 2013). The role of habitat in protecting human communities, as well as ecological ones, from storm tides, high winds, and flood surges was not appreciated, nor were sensitive plants, animals, and ecosystem (Maslo and Lockwood 2014). The concept of "ecological services" had not yet been developed (Costanza et al. 1997).

With increasing development in the last half of the twentieth century, many beach-nesting birds moved to salt marsh islands to nest, abandoning the sandy beaches of barrier islands (gulls, Common Terns. Black Skimmers), as well as some small coastal forest patches (egrets, herons) (Burger 1980a; Burger and Shisler 1980a,b). The salt marsh islands were largely ignored by people except for mosquito control personnel. Then sea level rise and land subsidence began to erode salt marsh islands, first washing away the small patches of sand that were used by nesting birds. As more and more people owned boats, salt marsh creeks and pools were invaded by boats and later by jet skis that were able to operate in shallow water. Jet skiers could reach every island, often apparently delighting in flushing birds nesting on salt marsh islands (Burger 2002b). The public and managers alike responded to the erosion of ocean front beaches by having the U.S. Army Corp replenish them (Psuty and Ofiara 2002). Beach replenishment (or nourishment), whereby sand dredged from the ocean is pumped onto beaches along the 208 km of New Jersey's Atlantic coast, has cost taxpayers $475 million since 1990 (Pffaf 2012). Such replenishment builds beach but destroys the natural coastal processes. Preserving the natural structure and function of the beach/dune/marsh habitat provides protection and resiliency for human communities, as well as creating and protecting natural habitat for fish and wildlife (see Chapters 2 and 4).

ENVIRONMENTAL CONTAMINANTS

We provide a brief overview of contaminants here because they form an important part of the book, and are one of the potential stressors that birds face; however, they are not discussed in detail until Part III. Much of the existing contamination of coastal ecosystem is historic (nineteenth and twentieth centuries). Current contamination is partly controlled by environmental regulations and enforcement, which does not eliminate effluent and emissions. Contaminants, such as metals, are persistent, ubiquitous, and easily transported globally in the atmosphere. Environmental policy makers, managers, scientists, and the general public are concerned about increasing accumulation of contaminants in the environment resulting from the concentration of people in cities (Lee et al. 2006). Contaminants derived from natural geological and oceanic processes amount to less than one-fourth of the annual circulation of metals, and these are augmented by anthropogenic sources, including urban, industrial, mining, and agricultural emissions, waste incineration, and electricity generation (Mailman 1980; Fitzgerald 1989; Hoffman et al. 2011).

Atmospheric transport results in some contaminants, particularly mercury, being transported globally, and the land–margin interface is particularly vulnerable to pollutants, fertilizers, and wastes that flow from associated watersheds (Steinnes 1987; Greenberg et al. 2006; Pacyna et al. 2006). For example, mercury has been found in organisms in the Arctic and other isolated places (Hammerschmidt et al. 2006). Once in the sediment, atmospheric organic mercury is converted to methylmercury by bacteria (Bertilsson and Neujahr 1971; EPA 1997b), and then moves up the food chain, undergoing bioconcentration at each node (see Figure 1.6). Chapters 8 and 9 have a more complete discussion of the fate, transport, and effects of metals on birds. Although we focused on metals, it should be remembered that there are organic pollutants (PCBs, PAHs, pesticides, etc.) in the surface sediments of Barnegat Bay and in the other Northeastern bays and estuaries (Vane et al. 2008), and these can have serious impacts on bird reproduction.

Once in aquatic systems, different metals enter the food chain in different ways, are taken up by various organisms through different routes (ingestion, gills, skin), and are distributed among tissues or are excreted. With each step in the food chain, there is the potential for bioaccumulation and bioamplification (Lewis and Furness 1991; Peakall 1992; Burger and Gochfeld 2001b; Evers et al. 2011; Lodenius and Solonen 2013). Long-lived birds that feed at the top of the food chain have the greatest potential for high contaminant levels because their exposure level is higher for a longer time (Furness 1993; Monteiro and Furness 1995; Burger 2002c; Seewagen 2010). Waterfowl can be exposed to high levels of lead if wounded by shot, or through the incidental ingestion of lead shot in sediment, leading to fatal lead poisoning. Similarly, the chicks of Albatrosses nesting near old buildings on Midway nibble on lead-containing paint chips, and develop potentially fatal lead poisoning manifested by weakness and droopy wings (Figure 1.11).

Most avian bioaccumulation and effects studies have examined fish-eating birds that are linked directly to aquatic systems, such as Common Loons (Burger et al. 1994a; Burgess et al. 2005; Burgess and Meyer 2008; Evers et al. 2008, 2010), egrets (Frederick and Spalding 1994; Frederick et al. 1999, 2004; Spalding et al. 2000a,b), raptors (Albers et al. 2007), and seabirds (Furness et al. 1995; Burger and Gochfeld 2001b), although accumulation in songbirds has also been examined (Tsipoura et al. 2008; Jackson et al. 2011; Evers et al. 2012). When exposed to a local source of mercury from mining, songbirds can accumulate 15–40 times the level of mercury as birds from control sites (Custer et al. 2007a). Because fish-eating birds bioaccumulate metals, they are often used as bioindicators of ecosystem exposure, especially for mercury (Fox et al. 1991; Burger 1993; Custer 2000).

Pollutants can cause chronic effects and population declines, as well as acute mortality or other impairments (Monteiro and Furness 1995; (Frederick et al. 1999; Rattner 2000; Lord et al 2002; Jackson et al. 2011). Adverse effects on birds in the wild have been found for other metals, including lead (Burger and Gochfeld 1994a, 2000a), selenium (Ohlendorf et al. 1986a, 1989; Eisler 2000a,b), and cadmium (Spahn and Sherry 1999). Similar effects may be found for other metals not yet tested experimentally. Females can sequester metals in their eggs (Burger and Gochfeld 1991a, 1996a; Nisbet et al. 2002), allowing females to rid their body of contaminants, which in turn provides opportunities to use eggs as bioindicators. Comparison of mercury in feathers of Great Egrets in museum specimens from the 1920s to the 1970s, with Everglades samples from the present, indicated that present mercury levels are 4–5 times higher now than in feathers collected before 1970, but also show a decline in the most recent samples (Frederick et al. 2004). Analyses of feathers of museum specimens of Black-footed Albatross (*Phoebastria nigripes*) spanning 120 years found increases in methylmercury levels in feathers in the twentieth century (Vo et al. 2011). Levels of metals in feathers of birds are generally correlated with levels in internal organs (Burger 1993) and in their eggs (Burger and Gochfeld 1991a, 1996b; Burger 1994c; Nisbet 2002), although this is not always the case for all metals in all species (Burger et al. 2014a). Looking across many studies, we could not find a consistent ratio between feather and liver concentrations of mercury, lead, cadmium, or selenium (Burger 1993). Furthermore, laying sequence may influence metal levels within eggs in a clutch (Brasso et al. 2010). Mercury is sequestered mainly as methylmercury (DesGranges et al. 1998). The individual metals we studied are discussed in Chapters 8 and 9.

HUMAN DIMENSIONS

The estuaries of the Northeast have been, and continue to be, influenced by people. The Bay systems of the Northeastern United States differ in latitude and climate, size and configuration, exposure to the sea, and in the surrounding density of industrial, agricultural, and residential towns, and the size and population densities of their watersheds and shorelines. The New York–New Jersey Harbor is a complex biological system of bays, estuaries, and rivers around and through one of the world's most populous cities. Massachusetts Bays and Delaware Bay have large urban areas. Peconic Bay and Barnegat Bay, on the other hand, are surrounded by both permanent residents and summer communities (Burger and Gochfeld 2005b). Many of the summer residents own these second homes, which often have private access to beaches. Chesapeake Bay is much larger than

(a)

(b)

Figure 1.11   (a) Laysan Albatross (*Phoebastria immutabilis*) chick on Midway Island with drooping wings, indicative of lead poisoning, (b) along with a building with flaking lead paint.

the other bays, is more complex with a number of bays and estuaries, and is bordered mainly by Maryland and Virginia.

Human dimensions have impacted the Northeastern bays and estuaries dramatically, and have several components, including: (1) how people and biota use the bay, (2) how people have changed the bay ecosystems, (3) how the physical and biological aspects of bays have influenced people, (4) how people perceive bays or want to change them, and (5) how these perceptions influence public policy (and therefore action). Each of the bays has environmental quality problems and advocacy groups for cleaning them up. We will provide an introduction to each of these aspects for Barnegat Bay in the following section.

Barnegat Bay as a Microcosm

Barnegat Bay has two inlets: Barnegat Inlet in the middle of the bay, and Beach Haven or Little Egg Inlet in the south. The Bay is long and narrow, whereas Little Egg Harbor itself is a small bay at the south end of Barnegat Bay. The Barnegat Bay–Little Egg Harbor estuary covers 279 km^2 of open water (Kennish 2001a–c). It has seagrass meadows, plankton communities, habitat for spawning fish, and shellfish beds (Kennish 2001a). The bay is relatively shallow, much of it inaccessible to large boats at low tide. The extensive mosaic of marsh islands has been largely ignored, except for small boat traffic and clammers. Most people simply cross the bay on bridges to the Atlantic beaches. The Intercostal Waterway winds through Barnegat Bay, ensuring that channel passageways are kept open, resulting in heavy, but largely channelized, boat traffic during the summer. Although not surrounded by towns larger than Toms River, the entire bay area is s part of the urbanized Boston-to-Washington corridor. Toms River, the biggest town bordering the bay, has a population of 92,000 (U.S. Census Bureau for 2012). Both the bay and surrounding communities are vulnerable to severe storms and hurricanes (Burger 2015b; Burger and Gochfeld 2014a,b), but these communities were less impacted by Superstorm *Sandy*'s storm surge (November 2012) than communities along the north Jersey coast in Monmouth County or the south coast of Delaware Bay.

How People and Biota Use the Bay

How birds and other biota use the bays is discussed in several chapters. In this section, however, we explore how people use and perceive bays and estuaries. The "beach" is not just for swimming or fishing, although both are very important (MacKenzie 1992; Piehler 2000). People who live in coastal states often have an emotional commitment to the beach, none more strongly than Jerseyans have for the Jersey Shore. It remains a very popular destination for days, weeks, or the entire summer. For example, in a 2015 "instant clicker" survey of more than a hundred people attending a "Geology Day" event at Rutgers, 30% said they went to the beach once a week in the summer, and another 7% said they went every day. An additional 13% went monthly throughout the year.

In surveys of people living along the major bays and estuaries in New York and New Jersey, "nature" was the highest-ranked reason for going to the beach. When asked how they use coastal areas, the highest rankings were for "communing with nature," followed by "for a place without people" (Burger 2003a,b). This latter struck us as ironic considering the density or humans on some of our interviewing days. Although the ratings varied, "communing with nature" was rated the highest at all sites (Figure 1.12). Fishing was rated higher in some bays than others. Even so, these data indicate that people value bays and estuaries for ecocultural values, as well as for consumptive (fishing) or recreational activities (swimming).

When asked on an open-ended question—"How do you use the shore?"—responders in New Jersey mainly said they "go to the beach" (Figure 1.13; Burger 2015b). Frequency of activities, however, varied as a function of where people lived. People living along the shore went to the beach

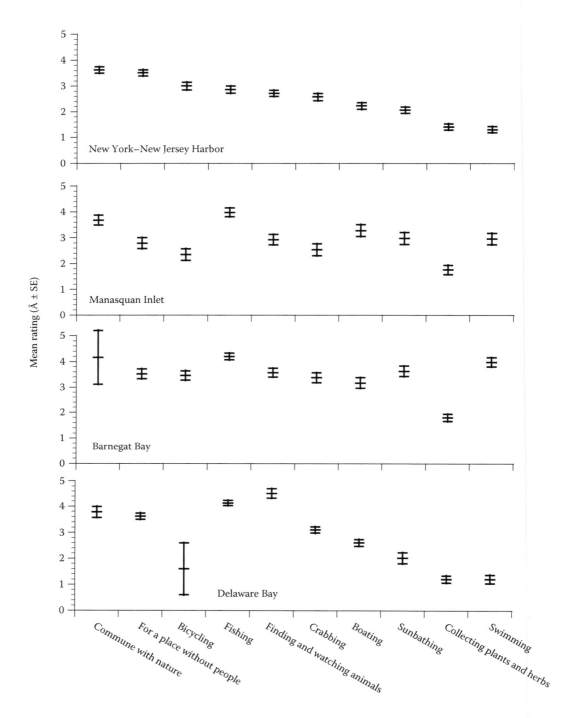

Figure 1.12 Mean rating people gave for the reasons they visit the shore and beaches. People rated these activities on a scale of 1 (low) to 5 (high). (After Burger, L., *J. Environ. Plann. Manag.*, 46, 399–416, 2003a; Burger, L., *Environ. Monit. Assess.*, 83, 145–162, 2003b; Unpublished data.)

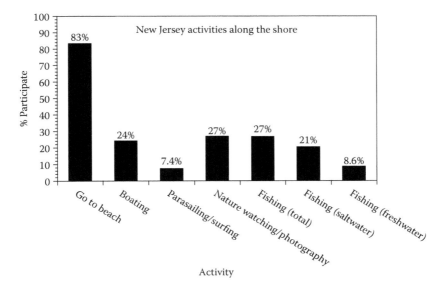

Figure 1.13   Percentage of people who participate in activities along the New Jersey Shore. (After Burger, J., *Urban Ecosyst.*, 18, 553–575, 2015b.) Given is the percentage of people who said they went for that reason. People could go for more than one reason.

significantly more often than those living in central New Jersey (Figure 1.14) and engaged more in nature watching and photography. Not surprisingly, shore people engaged significantly more often in saltwater fishing, whereas inland dwellers engaged in more freshwater fishing days/year (Figure 1.14; Burger 2015b). It is worth noting that although the number of anglers increased from the mid-1950s to 1985, numbers have remained similar or decreased slightly since then (Figure 1.15).

These findings are predictable—if you live along the shore you can go to the beach as often as you like, but if you live farther away, going to the beach involves travel time. There are, of course, many commercial and industrial interests along the shore, including marinas, party boat operations, fishing, shellfishing, and boat making. A century ago, duck hunting fostered a lively decoy industry, and both hunting and decoys remain characteristic of Barnegat Bay culture. The varied recreational activities contribute markedly to the economy of New Jersey, and other states along the coast.

How People Have Changed the Ecosystem of the Bay

Balancing the needs of people who continue to move to coastal areas with the sustainability requirements of natural ecosystems, is an ongoing issue (Thom et al. 2005; Wilson and Fischetti 2010). There are several ways that people have changed the Atlantic coast, many of which were mentioned earlier. The major changes include: (1) dredging channels and keeping the Intracoastal Waterway open for boat transportation; (2) performing beach nourishment or sand replacement on beaches; (3) building groins and jetties on the ocean side to "stabilize" the beaches; (4) filling in marshes to "reclaim" the land; (5) cutting the salt hay on *Spartina patens* marshes; (6) ditching and draining the marshes to prevent mosquito breeding; (7) building houses, marinas, and other structures on the barrier islands, some salt marsh islands, and the mainland; (8) building bulkheads or using rip-rap and other structures to stop the natural erosion of shorelines; and (9) developing shipping ports. For several estuaries, their ports are a prominent feature, with many port calls each day (Figure 1.16), numbers that have remained relatively constant for a decade (2002–2012).

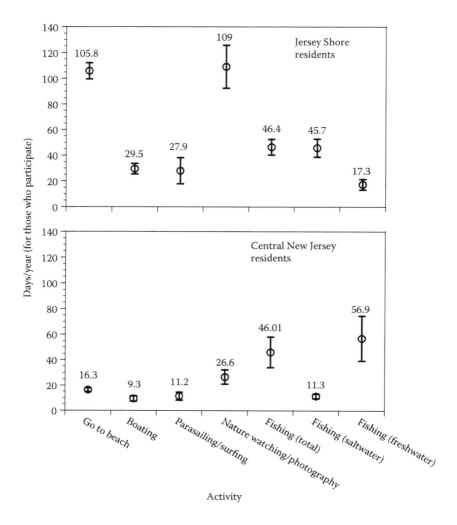

Figure 1.14 Differences in participation rates for people living in New Jersey depending on where they live. People living along the shore have higher participation rates than those living inland. (After Burger, J., *Urban Ecosyst.*, 18, 553–575, 2015b.)

All of these regional anthropogenic processes contributed to halting and redirecting the dynamic natural changes of the shore, beaches, dunes, marshes, bays, and estuaries (Nordstrom 2008). In all cases, the physical and biological aspects of East Coast bays and estuaries have changed dramatically.

How the Physical and Biological Aspects of the Bay Have Influenced People

The physical structure of each bay from Boston Harbor to the Chesapeake Bay has influenced how people use the bay. Bays that are deep, connected to rivers, or wide enough have become commercial ports and the lifeblood of the region. But ports, in turn, require maintenance, and as ships have gotten larger, channels have had to be deepened to remain competitive. Towns and cities have grown up around the ports. Barnegat Bay is not a major port because there are no rivers that could have once served as passageways to interior North America, as the Hudson River does for the New

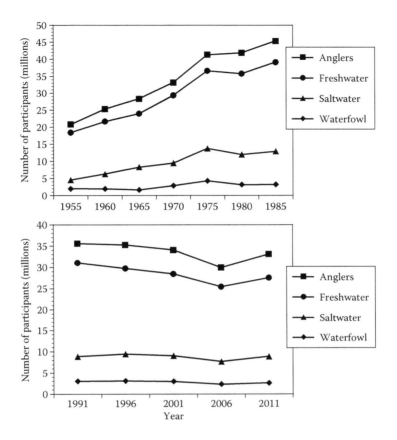

Figure 1.15 Changes in the number of anglers and waterfowl hunters since 1955. There are two panels because the U.S. Fish and Wildlife Service changed the way they count participants. Still, within each period methods were the same (USFWS 2011a). The number of saltwater anglers and fresh-water anglers is the number that engage in each, but because some people do both, the number of total anglers is not a sum of the two.

York–New Jersey Harbor Estuary, the Delaware River does for Delaware Bay, or the Potomac River does for Chesapeake Bay. Barnegat Bay is not deep enough to carry large boats or oil tankers. On the other hand, the shallow nature of Barnegat Bay has meant that there are extensive sea grass beds that serve as spawning areas and nurseries for fish and shellfish and wintering areas for waterfowl.

All of the bays and estuaries have received nutrients and contaminants from runoff, effluents from sewer overflows and other human activity, and oil from boats and ships. These aspects of the bays will be described further in Chapter 2. The Environmental Protection Agency's (EPA) Harbor Estuary Programs have contributed markedly to bringing together information known about each bay, and developing indicators that can be used in the different bays and estuaries to assess health and well-being (EPA 2007a–d). However, there never seem to be enough resources for actual comprehensive monitoring.

The biota of the bays have influenced the industries and activities that each bay experiences. Fish and shellfish are harvested both recreationally and commercially in all the bays and estuaries. The yearly migrations of species such as Bluefish, Striped Bass, and Shad influenced the activities of recreationists, commercial fisheries, and the industries that depend on them. Overfishing of Striped Bass required closing the fisheries for a decade to rebuild the stock (NOAA 2013). Migratory ducks that use the estuaries gave rise to market hunting in the 1800s. In the late 1800s, terns, gulls, and egrets were killed for their plumes for the millinery trade, and whole terns were mounted on ladies'

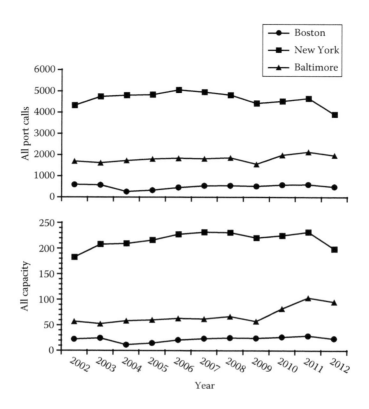

Figure 1.16 Use of the three main commercial harbors—Boston, New York, and Baltimore. Given is the number of port calls and capacity of the ships. (Data from United States Maritime Administration [USMA], 2015, 2002–2012 Total vessel calls in U.S. ports, terminals, and lightering areas report, http://www.marad.dot.gov/resources/data-statistics/.)

hats (Doughty 1974; Moore-Colyer 2000). Duck hunting was once a major commercial practice with hundreds killed with a single discharge of huge bore shotguns. Waterfowl hunting remains popular still, fostered by the National Wildlife Refuges along the Jersey Shore. At the same time, the different bays developed a history and reputation for carving different types of decoys, which remains a popular regional craft.

The physical and biotic aspects of the bays and estuaries also influence how people use them. Dunes and salt marshes buffer the mainland from the ravages of storms (Feagin et al. 2009). The protective role of dunes was particularly evident during *Sandy* in some communities (BBB 2012). Contamination of sediment and water leads directly to people through the food chain (refer to Figure 1.6). Not only do people enjoy the wildlife of the bay, engage in photography and kayaking to see wildlife, but they also catch fish and crabs from the bay, and hunt ducks for consumption. The levels of mercury and PCBs in some fish are sufficiently high to potentially cause adverse health effects if consumed often (IOM, 1991, 2006), leading to state and federal advisories.

Blue Crabs, Bluefish, and Striped Bass are popular species, both recreationally and commercially in the Northeast (Gobeille et al. 2006; Burger 2009b,c, 2013b). A large percentage of people eat them (Figure 1.17), yet there are advisories for all three, and in some places (e.g., Passaic River, Newark Bay) it is illegal to catch Blue Crabs (also called Blue-claw Crabs) because of dioxin levels (Belton et al. 1982; Pflugh et al. 2011). It is an environmental justice issue because some ethnic groups rely on subsistence fishing for protein, eat more of these species, and are less aware of the warnings (Figure 1.18).

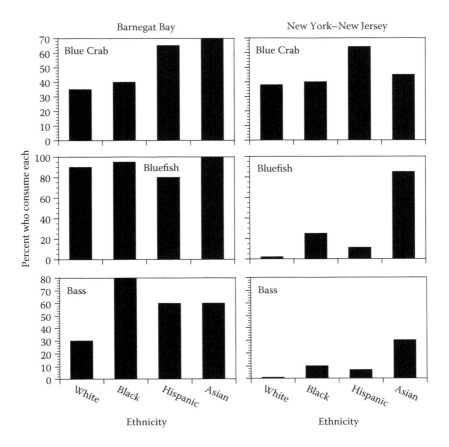

Figure 1.17 Percentage of anglers who consume the Striped Bass, Bluefish, and Blue Crabs they catch in the New York–New Jersey Harbor and in Barnegat Bay as a function of ethnicity. (After Burger, J., *Environ. Res.*, 109, 803–811, 2009c; Burger, J., *J. Risk Res.*, 16, 1057–1075, 2013b; Unpublished data.)

Advisories are usually issued based on the levels of contaminants in fish and the expected consumption rates. The U.S. EPA has an ambient water quality standard for methylmercury based on the concentration in fish tissue and the risk to humans consuming the fish. "The criterion is a concentration of methylmercury in fish that EPA calculated to protect human health." The standard is 0.3 μg/g (300 ng/g wet weight; EPA 2001). There is a great deal of variation in the mean levels of mercury among species within a bay. Some fish, such as Summer Flounder (also called Fluke, *Paralichthys dentatus*), are clearly safe for frequent consumption (Burger and Gochfeld 2011a,b; Burger 2013b), whereas others such as Striped Bass and Bluefish are not (Gochfeld et al. 2012). Fluke, for example, is quite low in mercury whereas Striped Bass can have mean levels exceeding 0.3 ppm (Figures 1.18 and 1.19). The potential risk to people and their offspring from eating fish varies, and people need to be aware of these differences to make wise choices about which fish to eat or release (see Chapter 14).

Whether or not people choose to eat the fish they catch depends on whether they know about any advice or advisories, *and* whether they choose to follow them, or whether they believe the warning applies to them personally. In general, people are aware of advisories, but this knowledge varies by ethnicity (Figure 1.18). Again, this is an environmental justice issue, and requires agencies and organizations to do a better job of spreading the word about advisories, particularly for Asians who might have recently migrated to the United States. According to a New York City blood survey of a representative sample of 1811 women, this is potentially a significant exposure problem, particularly

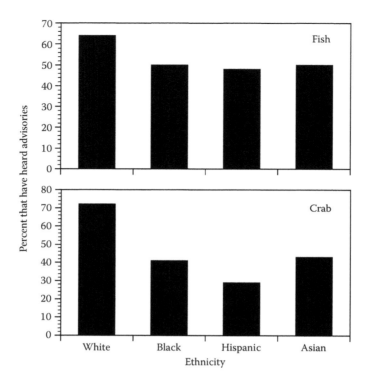

Figure 1.18 Awareness of consumption advisories for fish and for crabs of anglers in the New York–New Jersey Harbor Estuary and in Barnegat Bay as a function of ethnicity.

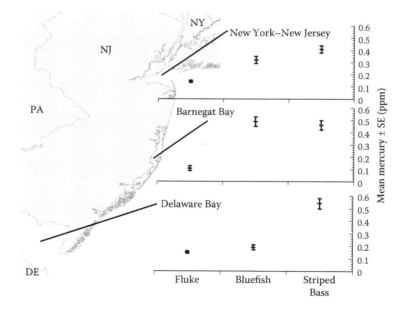

Figure 1.19 Mean (±SE) levels of mercury in fillets of Fluke (Summer Flounder, *Paralichthys dentatus*), Bluefish and Striped Bass from New York and New Jersey bays and estuaries. (After Burger, J., *Environ. Res.*, 109, 803–811, 2009c; Burger, J. et al., *J. Toxicol. Environ. Health*, 72, 853–860, 2009b; Gochfeld, M. et al., *Environ. Res.*, 112, 8–19, 2012.)

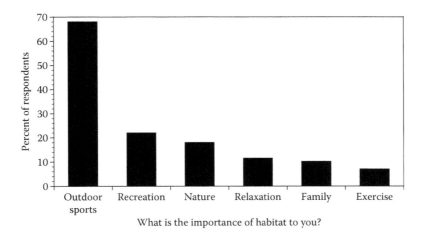

Figure 1.20   Importance of the coastal habitat for people from the New York–New Jersey Harbor to Delaware Bay, New Jersey. (After Burger, L., *Environ. Monit. Assess.*, 83, 145–162, 2003b; Unpublished data.)

for foreign-born Chinese women, two-thirds of whom have blood mercury levels exceeding the reporting threshold of 5 µg/dL (McKelvey et al. 2007).

How People Perceive the Bay and Want to See Changes

People living along the coast from the New York–New Jersey Harbor to Delaware Bay perceive bays and estuaries as providing a range of values for them. When asked what coastal habitats provide for them, most people mention outdoor sports, recreation, and nature (Figure 1.20). They perceive that the most important problems in the bays and estuaries are pollution (90%), followed by population increases (11%), and societal issues (about 5%) (Burger 2003b). Because the question was open-ended, people could list more than one option (thus, the data add up to more than 100%). An emphasis on pollution is not surprising, as contaminants in fish and shellfish have led to consumption advisories, or even outright fishing bans (Belton et al. 1982). Levels of many contaminants have declined, but they remain a problem (see Chapter 2).

How Perceptions Influence Management and Public Policy

Perceptions have the power to influence managers, elected officials, and public policy makers. People support restoration and management of places that they value and use (Shrestha et al. 2007; Burger et al. 2011a). People engaged in activities along the shore of the New York–New Jersey Harbor, Barnegat Bay, Manasquan Inlet, and Delaware Bay, were asked an open-ended questions— "What improvements would you like to see in bays and estuaries?" (Burger 2003b). They responded by mentioning pollution prevention first, followed by "other remediation," and then biological improvements (Figure 1.21; Burger 2003b). In the category "other remediation," people mentioned removing old cars and junk, fixing bulkheads, and cleaning up the beaches and marshes. Biological improvements included creating more fish spawning and bird breeding habitats (Figure 1.21).

When asked to rate the importance of a number of possible ecosystem and societal improvements from a list that they were given, people from several bays listed creating fish breeding habitat and building fishing piers as important (Figure 1.22; Burger 2003b). These perceptions acknowledge that to have better fishing, there needs to be both more fish breeding habitats to increase the number

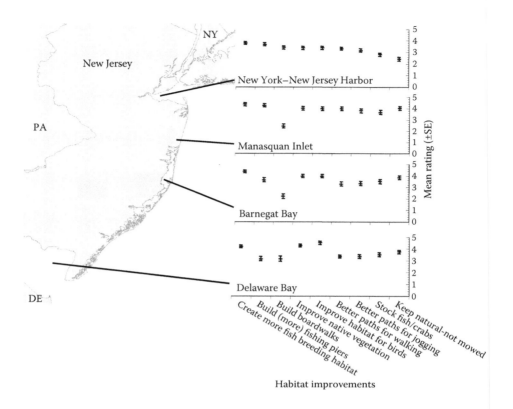

Figure 1.21 Habitat improvements people mentioned on open-ended questions. People were surveyed from New York–New Jersey Harbor to Delaware Bay. (After Burger, L., *J. Environ. Plann. Manag.*, 46, 399–416, 2003a; Burger, L., *Environ. Monit. Assess.*, 83,145–162, 2003b; Unpublished data.)

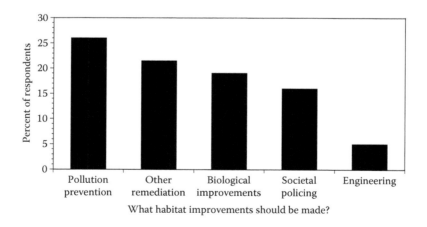

Figure 1.22 Habitat improvements people would like to see in bays from New York–New Jersey Harbor Estuary to Delaware Bay. (After Burger, L., *Environ. Monit. Assess.*, 83, 145–162, 2003b; Unpublished data.) People were asked to rate a list of possibilities on a scale of 1 (no interest) to 5 (high interest).

of fish, and more safe places from which to fish. The list contained several biological improvements (create more fish habitat) and several societal ones (build boardwalks). People could only rate the list they were given, but the data clearly indicate that people care about pollution prevention, mitigation, and biological improvements, as well as increases in amenities to improve their recreational opportunities. Public policy makers respond to public opinion and to the issues that people raise. Thus, knowing what issues people care about is critical to developing sound educational program and influencing public policy.

SUMMARY AND CONCLUSIONS

The health and well-being of avian communities within bays and estuaries, and the ecosystems they are part of, are driven by natural and anthropogenic stressors, both short term and long term. These include predators and competitors, invasive species, contaminants, human disturbance, habitat loss, sea level rise, climate change, and subsidence of the land. The pressures these stressors exert, and the interactions between them, result in the diversity and complexity of avian communities in the bays from Cape Cod to the Chesapeake Bay. The human dimensions have both shaped the bays, used the bays, and continue to have major effects on the habitats and overall ecosystems. These interactions are discussed further in the later chapters. Indicators of changes in avian populations need to be developed for adequate public action and protection.

Although maintaining the physical and biological integrity of bays and estuaries is critical for the success of birds and other wildlife within them, human dimensions are equally critical. People have developed the bays, and used the surrounding environments for commerce, recreation, and living, but the physical and biological aspects of the bay also influence people. Additionally, people have concerns and perceptions about the bay, what the problems are, and what should be done about them. Most of the bays and estuaries have special interest groups to support keeping them clean and productive (EPA 2007a–d). Over time, people's voices influence public officials, managers, and public policy makers responsible for preservation, active management, and restoration.

CHAPTER 2

Barnegat Bay and Other Northeast Estuaries

INTRODUCTION

This chapter describes the bays and estuaries covered in this book from Cape Cod, Massachusetts, to Cape Henry, Virginia, including Boston Harbor, Buzzards Bay, Long Island Sound, New York–New Jersey Harbor, Barnegat Bay, Delaware Bay, and Chesapeake Bay. The estuaries and bays vary in size and configuration, in their watersheds and airsheds, in their mix of developed and undeveloped land, in their roles in commerce and industry, and in their commercial and recreational fishing. They also differ in water quality and in the species compositions of their aquatic and coastal ecosystems. Most of the birds we discuss occur in all of these estuaries, but the numbers vary, partly along the north–south gradient.

As different as they are, the bays and estuaries share in common a history of growing populations, exploitation and overexploitation, ecosystems modified or eliminated, and domestic and industrial wastes disposed of, with little regard to the capacity of the system or the impact on its resources (EPA 2007b). The estuaries share a common theme of human impacts on fragile coastal habitat. They also share the challenges of maintaining biodiversity, achieving clean water compliance, and providing a full range of ecological services in a sustainable manner for the inevitably growing populations that will rely on them for future work, food, or fun. These estuaries provide a multitude of ecological, economic, and social benefits, including flood control, recreation, commercial and recreational fisheries, and fish and wildlife habitats (EPA 2007c). To varying extents, the estuaries receive runoff and/or effluent from urban point and nonpoint sources, from sewage treatment plants, from agriculture and industry, and even from mining and quarrying.

NATIONAL ESTUARY PROGRAM

The study and protection of the United States estuaries has been enhanced by the National Estuary Program (NEP), established by the U.S. Congress in a rare moment of clarity in the 1987 amendments to the Clean Water Act. The objectives were to "identify, restore, and protect nationally significant estuaries," and the program focused on the "integrity of the whole system—its chemical, physical, and biological properties, as well as its economic, recreational, and aesthetic values" (EPA 1997c). The NEP encourages and empowers stakeholder communities to play a role in managing their estuary. Our studies on Barnegat Bay benefited from its incorporation into the NEP, and the 28 estuaries embraced by NEP include Boston Harbor, Buzzards Bay, Long Island Sound, New York–New Jersey Harbor, Barnegat Bay, and Delaware Bay. The Chesapeake Bay has its own estuary program to protect and restore the estuary. Established in 1983, the Chesapeake Bay Program (CBP) became the prototype for the NEP (EPA 1997c). NEP is administered by the Environmental Protection Agency (EPA), but each estuary has a decision-making committee of stakeholders to prioritize issues for the estuary. Some of our research has been supported by NEP, and the many published reports from the different estuary projects provide background information for this chapter.

For the estuary programs, major concerns were toxic contamination (industrial chemical, agrochemicals, metals, and organics), pathogens from untreated sewage, and wetland loss (EPA 2007a–f). More recent concerns include pharmaceutical and endocrine active substances, which were not on the radar screen in 1988. In addition to the NEP, the National Oceanic and Atmospheric Administration (NOAA) conducted sediment toxicity testing as part of the National Status and Trends Program (Hartwell et al. 2001; Hartwell and Hameedi 2007). This included chemical analysis of sediment, Blue Mussel (*Mytilus edulis*) tissue, and a variety of standardized bioassays (survival of amphipods, mortality of clam larvae, and the Microtox™).

In addition to our study area of Barnegat Bay, the other Northeast coast estuaries also have significant waterbird populations, allowing for comparison of population trends, habitats, and metal contamination. The NEP and other sources provide a general overview of the characteristics of the bays and estuaries we discuss in this book (Table 2.1). For each of the estuaries, there are a variety of sources that report statistics in a variety of ways: entire watershed, estuary extent, open water area, and area within each state. Thus, the data in Table 2.1, assembled from publications and websites, are not as comparable across estuaries as one would like. In some cases, ranges of size measurements are given. Most habitats—sandy beaches, salt marshes, and tidal flats—are shared across all. The Massachusetts estuaries and Connecticut shore include rocky intertidal and shorelines. The Delaware and Chesapeake include oyster shell reefs.

From Boston to the Chesapeake, there are several general environmental trends with latitude. The water is generally deeper and clearer northward (EPA 2007f). Water quality is generally "good" in the north and "poor" in the south. Sediment input is generally greater southward, whereas tidal range is generally greater in the north (New England) than from New Jersey southward. Sediment quality did not show a latitudinal gradient, being poorer closer to large urban areas, mainly because of metal and organic contaminants (EPA 2007f). Sampling for sediment toxicity appears spatially uneven in the different estuaries. Toxicity was rated "good" for Barnegat Bay and "poor" for Massachusetts, New York–New Jersey, and Delaware. The methods employed for assessing the benthic index of biodiversity were not uniform. Most of the estuaries were judged using a composite diversity index (Paul et al. 2001), but in the north (Maine) and in the south (Chesapeake), the more familiar Shannon–Weiner index of biodiversity was used. This captures the number of species present as well as the evenness of the population numbers, such that overgrowth of a single species lowers the index (Spellerberg and Fedor 2003). One of the quality measures used by the EPA (2007f) is "Fish toxicity," but although concentrations of contaminants are reported, this is mostly based on very few samples. Fish contamination was rated "good" for the part of Delaware, "good to fair" for Barnegat and Massachusetts, and "poor" for the New York–New Jersey Harbor and most of Delaware Bay. In 2010, the CBP reported that 72% of the bay's tidal-water segments are fully or partially impaired as a result of the presence of toxic contaminants (EPA 2012a), and many individual waters have fish consumption advisories based on polychlorinated biphenyls (PCBs) and mercury.

When these metrics were combined into a single score overall, the Northeast estuaries achieved lower scores than those along the Southeast coast, Gulf coast, or West coast. EPA (2007f) concluded that this comparably low score was not unexpected and that overall condition scores correlated fairly well with population density for the individual Northeast coast NEP estuaries.

Water Quality Index

The EPA and participants in the various estuary programs use a water quality index that incorporates chemical and biological attributes, including dissolved inorganic nitrogen and phosphate content of the water, dissolved oxygen, chlorophyll *a* as a measure of phytoplankton density, and water clarity (usually measured by a black and white Secchi disk). Water quality varies

Table 2.1 Characteristics of Major Bays and Estuaries Discussed in This Book

	Boston Harbor and Massachusetts Bays	Buzzards Bay	Long Island Sound	New York–New Jersey Harbor	Barnegat Bay	Delaware Bay	Chesapeake Bay
References	EPA 2007a	BBNEP 2015; BBS 2015; EPA 2007e	EPA 2007g; LISS 2015a,b	EPA 2007b,c; NYNJ-HEP 1996	EPA 2007c	EPA 2007d; Hartwell et al. 2001	Brown and Erdle 2009; CBF 2015; CBP 2015; EPA 2012a, 2015b; Hartwell and Hameedi 2007; Phillips 2007; USGS 2012
Designated	1988	1987	1987	1988	1995	1996	1983
Watershed area	Ca. 1140 mi² 2950 km²	435 mi² 1209 km² relatively small	16,000 mi² 41,400 km²	8400 mi² 21,750 km²	660 mi² 1710 km²	14,119 mi² 36,570 km²	64,000 mi² 165,000 km²
Length of coastline	800–1200 mi 1280–1920 km	350 mi 560 km	ca. 600 mi 950 km	Not given. Estimate 135 mi (215 km) to Washington Bridge	Not given. Estimate 110 mi 180 km	170 mi (from Trenton) 272 km	11,684 mi 18,700 km
Estuary area	1650 mi² 4273 km²	233–279 mi² 603–722 km²	1300 mi² 3370 km²	Not given. Core area measures 1445 mi² (3750 km²)	75 mi² 195 km²	6747 mi² 17,475 km²	4480 mi² 11,600 km²
Includes	Cape Cod Bay, Massachusetts Bay, Boston Harbor, Merrimack River, N and S shores, Ipswich Bay (Massachusetts only)	Inner harbors of Elizabeth Islands, portions of the Cape Cod Canal	Connecticut, parts of New York, Massachusetts, New Hampshire, Rhode Island, and Vermont	Tidal waters of Hudson (from Troy Dam) and Raritan R. to Sandy Hook, NJ, and Rockaway Point, NY, to the mouth of the harbor	Point Pleasant to Little Egg Harbor Inlet, including Manahawkin Bay and Great Bay, all in New Jersey	Falls at Trenton and Morrisville, PA, south to the mouth of Delaware Bay between Cape May (New Jersey) and Cape Henlopen (Delaware)	Parts of Delaware, Maryland, Virginia, and West Virginia

(Continued)

Table 2.1 (Continued) Characteristics of Major Bays and Estuaries Discussed in This Book

	Boston Harbor and Massachusetts Bays	Buzzards Bay	Long Island Sound	New York–New Jersey Harbor	Barnegat Bay	Delaware Bay	Chesapeake Bay
Major freshwater inputs	Charles R., Merrimack R.	Agawam, R., Wankinco R., Wewantic R., Mattapoisett R., Acushnet R., Paskamanset R., Westport R.	Connecticut R., Housatonic R., Quinnipiac R., Peconic R., Mianus R., Bronx R., Mamaroneck R.	Hudson R., Raritan R., Passaic R.	Metedeconk R., Toms R., Forked R., Cedar Creek, Oyster Creek, Tuckerton Creek, Mullica R.	Delaware R., Schuylkill R., Murderkill R., Salem R., Cohansey R., Maurice R.	Susquehanna R., Potomac R., James R., York R., Rappahannock R.
Number of fish species	141	250+	250	200+	200+	200+	350+
Harvested fish species	Bluefin Tuna (*Thunnus thynnus*), Atlantic Cod (*Gadus morhua*), Winter Flounder (*Pseudopleuronectes americanus*), Atlantic Herring (*Clupea harengus*), Striped Bass (*Morone saxatilis*)	Alewife (*Alosa pseudoharengus*), American Eel (*Anguilla rostrate*), Blueback Herring (*Alosa aestivalis*), Bluefish, Mummichog (*Fundulus heteroclitus*), Striped Bass	Winter Flounder (*Pseudopleuronectes americanus*), Tautog (*Tautoga onitis*), Bluefish, Scup (*Stenotomus chrysops*), Striped Bass	Shad, Blueback Herring, Alewife	Striped Bass, Bluefish, Weakfish, Fluke, Summer Flounder	Striped Bass, Shad (*Alosa sapidissima*), Sturgeon (*Acipenser oxyrhynchus oxyrhynchus*), Amerian Eel (*Anguilla rostrate*), Blueback Herring, Atlantic Menhaden (*Brevoortia tyrannus*), Alewife, Bluefish, Weakfish	Atlantic Menhaden, Bluefish, Shad, American Eel, Atlantic Croaker (*Micropogonias undulates*), Atlantic Spot (*Leiostomus xanthurus*), Black Drum (*Pogonias cromis*), Red Drum, Striped Bass, Black Sea Bass (*Centropristis striata*), Spanish Mackerel (*Scomberomorus maculatus*), Summer Flounder (*Paralichthys dentatus*), Weakfish (*Cynoscion regalis*), Atlantic King Mackerel (*Scomberomorus cavalla*)

(Continued)

Table 2.1 (Continued) Characteristics of Major Bays and Estuaries Discussed in This Book

	Boston Harbor and Massachusetts Bays	Buzzards Bay	Long Island Sound	New York–New Jersey Harbor	Barnegat Bay	Delaware Bay	Chesapeake Bay
Harvested shellfish and crabs	Soft shell clams (*Mya arenaria*), oysters, Bay Scallops (*Argopecten irradians*), American Lobster (*Homarus americanus*), Blue Mussels (*Mytilus edulis*)	Soft shell clams, Quahogs (*Mercenaria mercenaria*), Scallops, Oysters, American Lobster	Oysters, American Lobster	Blue Crab	Clams, Scallops, Mussels, Blue Crab Horseshoe Crabs for medical uses and bait (Delaware only)	Blue Crab, oysters	Blue Crab, oysters, clams, mussels
Number of people in watershed	3.8 million + (2000)	250,000 (2000)	14.6 million (2000)	16.9 million (2000); 3097 persons/mi^2	560,000 to 1.55 million (2000); 1,055 persons/mi^2	9.4 million (2000); 772 persons/mi^2	17.8 million (2013)[3]
Population growth	23% in 40 years	72% during 40 years	14% during 40 years	13% during 40 years	132% during 40 years	35% during 40 years	113% in 60 years[3]
Overall condition (see Table 2.2)	Fair	Fair	Poor	Poor	Fair	Poor	Poor
Environmental concerns	Stormwater runoff, sewage-related pollution, effects of human development	Toxic contamination, bacterial contamination, nonpoint source pollution, habitat loss, nitrogen loading, coastal eutrophication, population growth	Hypoxia, toxic substances, and land-use changes, development, and urban sprawl has increased pollution and stormwater runoff, altered land surfaces, decreased natural areas, floatables	Toxics, pathogens, wetland loss, raw sewage	Nonpoint pollution, habitat loss and alteration, human activity and competing uses, water supply protection, fishery decline	Pollution, storm water runoff, past industrial effluent	Agricultural runoff, air and water pollution, chemical pollution, land use

(Continued)

Table 2.1 (Continued) Characteristics of Major Bays and Estuaries Discussed in This Book

	Boston Harbor and Massachusetts Bays	Buzzards Bay	Long Island Sound	New York–New Jersey Harbor	Barnegat Bay	Delaware Bay	Chesapeake Bay
Biological issues	Overfishing; invasive species	Over harvest, habitat degradation	Parasitic disease, habitat loss, human and predator intrusion, increased development	80% of tidal wetlands are gone; most of freshwater wetlands gone	Land use changes	Land use changes; declines in living resources	Fisheries, dead zones, habitat degradation
Pollutants issues	PAHs, copper, arsenic, lead, cadmium, mercury, zinc, PCBs, pesticides	Bacterial contamination, 2003 oil spill, excess nutrient levels	PCBs, nitrogen and phosphorus, sewage and pathogen contamination	PCBs, dioxins, PAHs, mercury, and cadmium	Mercury, dieldrin, PCBs	Advisories for PCBs and pesticides. Mercury. Also PAHs	Nitrogen and phosphorus, natural gas, polluted runoff, sewage
Fish tissue contamination	80% of samples had moderate or high PCBs; statewide marine fish advisories because of mercury ($n = 20$)	83% of samples exceeded EPA Advisory Guidance values for at least one contaminant	PCBs, metals, PAHs	Fish advisories issued for mercury ($n = 14$)	31% had elevated levels of mercury, PCBs or dieldrin ($n = 13$)	Mercury advisories for fish consumption; 62% of fish tissue were rated poor for contaminants (PCBs, DDTs, PAHs, dieldrin; $n = $ ca. 44)	Mercury, PCBs, PAHs

Note: Each bay has somewhat different criteria for reporting areas of the bay, the estuary, and the watersheds and coastline. Where conflicting estimates are identified, the larger value is given.

spatially and temporally, influenced by rainfall, runoff, temperature, and pollution. Good water quality has low nitrogen, low phosphorus, low chlorophyll, high dissolved oxygen, and high water clarity.

Sediment Quality Index

The EPA National Coastal Assessment methods measure contamination, toxicity, and total organic carbon in sediment samples. Contaminants include polycyclic aromatic hydrocarbons (PAHs), PCBs, pesticides, volatile organic compounds, and metals. These pollutants are discharged into waterbodies from towns and cities, industries, agriculture, and mining. Much of the contamination is historic, occurring before the 1970s and the origin of the EPA, and before many key environmental regulations. But some contamination, particularly urban runoff and combined stormwater outfalls (CSO), continue. Contaminants may be dissolved or particulate, and may undergo complexation with organic matter or chemical alteration (i.e., methylation of mercury). Contaminants can be directly toxic to microorganisms and benthic organisms living in the sediment, or can bioaccumulate and bioamplify in the food chain, resulting in harm throughout the food web, but particularly to higher trophic-level organisms (LISS 2015a; Figure 2.1).

Quality Index Comparisons

Water quality is based on dissolved inorganic nitrogen (DIN), dissolved phosphorus (DIP), chlorophyll a, clarity, and dissolved oxygen. Sediment quality is based on toxicity, contaminant levels, and total organic carbon. The benthic index combines the diversity of all benthic species, reduced by excess abundance of oligochaete and polychaete organisms (Paul et al. 2001). Fish tissue contamination includes PCBs, DDT compounds, and mercury, and more recently, pharmaceutical agents. The sample size for these analyses is very small. Each of the individual components is assessed separately

(a) (b)

Figure 2.1 Higher trophic-level birds in Delaware Bay include Bald Eagles (a) and Ospreys (b). Both catch fairly large fish. Eagles may pirate fish from Ospreys and also catch waterfowl.

Table 2.2   Ratings of Environmental Quality for Several Bays and Estuaries in the Northeast

Estuary	Water Quality	Sediment Quality	Benthic Biodiversity	Fish Tissue Contaminants
Massachusetts	Good	Poor	Fair to poor[a]	Fair ($n = 20$)
Buzzards Bay	Good	Fair	Good to fair	Poor (n = not given)
Long Island Sound	Fair	Poor	Poor	Poor ($n = 13$)
New York–New Jersey Harbor	Poor	Poor	Poor	Poor ($n = 14$)
Barnegat Bay	Good to fair	Good to fair	Fair	Fair ($n = 13$)
Delaware Bay	Poor	Good to fair	Poor	Poor ($n = $)[b]
Chesapeake Bay[c]	Poor	Sediment quality triad = variable	Variable	Poor ($n = $)[b]

Sources: EPA, *Northeast National Estuary Program Coastal Condition, Massachusetts Bays Program*, Environmental Protection Agency, Washington, DC, http://www.epa.gov/owow/oceans/nepccr/index.html, 2007a; EPA, *Northeast National Estuary Program Coastal Condition, New York/New Jersey Harbor Estuary Program*, Environmental Protection Agency, Washington, DC, http://www.epa.gov/owow/oceans /nepccr/index.html, 2007b; EPA, *Northeast National Estuary Program Coastal Condition, Barnegat Bay National Estuary Program*, Environmental Protection Agency, Washington, DC, http://www.epa.gov /owow/oceans/nepccr/index.html, 2007c; *Northeast National Estuary Program Coastal Condition, Partnership for the Delaware Estuary*, Environmental Protection Agency, Washington, DC, http://www .epa.gov/owow/oceans/nepccr/index.html, 2007d. Data from the Chesapeake is from EPA, Toxic Contaminants in the Chesapeake Bay and Its Watershed: Extent and Severity of Occurrence and Potential Biological Effects, U.S. EPA Chesapeake Bay Program Office, Annapolis, MD, http://execu tiveorder.chesapeakebay.net/ChesBayToxics_finaldraft_11513b.pdf, 2012a. With permission.

[a]   Based on Shannon–Weiner diversity index (Spellerberg and Fedor 2003).
[b]   Sample size not found.
[c]   Based on Hartwell and Hameedi (2007).

in the relevant Harbor Estuary Program (HEP) documents (EPA 2007a–e), and related to the proportion of sampling sites in each of the bins: "good, fair, and poor." For example, the water quality index is rated "good" for 93% of sites (mainly in Cape Cod Bay) and "poor" for only 7% of the Massachusetts bays (EPA 2007a,f). Direct comparisons among the bays proved challenging. Conditions of the main bays are summarized in Table 2.2. Information for the Chesapeake Bay, obtained from Hartwell and Hameedi (2007), is not available in the same format because it was not part of the NEP program.

ECOREGIONS

One important method of mapping ecological conditions is with use of Ecoregions. These are regions of the United States that are defined on the basis of geology, soils, physiography, climate, vegetation, wildlife, and land use (Omernik 1987, 2004). It is useful to examine these regions from Maine to Florida to place the Northeast estuaries in an Atlantic coastal context. There are many levels of distinction for Ecoregions, but herein we provide the basic overview (Figure 2.2). These are updated periodically, and with global warming and sea level rise, the regions will change slightly as climate, vegetation, and wildlife change.

The following sections introduce the bays and estuaries to be discussed, including geography, social history, uses, and abuses. Our research focus has been on Barnegat Bay; hence, we describe it first and in most detail. The other estuaries are covered from north to south.

BARNEGAT BAY ECOSYSTEM

Barnegat Bay, sheltered by barrier islands that extend from Point Pleasant to Little Egg Harbor and Great Bay, has two inlets to the Atlantic Ocean: Barnegat Inlet in the middle of the bay and

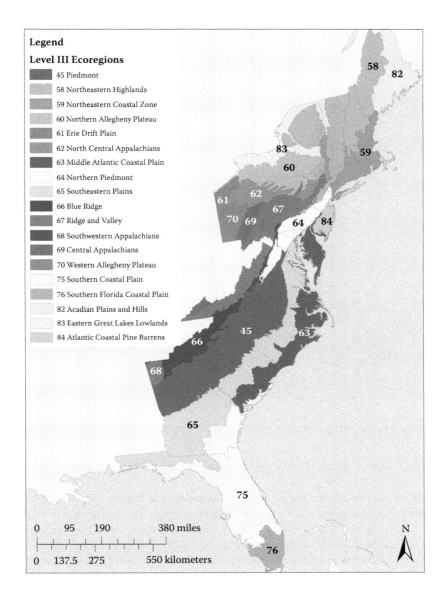

Figure 2.2 Major ecoregions from Maine to Florida. (From EPA 2014c, Cleaner Powerplants, http://www.epa.gov/mats/powerplants.html. With permission.)

Beach Haven Inlet (Little Egg Harbor Inlet) in the south (Figures 2.3 and 2.4). The bay is mostly narrow (2–6 km wide), long (67 km), and shallow (maximum 6 m deep) with a small tidal swing. On the east, Barnegat Bay is bordered by Island Beach and Long Beach Island (barrier islands), most of which are developed for residential communities. It is bordered on the west by salt marshes and by the mainland, extending into the New Jersey Pine Barrens. There are more than 250 small islands in the bay, as well as extensive salt marshes that fringe the mainland and, to a lesser extent, the barrier island (Burger and Lesser 1978). Two bridges provide access to the barrier islands—the Thomas A. Mathis Bridge from Toms River to Island Beach (replacing the first bridge built in 1914), and the Dorland J. Henderson Bridge, better known as the Manahawkin Bay Bridge (1958), to Long Beach Island. Long Beach Island is 29 km long, ranging from 300 to 800 m wide, with

Figure 2.3   View of Barnegat Lighthouse from Barnegat Inlet showing low coastal forest habitat and bulkheads.

a resident population of about 10,000, and swelling to 150,000 in the summer (NJCOM 2015) (Figure 2.5).

Most of the visitors flock to the sandy beaches of the Atlantic, heedless of the marshes with biting Greenhead Flies, mosquitoes, and nesting birds (Burger 1996a). The inland waterway goes through Barnegat Bay, ensuring heavy boat traffic in the summer. However, the bay is shallow so that most boaters keep to the marked channels, keeping them several hundred meters away from most bird nesting islands. The salt marshes (both islands and those attached to the mainland) were largely ignored by people, at least prior to the advent of shallow-draft Personal Watercraft capable of cruising across tidal islands. Like the other bays and estuaries, Barnegat Bay harbors a rich diversity of nesting and migratory birds. The beach, barrier islands, bay, and salt marsh islands form a highly productive system for invertebrates, fish, and birds (Burger 1996a; Burger et al. 2001b; Able 2005; Able and Fahay 2014). Barnegat Bay is part of a coastal Ecoregion. A map of the ecoregion is shown in Figure 2.6.

Like all Northeastern U.S. bays, Barnegat Bay has a long history of human exploitation, and is heavily influenced by human activity. However, most of the other bays considered in this book are also heavily influenced by large urban core areas and busy shipping ports and channels. Barnegat Bay's many channels are occupied by pleasure boats and small fishing boats, and its economies today are driven by the underlying and diverse coastal ecology, mainly tourism and fishing. The only cities of note, Toms River (population of 92,000) and Atlantic City (population of 40,000), lie at the ends of the bay rather than at the core.

The Barnegat Bay–Little Egg Harbor (also known as Great Bay) estuaries became part of the U.S. EPA–designated NEP in 1995. At the time, three priority problems were identified: (1) nonpoint pollution and water quality, (2) habitat loss and alteration, and (3) human activities and competing

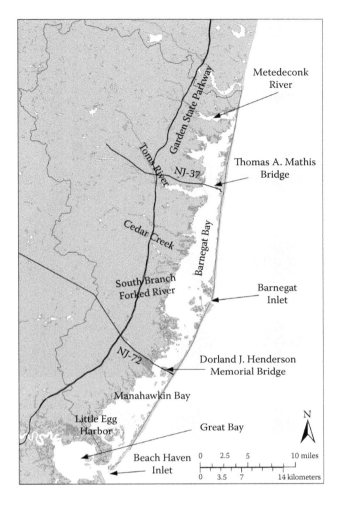

Figure 2.4 Map of Barnegat Bay showing major freshwater inputs (rivers and creek), ocean inlets, and access bridges.

uses (Kennish 2001b; Able 2015). The water quality of estuaries is partly determined by anthropogenic inputs from atmospheric deposition, industrial discharges, river discharges, and surface runoff, which in turn results from farms, industry, and dwellings in the watershed. Further details on the physical attributes of the Bay are described by Kennish (2001c). A history of Coast Guard Station 119 at the southern end of the Bay can be found in the work of Able (2015); the station later morphed into the Rutgers Marine Field Station.

As is true for the other areas, many local residents and policy makers believe that Barnegat Bay is impaired because of overdevelopment, excess nitrogen loading, and contaminants (de Camp 2012). Thus, urban planners and managers are concerned with ongoing development and the ecological consequences of land-use policies. These concerns apply throughout the urban areas from Boston to Washington, DC. During the twentieth century, the bay experienced great development, including bulkheads, infilling, residential development, marinas, mosquito ditching, and lagoon construction. Most of the development occurred before the 1970s (Kennish 2001b), but residential growth has continued. The area developed in the bay's watershed increased from 18% in 1972, to

Figure 2.5   Atlantic City beach with boardwalk in early summer. This use of the beach leaves no place for beach-nesting birds, such as Piping Plovers, Least Terns, and Black Skimmers. (Photo from https://commons.wikimedia.org/wiki/File:3813GrandChpeau1911.png.)

21% in 1984, 28% in 1995, and about 70% by 2010 (Ocean County 2012), although the estimates come from different sources. About 45% of the bay shoreline is bulk-headed (Lathrop and Bognar 2001). Most of the bay lies in Ocean County, which showed the fastest growth of any New Jersey County, increasing 12.8% to a population of 576,000 between 2000 and 2010 (Figure 2.7). Some of the townships bordering the bay, such as Barnegat, grew 25%, in the same period (Ocean County 2010).

Only 29% of the upland-bay buffer area has natural land cover; the rest is developed (Kennish 2001b). More than 33% of the watershed is publicly owned, although this does not guarantee sound management. Halting or slowing development requires sound and enforceable land-use policies. A modeling effort for the ecological consequences of land-use policies indicated that the four generally accepted policies (downzoning, cluster development, wetland/water buffers, open space protection) were unable to alter future land-use patterns (Conway and Lathrop 2005). It is not only natural areas that have decreased in the bay, but public access has as well (Kennish 2001b). The question of public access is only one of the competing claims. There is competition between commercial and recreational fishermen, between homeowners and tourists, between land users and developers, and between commercial and personal recreation, as well as conflicts over the use of boats and personal watercraft (Burger 1998b). Resolving conflicts between stakeholders has always been challenging, and will become more difficult with increasing populations and additional demands on declining resources.

Protecting and enhancing estuaries, such as Barnegat Bay, requires adaptation to natural forces and decreasing or mitigating anthropogenic stressors. Natural and anthropogenic factors interact,

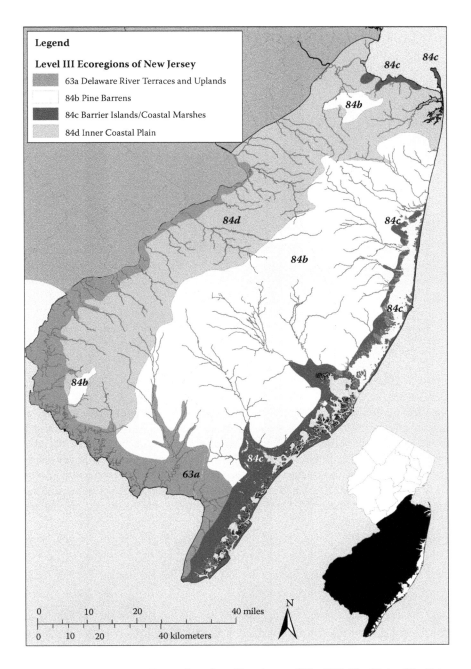

Figure 2.6 Ecoregions of Barnegat Bay and southern New Jersey. (After EPA, The State of the Estuary—2012, NY–NJ Harbor Estuary, http://water.epa.gov/type/oceb/nep/upload/New-York-New-Jersey-SOE _Rprt.pdf, 2012c.)

and governmental responses are often modified by the perceptions and competing concerns of residents and other concerned citizens. All of this plays out against the specter of climate destabilization, associated with sea level rise and increased frequency of superstorms. Despite the severity of several recent storms and hurricanes, and the damage caused by these storms (Freedman 2013), many people in New Jersey and elsewhere in the Northeast do not recognize the role of

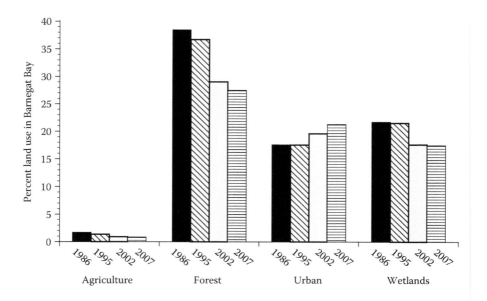

Figure 2.7   Decline in forest and marsh acreage of the Barnegat Bay watershed with increases in urban development. (Data from New Jersey DEP Land Use/Land Cover Maps.) Percentages do not add up to 100 because water and barren lands are not shown.

human activities fostering global climate change and sea level rise (Nordstrom and Mitteager 2001; Burger 2015b).

The management of coastal dune and beach restoration has rested largely with the U.S. Army Corps, which has engaged in large-scale beach nourishment (placement of sand on beaches; Nordstrom 2008). Despite the massive efforts, sand was removed from many beaches in the Northeast by Superstorm *Sandy* (BBB 2012). Before, during, and after *Sandy*, people in New Jersey were more worried about property damage (houses, cars, boats) than they were about their own or community health (Burger and Gochfeld 2014a,b, 2015), and few recognized the importance of dunes and salt marshes in protecting their property (Burger 2015b) (Figure 2.8). After *Sandy*, New Jersey initiated a federally

Figure 2.8   Superstorm *Sandy* (October 29–31, 2012), the second costliest of all hurricanes, devastated communities and habitats from Cape Cod to Virginia, with the heaviest hit areas in northern New Jersey and western Long Island. Shown here are houses destroyed along the Jersey Shore.

funded voluntary buyout of some flood-prone properties—the Blue Acres program. However, most purchases have been homes in river floodplains, rather than on the barrier beaches (Climate Central 2014) where homes have been rebuilt rather than removed.

MASSACHUSETTS BAYS AND BOSTON HARBOR

This estuary covers more than 1280 km (800 mi) of coastline from the New Hampshire border out to the tip of Cape Cod and embraces an area of 4200 km² (1650 mi²). As defined, this estuary is bordered on the north by the cold water Gulf of Maine. It includes Massachusetts Bay, Boston Harbor, Ipswich Bay, and Cape Cod Bay, as well as the Charles and Merrimack Rivers (EPA 2007a). From Salisbury (Maine) to Provincetown on the tip of Cape Cod is 1760 km (1200 mi) of shoreline. The watershed covers more than 17,900 km² (7000 mi²) with a population of 3.8 million. Beyond the immediate influence of Boston City and the Harbor, the bay supports a healthy marine ecosystem of barrier beaches, tidal flats, submerged aquatic vegetation (eelgrass), as well as freshwater and saltwater marshes (EPA 2007a, Figure 2.9). Commercial and recreational fishing are important sources of revenue as well as tourism, ecotourism (particularly whale watching), marinas, and commercial shipping.

Figure 2.9 Cape Cod, a natural beach is part of a long coastline with diverse habitats. (Photo from https://commons.wikimedia.org/wiki/File:SandyNeckDunes.JPG.)

BUZZARDS BAY AND NEARBY WATERS

Buzzards Bay became part of the NEP in 1987. It encompasses coastline extending west to the Rhode Island border, and east along the South Shore to Cape Cod and to Nantucket Sound (EPA 2007e). Officially, the bay is 13×45 km, with a mean depth of 11 m and covers about 600 km². The south shore of Massachusetts is indented with many small bays, resulting in a 563-km coastline on the mainland and Elizabeth Islands. The watershed covers 1123 km² (434 mi²), and is occupied by about 260,000 persons (2000 census). Except for the industrialized New Bedford Harbor (Figure 2.10), most of the boating activity is recreational within a portion of the Intracoastal Waterway. To the east, the 10-m-deep Cape Cod Canal carries very heavy commercial shipping (tankers, barges, ferries, fishing vessels, cruise ships, and even container vessels; Figure 2.10).

The fishing fleet of New Bedford Harbor is one of the largest in the country, ranking number one in the United States in terms of "dollar value landed" (NOAA 2012b), mainly based on scallops. On the other hand, New Bedford has the dubious distinction of being one of the most polluted harbors in the nation (Pesch et al. 2011). The main contaminants of interest are PCBs, the legacy of improper waste disposal by large electric device companies. The harbor itself occupies about 18,000 acres and is highly contaminated with metals and PCBs, over an area of about 10 km from Acushnet River into Buzzards Bay. The harbor was placed on EPA's National Priorities List in 1982, and continues to require significant time and funding to clean up, and hydraulic dredging of contaminated sediment has continued since 2003 (EPA 2011).

Ian Nisbet's studies on the Common and Roseate Terns nesting on Bird Island at the mouth of Sippican Harbor represents one of the longest term studies of birds on a single island in the world (Nisbet 1994, 2002; Nisbet et al. 2011, 2014). Our joint studies are mentioned later in Chapters 6 and 10 through 13 (Figure 2.11). Other major tern nesting islands include Ram Island and Penikese Island (Figure 2.12).

Figure 2.10   New Bedford Harbor in Buzzards Bay with a large fishing fleet of some 500 commercial vessels is among the leading fishing ports in the United States. (Photo from http://www.newbedford-ma.gov /wp-content/uploads/slideshow/fishing-vessels-no-gradient-1024x1024.jpg.)

(a)

(b)

Figure 2.11 Roseate Terns have been studied extensively by I. Nisbet, J. Spendelow, and others for many years, and their populations are small enough to have the major colonies all under study, such as this one at Falkner Island (a). Common Terns nest in the more open spots, whereas Roseate Terns typically nest partly under the rocks or in artificial structures (b).

LONG ISLAND SOUND AND PECONIC BAY

Long Island Sound borders New York and Connecticut, extending 176 km (110 mi) and 34 km (21 mi) across, with 950 km (600 mi) of coastline (EPA 2007g). It has a water surface area of about 3370 km^2 (1300 mi^2), and it extends from the East River and Westchester County, into Block Island Sound. The population of the watershed exceeds 8 million. The Connecticut shore is highly urban with major cities from Stamford east to New Haven. The Long Island shoreline has a mix of residential communities and former estates, some of which are now public parks. Population density

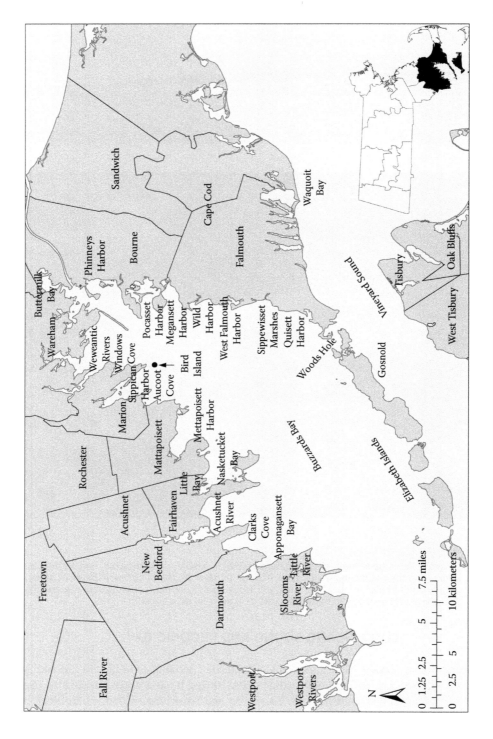

Figure 2.12   Map of Buzzards Bay Estuary. The arrow points to Bird Island, the site of Ian Nisbet's long-term tern studies.

has been constrained by the huge estates and extensive private land holdings, and by wealthy communities that border the water (NYDEC 2015). A dozen rivers enter the Long Island Sound from the north, the largest and longest of which is the Connecticut River, arising in Quebec and draining much of New England. It accounts for 70% of Block Island Sound's freshwater. Other major rivers are the Mianus, Saugatuck, and Housatonic in Connecticut and the Hutchison (Bronx) and Mamaroneck River (Westchester; EPA 2007g). Long Island accounts for very little freshwater input to the Long Island Sound, with no major rivers; most of the islands' surface is in the Atlantic Ocean watershed.

Long Island became part of the NEP in 1987. The Long Island Sound Study project tracks the status and trends of five environmental indicators: climate change, habitats, land use and population, marine and coastal animals, and water quality (EPA 2007g). Its Comprehensive Conservation and Management Plan (CCMP) identifies several priority problems: toxic substances, hypoxia, pathogens, floatable debris, management and conservation of living resources and their habitat, and land use and development. Some of the environmental quality metrics vary spatially (from west to east), and others vary temporarily (e.g., algal blooms). The CCMP has evolved to keep in touch with emerging issues facing all the estuaries: pharmaceuticals among the toxics and climate change (NYDEC 2015). The program also tracks approved versus closed shellfish bed acreage and beach closures or advisories. Toxic contaminants in Blue Mussel are analyzed as part of NOAA's National Mussel Watch (NOAA 2011; Figure 2.13). A variety of point source measurements are made to track industrial chemical discharges. The water quality is generally "poor" in the western basin of the Long Island Sound from Stamford to Bridgeport, Connecticut, and improving to "good" in the eastern basin well to the east of New Haven. Sediment quality was rated "poor" for almost half of the western basin, "poor" for only about 10% of the central basin, but "poor" again for about a third of the eastern basin (NYDEC 2015).

Floatable debris washing up on beaches is the most conspicuous contaminant. Over the past 10 years, for example, volunteers have recovered between 600 and 1600 yd³ of floatables per year (LISS 2015b). However, hypoxia is probably the most serious problem. During the late 1980s, Long Island Sound gained widespread recognition for huge "dead zones" of hypoxia (dissolved oxygen below 3 mg/L) and even anoxia attributable to excess inputs of nitrogen (Figure 2.14). This resulted in a tristate plan to reduce nitrogen inputs into the Long Island Sound. Chlorophyll *a* was high in

(a) (b)

Figure 2.13 Blue Mussels (a) are bioindicators for the NOAA's National Mussel Watch Program. (Photo from http://www.mass.gov/envir/massbays/images/bhha_shellfish_figure2.jpb.) Ribbed Mussels (b) are also common on the edges of salt marsh creeks.

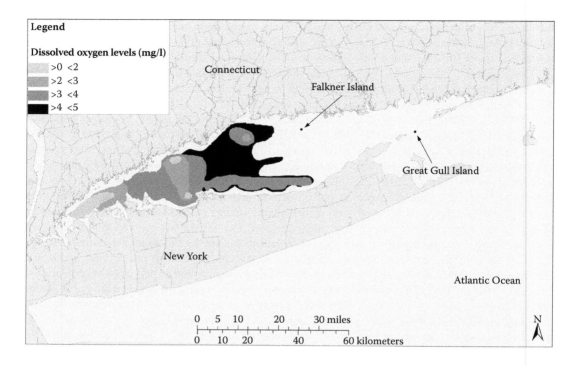

Figure 2.14   Hypoxic areas of Long Island Sound; the legend shows areas below 5 mg/L dissolved oxygen. (From http://longislandsoundstudy.net/about/our-mission/management-plan/hypoxia/.) The map also indicates locations of Falkner Island and Great Gull Island tern colonies.

the early 1990s, declined in 2000, and increased somewhat through the decade, peaking in 2012 (LISS 2015b).

The Peconic Estuary is at the east end of Long Island between Long Island's north and south forks. The watershed begins at Brookhaven National Laboratory, the headwaters of the Peconic River, and spans the several bays from Flanders to Gardiners Island, and joins Long Island Sound in Block Island Sound between Plum Island and Montauk Point. More than 125,000 land acres and 158,000 surface water acres are included in the Peconic Study Area (Peconic 2015). Nearly 25% of watershed area is open space; the year-round human population is just 100,000 (Peconic 2015). However, the Peconic is vulnerable to urban sprawl and growing population. Starting in 1985, a brown tide algal bloom devastated the local scallop industry, a result of excessive nitrogen and hypoxia. Eastern Long Island was highly agricultural and experienced extensive use of pesticides, including aldicarb contamination of its aquifers (Jones and Marquardt 1987). The U.S. Department of Energy's Brookhaven Laboratory has been a significant source of mercury contamination to the Peconic River (Burger et al. 2008b), resulting in conflicting information on the safety of consuming fish from the river (Burger and Gochfeld 2005b), and in the Department of Energy's extensive cleanup program.

Jeff Spendelow's studies of the terns nesting on Falkner Island and Helen Hays and colleagues' studies of Great Gull Island tern colonies represent important long-term studies, particularly for the endangered Roseate Tern. These are some of the largest tern colonies in the United States, and the long-term studies have contributed markedly to our understanding of avian ecology (Spendelow et al. 1995) (Figures 2.11 and 2.15). Falkner Island, about 2 ha, is subject to erosion and requires restoration to maintain tern breeding habitat (USFWS 2011b). Great Gull Island (about 7 ha), owned by the American Museum of Natural History, is one of the major tern colonies in the Northeast.

Figure 2.15   Falkner Island hosts a mixed-species Common and Roseate Tern colony on Long Island Sound. This colony is one of the major Northeast Roseate Tern colonies.

NEW YORK–NEW JERSEY HARBOR

With New York City as its core, this estuary embracing the Hudson River and Raritan River watersheds is "one of the most vibrant and economically important in the country" (EPA 2007b). Its components include the Hudson River up to the limit of tides at Troy dam, near Albany, New York Bay at the mouth of the Hudson, Newark Bay and the Passaic River in the Northwest, and Long Island Sound in the Northeast (Figure 2.16). Jamaica Bay and Sandy Hook Bay are the east and southern extent of the Harbor, whereas Raritan Bay is fed by the Raritan River system and by the Kill Van Kull. Other rivers with significant inputs to the estuary are the Hackensack, Shrewsbury, Navesink, and Rahway Rivers. The watershed is about 16,300 mi^2 (42,450 km^2) and is highly urbanized (EPA 2007b). Agriculture and industry, responsible for a great deal of the historic pollution, still flourish in the region. However, both agriculture and industries have declined because of suburbanization and gentrification, whereas pollution controls have substantially reduced the chemical inputs and effluent over the past 40 years. However, increased development with increases in pavement and impermeable surface, coupled with increased population and vehicles, decreased recharge and increased runoff, resulted in significant ongoing inputs to the Harbor (Steinberg et al. 2004). Sewage is still an intermittent problem in heavy rain events that overcome the combined sewer overflows. Activities specifically identified as "uses" include fishing, boating, swimming, bird watching, transportation, and commerce—all of which benefit from clean water, uncontaminated sediments, and biodiversity (EPA 2007b).

Sandy Hook, Staten Island, Breezy Point, and Jamaica Bay are part of the Gateway National Recreation Area administered by the National Park Service. Sandy Hook, in particular, sees heavy recreational visitation throughout the year. A priority of the Park Service is to maintain the

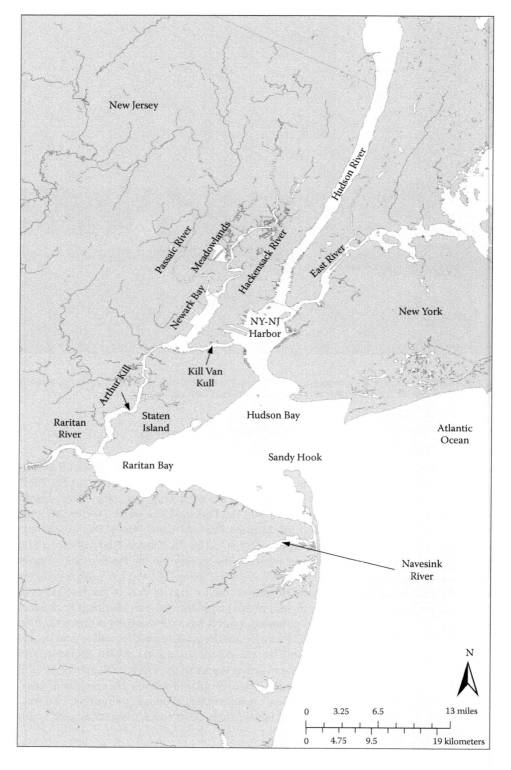

Figure 2.16 Map of the lower New York–New Jersey Harbor Estuary.

recreational opportunities, including wildlife conservation for Gateway visitors. The *Comprehensive Conservation Management Plan* (NYNJ-HEP 1996) outlines five priority areas of concern: (1) habitat loss and degradation, (2) toxic contamination and management of dredged material and sewage overflows, (3) pathogen contamination affecting public beaches and shellfish beds (particularly after heavy rains), (4) floatable debris, and (5) nutrient and organic enrichment supporting toxic algal blooms resulting in low levels of dissolved oxygen, in turn impairing fish and shellfish production (NYDEC 1996). About 80% of the harbor estuary's tidal wetlands have been developed. The urban core once had 224,000 acres of freshwater wetlands, which are now mostly gone (EPA 2007b). Chief among these are the Hackensack Meadowlands, of which more than 70% has been drained (EPA 2012c), and much of the remainder was converted to impoundments or invaded by *Phragmites*. Under the scrutiny of many agencies through the NEP, the rate of loss has declined. There are 21 coastal counties that intersect the New York–New Jersey Harbor, and they were already heavily urbanized by the 1970s (NYDEC 1996). The population increased from 15 million in 1960 to 16.9 million in 2000 (or 13%), which is slower than that in other estuary areas.

Pollution Prevention and Industrial Ecology in the New York–New Jersey Harbor

In the mid 1990s, the New York–New Jersey Port Authority had many grand plans for the new millennium, including the plans to dredge and deepen the New York–New Jersey Harbor. Maintaining the economic viability of the Port of New York relies on periodic dredging to maintain a 50-ft-deep (15 m) channel in a waterway that is naturally less than 10 m deep. In the 1990s, dredging became highly controversial because of resuspension of contaminants and proposed disposition of the sediment contaminated by metals, organics (PCBs, dioxins), petroleum products, and probably pathogens (Keegan 2011). Resuspension was predicted to cause local impacts on biota in the harbor, whereas continued disposition at sea impacted the marine ecosystem. In 1997, the U.S. EPA approached the New York Academy of Sciences about the feasibility of using industrial ecology approaches to investigate sources, fate, transport, and consequences of several significant pollutants in the Harbor. In addition to standard environmental science questions—What is in the sediment? How much is there?—industrial ecology could approach the following issues: Where does it come from? What happens to it once it gets in the harbor? Where does it go? What is the timeline of such processes? And what can be done to prevent further pollution?

Under the auspices of the academy, Charles "Chuck" Powers organized the New York–New Jersey Harbor Consortium comprised of science and policy experts and diverse institutional and advocacy stakeholders, to explore the historical and current sources of pollution and the approaches to effective pollution prevention. It was our privilege to participate in the frequent workshops as members of the Harbor Consortium, and a pleasure to see the results through to publication, and to see them impact decisions made ultimately by the Port Authority, state legislatures, and diverse regulators. Chuck Powers describes the results as "audacious in scope, rigorous in its scientific and analytic conclusions, and bold in its recommendations affecting a wide variety of institutional interests and practices" (Powers Preface to de Cerreño et al. 2002, p. 4). The results were reported and ratified in multiple workshops, published in six monographs under the rubric of Industrial Ecology, Pollution Prevention and the New York–New Jersey Harbor Project. The reports addressed mercury (de Cerreño et al. 2002), cadmium (Boehme and Panero 2003), PCBs (Panero et al. 2005), dioxins (Muñoz and Panero 2006), PAHs (Valle et al. 2007) and suspended solids (Muñoz and Panero 2008). Each of the investigations had unique lessons and each contributed to the field of industrial ecology, as well as influencing Port Authority management decisions and both federal and state regulations, particularly for mercury.

The Estuary has had some dramatic pollutant inputs from the Hudson River. PCBs from a General Electric plant contaminated a large stretch of the Upper Hudson River, becoming one of the nation's most famous and contentious Superfund Sites. Although the source was near Fort

Edward, upriver from the Troy Dam, this source contributes more than 50% of the PCB load to the harbor (Panero et al. 2005). The former Marathon Battery Plant at Cold Spring polluted Foundry Cove and the adjacent Constitution March and the Hudson River. This Superfund Site, one of the most significant cadmium-pollution sites in the world (see Chapter 9), was remediated in the 1990s. The Industrial Ecology Consortium concluded that, by 2000, the site was not a significant source of cadmium for the harbor.

DELAWARE BAY ESTUARY

Between New Jersey on the east and Delaware on the west, Delaware Bay is one of the busiest navigational waterways in the United States. The watershed extends from the western flanks of the Catskills, through eastern Pennsylvania and western New Jersey to most of Delaware (EPA 2007d). The Delaware River is the chief input to the Bay. For much of its length, it is the boundary between New Jersey and Pennsylvania. The Delaware River is listed as the last undammed river in the eastern United States, although proposed construction of the Tocks Island Dam in the 1970s, at what is now the Delaware Water Gap National Recreational Area, was narrowly averted. The Upper Delaware is designated a "Scenic and Recreational River" (NPS 2015a). The Lower Delaware is part of the National Park Service Wild and Scenic River System. Other tributaries to the river include the Lehigh River at Easton, Pennsylvania. The river and bay extend 419 mi (674 km), draining 14,119 mi^2 (36,570 km^2) of watershed. The mean freshwater discharge of the Delaware River into the estuary of Delaware Bay is 11,550 ft^3/s (330 m^3/s) (EPA 2007d; Figure 2.17).

The Trenton Falls, although only 2.5 m high, marks the beginning of the estuary. From Trenton, the river flows between Camden, New Jersey, and Philadelphia, past Marcus Hook and Wilmington, before broadening out into the bay, as it approaches Cape May (New Jersey) and Cape Henlopen (Delaware). Several rivers enter the bay, including Schuylkill near Philadelphia, the Salem, Stowe Creek and Cohansey River, and the Maurice River. The Maurice River is designated as Wild and Scenic largely because of its rich fishery history (NPS 2015b; Figure 2.18). The Maurice River and its tributaries of the Manumuskin Watershed drain the Pinelands. However, the Maurice potentially contributes more than freshwater, for it passes through the industrial cities of Vineland and Millville, where it widens into the arsenic-contaminated Union Lake. Vineland Chemical manufactured arsenic-based herbicides and polluted the river and the sediment of Union Lake, earning an early slot on the National Priorities or "Superfund" List (EPA 2013a), requiring extensive remediation during the past decade, habitat restoration, and ongoing monitoring of water and sediment (EPA 2013a).

Delaware Bay is a huge port complex, covering not just Philadelphia and Wilmington, but also Salem, Camden, and Gloucester City, and all contribute to bay shipping traffic. Despite the de-industrialization of the Northeast, Delaware Bay remains a major petrochemical port. Population and pollution go hand in hand. The densely populated upper reaches of the Bay contribute both industrial pollutants and combined stormwater (CSO) overflow, somewhat curtailed since the Sutton et al. (1996) report. Cities along the river and bay include Port Jervis, New York (population 8700); Easton, Pennsylvania, (27,000); Trenton, New Jersey (84,000); Camden, New Jersey (77,000); Philadelphia, Pennsylvania (1.5 million); and Wilmington, Delaware (71,000). The latter two are huge ports that make Delaware Bay a critical economic entity for the nation. Boat traffic from Delaware Bay has access to Chesapeake Bay through the Chesapeake and Delaware Canal. Below Camden and Philadelphia, the towns are small, and on the New Jersey side, vulnerable to erosion and flooding.

Three major ecological zones differing in salinity, turbidity, and productivity can be distinguished in Delaware Bay. The upper zone (zone I) of tidal freshwater wetland is from Trenton to Marcus Hook, a highly developed and industrialized zone that includes the biggest cities (Philadelphia and Camden). From Marcus Hook to Artificial Island, there is a wide salinity range

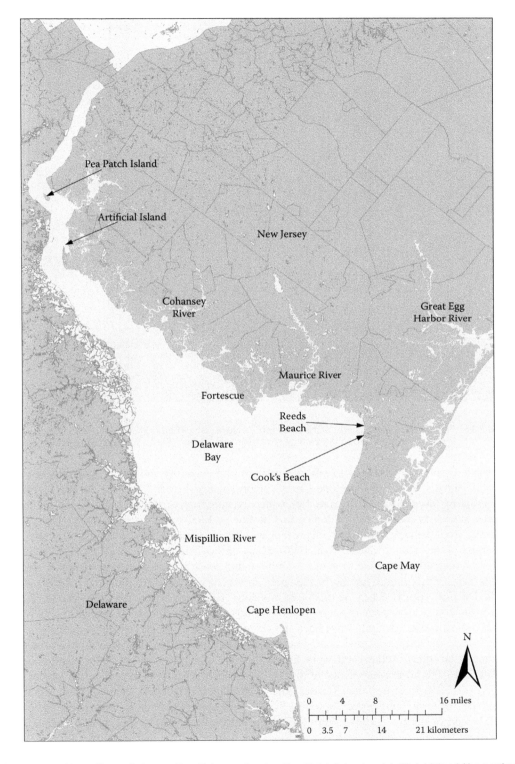

Figure 2.17 Map of lower Delaware Bay Estuary, showing Pea Patch Island and Artificial Island (the northern extent of Delaware Bay itself). The estuary begins at the falls at Trenton (not shown).

Figure 2.18   The Maurice River is one of the main tributaries leading into Delaware Bay. There are still extensive salt marshes along the banks of the river and creeks. (Photo courtesy of Jane Galetto.)

from fresh (0 ppt) to brackish at 15 ppt, in a zone of high turbidity and low productivity (zone II). Artificial Island is home to the Hope Creek–Salem Nuclear Power Plant. Below there, the river widens rapidly into the open Bay (zone III), which extends from Artificial Island to the ocean. The bay is saline, fairly shallow (<9 m depth), and highly productive (Sutton et al. 1996). Diversion of freshwater from the Delaware River, channel deepening, and sea level rise all contribute to increasing salinity of the estuary (Sutton et al. 1996). Many of the Delaware Bay marshes have been overgrown by a nonnative, invasive form of *Phragmites*, and control efforts have been in place since for decades, mainly using multiyear herbicide applications (Able et al. 2003; Teal and Peterson 2005).

The Delaware Bay beaches are the major breeding ground for Horseshoe Crabs and the feeding grounds for tens of thousands of migratory shorebirds of a dozen species (Clark et al. 1993; Niles et al. 2008). However, the crabs have been subjected to intensive overharvesting, and the shorebird numbers have declined 80% over a 30-year period (WHSRN 2015). The Ramsar Convention on Wetlands is an international framework for the conservation and wise use of wetland resources (Ramsar 2015). Delaware Bay is one of 36 U.S. sites identified under the convention (since 1992), referring to 51,252 ha of wetland (Ramsar 2015). Delaware Bay was also the pioneering site identified for the Western Hemisphere Shorebird Reserve Network (http://www.whsrn.org), encompassing 21,208 ha of the bay and bayshore (WHSRN 2015) (Figure 2.19). Delaware's Coastal Zone, comprising 270,000 acres of wetlands and uplands, has been designated as an "Important Bird Area of Global Magnitude," a list sponsored by the National Audubon Society and the American Bird Conservancy. The bay beaches, with abundant Horseshoe Crab eggs, are a critical migratory stopover for shorebirds.

Figure 2.19 Semipalmated Sandpipers (*Calidris pusilla*) and Laughing Gulls forage on eggs of the retreating Horseshoe Crabs on the Delaware Bay shore.

The dense shorebird flocks, avidly probing the muddy shoreline of Delaware Bay, provide conditions for the exchange of infectious agents, such as avian influenza virus. The prevalence of infection (25% determined by oropharyngeal swabs) and resistance (66% elevated antibody levels) is very high among Ruddy Turnstones trapped on the bayshore. Laughing Gulls are susceptible to avian influenza, but swabs from Laughing Gulls revealed a low rate of infection and a high prevalence of antibodies (Maxted et al. 2012). During the several weeks they spend on the bay, Turnstones show an increase in virus infection.

The bay is also an important wintering area for waterfowl. For example, the 10-year averages for species were as follows: Black Duck, 27,822; Mallard, 15,449; Canada Goose, 71,333; and Snow Goose, 14,314 (Sutton et al. 1996). Unfortunately, Black Ducks have declined steadily in the bay from the 1960s to the present; they face problems from interbreeding with Mallards.

Important breeding colonies include the Pea Patch Island heronry, as well as breeding areas for endangered or "Watch-listed" species such as Piping Plover (*Charadrius melodus*), Least Tern, and Saltmarsh and Seaside Sparrow (*Ammodramus caudacutus*, *Ammodramus maritimus*), and the migratory shorebird concentrations (Sutton et al. 1996; Delaware Audubon Society 2002). Ospreys, representing a major conservation success story, breed in Delaware Bay. Of the 542 nests identified in New Jersey in 2013, two were on the Delaware River itself, 63 in the Maurice Estuary, and 26 in the Salem Artificial Island area (Clark and Wurst 2013). Kerlinger and Wiedner (1991) estimated the economic value of sport fishing in the Bay at $25 million per year. Birding ecotourism also contributed more than $5 million per year in the late 1980s (Kerlinger and Wiedner 1991), and is much higher today.

CHESAPEAKE BAY

Chesapeake Bay is the largest of the 100+ estuaries in the United States in terms of surface area, shoreline, and watershed extent (CBP 2015). It drains most of southcentral New York, Central Pennsylvania, all of Maryland (including the District of Columbia), and much of Virginia (Figure 2.20).

Chesapeake Bay and its various projections lie between the DelMarVa Peninsula on the east and the mainland of Maryland and Virginia on the west. It extends more than 300 km north to south, ranging in width from 6 to 50 km. The northern extremity of the bay is in the Northeastern corner of Maryland where the Susquehanna River enters the Bay, delivering about half of all the freshwater input to the Bay. The Susquehanna, arising in central New York and meandering across Pennsylvania and through Harrisburg, enters the bay just north of the Aberdeen Proving Ground, home of Edgewood Arsenal, where toxic gases were manufactured for World War I, and other warfare chemicals thereafter. The site housed the U.S. Army Chemical Corps. There are many areas of contamination (metals, PCBs, organics, pesticides, chemical warfare agents, volatile organics), and unexploded ordnance in areas adjacent to the streams, wetlands, and bay, and the site is on the National Priorities List (EPA 2015a) (Figures 2.21 and 2.22). An analysis of environmental contaminants indicates that some pose potential hazards (Rattner and McGowan 2007).

Another threat to birds are disease and toxins. Bird die-offs occur periodically in many wetlands, and often the cause goes unidentified. In 2014, Poplar Island experienced a botulism outbreak that killed several hundred waterfowl, and later in the summer, a second outbreak killing 262 cormorants was traced to a paramyxovirus, related to Newcastle Disease (Kobell 2014).

There are no large urban areas on the immediate eastern shore of the bay, but several rivers drain from the eastern shore into the bay, including the Choptank, Nanticoke, Wicomico, and Pokemoke. The western shore is part of the urban megalopolis from Baltimore (population 622,000) and Annapolis (38,000) to Washington, DC (649,000), and Alexandria (150,000), drained by the Potomac River (EPA 1983). The rest of the western shore is rural agriculture, until one reaches the Hampton Roads military complex, with more than a million residents in the cities of Newport News and Hampton to the north of the mouth of the Bay, and Norfolk and the Chesapeake to the south. The James River, flowing through Richmond, Virginia, is the major freshwater input to the southern part of the Bay. The mouth of the bay is spanned by the Chesapeake Bay Bridge–Tunnel, which was built in 1964, and spans 23 mi from the Cape Charles area of southern DelMarVa to the Cape Henry area of Virginia Beach.

Phillips (2007) and colleagues describe the physical, hydrologic, biologic, and ecotoxicologic features of the Chesapeake Bay area. The Chesapeake Bay Estuarine Complex (38°N × 76°20′W) is listed on the Ramsar Convention (1987) (USFWS 2015). In 2010, the CBP reported that 72% of the Bay's tidal-water segments were fully or partially impaired as a result of the presence of toxic contaminants (EPA 2012a). This has necessitated fish-consumption advisories for some areas.

Collaboration between the U.S. Fish and Wildlife Service and the Army Corps of Engineers has resulted in using dredge "spoil" from deepening the shipping channels to augment islands in the Bay. Several islands, such as Poplar Island and Hart Miller Island, have had substantial increase in size, providing wildlife and recreation habitat, while at the same time enhancing the viability of Baltimore Harbor. The Poplar Island restoration project will result in a mix of upland and wetland habitat. To avoid contamination, the island augmentation did not use dredge material from close to Baltimore. Poplar Island is a breeding ground for colonial waterbirds, Ospreys, and Diamond-back Terrapin (*Malaclemys terrapin*). The absence of terrestrial predators resulted in very high breeding success for terrapins. Poplar Island is the only Common Tern nesting colony in the Maryland portion of Chesapeake Bay. However, recently arrived terrestrial predators now play havoc with the terns (Michael Erwin, personal communication, July 2015).

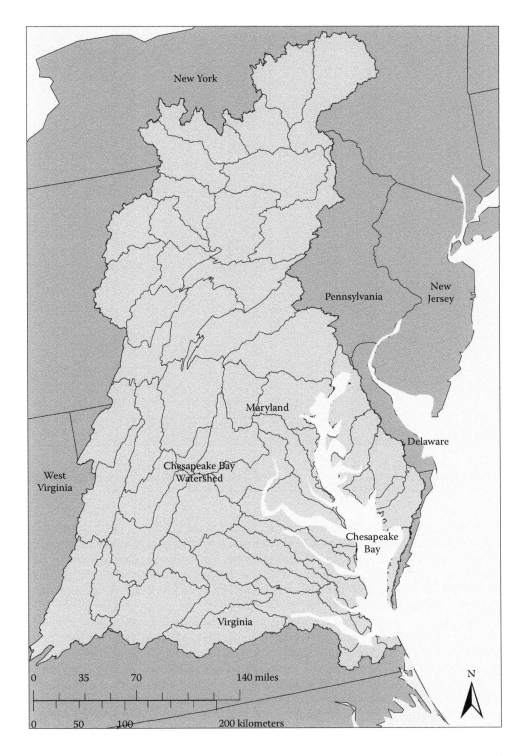

Figure 2.20   Chesapeake Bay and its watershed. (From EPA, http://www.epa.gov/chesapeakebaytmdl/.) The watershed, shown in light gray, extends up into New York State.

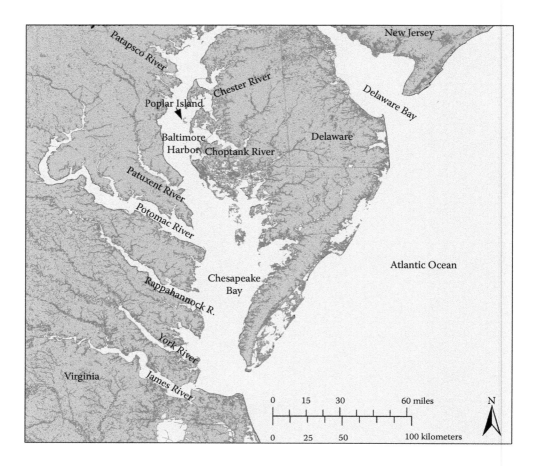

Figure 2.21   Map of the Chesapeake Bay, showing the intricate system of bays and drainage systems. The Susquehanna River (not shown) enters the northernmost reach of the bay.

The EPA monitors the progress that regions make toward achieving the goals of the Clean Water Act. For example, the EPA (2014a) reported that the states bordering the Chesapeake Bay, committed to reducing nutrient and sediment pollution, had achieved goals in reducing nitrogen and phosphorus, but not sediment inputs. The U.S. Geological Survey (USGS), the scientific arm of the U.S. Department of Interior, conducted extensive investigations of environmental quality of the Chesapeake (Phillips 2007). As with other estuaries, the Chesapeake has been affected by human population increases, which resulted in degraded water quality, loss of habitat, and declines in populations of biological communities. Multiple agencies have partnered to "restore" the bay ecosystem with respect to stewardship of the living resource and the protection of vital habitats. The prospect of continued human population growth (Claggett 2007), coupled with sluggish improvement in water quality, was identified as a concern (Phillips 2007). The USGS report (Phillips 2007) covered a broad range of topics to address governmental and public concerns that ecological conditions in the Bay and its watershed had not significantly improved. Phillips identified "four primary themes: (1) causes and consequences of land-use change; (2) factors affecting water quality and quantity; (3) ability of habitat to support fish and bird populations; and (4) a synthesis and forecasting to improve ecosystem assessment, conservation, and restoration."

(a)

(b)

Figure 2.22 Typical views of marsh habitats in the Chesapeake Bay area, showing a salt marsh creek (a and b).

SUMMARY AND CONCLUSIONS

The birds and other biota we examine in this book are components of the Northeast estuaries from Cape Cod to Chesapeake Bay. The bays differ in size, shape, number of rivers, components, biota, and human activities. Most include large seaports with complex infrastructures connected to transportation of goods, whereas others (e.g., Barnegat Bay) do not have tankers and seaports. All of the bays are extensively developed with industry, residential, and tourism. All except Chesapeake Bay are part of the EPA's extensive HEP (EPA 2007a–f); the Chesapeake Bay already had a program that was used as a model for the EPA HEP. The advantage of the HEP was that it provided a uniform method of evaluating the status of the bays, which allows comparisons. The overall conclusions were: (1) water quality varies from "poor" (New York–New Jersey Harbor, the Chesapeake) to "good" (Barnegat Bay; Massachusetts Bays); (2) sediment quality varies from "poor" (Massachusetts, New York–New Jersey Harbor) to "good" to "fair" (Barnegat, Delaware Bays); (3) benthic biodiversity was "poor" in New York–New Jersey Harbor and Delaware Bay, and better in the other bays; and (4) fish tissue contaminants vary, but were based on few samples (EPA 2007a–f). Barnegat Bay is the smallest of the estuaries and the least impacted by urbanization. Although hardly pristine, its water quality and sediment quality are better than those in other areas in that none of the four metrics—water quality, sediment quality, biodiversity, or fish toxicity—are rated "poor."

CHAPTER 3

Species, Methods, and Approaches

INTRODUCTION

The main emphasis of this book is on habitat, population dynamics, and metal (including the metalloids selenium and arsenic) levels in birds in Barnegat Bay, with comparisons to other estuaries along the Northeast Atlantic coast. The study spans the years 1970 to 2015 in Barnegat Bay, with data and reference from other bays. The primary species identified in this chapter are ones that we targeted for observations, monitoring, and sample collection in Barnegat Bay. We also identify a list of secondary species of ecological importance in the estuaries, including many species that are classified on state or federal lists as threatened or endangered. There is a rich literature on the colonial birds of the Massachusetts area (Nisbet and colleagues), Long Island Sound (Spendelow, Hays, and colleagues), the New York–New Jersey Harbor (Gochfeld, Parsons, Elbin, Craig, Tsipoura, and colleagues), and Chesapeake Bay (Erwin, Watts, and colleagues), and we use these data for comparative purposes. There are few breeding colonies of waterbird species in Delaware Bay; the emphasis in Delaware Bay is on populations and contaminants in migrating shorebirds (Niles, Dey, and colleagues).

Our overall approach for Barnegat Bay was to census birds on salt marsh islands every year and to conduct a variety of behavioral, reproductive, and ecological studies (Figure 3.1). For our ecotoxicologic research, we collected eggs of several species, and feathers from fledging young. To provide information on nodes in the food chain, we also collected and analyzed fish representing several trophic levels, including larger fish that are eaten by larger colonial birds, as well as by humans. For other bays from Cape Cod to Chesapeake Bay, we obtained data on populations and heavy metals to provide a regional comparison. In addition, we sampled Horseshoe Crab (*Limulus polyphemus*) tissues (eggs, internal) from Maine to Florida and Herring Gulls (feathers) as indicators of metal exposure from New England to Virginia.

Methods unique to a particular topic or species are discussed in each section. Experimental studies are mentioned in Chapter 9 (Burger and Gochfeld 1990a). The main avian study species are the Great and Snowy Egrets; Black-crowned Night-Herons (*Nycticorax nycticorax*); Common and Forster's Terns; Black Skimmers; Great Black-backed, Herring, and Laughing Gulls; and Red Knots; with selected data on other species (Table 3.1).

There are other species characteristic of the estuary and its salt marsh islands, which are secondary to our studies. We mention them in various contexts related to habitat, behavior, and contaminants.

Figure 3.1   Herring Gulls at Captree colony (Suffolk County, Long Island, New York). Adults from adjacent territories fighting over chicks that have wandered into a neighboring territory.

Table 3.1   Avian Species with Types of Data Presented in This Book for Barnegat Bay

Species	Habitat	Population	Contaminants
Great Egret (*Egretta alba*)	General and site-specific	1977–2015	Eggs: 2003–2015 Feather data: 1989–2014
Snowy Egret (*Egretta thula*)	General and site-specific	1977–2015	Eggs: 1994–2009 Feathers: 1989–2010
Black-crowned Night-Heron (*Nycticorax nycticorax*)	General and site-specific	1977–2014	Eggs: 1991–2009 Feathers: 1989–2014
Great Black-backed Gull (*Larus marinus*)	General data, many islands	1977–2015	Eggs: 1995–2014 Feather: 1995–2014
Herring Gull (*Larus argentatus*)	General, many islands	1976–2015	Eggs: 1989–2014 Feathers: 1989–2006
Laughing Gull (*Larus atricilla*)	General	1976–2015	Eggs: 1995–2011 Feathers: 1992–2014
Common Tern (*Sterna hirundo*)	Site specific	1976–2015	Eggs: 1970–2015 Feather: 1978–2014
Forster's Tern (*Sterna forsteri*)	More than 30 islands in Barnegat Bay	1976–2015	Eggs: 1990–2012 Feathers: 1989–2005
Least Tern (*Sterna antillarum*)	NJ and NY colonies	1977–1999	None
Roseate Tern (*Sterna dougallii*)	NE colonies	Recovery team data	Long Island eggs: 1989–1994
Black Skimmer (*Rynchops niger*)	NY and NJ colonies	1976–2015	Eggs: 1989–2000 Feathers: 1989–2004

Note: "General" refers to information about the topic. The table identifies some topics where site-specific data were obtained, and species for which egg and feather sampling was done. Over the course of 40 years of our Barnegat Bay research, the scientific names of some species have changed (see species accounts). Temporal patterns are from Barnegat Bay unless otherwise stated.

ETHICAL ISSUES IN FIELD STUDIES

Throughout our studies, we were mindful of the health and well-being of the birds and other animals we studied. We were attentive to minimizing disruption of the breeding birds during observations and sampling (Burger and Gochfeld 1983a). We adhered to the values and principles eloquently articulated by Nisbet and Paul (2004), recognizing the need to balance disturbance against the conservation value of the research results. All the bird species are protected by the Migratory Bird Treaty Act of 1918, with subsequent amendments (USFWS 2013). Some of the species have special status in New Jersey and/or elsewhere (Table 3.2). Except for the Arctic-nesting Red Knot, all the species in Tables 3.1 and 3.3 breed in most states along the Atlantic coast. The Red Knot migrates from its wintering grounds in the Neotropics (from southern United States to Tierra del Fuego), pausing to refuel on New Jersey's Delaware Bay shore and estuaries on its way to the Arctic (Niles et al. 2008, 2010). All of the fieldwork, including observations, monitoring, and collecting, was performed under the necessary Federal, State, Refuge, and Park Permits, and all protocols were approved by the Rutgers University Institutional Animal Care and Use Committee.

CONSERVATION STATUS DEFINITIONS

The federal government has established lists of federally "Endangered" and "Threatened" species under the Endangered Species Act (1973), which also requires that states do the same. States must evaluate their wildlife resources and designate as "Endangered" those species that are in danger of disappearing from the state for a variety of reasons (habitat change, overexploitation, predation, competition, disease, disturbance, or contamination). Endangered species require active assistance and protection to assure survival. Assistance is needed to ensure continued existence as a viable component of the state's wildlife community. "Threatened" species are those that may become endangered if their populations continue to decline. The process for listing, the enthusiasm for listing, and the opposition to listing by vested interests varies from state to state and has varied by the political philosophies of "Washington." A category of "Special Concern" applies to species that warrant special attention because of inherent vulnerability to environmental deterioration or habitat modification that would result in its becoming threatened if conditions surrounding the species begin or continue to deteriorate (NJDEP 2012). There are also provisions for downlisting when species have recovered sufficiently to thrive without active or legal protection. For example, the Eastern Brown Pelican, formerly listed as endangered, was downlisted in 1985 (Shields 2014; Figure 3.2).

The officially designated conservation status (federal list, states from Maine to Virginia) for the species discussed in this book are given in Table 3.2. Several states adopted the federal listing for species such as the Roseate Tern and Piping Plover. Nonetheless, all species in Table 3.2 are protected under the Migratory Bird Treaty Act (USFWS 2013), and permits are required to implement controls. Most of the species in these tables were historically egged both for subsistence and commercial purposes, and during the late nineteenth century were wantonly slaughtered for the plume trade period of the late nineteenth century (Ehrlich et al. 1988), which led to the formation of the National Audubon Society and protections in the early twentieth century.

TAXONOMY AND NOMENCLATURE

The understanding of what constitutes a species is constantly evolving. In the mid twentieth century, there was a strong tendency for taxonomists to recognize the similarities among different groups by "lumping" them as a single species (with subspecies). Extensive studies of behavior,

Table 3.2 Conservation Status of Species Discussed in Our Studies Derived from Recent Lists on State Web Sites (July 2015)

Species	USA	Maine	New Hampshire	Massachusetts	Rhode Island	Connecticut	New York	New Jersey	Delaware	Maryland	Virginia
Brown Pelican	D										
D.C. Cormorant											
Great Egret					C	Th					
Snowy Egret					C	Th					
Black-crowned Night-Heron		Th			C			Th	End		
Little Blue Heron					C	SC		SC		Rare	
Tricolored Heron								SC		Rare	
Glossy Ibis					C	SC		SC			
Osprey			Was Th[a]		C		SC	Th			
Clapper Rail					C						
American Oystercatcher (*Haematopus palliates*)					C	Th		SC	End	Rare	
Piping Plover	End and Th[b]	End	End	Th	Fed End	Th	End	End	End	End	Th
Willet					C						
Red Knot	TH[c]						Fed	End	End		Not listed
Great Black-backed Gull											
Herring Gull											
Laughing Gull										Rare	
Common Tern			Th[a]	SC		SC	TH	SC	End		
Forster's Tern							Marg	End	End		

(Continued)

Table 3.2 (Continued)   Conservation Status of Species Discussed in Our Studies Derived from Recent Lists on State Web Sites (July 2015)

Species	USA	Maine	New Hampshire	Massachusetts	Rhode Island	Connecticut	New York	New Jersey	Delaware	Maryland	Virginia
Roseate Tern	End and Th[b]	End	End	End	Ext	End	End	End	End	Ext	End
Least Tern	End	End	End	SC	Th	Th	Th	End	End	Th	
Black Skimmer							SC	End	End	End	
Sedge Wren		End	End	End		End	Th	End	End	End	
Seaside Sparrow					C	Th	SC				
Saltmarsh Sparrow						SC				Rare	
Henslow's Sparrow				End	Ext	SC	Th	End	End	Th	Th

Note: C, concern; D, delisted refers to the Federal Endangered and Threatened Species list; End, endangered; Ext, extirpated from the state; Marg, identified as a marginal inhabitant; Rare, species is listed on website without designation; SC, special concern; Th, threatened.

[a] Species has been downgraded from earlier listings because of improved status.
[b] Some populations of Least Tern and Piping Plover are listed as federally endangered, others as threatened.
[c] Red Knot was added to the Federal Endangered and Threatened Species list in January 2015.

Table 3.3   Other Salt Marsh Species in Bays and Estuaries

Brown Pelican (*Pelecanus occidentalis*)	Osprey (*Pandion haliaetus*)
Double-crested Cormorant (*Phalacrocorax auritus*)	Clapper Rail (*Rallus longirostris*)
Little Blue Heron (*Egretta caerulea*)	Piping Plover (*Charadrius melodus*)
Tricolored Heron (*Egretta tricolor*)	Willet (*Catoptrophorus semipalmatus*)
Glossy Ibis (*Plegadis falcinellus*)	Osprey (*Pandion haliaetus*)

Figure 3.2   Brown Pelicans were once endangered owing to reproductive failure caused by the endocrine disruptive chlorinated hydrocarbon pesticides, particularly DDT, which caused eggshell thinning. The species has recovered substantially, and now occurs regularly as a visitor to the Barnegat Bay coast.

vocalizations, biochemistry, and most recently genetics, have changed the way taxonomists think about species. There is an increasing tendency to emphasize differences by recognizing certain groups, formerly called subspecies, as species. This is happening with the Herring Gull, which has been split by some authorities so that the American Herring Gull is *Larus smithsonianus* rather than *Larus argentatus smithsonianus*. In the past 25 years, a half-dozen additional "species" have been recognized among the Herring-type gulls of Eurasia. Because that split occurred relatively late in our studies, after decades of referring to our Herring Gulls as *Larus argentatus*, we have retained that name.

Genetic studies have also shown that species formerly considered closely related and placed in the same genus are only distantly related, and conversely some species have been moved to other genera to reflect hitherto unsuspected close relationships. Some of the name changes were made in the American Ornithologists' Union, *Checklist of North American Birds* (1983, 6th edition). That volume reassigned the Little Blue Heron (formerly known as *Florida caerulea* in some of our earliest papers) and the Tricolored or Louisiana Heron (once known as *Hydranassa tricolor*) to the genus *Egretta*, and we have used the names *Egretta caerulea* and *Egretta tricolor* in recent papers. On the other hand, among our key species, the Great Egret has had its genus recently reassigned, so that it is now considered more closely related to the Great Blue Heron genus *Ardea* than to other egrets in *Egretta*. However, we have referred to it as *Egretta* in most publications. Likewise, the

Willet, formerly in a monotypic genus *Catoptrophorus*, is now recognized to be a member of the large shorebird genus *Tringa*—hardly a surprise to those familiar with the Willet and other large shorebirds such as yellowlegs in the field. However, that change is very recent, so we continue to use the awkward *Catoptrophorus*.

In the world of biodiversity, each species is considered a unique entity or a proper noun. Therefore, throughout the book we capitalize the names of species to indicate that they are unique entities and to avoid confusion. This is contrary to the style manuals that journalists follow, but is standard in such ornithology journals as *The Auk*. Capitalization helps us distinguish the name of a species from a string of adjectives. Thus, "Little Blue Heron" is the name of a species, and is not merely descriptive of a heron that happens to be small and blue. We do not capitalize groups of species such as "gulls," "terns," or "herons." On the other hand, we do capitalize "Skimmers" or "Night-Herons," because those terms refer to a single unique species in the Northeast.

PRIMARY SPECIES DESCRIPTIONS

Great Egret (*Egretta alba*)

The 1-m-tall, stately, all-white Great Egret is a striking and conspicuous species along the Atlantic coast shores and estuaries. As a stand-and-wait predator, its shape is a familiar iconic image. In the breeding season, it sports large plumes that made it a prime target for plume hunters, and it was essentially wiped out by millinery hunters in the 1800s. Its long, stout beak is bright yellow. The shift of the Great Egret to the genus *Ardea* is relatively recent, and we have published studies under its former generic designations of *Casmerodias albus* and *Egretta alba*. It breeds along both Atlantic and Pacific coasts from southern New England (rarely) and Washington State through Mexico and Middle America and most of South America (McCrimmon et al. 2011). It is also widely distributed in the Old World. Birds of the Atlantic coast are mostly migratory to the southern states, the Caribbean, and Mexico. The species showed great increase from a nadir in the early 1900s to peak numbers in the 1970s, and it is again increasing. It breeds in mixed-species heronries, typically occupying the highest positions (Burger 1978b,c, 1979c, 1981a; McCrimmon et al. 2011). It feeds primarily on fish, often fairly large-sized, but may hunt for crabs, frogs, snakes, and mice (Figure 3.3).

(a)   (b)

Figure 3.3 A Great Egret (a) illustrating its "stand-and-wait" fishing pose and a Snowy Egret (b) perched above the water also waiting for fish in striking range.

Snowy Egret (*Egretta thula*)

The Snowy Egret is a small, elegant, all-white heron, with bright yellow feet that contrast with its black legs (Figure 3.3). Its bill is black. Although all colonial waterbirds were exploited almost to extinction in the 1800s, the Snowy was "among the most sought-after of all herons" (Hancock and Kushlan 1984). Like the other species, population numbers increased when persecution stopped. Numbers continued to increase into the 1980s, and the species spread northward as a breeding bird, to New England, beyond its historical range (Parsons and Master 2000). Subsequent declines have been documented, partly attributed to the decline in coastal wetland acreage. Unlike the watch-and-wait tactics of the Great Egret, the Snowy Egret is an active hunter, dashing one way or another using its feet to stir the bottom and then darting its head forward to seize small fish or, occasionally, crustaceans. This active pursuit feeding is considered more specialized than waiting, and is energetically costly (Kushlan 1978; Kent 1986). This species may be more vulnerable to changes in prey density and accessibility (Parsons and Master 2000). Egrets gather by the dozens in areas where fish are abundant or available, as in drawdowns or tidal pools in salt marshes. The Snowy Egret nests in mixed-species colonies, and often it is the most numerous member (Parsons and Master 2000). Before 1986, this species was called *Leucophoyx thula*.

Black-Crowned Night-Heron (*Nycticorax nycticorax*)

This species also occurs in both the Old World and New World. It is a familiar breeding bird across North America from the Maritime Provinces to British Columbia, with both coastal and inland populations (Figure 3.4). There are breeding populations in Middle America and South America, as well as in the Old World (Hothem et al. 2010). Compared to the egrets, this is a stockier-built, shorter-necked species. Although often seen in daytime, it is primarily nocturnal. It leaves its daytime roosts at dusk, uttering a load, characteristic croak. It feeds in both marshes and meadows. The Night-Heron has the broadest spectrum of foods of any heron or egret, including worms, insects, crustacea, shellfish, squid, fish, amphibians, mammals, birds, eggs, carrion, and garbage (Hothem et al. 2010). They are much less reliant on fish than other similarly sized herons. A few Night-Herons can play havoc at a tern colony, stalking through it at night and swallowing several chicks to take to its own nestlings (Collins et al. 1970; Burger 1974). It is not averse to eating untended Night-Heron chicks in neighboring nests.

Figure 3.4   A Great Egret and Black-crowned Night-Heron at the edge of a salt marsh, looking up after searching for food.

Great Black-Backed Gull (*Larus marinus*)

This is one of the largest gulls in the world, weighing up to 2.5 kg. Its breeding range extends from northern Europe to the Maritime provinces of Canada and to the mid-Atlantic states (Figure 3.5). It has spread westward to the Great Lakes and southward to the mid-Atlantic states, where it nests on salt marsh islands. Great Black-backed Gulls begin breeding mainly at 5 (some at 4) years of age. There are few natural predators on adults, although the young are vulnerable to conspecific aggression and predation (Good 1998). It is a large, voracious predator on eggs, preflying young, and on fledgling birds (Burger and Gochfeld 1984a). At a nesting colony of Black-legged Kittiwakes

(a)

(b)

(c)

Figure 3.5 A Great Black-backed Gull (a) and Herring Gull (b) in a salt marsh colony; and a Laughing Gull (c) with nest material.

(*Rissa tridactyla*) in northern Norway, we watched several Great Black-backed Gulls kill the young Kittiwakes by forcing them to the water, seizing their heads and shaking them vigorously until the fledgling was dead (Burger 1978a). In addition to feeding on fish and crustacea, they opportunistically follow fishing boats, attend landfills, kleptoparsitize other birds, and patrol beaches for dead and dying foods, including stranded Horseshoe Crabs. As these gulls have colonized southward, they have preempted higher island habitats, and by remaining in the area through the winter, they occupy the most favorable sites formerly used by terns. The downy chicks readily take to the water. Like most of the colonial nesting species, the Great Black-backed Gull was extirpated from much of its range by egging and plume hunting in the late 1800s. When these practices were banned in the early 1900s, the survivors in the less accessible northern parts of its range were able to recolonize. In the mid 1900s, the populations grew and the species spread southward abetted by abundant food at garbage dumps (Good 1998). By 1972, the species was found nesting in North Carolina. Burger (1978a) found the first successful New Jersey nest on salt marsh islands in 1976. In Newfoundland, a decline in the 1990s was attributed to the crash of its main fish food, the Capelin (*Mallotus villosus*; Good 1998).

Herring Gull (*Larus argentatus smithsonianus*)

This is the familiar "seagull" of the Northeastern United States, occurring in large numbers along coastal shores and inland lakes throughout the year (Figure 3.5). The Herring Gull complex is one of the most studied species of bird. Like the other colonial birds, Herring Gull numbers were greatly reduced in the late 1800s, and they also made a comeback in the mid 1900s, spreading southward some years ahead of the more northern Great Black-backed Gull, often defending large territories (Burger 1981b; Pierotti and Good 1994). The increase in the number of garbage dumps in the post–World War II period, and a growing and wasteful fishing industry, provided abundant food for recently fledged gulls, allowing them to survive the early independent life phase that typically has the highest mortality (Burger 1981c; Burger and Gochfeld 1981a). Gull control has been practiced at various times for various reasons (Gross 1951). In the 1960s, gulls—particularly the Herring Gull—were considered by field biologists as a pest species, threatening other coastal nesting species, particularly terns.

In the 1980s, large, thriving, and expanding gull colonies exceeding 1000 pairs occurred on several parts of Long Island, and there were many colonies of greater than 100 pairs scattered through Barnegat Bay (Burger 1977a). Once gulls occupied a colony, terns were likely to abandon or to nest as far as possible from the gulls (Erwin et al. 1981). Gulls displaced other species from high ground with *Iva* bushes, relegating them to the flood-prone margins of islands (Burger and Shisler 1978a). Although Great Black-backed Gulls could displace Herring Gulls, they were usually outnumbered by the smaller species, resulting in coexistence in many colonies both on *terra firme* and on salt marshes (Burger 1983). In the 1980s, laws required that garbage dumps be covered by earth each day, thereby greatly reducing the abundant food source. The loss of this food resource is believed to be one of the reasons that Herring Gull breeding populations declined, reaching less than 50% of the peak years by the end of the 1990s, and showing a 78% decline between 1970 and 2010 (Cornell 2014). Herring Gulls in Britain declined by 48% from 1970 to 1988 (JNCC 2002).

Laughing Gull (*Larus atricilla*; Now Placed in Genus *Leucophaeus*)

Laughing Gulls breed from southern New England, along the Atlantic and Gulf Coasts, the Caribbean, and northern South America (Figure 3.5). This is very much a coastal species throughout its range, and New Jersey is a stronghold for the species (Burger 1996a,b, 2015a). This gull is one of the most widely recognized of New Jersey's coastal birds, where its loud, strident laughing calls are a familiar sound. In breeding plumage, the black head and slate gray back are a stark contrast to the white underparts and underwing. A closer look reveals a dark red bill. This species is much more migratory than the other gulls, and few remain in the Northeast during the winter. It nests in a variety of salt marsh habitats in the bay (Burger and Gochfeld 1985a). Nests are usually separated by several meters, such that incubating adults have only a partial view of their neighbors (Burger 1977b, 2015a). This is one of the species we have studied very extensively to understand their activity patterns (Burger 1976), and their response to flooding (Burger 1978d), as well as their relationships with larger gull competitors (Burger 1979a, 1981c). The Laughing Gull increased dramatically during the late 1900s. The breeding population in Jamaica Bay, New York, near John F. Kennedy (JFK) International Airport increased from 2 to 2000 pairs in just a few years, resulting in an extensive culling program (Buckley et al. 1978; Brown et al. 2001a,b).

Common Tern (*Sterna hirundo*)

This small, 120 g species has a streamlined profile with a slender, sharp beak, and long pointed wings and tail. It is whitish below and pale gray above with a black cap and red bill tipped with black (Figure 3.6). The species occurs in North America and across Eurasia, breeding in the North Temperate Zone and migrating to the South Temperate Zone (Nisbet 2002). It is vulnerable to extensive hunting for food on its wintering grounds in South America and Africa (Nisbet 2002). Like most of our colonial species, Common Terns were nearly wiped out by egging and the millinery trade. The birds were killed for their feathers and wings, and even whole birds were mounted on women's hats (Nisbet 2002). Terns were gone from most of their historic range by 1900. Under protection, the species showed great resurgence into the mid 1900s, although sometimes it encountered severe competition from the growing gull population. Toxic chemicals also reduced breeding success (Nisbet 2002). Pressure on its prey base may account for recent declines, and loss of nesting habitat is the primary problem in marshes (Erwin et al. 1981; Burger and Gochfeld 1991b). The Common Tern has both coastal and inland populations (e.g., Great Lakes). It is most often noticed in its noisy feeding flocks, plunge-diving for small fish near inlets, or heading back to the colony carrying a single fish in its beak. A person approaching a colony results in a swirling flock of noisy birds, swooping overhead, and even diving and defecating to discourage intrusion. When we started our studies, Alfred Hitchcock's 1963 terrifying movie, *The Birds*, was still familiar, and casual visitors were easily dissuaded from entering nesting colonies by the noisy mobbing terns. As memory of the film fades, human disturbance at colonies has increased.

Historically, the Common Tern has been a beach or island nesting species, building nests on sand or gravel. In New Jersey, beaches have been usurped by people, and we were surprised at the beginning

(a)   (b)

Figure 3.6   A Common Tern (a) in a sandy beach colony and a Roseate Tern (b) waiting by its nest under the log.

of our studies to find most Barnegat Bay Common Terns nesting on salt marsh islands in a variety of habitats. Nisbet (2002) noted that maintaining Common Tern populations, even at current modest levels, requires intensive management. Common Terns habituate well to researchers, and even recognize individuals (Burger et al. 1993a), attacking familiar ones by dive-bombing and striking them. Most Common Tern colonies are readily accessible and their availability to researchers in the United States and Europe has resulted in numerous studies of behavior and ecology, including Marples and Marples (1934), Austin and Austin (1956), Nisbet (1978, 2002), Burger and Gochfeld (1991b), Hume (1993), Becker (2003), and Palestis (2014). We summarized many of our results on Common Terns in *The Common Tern: Its Breeding Biology and Social Behavior* (Burger and Gochfeld 1991b).

Forster's Tern (*Sterna forsteri*)

Forster's Tern, a species confined to North America, is superficially similar to the Common Tern in appearance (but upper surface of the wing is whiter) and in many other attributes. The main breeding areas are in the northern Great Plains, the northern Great Basin, and coastal marshes from Long Island to South Carolina, and then along the Gulf Coast from Alabama to Mexico. Most research has been done on inland populations (McNicholl et al. 2001). It is more of a marsh species than the Common Tern, and where they occur on the same salt marsh islands, nests of Forster's are more likely to be in wetter parts of the marsh where they are built up with dead grass stems. The nest characteristics, subtle difference in egg color (browner) and spotting (denser), and their higher pitched mobbing calls, tell us when we have arrived at a Forster's Tern colony (or subcolony). Forster's Terns feed on small fish, mainly from waters on or around the salt marsh islands or in nearby impoundments, and they also opportunistically feed on invertebrates including insects (McNicholl et al. 2001). At Forsyth Refuge, we watched them capture Polychaete (nereid) worms by shallow diving.

Roseate Tern (*Sterna dougallii*)

Roseate Terns closely resemble Common Terns, and are usually distinguished first by their high-pitched "keek" note, unlike anything in the Common Tern vocabulary (Figure 3.6). Roseate Terns are pure white above (arriving on the breeding ground with a faint salmon-pink tinge to their breast) and are overall paler than Common Terns. The bill is all black at the start of the season, gradually reddening toward the base. It is shorter winged and longer tailed than Common Terns. Roseate Tern breeds in New England and Long Island, Florida and the Caribbean, Europe and Africa, and Australasia (Nisbet et al. 2014). It is mainly a tropical species that has extended its range northward on both sides of the Atlantic. The U.S. Fish and Wildlife Service listed the Northeastern population as endangered and the Florida–Caribbean population as threatened (USFWS 2010). It is considered a former breeder in Virginia, Maryland, and New Jersey (Nisbet et al. 2014). Roseate Terns in the Northeast are concentrated in three to four major colonies, forming a metapopulation (Spendelow et al. 1995). The European colonies are also considered a metapopulation, and in both cases adult survival and breeding dispersal are important aspects of their population biology (Spendelow et al. 1995; Ratcliffe et al. 2008). Early in our studies, we found a Roseate Tern paired with a Common Tern attending a nest in Barnegat Bay, but we have not seen or heard it since, and the Roseate is now only a rare migrant in New Jersey. Our metal studies in Roseate Terns were conducted with birds from Long Island, Connecticut, and Massachusetts.

Black Skimmer (*Rynchops niger*)

The Skimmer is one of North America's most distinctive coastal waterbirds, noted for its unusual voice, unique bill, and feeding behavior (Gochfeld and Burger 1994). It has a laterally compressed, almost flat bill, which is used to "slice" the water surface as the bird skims low over tidal channels and bays, bill open, waiting to contact a surface-dwelling prey item, such as a sand eel. The

Skimmer is black above and pure white below. The long bill, its most conspicuous feature, is black with a red base (Figure 3.7). In flight, it is long winged and graceful, and while feeding it glides effortlessly along the surface, flapping only occasionally. It has a high degree of sexual dimorphism; males weigh 35% more than females. It nests in dense monospecific colonies or in association with Common Terns, benefitting from the terns' aggression (Burger and Gochfeld 1990b; Pius and Leberg 1998). In most places it is a beach-nesting species, and even in the backwater areas it prefers sandy substrates. But in Barnegat Bay, its colonies are mostly on wrack strewn up by winter storms on the edge of the salt marsh islands. When flushed from the colony, Skimmers fly away, circling out over the water, or may perform distraction displays and injury-feigning, rather than diving and attacking the intruder (Gochfeld and Burger 1994). We summarized our early studies of Skimmers on Long Island and in New Jersey in a book, *The Black Skimmer: Social Dynamics of a Colonial*

(a)

(b)

Figure 3.7 A Red Knot with a leg flag number L77 (a) feeding on Horseshoe Crab eggs on Delaware Bay. Black Skimmers flying near their colony in Barnegat Bay (b).

Species (Burger and Gochfeld 1990b). The Black Skimmer was declared endangered in New Jersey, and we turned our attention to studying vulnerability to disturbance by boaters (Burger et al. 2010a). It is also listed as endangered in Delaware and Maryland.

Red Knot (*Calidris canutus*)

This 120 to 200 g shorebird is an iconic species of the Delaware Bay coast of southern New Jersey. Beachgoers, photographers, and bird watchers visit the bayshore to see the throngs of shorebirds each spring (Burger and Niles 2013a,b, 2014; Burger et al. 2015b). On their north- ward migration in May, Knots and other shorebirds feast on the eggs of Horseshoe Crabs in a rapid quest to accumulate sufficient fat to fuel the final leg of their migration (Niles et al. 2008, 2009). Estimated at hundreds of thousands when the South Jersey shorebird concentration was discovered in the early 1980s, the migratory shorebird populations of Delaware Bay have declined drastically in only 30 years (Morrison et al. 2004; Niles et al. 2012). Overharvesting of Horseshoe Crabs is a major culprit as is hunting for food by people in the South American win- tering grounds (Baker et al. 2013). Red Knots breed in the Arctic, making a nonstop 4000-km flight north from Delaware Bay (Figure 3.7). During their 2 to 3 weeks in New Jersey, they must almost double their weight, storing enough energy in their fat to propel them northward for sev- eral nonstop days, and arriving with sufficient reserves for the female to begin egg-laying and start their short-breeding season (Baker et al. 2004). The development of lightweight geolocators, capturing sunrise and sunset data, has revolutionized migration studies. Once retrieved after a year, these devices allow more precise tracking of individual birds day by day over a yearlong cycle (Niles et al. 2010; Burger et al. 2012b–d). The Red Knot is discussed extensively in the book because of its recent federal listing, and because it is emblematic of the problems faced by migrant shorebirds in the Northeast.

SECONDARY SPECIES DESCRIPTIONS

Brown Pelican (*Pelecanus occidentalis*)

The Brown Pelican is a permanent resident of the Atlantic, Gulf, and Pacific Coasts (Shields 2014). Large, conspicuous, and familiar where it occurs (Figure 3.2), the almost complete dis- appearance of Pelicans from the Gulf Coast and California in the 1950s and 1960s because of dichlorodiphenyltrichloroethane (DDT) and other pesticides, was readily noticed (Jehl 1973). The disappearance of the flotillas of Pelicans gliding on thermals along a beach or plunge-diving for fish was one of the most dramatic impacts of pesticides. Although there were only 11 New Jersey records before 1980 (NJAS 1999), the species became a regular summer visitor, with peak numbers in 1992, when we found nests (without eggs) on an island in Barnegat Bay (Burger et al. 1993b). The anticipated colonization has not materialized, although Pelicans continue to appear annually. As a ground nester, islands free of terrestrial predators are important (Shields 2014)— hence, our optimism when nests were discovered on an island in the bay, far from the mainland. The Brown Pelican was placed on the Federal Endangered and Threatened Species list in 1970, and its decline was a contributing factor to the banning of DDT and reduction of endrin use in the United States (Shields 2014). Pelican numbers increased, and the species was downlisted in 1985.

Double-Crested Cormorant (*Phalacrocorax auritus*)

Double-crested Cormorants are widespread in both coastal and inland areas of North America. They were persecuted by fishermen and the breeding range shrank (Dorr et al. 2014) until Lewis

(1929) showed that Cormorants were not directly competing for commercial fish species. DDT impacted this piscivorous species, and banning DDT was followed by population increases. Adding the Cormorant to the Migratory Bird Treaty Act occurred in 1972, preventing uncontrolled harassment and killing. The breeding population spread southward out of northern New England, into Massachusetts (first nesting in 1940), the New York–New Jersey area (first nesting in 1977), and New Jersey where the first known breeding was documented in 1987 (Parsons et al. 1991). However, Cormorants were breeding in Chesapeake Bay in 1978 (Watts 2013). Cormorants are reported to kill the trees on which they nest (Craig et al. 2012). They have been subject to intensive research (Weseloh et al. 2011). They are numerous in Barnegat Bay, but do not breed there (Figure 3.8). They are decreasing in Maine and in the Great Lakes.

Little Blue Heron (*Egretta caerulea*)

This is a relatively uncommon, all dark heron of coastal marshes, streams, and lakes. It feeds solitarily, but nests in mixed-species heronries (Rodgers and Smith 2012). It feeds mainly on small fish, but like other herons, it can consume invertebrates, frogs, and crabs. We studied the competition and aggression between Little Blues and Cattle Egrets and their use of nest sites (Burger 1978b,c). Previously, this species was called *Florida caerulea*.

Tricolored Heron (*Egretta tricolor*)

This medium-sized, slender, egret stands about 70 cm tall and weighs about 400 g. Although it was not primarily targeted for the millinery trade (Frederick 2013), it suffered collateral damage as the plume hunters invaded nesting colonies and slaughtered birds indiscriminately. It is primarily a southern species that extended its breeding range tentatively northward in the twentieth century. It was first recorded breeding in New Jersey in 1951, New York in 1955, and Massachusetts in 1976. In the 1970s, it was a fairly common breeding bird in the heron colonies of Long Island and New Jersey, but numbers declined subsequently (Frederick 2013; Gochfeld, Unpublished data). Small fish make up more than 90% of the diet, but it can exploit frogs and invertebrates if they are locally abundant (Frederick 2013). Before 1986, this species was called *Hydranassa tricolor*.

Glossy Ibis (*Plegadis falcinellus*)

It occurs in both the New World and Old World. These dark brown birds, with irridescent green highlights with strongly curved bills offer striking silhouettes (Figure 3.9). They are tactile feeders, swiping their curved bills back and forth through the mud to capture mainly insects and crustaceans. The Glossy Ibis went from a rare and local Florida species in the 1940s to a locally common breeder as far north as New England, a remarkable range expansion (Davis and Kricher 2000). The species first bred in New Jersey in 1957, Long Island in 1961, and Maine in 1971. Glossy Ibis nest in mixed-species colonies in a variety of vegetation types including *Phragmites* and in *Iva* bushes (Burger and Miller 1977).

Osprey (*Pandion haliaetus*)

The Osprey is a large, brown and white fish-eating eagle with a wing span of 5–6 ft (Figure 3.8). The Osprey was one of the prominent victims of the DDT era. DDT was the first endocrine-disrupting chemical identified in wildlife (Poole et al. 2002). After the banning of DDT, Osprey populations in New Jersey were augmented by hacking. Pesticide use on the wintering grounds continues to be a problem, but this species has bounced back in the past 40 years and has once

(a)

(b)   (c)

Figure 3.8   (a) Double-crested Cormorants on posts near a nesting colony in the New York–New Jersey Harbor Estuary, (b) Osprey incubating on a pole nest, and (c) with its nearly full-grown chicks 2 months later (Delaware Bay).

again become a familiar sight along the Intracoastal Waterway. The downlisting of the Osprey from "endangered" to "threatened" in New Jersey was a landmark in conservation. Today, more than 500 pairs nest in New Jersey (Clark and Wurst 2013). Ospreys feed by plunge-diving for fish, often from above 10 m, emerging with large fish in their talons, gripped by the tiny spines along their toes (Poole et al. 2002).

(a)   (b)

Figure 3.9   Glossy Ibis feeding (a) and Glossy Ibis chicks in a nest (b).

Clapper Rail (*Rallus longirostris*)

This chicken-sized marsh bird occurs in *Spartina alterniflora* marshes (Figure 3.10). They nest solitarily on high ground in the marsh, particularly spoil mounds along historic mosquito ditches (Burger and Shisler 1979; Rush et al. 2012). The loud, clattering call was once a familiar sound in most of the New Jersey salt marshes (Kozicky and Schmidt 1949). Clapper Rails are still fairly common in some of the Delaware Bay marshes, but there has been a steep decline along the Atlantic coast. Rails feed on mudflats at the edge of the salt marsh and along channels, where its main food (Fiddler Crabs, *Uca* spp.) is abundant (Lewis and Garrison 1983; Rush et al. 2012).

(a)   (b)

Figure 3.10   Clapper Rails are difficult to spot in salt marshes (a), whereas American Oystercatchers are conspicuous in the same habitat (b).

American Oystercatcher (*Haematopus palliatus*)

This large, strikingly patterned shorebird extended its range northward in the mid twentieth century, establishing itself as a breeding bird in New Jersey and Long Island in the 1960s, eventually reaching Nova Scotia (Figure 3.10). Since then, it has become very common and conspicuous in Barnegat Bay, and most salt marsh islands now have at least one or two breeding pairs. It feeds on small mussels, crustaceans, and bird eggs (American Oystercatcher Working Group 2012).

Piping Plover (*Charadrius melodus*)

Once a familiar sight and sound on Atlantic coast beaches, the Piping Plover decline began in the 1960s. The Piping Plover is listed as endangered in most states, including New Jersey, and is on the Federal Endangered and Threatened Species list (Conserve Wildlife 2014; FWS 2014a,b). It is a species of open beaches, nesting solitarily or sometimes among Least Terns for protection (Burger 1984a), far from the water (Figure 3.11). Its nest, usually with four eggs, is also difficult to find (Elliot-Smith and Haig 2004), and the adults give broken wing distraction displays when an intruder is close. The chicks are precocial (e.g., able to run as soon as they hatch), and they must gain access to the intertidal zone to find food (small invertebrates and insects; Elliott-Smith and Haig 2004).

Willet (*Tringa semipalmata*)

The Willet is our largest sandpiper and one of the noisiest (Figure 3.12). The eastern subspecies is a coastal bird from the Canadian Maritime Provinces to the Gulf Coast of Mexico (Lowther et al. 2001). It is a common breeding bird of the salt marsh islands. Willets nest solitarily or in loose colonies, hiding their clutch of typically four eggs among the stems of *Spartina* (Burger and Shisler 1978b). The nest, eggs, and incubating adult are very cryptic, rarely seen until the adult bursts from almost underfoot, and circles away with loud distraction calls (Vogt 1938; Burger and Shisler 1978b). Willets feed on beaches and mudflats, both tactilely and visually, on a variety of insects, crustaceans, mollusks, polychaetes, and occasionally small fish (Lowther et al. 2001). The Willet is now known as *Tringa semipalmata*.

(a)   (b)

Figure 3.11   A Piping Plover giving a distraction display to draw predators (humans) away from its nest (a), and a Least Tern brooding a chick (just behind the clam shell) (b).

(a)   (b)

Figure 3.12 Willet chicks (a) rely on cryptic coloration for camouflage while in the nest. Chicks are semi precocial. Even the adults (b) blend with the background. They are semicolonial but with large internest distances.

Least Tern (*Sternula antillarum*)

The Least Tern is one of a worldwide group of small terns now assigned to the genus *Sternula* (Figure 3.11). It is a familiar beach-nesting bird that has declined dramatically in the past 50 years owing to a combination of habitat loss, human disturbance, and predation (Thompson et al. 1997). It is listed as threatened or endangered in almost every coastal state in the Northeast. Least Terns prefer open beaches with little or no vegetation for nesting (Gochfeld 1983; Burger 1988b), and often have to shift colony sites because of changes in habitat (Burger 1984b; Burger and Gochfeld 1990c). Larger colonies on the mainland suffer higher predation than smaller colonies on islands. Human disturbance also contributed to reproductive failure (Burger 1984b). Least Tern recovery has involved extensive management of habitat, fencing, wardening, and—in some cases—predator control (Burger 1989).

Four songbirds—Saltmarsh Sparrow, Seaside Sparrow, Henslow's Sparrow, and Sedge Wren— are or were residents in marshes in the Northeast region, and all four are declining, listed as threatened or endangered in some states. We have personally noted the decline of the first two in New Jersey salt marshes. The latter two are primarily wet grassland species. We do not provide data on these species.

BARNEGAT BAY METHODS

Barnegat Bay has been the focus for our studies for more than four decades. Data on breeding colonies, habitat, and nesting were collected every year from 1976 to 2015 on some species. Early in the study, we documented 259 islands in the bay, of which 34 were deemed suitable for nesting birds based on distance from the mainland shore, size, and vegetation (Burger and Lesser 1978). We subsequently tracked colony occupancy and species composition. For ecotoxicology, we collected eggs and feathers for metal analysis from 1970 to 2015 for Common Terns, and from the mid 1980s to the 2015 for other species (Figure 3.13), although we could not always find eggs or chicks when no chicks fledged. Sampling design depended somewhat on avian reproduction timing (phenology), weather, and logistics. That is, over the course of the study, (1) there were changes in species breeding numbers, distribution, and phenology; (2) some species increased markedly, whereas others declined; (3) in some years, reproductive success was zero because of heavy rainstorms and high

Figure 3.13   David Wiser holding an adult Common Tern for processing (a), Fred Lesser and Joanna Burger with egret chicks captured for feather collection (b), and Brian Palestis measuring a young Common Tern (c) in Barnegat Bay.

tides, making it impossible to collect feathers from fledglings; (4) in some years, severe storms made it impossible to get out to some of the colonies to check the number of pairs, eggs, or chicks; and (5) in some years, birds abandoned colony sites because of severe storms or wash-over tides before the first census.

Weather events had substantial impact on our collection of data. Barnegat Bay is wide and long (67 km), and severe storms made it impossible to always check nests on a given schedule. Excessively high tides and winds made travel between some islands difficult, because it was hard to negotiate the narrow channels. Very low tides made it challenging to reach some islands without long walks in thigh-deep water over muddy bay sediments. Weather and work schedules sometimes contrived to keep us out of the bay for days at a time. We did not check nests during consecutive days of rain as some adults (particularly Black Skimmers) fail to incubate or brood during heavy rains (Gochfeld and Burger 1994), and any additional disturbance could increase chick mortality.

Habitat Availability, Use, and Selection

Information on habitat use and selection for birds nesting in Barnegat Bay are partly derived from previous work, but are synthesized in this book to provide an overall picture of habitat use by the different species in Barnegat Bay (Chapter 4). Barnegat Bay lies between the barrier beach and the mainland, the former being covered with dune vegetation, the latter fringed with salt marshes. The bay has numerous salt marsh islands with low (*Spartina alterniflora*) areas subject to daily flooding and higher (*Spartina patens*) marsh, with some areas of higher shrubs (*Iva frutescens* and *Baccharis halimifolia*), and in some places Common Reed (*Phragmites communis*), Poison Ivy (*Toxicodendron radicans*), and small Cherry trees (*Prunus* spp.). We described each island in terms of its vegetation cover by these main types. Some islands in Barnegat Bay are essentially a monoculture of *Spartina alterniflora*. Other islands, created by dredge spoil deposition, have much greater altitudinal gradient from marsh, to small dunes, with more than 50 species of plants, some shrub diversity, and even a semblance of maritime forest. This combination of islands provides habitat diversity for nesting birds. Like many areas of the Atlantic coast, the salt marsh islands in Barnegat Bay were ditched for mosquito control in the early 1900s, and the effect of mosquito control on water management, vegetation, and habitat availability is also discussed.

Four concepts need to be distinguished: habitat suitability, habitat availability, habitat use, and habitat selection. In our work, we refer mainly to nesting habitat, but similar approaches apply to foraging habitat. The methods for examining habitat use and selection are described separately for each species or species group because the methods varied. Suitable habitat encompasses the specific needs of each species. Habitat suitability may range from optimal, meeting all needs, including security from water and predators, to marginal. Moreover, not all suitable habitat is available to that species. Some habitat has been altered, and some has been preempted by other species including humans. Habitat availability indicates space that birds can use currently, without regard to suitability. Coastal habitats include the open ocean, surf zone, beaches on barrier islands, sandy or rocky islands (or combinations), salt marsh islands, and mainland salt marshes. These are major categories, however, and each can be subdivided into others.

Habitat use reflects where we find the birds nesting regardless of what we think of as available or suitable. For example, Common Terns nest on wrack (dead stems of grasses) strewn on the high places of *Spartina alterniflora* and *Spartina patens* salt marsh islands (Figure 3.14). Herring Gulls nest on *Spartina patens* and near *Iva* bushes. These are descriptions of where birds nest, but not whether they have selected particular habitat types from those available. Habitat selection requires knowing not only where birds nest, but what habitat is available that they can choose from.

(a)

(b)

Figure 3.14 Common Terns prefer to nest on wrack (a), whereas Black Skimmers choose sand patches (b).

In general, habitat availability was determined by estimating (with transects) the percentage of salt marsh islands that were composed of a particular vegetation type. Habitat use was determined by recording the habitat location of nests, either all of them in a colony, or a sample, depending on colony size. Habitat selection was examined by comparing the habitat (or vegetation) used with the habitat available. Thus, we compared occupied sites with those that were unoccupied (determined by random selection of points). Although birds exhibit flexibility in their choice of nesting sites, they prefer particular types of habitats (Wilson and Vermillion 2006). Our approach was comparative among species.

Population Numbers

Since 1976, we have conducted yearly censuses of Common Terns, Forster's Terns, and Black Skimmers on the salt marsh islands of Barnegat Bay. This has involved traveling by car, boat, and helicopter from the north end of the bay (Lavallette area) to the south end at Tow Island, stopping at all islands with birds to count the number of adults, nests, and young. This usually resulted in us being mobbed by terns or gulls defending their colony (Figure 3.15). Islands were visited several times per month, depending on the year and the questions being addressed. In some years, islands

(a)

(b)

Figure 3.15 To defend their nests, Common Terns mob M. Gochfeld, who is counting nests (a). Often, the terns actually strike the heads of scientists walking through the colony (shown here are Fred Lesser and Joe Jehl) (b).

were visited more often to determine tidal or predator effects, or laying sequences. Ground surveys consisted of surveying all available habitat and sites known to be used previously (Burger and Lesser 1978; Burger and Gochfeld 1991b). Once colonies were located, we observed from the edge at a distance that did not unduly disturb the birds, then counted the number of adults present, and then estimated population size and phenology by making partial or complete counts of nests, eggs, and chicks. Aerial surveys by helicopter allowed estimations of available habitat, and assured that new colony sites were located.

We performed two types of nest checks: in-depth checks and routine checks. Routine checks involved walking through the colony and recording the following information: nest contents, age of chicks if present (alive or dead), and evidence of predation, starvation, abandonment, or flooding out of eggs or chicks. These nest checks provided us with information on colony size, clutch size, and fledging success. In-depth nest checks involved marking each nest with a numbered flag, marking eggs with a magic marker indicating laying order, banding chicks, and following them until hatching.

In-depth nest checks occurred every 3–4 days throughout the season. Because colonies of terns and Skimmers contained fewer than 500 pairs, we could count all nests. Eggs that were washed out of nests generally had a chalky white color and were scattered away from nests, often in a windrow of eggs at the edge of *Spartina* wracks or against dense *Spartina*. Abandoned eggs usually remained in the nest, and were cold to the touch. We also tested whether a nest was abandoned by separating the eggs and leaving a space of about 2 cm between them. Returning adults quickly gathered the eggs together for incubation, and on our next visit, the eggs were again touching one another. If eggs remained in the displaced position for 24 h, we assumed they were abandoned. Eggs preyed upon by American Oystercatchers had a characteristic slit. Larger peck holes usually indicated predation by gulls, although Herring Gulls usually removed eggs to eat. Starved chicks were underweight; flooded chicks looked bedraggled with wet feathers and were usually some distance from the nests. Chicks that died from the attacks of neighbors had peck marks on their heads, were often bloody, and were usually some distance from the nest. Productivity was determined by dividing the number of fledgling chicks (fully feathered preflying and flying) by the number of pairs present (sometimes estimated by the maximum number of nests present).

Data on number of birds present were also collected on Forster's Tern, Laughing Gulls, Herring Gulls, and Great Black-backed Gulls during our censuses of Common Terns. These species nested on many of the same islands as the terns, although they also nested on islands without the terns. Data on herons and egret populations were obtained from the Endangered and Nongame Program of the New Jersey Department of Environmental Protection's aerial census (NJDEP, Unpublished data), as well as our own data. Although we entered the heronries to collect eggs and feathers from fledglings, we did not enter regularly because of disturbance to chicks that leave their nests. The gull, tern, and skimmer chicks hid in the *Spartina* near their nests during our nest checks, and usually remained only a few centimeters from their nests, whereas the older heron and egret chicks tended to climb or run longer distances from their nests when disturbed.

COLLECTION OF DATA FROM OTHER BAYS AND ESTUARIES

Information from other bays and estuaries comes from data we collected, from samples collected by others for us, from data provided by others, and from published data (Figure 3.16). Much of the background information for the Northeastern bays and estuaries comes from the Environmental Protection Agency National Estuary Program Coastal Condition reports (EPA 2007a–g). Although each report differs somewhat, they all contain some of the same material about the condition of the bay, including toxic contaminants, pathogens, floatable debris, nutrients and organic enrichment, and biological resources, except for edible fish. Other information comes from published studies (our own and others).

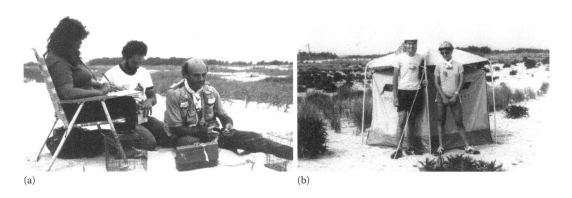

(a)   (b)

Figure 3.16 Some of our early work on Common Terns was at Cedar Beach on Long Island. Here, Mike Gochfeld is collecting feathers from a Common Tern; Carl Safina is placing feathers in envelops for later analysis, and Joanna Burger is recording measurements (a). Blinds were used for observing nesting birds. Shown here are field assistants, David and Debbie Gochfeld in front of a typical blind (b).

We also benefited from interacting with colleagues conducting long-term studies of waterbirds in other estuaries. For example, Common and Roseate Terns have been studied by Jeff Spendelow in Connecticut colonies, Helen Hays on Great Gull Island (eastern Long Island), and Ian Nisbet (Bird Island in Buzzards Bay). Their long-term studies on both Roseate and Common Terns have contributed greatly to the literature (Hays 1970; Hays and Risebrough 1972; Nisbet 1994, 2000, 2002; Spendelow et al. 1995; Figure 3.17).

Wading birds have also been under long-term study. Katherine Parsons studied wading birds in New England, in the New York–New Jersey Harbor Estuary, and Delaware Bay, and Susan Elbin, Nellie Tsipoura, Elizabeth Craig and others have continued the monitoring and management of the "Harbor Herons," including educating the public about the importance and value of this resource so close to millions of people (Figure 3.18). Michael Erwin (Erwin et al. 2007a) and Bryan Watts (2013) have studied colonial birds of Chesapeake Bay. There are also long-term studies of colonial birds on the Great Lakes (Weseloh et al. 2011).

COLLECTION OF SAMPLES FOR METAL ANALYSIS

For our ecotoxicological studies, we sampled invertebrates, fish, and birds. For birds, we collected mainly eggs and feathers, which we analyzed for several metals and metalloids to determine a possible link with population changes. Such data are valuable for establishing trends and as a baseline for future comparisons (Jerez et al. 2011), as well as to examine individual variations (Barata et al. 2010). We collected Horseshoe Crab egg masses from Crab nests (Figure 3.18), and tissue from recently dead adults stranded on the beach or from bait collections.

All collecting was done under the necessary Federal and State Permits and under protocols approved by Rutgers University. For birds, we generally collected eggs and feathers from fledglings (which represent local exposure because parents feed locally, usually within 10–20 km of the colony). The birds we studied typically lay three to six eggs per clutch (varying by species and food availability), but they rarely raise all the chicks to maturity (Pierotti and Good 1994; Parsons and Masters 2000; Nisbet 2002; McCrimmon et al. 2011). Thus, our practice of taking one egg/clutch would have little impact on populations.

Figure 3.17   Researchers with long-term studies on Roseate and Common Terns in Northeastern bays. (a) Mike Gochfeld works on Cedar Beach in New York with a fenced Roseate Tern nest (top right). (b) Jeff Spendelow, Joanna Burger, and Carl Safina (left) check Roseate Tern nests in tires on Falkner Island. (c) Researchers at Great Gull Island check Roseate and Common Tern nests. (d) Ian Nisbet and an aerial view of Bird Island.

(a)   (b)

Figure 3.18 Long-term studies for the New York–New Jersey Harbor Estuary have been conducted by the New York City Audubon, under the Harbor Herons Project. Here, Susan Elbin bands a Great Egret (a). Similar long-term studies in Delaware Bay involved both Horseshoe Crabs and shorebirds. Nellie Tsipoura and Joanna Burger excavate nests of Horseshoe Crabs to gather their eggs for analysis (b).

Feathers can be collected noninvasively (Kim and Oh 2014), making them ideal samples because their collection does not impact reproduction or survival. We captured preflying or recently flying chicks by hand or by net. We plucked a pinch of breast feathers (about a dozen feathers) from each side (see below), and released the birds. Feathers were stored in envelopes. Although data are based on a convenience sample within colonies, there is no reason to assume that samples were biased.

Biomonitoring Metals in Eggs

Eggs were collected at the beginning of the breeding season to ensure that eggs were from first clutches and not from females that had renested. Usually, the first egg was collected from each nest to ensure comparability among nests and colonies. Several studies have shown that contaminant levels vary by egg-laying order (Kennamer et al. 2005; Akearok et al. 2010; Custer et al. 2010), although this is not always the case (Brasso et al. 2010). Thus, it is desirable to use the same procedure over space and time. Freshly laid eggs were collected from widely separated locations within a colony. If we could not get to a colony at the very beginning of egg-laying owing to the inclement weather, eggs were floated to determine if they were undeveloped (Hays and LeCroy 1971). Recently, Ackerman and Eagles-Smith (2010) showed that the precision of flotation declined during incubation, but because our purpose was to identify eggs with minimal embryo development, we used the Hays and Lecroy method. Eggs were taken back to the laboratory and stored in a refrigerator for immediate analysis; some eggs were frozen for archival purposes.

Biomonitoring Metals in Feathers

Feathers play an important role in the toxicokinetics of metals (see Chapter 8). Metals absorbed into the bloodstream are delivered to target organs, excreted, or sequestered in feathers (Lewis and

Furness 1991) or other tissues. Growing feathers serve as a sink for methylmercury (Spalding et al. 2000a) and other metals. The mercury in feathers is close to 100% methylmercury (Thompson et al. 1991; Bond and Diamond 2009). In young birds, a substantial part of the body burden for several metals, particularly mercury, is stored in the feathers (Braune and Gaskin 1987). Down and growing plumage of young birds play a key role in eliminating metals (Burger 1993; Spalding et al. 2000a). Metals enter feathers during the 2–3 weeks it takes for them to grow, after which time the blood supply atrophies, and there is no further uptake of metals (Monteiro 1996; Thompson et al. 1998a,b). Although some seabirds may travel thousands of kilometers from the colony to find food for chicks (Schreiber and Burger 2001; Safina 2002), most of the species we studied are feeding within a few kilometers of the colony. The food brought back to nestlings is representative of the vicinity of the colony (Custer et al. 2007b). The feathers of adult birds reflect circulating metal levels at the time feathers were grown, but where this happened depends on molt cycles.

Breast feathers, rather than wing feathers, were selected for our studies because they are considered to be more representative of exposure to metals than wing or tail feathers (Furness et al. 1986; Bortolotti 2010), and metal concentrations in breast feathers are correlated with internal tissues (Gochfeld 1980b; Burger 1993; Zamani-Ahmadmahmoodi et al. 2014). We captured chicks by hand or net, and plucked about two dozen feathers per bird. The areas of pulled feathers were not noticeable, and breast feathers grew back in about 3 weeks (Burger and Gochfeld 1992a; Burger et al. 1992a), reducing any impact on the bird. Palestis and Stanton (2013) found no differences in survival of Common Tern chicks as a function of the age of feather plucking. Breast feathers are easy to collect noninvasively and can be stored indefinitely in sealed envelopes, making them especially useful for establishing contamination in threatened or endangered species, without impacting populations. The ease of storage facilitates the use of feathers for establishing time trends. To minimize analytic variability, feathers collected in different years were analyzed at the same time, using the same equipment with the same detection limits (e.g., Frederick et al. 2004). We aimed to collect about 12 to 20 feather samples per colony. Breast feathers were collected from prefledging and recently flying young, rather than when they were younger (Frederick et al. 2002) (Figure 3.19). We did not pool samples.

Tissue Samples

For some birds recently predated or found with severe injuries and euthanized, we examined internal tissues (e.g., shorebirds). We dissected birds in the laboratory, and removed liver, kidney, heart, pectoral muscle, and brain (Burger et al. 2013f). For invertebrates (e.g., Horseshoe Crabs) internal tissue samples were also taken. Small prey fish were digested whole because birds eat the whole fish, but for larger fish, we took a small sample from the midline of the fish (a fillet), which is equivalent to what gulls eat as carrion and also people eat.

METAL ANALYSIS

All samples were analyzed in the Elemental Analysis Laboratory of the Environmental and Occupational Health Sciences Institute of Rutgers University, in Piscataway, New Jersey. Mercury was analyzed by cold vapor atomic absorption spectrophotometry, and other metals were analyzed by graphite furnace (flameless) atomic absorption (Figure 3.20; Burger and Gochfeld 1993a). A quality assurance protocol was followed. Instruments were calibrated each day with a calibration curve that included a blank, four standards, a spiked sample, and a reference sample (NIST fish tissue, DORM-2). The correlation coefficient for the calibration curve exceeded 0.99. The variation from the known standards and the coefficient of variation on replicate samples was within 15%. Instrument detection limits were 0.02 ppb for arsenic and cadmium, 0.08 ppb for chromium,

(a)   (b)

Figure 3.19 It is important to collect feathers from chicks with fully formed breast feathers. (a) Burger is hold-ing a Skimmer chick with full breast feathers, whereas (b) Taryn Pittfield is holding a chick that is too young for feather collection (note the down on the throat).

(a)   (b)

Figure 3.20 Christian Jeitner collecting feathers from a fledgling Herring Gull from Dredge Island in Barnegat Bay (a). Here, Jeitner is in the laboratory working with the atomic absorption (AA) spectropho-tometer to analyze heavy metals (b).

0.15 ppb for lead, 0.09 ppb for manganese, 0.02 ppb for mercury, and 0.7 ppb for selenium, but matrix detection limits were an order of magnitude higher.

All of our analyses were for total metal. The mercury in avian tissues is usually 90–100% methylmercury (Thompson et al. 1991), with feathers containing essentially 100% methylmer-cury. All concentrations are expressed as parts per billion equivalent (ppb) equal to nanograms per gram (ng/g) or micrograms per kilogram (μg/k) on a dry weight basis. In discussions and tables, we converted published data from parts per million (ppm) to parts per billion (ppb). We converted wet weight values to dry weight values by multiplying by the average moisture content of eggs.

Analysis of Eggs

In the laboratory, egg contents were emptied into acid-washed weigh boats, weighed, and then dried to constant weight. Results are presented on a dry weight basis. Whole egg contents were homogenized and digested individually in 70% nitric acid within microwave vessels for 10 min at 150 lb/in² (10.6 kg/cm²) and subsequently diluted with deionized water. Mercury was analyzed as total mercury, of which about 90% or higher is assumed to be methylmercury (Grieb et al. 1990).

Because we found substantial variability in the moisture content of fresh eggs, we report our egg results on a dry weight basis. Dry weight concentrations are 3× to 5× higher than wet weight concentrations on the same sample. Table 3.4 summarizes the mean and range of the moisture content of eggs for egrets, herons, ibis, gulls, and terns.

The wet to dry conversion ratio depends on the moisture content of the eggs. However, published data on moisture content of bird eggs are sparse. For example, few of the authoritative *Birds of North America* species accounts provide such information. The water content of Common Tern eggs from Massachusetts was 74.7–77.6% (Nisbet 1978), close to our Common Tern average of 75.8%. We determined the moisture content of egret, heron, gull, and tern eggs (Table 3.4) by pouring the homogenized contents into a plastic weight boat, and evaporating them at 50°C for 48 h to a constant weight, usually achieved by 24 h. Moisture content was not very variable for most species, with the coefficient of variation ranging from 1% to 4%, except for 8% for Herring Gull eggs and 11% for Great Egret eggs. Great Egret eggs had the higher moisture content (mean 83.5%) compared with Snowy Egrets eggs (mean 80.5%, $t = 1.81$, $P = 0.07$), and the difference approached statistical significance. Likewise, the difference between Great Black-backed Gull eggs (moisture content average 78.3%) and Herring Gull eggs (average 76%) approached significance ($P = 0.07$).

Whether data are reported as wet weight or dry weight makes a substantial difference, and omission of that detail has been a recurrent problem in interpreting older published results. Contaminants in eggs have been published on both wet weight (ww) and dry weight (dw) bases, and few papers are helpful in including both values (Champoux et al. 2002). The conversion ratio between wet weight and dry weight is the reciprocal of 1 minus the percent moisture or 1 divided by the dry matter

Table 3.4 Moisture Content of Bird Eggs from Barnegat Bay

Species	Sample Size	% Water Mean ± SD	% Water Range	% Dry Matter (Mean)	Conversion Ratio Used
Great Egret	259	83.8 ± 9.6	79.7–87.4	16.2	5[a]
Snowy Egret	28	80.5 ± 1.6	78.4–83.4	19.5	5
Black-crowned Night-Heron	27	81.0 ± 3.1	79.1–83.1	19	5
Glossy Ibis	11	81.0 ± 0.7	80.2–82.9	19	5
Great Black-backed Gull	25	78.3 ± 3.0	74.5–87.9	21.7	4[a]
Herring Gull	57	76.0 ± 6.0	65.7–86.1	24	4
Laughing Gull	52	75.5 ± 2.2	72.0–88.0	24.5	4
Common Tern	300	75.8 ± 1.7	67.0–84.0	24.2	4
Forster's Tern	100	76.7 ± 1.0	73.9–82.9	23.3	4

Note: These data can be used for converting concentrations between wet weight (ww) and dry weight (dw) for studies that published concentrations in eggs on wet weights.

[a] May result in about a 10% underestimate of the concentration on a dry weight basis.

percent): Conversion ratio = 1/[1 − moisture%]. Thus, if the wet weight of an egg's contents is 100 g and the dry weight is 20 g, the moisture content would be 80% and the dry content would be 20%. The appropriate conversion ratio would be 1/0.2 = 5. If the results were reported as wet weight, we would multiply by 5 to obtain a comparable dry weight value.

Table 3.4 shows that most gulls and terns have a ratio about 4 and most herons have a ratio about 5. The Great Egret data would actually result in a ratio closer to 6, but we have been conservative in using 5 for that species as well. For the gulls, a ratio of 4.5 would be appropriate for the Black-backed compared with 4.2 for the Herring Gull. However, in the absence of data on moisture content of the eggs in other published studies, we decided it was simpler and slightly conservative to round off the ratio to 4.0 for gulls and terns and 5.0 for herons, egrets, and ibis. This may underestimate the converted value for Great Egrets and Great Black-backed Gulls by about 10%, which is within the range of analytic error.

Analysis of Feathers

Feathers were washed three times with acetone and then with deionized water to remove surface contamination. The feathers from an individual bird were pooled, but each bird was analyzed individually. We did not pool across birds. Feathers were digested in a microwave in warm nitric acid mixed with the addition of 30% hydrogen peroxide, and subsequently diluted with deionized water (Burger and Gochfeld 1991a). All concentrations are expressed in ng/g (ppb) on a dry weight basis using weights obtained from air-dried specimens. Detection limits were the same as described earlier. All specimens were analyzed in batches with reference material (NIST fish matrix, DORM-2), calibration standards, and spiked specimens. Recoveries generally ranged from 88% to 112%. Batches with recoveries of less than 85% or more than 115% were reanalyzed. The coefficient of variation on replicate, spiked samples ranged up to 10%.

Organs and Tissues

For invertebrates (crabs, mollusks), fish, and some birds, we analyzed internal tissues. At Environmental and Occupational Health Sciences Institute, a 2-g (wet weight) sample of tissue was digested in Ultrex ultrapure nitric acid in a microwave (CEM, MDX 2000), using a three-stage digestion protocol of 10 min each under 50, 100, and 150 lb/in^2 (3.5, 7, and 10.6 km/cm^2) at 70% power. Digested samples were subsequently diluted to 25 ml with deionized water. All laboratory equipment was washed in 10% HNO_3 solution and rinsed with deionized water before use.

STATISTICAL ANALYSIS

We present the mean ± 1 standard error unless otherwise specified. In some cases, we used log transformation and parametric methods (SAS 2005). Many of the characteristics we examined are not normally distributed and do not lend themselves to parametric analysis. We used nonparametric statistical procedures as appropriate (Siegel 1956), Chi square tests on Contingency Tables, and Goodness of Fit (Sokal and Rohlf 1995). Correlational analyses were performed using the Kendall tau rank correlation method. We used nonparametric statistics because the Kruskal–Wallis chi square test has 95% of the power of analysis of variance (ANOVA), without having restrictive assumptions (Siegel 1956).

The graphs on temporal trends report the nonparametric Kendall tau correlation between each sample and the year. Thus, although there may be only six sample years on some graphs, the *n* for the analysis is the total number of samples (shown in the tables). The curved lines indicate the curvilinear best fit to the data using third-degree polynomial curves created in DeltaGraph 7. This function estimates the coefficients a, b, c, d that best describes the data by minimizing the differences between the equation and the data and plotting the function: $f(x) = ax^3 + bx^2 + cx + d$.

In many of our published studies, we used multivariate analysis to explore the effect of independent variables using stepwise multiple regression, and then used those variables that entered significantly in a general linear models procedure that gives a more valid estimation of the significance level for each variable. Data were log-transformed and analyzed using ANOVA to determine differences among metals (SAS 2005). We used the Duncan multiple range option with ANOVA (SAS 2005) as a post hoc test of the significance of the differences among years and metals. We usually present the F value for the model, as well as the degrees of freedom and the P value for each independent variable. We consider $P < 0.05$ as significant, but sometimes present $P < 0.10$ where it provides additional information relevant to management.

Statistical Considerations

For both the population data and the contaminant data, we were interested in spatial and temporal trends. Examples of the null hypotheses we formulated are as follows:

1. There are no differences in number of breeding terns between years.
2. There are no differences in mercury levels in Common Tern feathers between the north and south ends of Barnegat Bay.

We looked to our data to either reject or fail to reject the null hypothesis in favor of alternative hypotheses (Table 3.5). We tried to sample enough individual birds to achieve at least 80% power to reject the null hypothesis of no difference between two samples. For population trends, it is important that the sampling has sufficient *sensitivity* to detect a change in number or an upward or downward trend. In this context, *specificity* refers to not reporting a trend, when there are mere fluctuations. There is much variability in contaminant levels and population numbers between years, and 2 or 3 years of sampling cannot indicate a real trend.

If we detect a trend that is really happening, that is a true positive. But if we detect such a trend, we may not know for several years whether it is real. Thus, we are interested in the *sensitivity*, which is the ability to detect a trend that is happening, and in the *positive predictive value*: If we detect a decline, how confident are we that it is real? Bird populations may vary erratically, with several years of growth or decline, followed by reversals. Many factors go into the year-to-year variation, as well as into the long-term population changes.

Table 3.5   Censusing: Sensitivity and Specificity for Detecting Population Trends

Possible Outcomes	There Is No Long-Term Trend in Population	There Is a Real Downward Trend in Population	
Our censusing indicates no change over time	TN TRUE NEGATIVE	FN FALSE NEGATIVE	Negative Predictive Value TN/(TN + FN)
Our censusing indicates a downward change over time	FP FALSE POSITIVE	TP TRUE POSITIVE	Positive Predictive Value TP/(TP + FP)
	Specificity = TN/(TN + FP)	Sensitivity = TP/(TP + FN)	

SUMMARY AND CONCLUSIONS

The indicator species, as well as other typical marsh species, are described to provide background for the rest of the book. The main methods for this study of birds nesting in the colonies in Barnegat Bay were frequent surveys of the number of birds/colony from 1976 to 2015, studies examining habitat use and selection in several species during the study period, and collection of eggs and fledgling feathers for metal analysis from the 1970s to 2015. Population studies for some species involved collecting data on number of adult birds present, number of nests, clutch size, hatching rate, and fledging rate. Habitat studies were conducted with the main indicator species to understand habitat use and competition for space. Studies of metals were conducted on several bird species, invertebrates, and fish from Barnegat and other estuaries from Maine to Florida. The methods were designed to examine colony and population dynamics, habitat use, and contamination by metals over the period from 1970 to 2015 in Barnegat Bay, and compare these aspects with birds in other Atlantic coastal bays and estuaries, particularly in the Northeast.

PART II

Habitat and Populations Dynamics

CHAPTER 4

Habitat

INTRODUCTION

The ecosystems that occur between the ocean and the land are some of the most productive on Earth. Similarly, they are very important sites for carbon sequestration, and they account for half of the carbon sequestration that resides in ocean sediments (Duarte 2009). Yet, they are also some of the most imperiled. Worldwide, there is about 1–2% loss of coastal salt marshes each year (Achard et al. 2002). Although the coasts of the world are extensive, the size of natural coastal ecosystems is shrinking at an alarming rate, mainly because of human development, nitrification, contaminants, and natural geophysical processes. The major drivers are land reclamation, aquaculture, excessive sedimentation, nutrient and organic inputs, coastal hypoxia, and a combination of sea level rise and subsidence. Even the ecosystem below the water surface of estuaries and bays is imperiled. Sea grasses—referred to generally as submerged aquatic vegetation—that support fish and shellfish nurseries are declining in both tropical and temperate waters (Green and Short 2003; Fertig et al. 2013). Conservation of these fragile environments is essential for the protection of avian communities and the other components of coastal communities (Boersma et al. 2001; Salafsky et al. 2008). Conservation, management, and restoration may be a decades-long process, but because of losses already suffered (especially given sea level rise), the effort is necessary (Warren et al. 2002).

The Atlantic coast of North America boasts an extensive mosaic of barrier islands, bays, estuaries, freshwater and saltwater marshes, other wetlands, and many islands, large and small, natural and artificial, within the bays and estuaries (Burger 1991b). Large rivers and small streams empty into the bays, and bring freshwater, nutrients, and contaminants. Although we think of the sea as having the major impact on coastal ecosystems, riverine flow is also extremely important, although declining. For example, the Mississippi–Atchafalaya River system represents about 80% of the freshwater inflow into the Gulf of Mexico each year, in contrast to other rivers and streams (Nyman and Green 2015). These carry sediment that creates a large river delta. The Hudson River, large and important for the New York–New Jersey Harbor Estuary, has an ebb and flow of fresh and saltwater. The Hudson, at 606 m³/s, delivers about 19 trillion m³/year of freshwater to New York Bay. The Delaware River flow, which is seasonally variable, is about 330 m³/s.

In this chapter, we briefly discuss some of the factors that affect the availability, suitability, and use of habitat by birds in the Northeastern bays and estuaries, although these factors apply to other habitats as well. This is not meant as a review of habitat use in birds, but rather as an introduction to the factors that birds nesting in bays and estuaries face. Throughout, it is clear that changes in species, communities, and ecosystems are a widespread, normal feature of ecosystems (Moore et al. 2009), and coastal ecosystems experience change sooner than others.

To the casual visitor, a salt marsh island seems to be a simple habitat—a large grassland, perhaps with brown patches of dead vegetation known as "wrack" and perhaps surrounding a core of bushes or tall reeds. The reality is somewhat more diverse (Teal and Teal 1969). Most of the "grassland" is salt marsh dominated by *Spartina alterniflora*, a coarse grass that tolerates daily tidal flooding. At a higher elevation of a few centimeters, the aspect changes to the much finer, *Spartina patens*. Higher and dryer, not subject to frequent flooding, is a zone of bushes, Groundel Bush (*Baccharis halmifolia*) and Marsh Elder (*Iva frutescens*). Reeds (*Phragmites*) create their own stable habitat by spreading with runners, until they encounter the daily tidal zone. Deposition of mud from historic mosquito ditching or dredging has added to the complexity of many islands, although areas able to support short woodlands of cherry trees and dense Poison Ivy occurs more on the back side of the barrier beach or on the mainland, rather than on islands. A few islands created or greatly augmented by dredge spoil are tall enough to support a very complex flora, including bushes and reeds. Gulls are the dominant nesters on the latter islands.

HABITAT LOSS

The extensive coastal mosaic along the Atlantic, close to the Boston–Washington megalopolis, is vulnerable to degradation. In the New England region, 50% of the salt marsh areas were lost by the mid 1970s (Nixon 1982). The losses have continued, but through management efforts and public awareness, the rate of decline has slowed down (Valiela et al. 2009). The attention and stakeholder empowerment brought by the National Estuary Program (EPA 2007a–f) has contributed to slowing the habitat loss throughout the region. In the continental United States, the Northeast had 16% of the total salt marshes and 14% of the tidal mudflats. In the late 1980s, New Jersey had more salt marsh area than any other state in the Northeast, as well as extensive tidal mudflats (Field et al. 1988; Figure 4.1). The Gulf of Mexico coastline had well over half of the eastern salt marshes in the 1980s, but areas of the Gulf Coast have been losing salt marshes at an alarming rate as well (Valiela et al. 2009).

In 1993, Kearney et al. (2002) categorized as "nondegraded" only 38–55% of Delaware Bay and only 28–31% of Chesapeake Bay. Likewise, less than 30% of Barnegat Bay is unimpacted. Although much of the loss is attributable to human activities (ditching, development), some is due to sea level rise, land subsidence, salt marsh die-back, and invasive species (Alber et al. 2008). The direct effects (ecological changes, fragmentation) can be seen visually from satellite photography

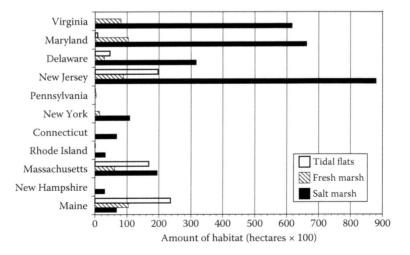

Figure 4.1 Amount of different habitat types from Virginia to Maine, in the late 1980s (hundreds of hectares). (After Field, D.W. et al., *Mar. Fish. Rev.*, 50, 40–46, 1988.)

and digital land cover data (Lathrop and Bognar 2001), but the indirect effects mediated by warming of the atmosphere leading to ice cap melt and sea level rise are more subtle (Valiela et al. 2009). Fragmentation of habitat is one of the chief ecological effects in a cascade leading to a broad range of other effects (Forman and Godron 1986; Andren 1994; Fahrig 2003). Sand washes away, dunes disappear, saltwater overwash occurs, coastal forests shift composition, salt marshes disappear in some places, and *Phragmites* monocultures appear.

In addition to ecological change, there is a range of goods and services that will be lost, including export of energy-rich materials to the food webs of oceanic waters, nutrients and nurseries for commercially important fishery stocks, loss of habitat for shell and finfish stocks, loss of aquaculture sites, contaminant interception, shoreline stabilization, sources of forage and hay, loss of waterfowl refuges, and loss of secure and productive migratory stopovers for waterbirds. These, in turn, impact ecotourism aesthetic and existence values (Costanza et al. 1997, 2014; Gedan et al. 2009). The impact of estuary degradation on fisheries is large and economically critical. Estuarine-dependent fish comprised 46% by weight and 68% of the value of U.S. commercial fish landings (2000–2004), and more than 80% of recreational fish harvest (Lellis-Dibble et al. 2008).

A similar picture of rapid coastal changes in habitats was developed by Erwin et al. (2004) for the period 1932–1994 for marsh morphology at Northeastern Atlantic coastal sites. They used aerial photographs from a variety of sources, and georeferenced them to examine changes in total salt marsh area at specific sites (Figure 4.2). Overall, Nauset Marsh in Cape Cod (Massachusetts) experienced a 19% loss of salt marshes from 1947 to 1994. Forsythe NWR ("Brigantine" New Jersey) experienced a 17% loss from 1932 to 1994. Curlew Bay (Virginia) experienced a 9% loss from 1949 to 1994. The losses for both Massachusetts and Virginia might have been higher if aerial photographs were available for 1921, as they were for New Jersey. Although these data are not for the entire coast from Virginia to Massachusetts, they provide an indication of rapid shifts in available salt marsh habitat. As the authors have noted, such large-scale changes in available salt marsh have implications for salt marsh specialists, such as Laughing Gulls, Forster's Terns, Clapper Rails, and Saltmarsh Sparrows (Erwin et al. 2004).

Suitability and availability are key features of habitat. Suitable habitat has the structural and biotic features to support a species during the breeding season. The suitability may range from

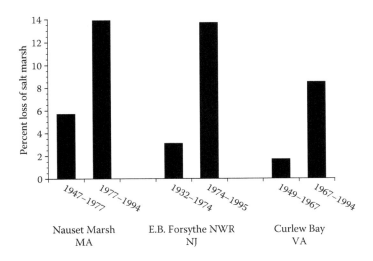

Figure 4.2   Percent of loss of three salt marshes located in Massachusetts, New Jersey, and Virginia. Most of the losses were in the mid 1970s to the mid 1990s. (After Erwin, R.M. et al., *Wetlands*, 24, 891–903, 2004.)

optimal to marginal. It may be rendered unavailable by environmental changes or human distur-bance, or made more available and suitable by restoration.

Not only has there been a loss of marsh land, but there has also been a degradation of the remaining marshes, which is not always readily apparent. In the post–World War I era, there was extensive ditching of salt marshes to remove shallow mosquito-breeding pools (Burger and Shisler 1978a, 1979; Niemi et al. 1999; Berg 2010). Ditching left spoil piles of higher elevation, while draining large areas of the marsh and changing the natural vegetation (Burger and Shisler 1979). The lack of ephemeral ponds in marshes reduced the available foraging space, as well as the avail-able prey for some salt marsh birds. Similarly, the general drying of the marsh shifted the species composition of marsh vegetation; there was less *Spartina alterniflora*, and more *Spartina patens* and bushes. The changes are dramatic in New Jersey and elsewhere along the coast. They reduce both the suitability and availability of habitat for birds. Unless they are protected by federal, state, or local agencies, the fate of salt marshes is usually ignored by people, even though the barrier islands are highly developed (Figure 4.3).

Like many management practices, however, ditching had some positive as well as negative impacts on the wildlife living in marshes. The early ditching of salt marshes resulted in rather straight-line ditches, with spoil placed in long piles up to about a half meter high along the ditches. Because this substantially reduced the tidal inundation of the spoil sites, Marsh Elder (*Iva fru-tescens*) and Groundsel Tree (*Baccharis halimifolia*) colonized the higher terrain. This slightly higher elevation created places for birds to nest, including Herring Gulls, Clapper Rails, and Willets (Burger and Shisler 1978b,c, 1979). The bushes provided shade for eggs and chicks, as well as decreasing the visibility of nests from avian predators above. Figure 4.4 shows that these species have a high preference for nesting in this habitat, selecting it disproportionately more often than its spatial availability. Some of these data on nest site selection were gathered in the late 1970s, sug-gesting that even then the amount of this higher habitat was limiting for marsh nesting birds. More recent practices for mosquito control, called Open-Marsh Water Management (OMWM), avoid draining the marsh, but rather drain pools that allow mosquito breeding (e.g., dry down between rains), while dredging ditches that meander and mimic natural marsh physiognomy (Berg 2010). The spoil is no longer just dumped in long mounds along the edges of the ditches, but is broadcast across the marsh. Natural, more permanent pools are left, and allow both prey fish and small preda-tory fish to move freely. OMWM has the effect of increasing salinities because there is more free-flowing water from the bay, which also affects fish populations (Talbot et al. 1986), creating habitat for terns foraging over these open pools.

Habitat has also been lost along the open ocean, both because of human development and a combination of sea level rise and natural sand movements (Nordstrom 2008; Pries et al. 2008). Groins and jetties accentuate some beach loss. Beach nourishment projects by the Army Corps of Engineers alters the substrate of the beaches, including changing the elevation and modifying the surface. Numerous residential communities now occupy the outer beaches in a nearly unbroken line along barrier beaches. Some coastal residents do not want dunes to spoil their view of the sea. Others build tall, stabilized dunes for protection, and these rapidly become too densely vegetated for beach nesting birds such as Least Terns and Piping Plover. These dunes provide a barrier against which winter storms can erode and narrow the beaches. Valuable mudflats, wide sandy beaches, and dunes have been lost, and the natural processes could not always replenish them. Such losses are problematic for the open beach nesting species (Thompson et al. 1997; Elliot-Smith and Haig 2004). Beaches, both public and private, are raked (daily in some cases), which is incompatible with plover and tern nesting. Habitat fragmentation is also apparent and a problem (Andren 1994; Fahrig 2015). Even where habitat is preserved, the size of each tract is small with respect to the perimeter, meaning that predators and people are closer to the nesting birds (Forman and Godron 1986).

Loss of coastal habitat results in fewer potential places for birds to nest, roost, and forage, and less productivity because there is less habitat for the base of the food chain (e.g., invertebrates, small

(a)

(b)

Figure 4.3 Typical salt marsh island in Barnegat Bay, with heavily developed Long Beach Island in the background (a: Clam Island). (b) Shows a pristine, undeveloped barrier island.

prey fish in sea grass beds). The overall available habitat is decreased, and some of the available habitat is no longer suitable for nesting by birds, the prey base shifts, and the suite of predators changes (Burger 1991b). Furthermore, the smaller the habitat, the more likely it is that prey depletion will occur (Mendonça et al. 2007). Smaller habitat patches have proportionally larger edge, which increases vulnerability to disturbance and predation.

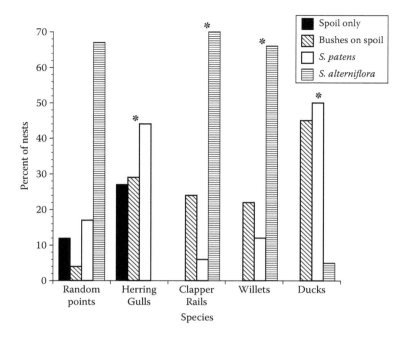

Figure 4.4 Nesting habitat use of Herring Gulls, Clapper Rails, Willets, and Ducks compared to available habitat, as determined from random points. An asterisk indicates significant differences from the available habitat represented by the random points. The species prefer different habitats for nesting but all show a preference for bushes on spoil. "Spoil" refers to dredged sand and mud. (After Burger, J., Shisler, J., *Am. Midl. Nat.*, 100, 54–63, 1978a; Burger, J., Shister, J., *Biol. Conserv.*, 15, 85–104, 1979; Unpublished data.)

HABITAT AND ACTIVITY

There are differences in avian habitat use as a function of activities (breeding versus migrating and overwintering; nesting versus foraging). In the nesting season, birds are tied to their nest site during the incubation period and chick rearing period, and must feed in the neighborhood of the colony, usually with distances of a few kilometers (Visser 2001). The chicks of precocial species, such as Willets, Oystercatchers, ducks, and rails, are covered with down at hatching and are able to leave the nest within hours after hatching to forage on their own, although they are accompanied by their parents. Gulls and terns are semiprecocial. At hatching, chicks are covered with down and their eyes are open, but they are dependent on parents for food for weeks or months until they are able to fly (Schreiber and Burger 2001). The other extreme, represented by herons and most passerine birds (wrens, sparrows) is altricial, where young hatch naked, with their eyes closed. These chicks must be brooded early on because they cannot regulate their body temperature, and they are guarded and fed until they are able to forage on their own. The Saltmarsh Sparrow has adapted by having a very short nestling period.

Although there are constraints on birds to select nest sites that are removed from the threat of tides, floods, inclement weather, and predators, the period that birds are place-based varies by both activity and stage in the life cycle (eggs, chicks, fledglings, adults). Although all the bird species that nest, migrate, or winter in the estuaries along the Atlantic coast can fly, their flight abilities and inclinations to migrate or disperse are variable, and are dependent on energetics (Ellis and Gabrielsen 2001; Whittow 2001). Seabirds are the most mobile, both seasonally and on a daily basis. Seabirds are likely to fly the greatest distances from their nest sites to obtain food for their chicks, with some foraging trips lasting more than a week (i.e., albatrosses; Safina 2002). Laughing Gulls, for example, will fly 30 km out to sea (Wickliffe and Jodice 2010), and 40 km inland to forage (Dosch 2003). Many seabirds nest on stable offshore islands far removed from predators, but in the bays and estuaries of the Northeast, the availability of such islands varies. Still, other species are resident and quite sedentary, and are not likely to fly long distances. This has the effect of isolating populations if the habitat is fragmented. Clapper Rails, herons, and egrets are short-distance migrants, moving to the Southeast states in winter (Rush et al. 2012), whereas Seaside Sparrows and Sharp-tailed Sparrows (Saltmarsh Sparrow) are generally nonmigratory (Greenlaw and Rising 1994; Post and Greenlaw 2009).

AVAILABLE HABITAT AND SUITABLE HABITAT

The Atlantic coastal ecosystem has a number of key features that provide sufficient habitat diversity to support diverse avian communities, including the open sea, a wide, relatively shallow continental shelf, a continuum from open water to barrier islands, tidal marshes, freshwater marshes and ponds, sandy beaches, mudflats, and uplands. In New England, there are rocky coves, cliffs, caves, and tidal pools not known from New York southward. The different habitats are intermixed, creating habitat for different mixtures of nesting, foraging, and roosting birds. We distinguish suitability and availability. The land that is available for birds, however, is also a function of how much land is protected. If the land is not protected, then it is available for sale or development rather than for birds.

Suitable habitat is the habitat that meets the needs of each species. Species have preferences and tolerances. We identify preferences by demonstrating that the birds are selecting nest sites more frequently than the availability of that habitat. However, where preferred habitat is limited, birds may nest in other habitats, perhaps suboptimal ones. Laughing Gulls, for example, may prefer to nest in salt marshes rather than sandy substrate, and there may be 100 ha of salt marsh available. However, if that salt marsh is attached to the mainland, it is not suitable because mammalian predators can easily get there to eat the eggs and chicks. Similarly, if there are low, sturdy bushes for egrets and herons to nest in (thus, there is available habitat), they will not nest there if it is attached to mainland or to large barrier beaches that allow tree-climbing Raccoons and other predators to have easy access. If the 100 ha of salt marsh is fragmented into a dozen very small patches separated by large distances, the habitat is not suitable for nesting by colonial species because they cannot have social interactions or exhibit group antipredator defense. Common Terns may prefer to nest on sandy beaches well above the tide line, but if these are connected to larger areas that harbor predators, the habitat is not suitable. Similarly, if the sandy area is used extensively for recreation, often crowded with people on weekends, the otherwise suitable habitat is rendered unsuitable.

The dichotomy between available and suitable habitat is perhaps sharpest on beaches, especially those highly prized by people. Although the habitat features (sand, vegetation, slope, shells, tidal regime) may be available, the habitat may not be suitable because of the presence of people, human commensals (dogs, cats), and predators (Red Foxes, Raccoons). That is, the availability and apparent suitability may be an ecological trap (e.g., Cook-Haley and Millenbah 2002; Battin 2004) in that birds may select appropriate appearing habitats that are actually "bad" for them because the presence of people or predators makes it very unsuitable. Least Terns typically nest on sandy beaches in colonies high enough on the beach to avoid storm tides, but far enough from heavily vegetated areas or dunes that may harbor mammalian predators. Species nesting in these habitats have been exposed to intense pressure from developers, increased human disturbance, increased predators, and increased competition, as well as sea level rise (Burger 1990a; Thompson et al. 1997; Nisbet 2002; Chapter 7). Even gulls nesting on some remote offshore islands are vulnerable to people in their colonies (Burger and Gochfeld 1981b).

In the 1980s, human activities (e.g., causing disturbances) accounted for more than half of the colony failures in Least Terns (Burger 1984b), and much more for Piping Plover nest failures (Burger et al. 2010c). Dividing the beach into thirds (near the ocean, the middle, and near the dunes; Figure 4.5) indicates that Least Terns prefer to nest in the middle third (Burger and Gochfeld 1990c), where they rely on their cryptic nests, eggs, and chicks for protection.

The endangered Piping Plover nests in the same habitats as Least Terns, and is threatened by the same stressors (Elliot-Smith and Haig 2004), but has the additional constraint that chicks need to get to feeding areas (Least Tern adults carry fish to the nest). Generally, Piping Plover space their nests along the beach, nesting about 100 m from another plover, and they often nest closer to dunes than to the water (less than 50 m to dunes, more than 150 m to water; Burger 1987a). Although they generally nest solitarily, they often nest within, or near Least Tern colonies (Burger 1987a). Compared to random points, for example, Piping Plovers nest closer to dunes and vegetation, farther from water, and closer to Least Tern nests (Burger 1987a). Figure 4.6 illustrates the horizontal stratification from beach-nesting Least Terns and Piping Plovers to birds nesting behind dunes and in upland habitats. Birds nesting in these coastal environments, however, are also threatened by habitat loss and change. The stratification results in increased competition when there is a shortening of the beach because of sea level rise or human development.

Interestingly, Least Terns have remained nesting on beaches, whereas some other typical beach-nesting species (e.g., Common Terns, Black Skimmers) have moved onto salt marshes in bays (Burger and Gochfeld 1990b, 1991b). Even Herring Gulls and Great Black-backed Gulls that traditionally nested on sandy or rocky beaches or islands in New England have moved into salt marshes (Burger 1977a, 1978a; Pierotti and Good 1994; Good 1998), allowing them to colonize Barnegat Bay. This is well illustrated by the Great Black-backed Gull. As its breeding range spread southward out of New England, it was first recorded breeding in 1976 in New Jersey. Finding no typical New England type habitat (Good 1998), the Great Black-backed Gulls began nesting on the salt marsh islands in the bay. They showed a preference for higher islands, associating with and then displacing Herring Gulls. These two large species displaced terns and Skimmers to the island edges (Burger 1978a) or to lower islands (Figure 4.6).

Habitat suitability must also take into account natural mobility patterns of each species. Suitable habitat must be available within the flight range for a species, and the matrix of habitats must allow

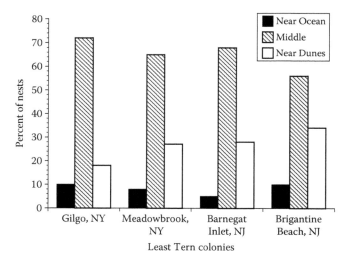

Figure 4.5 Location of Least Tern nests relative to the ocean and dunes. The habitat was divided into thirds, and we plotted the percent of nests in each third of the beach. Terns nested mainly in the middle third. (After Burger, J., Gochfeld, M., *Colon. Waterbirds*, 13, 31–40, 1990c; Unpublished data.)

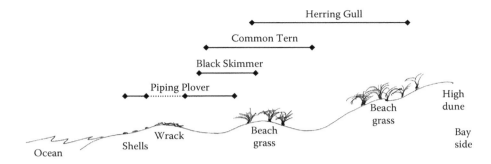

Figure 4.6 Nesting stratification of beach-nesting species from Long Island to the Chesapeake. The dotted line for the Piping Plover indicates it usually nests near the wrack line, and near beach grass or on the higher areas.

for easy movement between habitat patches for nesting and feeding. Habitat suitability, and the relationship to available habitat, leads directly to the concept of habitat selection for colony sites, nest sites, and foraging areas. Selection implies a nonrandom choice of a particular type of habitat from the available habitat (Garshelis 2000). Thus, colony site selection is the choice of a colony site from the habitat that is available, and nest site selection is the choice of a specific nest site from all the sites available within that colony (Burger 1985a).

HABITAT SELECTION

Species show preferences for particular habitats that meet their needs for foraging, roosting, nesting, migrating, and wintering, and that allow them to nest securely, avoid predators, tides, and other stressors, while being close to a food source. Recognition of suitable habitat is an evolved trait, modified by a bird's previous experience of being successful or unsuccessful in that habitat or location. In short, they nest where they nest because they have higher fitness by nesting there than in other habitats (Lack 1968; Ricklefs 1977; Burger 1985a; Bried and Jouventi 2001; Jones 2001; Cook-Haley and Millenbah 2002). Species composition varies along the coast, and in specific habitats. Species are most likely found in specific habitats for breeding, and particular places for foraging (Figure 4.6). Where habitat is limited, species often nest with other species that they otherwise might avoid. Thus, there are both passive and active colony assemblages, particularly where a specific habitat type is scarce.

There are really three levels of selection for breeding sites: (1) habitat or colony site, (2) territory (Burger 1980b, 1984c,d), and (3) nest site (Burger 1985a; Burger and Gochfeld 1987a). The first is the selection of the general area to nest, such as a beach, salt marsh, shrubs, or trees. The second is where to place a territory, including a display site within the general habitat, and the third is where to place the nest within the territory. Solitary-nesting species (e.g., Clapper Rail) select habitat, establish territories with respect to other conspecifics, and then may have many possible nest sites to choose among. Another salt marsh species, the Willet, may select a territory that is close to other Willets to facilitate social mobbing of predators. The internest distances are shorter than if the Willet nests were randomly distributed on the salt marsh. They are semicolonial. Truly colonial species (e.g., gulls, terns, egrets) must consider the proximity of both conspecific and often heterospecific individuals in nest site selection (Burger 1984c).

The range of characteristics selected may also be narrow (e.g., some vegetation, but not dense vegetation). Herring Gulls that nest in salt marshes prefer nest sites in the higher places where there are some small shrubs for cover and protection for the chicks, but they select high open places as display sites. On Clam Island in Barnegat Bay, in a section of island with 40% *Iva* bush coverage, Herring Gulls chose to nest in areas with only 22% *Iva*. On other parts of the island that averaged 8% *Iva* generally, they chose to nest in areas with 20% *Iva*. These choices reflect the selection of partly open places to allow visibility of potential predators, while having some protection and cover from predators, the sun, and heavy rains (Burger 1985a). Similarly, Herring Gulls at Captree on Long Island chose territories in areas with about 20% vegetation (grass clumps in a sandy habitat), but actually nested in areas with less vegetation (i.e., more open) than generally occurred in the habitat. Visibility and landmarks are also important in helping chicks find their siblings and nests (Palestis and Burger 2001a,b). Habitat selection varies somewhat, depending on the location along the Northeastern Atlantic coast (Erwin et al. 1981), but birds basically can choose among sandy beaches, dunes, salt or freshwater marshes, or rocky habitats farther north.

COLONIALITY

In any particular year, individual birds are likely to nest at a colony site where they have previously been successful (site tenacity). Although some birds may nest in the same places as other species passively, most birds that nest in colonies do so because of the advantages coloniality provides (Burger 1981a, 1984d; Coulson 2002; Mariette and Griffith 2013). The advantages of coloniality include increased early warning (more eyes to watch out for predators), predator swamping (one of many for a predator to choose from), and colony defense (more birds to dive or attack a predator) (Burger 1984a; Danchin and Wagner 1997; Colwell 2010). Colonies may serve as information centers to improve feeding success. The disadvantages are that a colony location is conspicuous visually and acoustically because of the many birds coming and going, and there is competition for nest sites and nest materials. Neighbors may kill and eat eggs or chicks, and there is competition for food near the colony (Furness and Birkhead 1984). The optimal colony size occurs where birds obtain the greatest advantages, while minimizing the disadvantages (Figure 4.7; Burger 1981a; Coulson 2002).

Niche separation in nest site selection and foraging may well have led to the evolution of mixed-species colonies (Burger 1981a). Birds can reduce the costs (competition for nesting space and food in surrounding habitat), while increasing the benefits (antipredator behavior, early detection of predators, social enhancement of foraging) by nesting in mixed-species colonies (Figure 4.7; Burger 1981a). The advantages of coloniality increase with colony size—the greater the number of birds, the more eyes there are to look out for predators or mob predators, and the more birds there are to enhance foraging opportunities. Similarly, the greater the number of birds, the higher the competition for nest sites and nest material, and the farther birds may have to fly to forage. Thus, in monospecific colonies, the maximum advantages occur with the least costs when colonies are relatively small (e.g., greatest separation between costs and benefits). However, in mixed-species colonies, the greatest benefits (and least costs) occur with far more nesting pairs. In other words, the benefits increase faster, and the costs increase slower in mixed-species colonies (Figure 4.8; after Burger 1981a).

Although the advantages and disadvantages exist for colonies located in coastal areas and on the mainland, they are not the same for remote, offshore islands where mammalian predators do not occur, and where there is normally not enough food year-round to support avian predators (Schreiber and Burger 2001; Coulson 2002). The advantages of nesting on remote islands disappeared once people landed on islands, bringing predators and, in some cases, leaving foxes for future trapping, and sheep or goats they could use for food on future sea voyages (Nettleship et al. 1994). Invasive rats and cats on seabird-nesting islands are among the greatest threats seabirds face

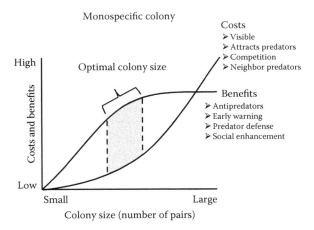

Figure 4.7 Nesting in colonies has both advantages (benefits) and disadvantages (costs). The optimum colony size (shaded area) occurs when the benefits (advantages) are greatest (before the benefit
curve plateaus) and the costs are lowest (before the cost curve rises steeply). (After Burger, J.,
Annu. Rev. Biol., 56, 1443–1467, 1981a.)

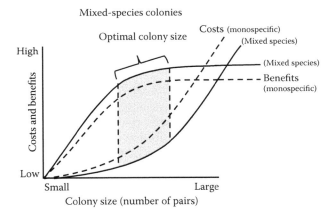

Figure 4.8 Advantages of nesting in mixed-species colonies are that competition is reduced, whereas antipredator behavior is enhanced. (After Burger, J., *Annu. Rev. Biol.*, 56, 1443–1467, 1981a.)

(Jones et al. 2008); many of these species are endangered, and some have been brought to extinction
by cats (Nogales et al. 2004).

FACTORS AFFECTING COLONY AND NEST SITE SELECTION

Nest site selection involves the ultimate, historic, or evolutionary factors that enhance success
in some habitats relative to others, and proximate factors that influence the choice at the beginning
of a nesting season. Several factors affect colony and nest site selection, including habitat stability,
nearby conspecifics or other species, competition with conspecifics and others, flooding/storms,
reproductive success, foraging, and human disturbance. Obviously, habitat loss is one of the most
important determinants of colony or habitat suitability and is a common cause of abandonment—
preferred suitable habitat becomes inundated or overgrown with invasive species or trees die.

Habitat Stability

Waterbirds that nest on coastal islands shift colony sites as conditions dictate, but have high site fidelity if colony sites remain unchanged and suitable (Buckley and Buckley 1980; Coulson 2002). In other cases, stability is reduced by erosion, invasive species, and loss of space. Herons, egrets, and ibises that nest in coastal colonies use the same colony sites as long as they remain safe from predators, parasites, and people, and are physically suitable. Over time, bird guano or salt-water intrusion may kill the trees, reducing suitability for some species. For many species, nest site requirements are driving their choice of colony site, and if these sites remain stable, they usually continue to nest on those islands. This is strongly illustrated for Common Tern and Roseate Tern colonies in Massachusetts (Nisbet 2002; Nisbet et al. 2014). By contrast, there were thriving colonies of Roseate and Common Tern at Jones Beach State Park and at Cedar Beach, Long Island, in the 1970s and 1980s (Burger and Gochfeld 1991b). Jones Beach colony was deliberately rendered unsuitable by the Park Commission, which planted dense Beach Grass in the partly vegetated areas preferred by terns. This was done to discourage terns from nesting close to the busy beach access roads. The Cedar Beach colony on the large barrier island was eliminated by Red Foxes and feral cats. Eventually, these colonies on Western Long Island disappeared before active predator control occurred. In contrast, colonies on Great Gull Island, New York (Helen Hays), Falkner Island, Connecticut (Jeffrey Spendelow), and Bird Island (Ian Nisbet) have remained for decades, but they are on islands where human access and predators can be controlled (Spendelow et al. 1995). For most species nesting in the bays and estuaries of the Northeast, the occurrence of stable colony sites invites long-term studies with marked birds, of which a few have been conducted.

Conspecifics and Other Species

The choice of a nest site is also dependent on nearby conspecifics, as well as available space. Birds usually defend a territorial space that is sufficient for their nesting needs, although sometimes they defend a much larger space, called a superterritory (Burger 1981b). In an expanding population of birds, the distance between nests may decline as the number of birds nesting in the colony increases. For example, when Herring Gulls were first establishing colonies in New Jersey, the distance between nests decreased significantly with the number of nests in colonies (Burger 1977a; Burger and Lesser 1980). In Barnegat Bay Common Tern colonies, nearest neighbor distance also decreased as the number of terns nesting on wrack increased. The wrack offered more suitable habitat than the surrounding salt marsh on which it lay, but space there was strictly limited (Burger and Lesser 1978; Palestis 2009). Wrack is the accumulation of dead vegetation, mainly *Spartina* and *Phragmites* stems as well as Eelgrass (*Zostera marina*), which is carried up onto islands by the highest tides and is then stranded there to dry out. It forms a platform on top of the salt marsh.

Nest site selection is also a matter of visibility of neighbors and predators. Territoriality involves aggressive displays toward neighbors, and birds nesting too close together may expend time and energy in aggressive displays. Birds can nest closer together when dense vegetation decreases visibility of neighbors and minimizes aggression (Burger 1977b; Burger and Gochfeld 1988a).

There are general observations that can be made about visibility from nests: some species prefer more open nest sites than others. For example, Piping Plovers and Least Terns prefer nesting in open places with visibility in all directions. At one extreme are the Royal Terns and related species that nest as close together as possible, just out of beak reach of their neighbors on all sides (Buckley and Buckley 1977). At the other end of the continuum, some species prefer nest sites that are more hidden; Glossy Ibis hide their nests even in a heronry, and Clapper Rail and Willets hide them in dense vegetation clumps. Incubating Willets sit tight until an intruder is almost overhead, and then burst from the nest explosively, and either flop weakly over the marsh with a distraction display or circle noisily, engaging nearby Willets in a weak mobbing effort, more symbolic than threatening.

Even within closely related species there are differences in nest site selection. On Long Island, Common Terns nesting in a dry land sandy colony always had 90% visibility above the nest, whereas Roseate Terns averaged 30% visibility from above (Gochfeld and Burger 1987a). Thus, there is a balance between visibility of neighbors and visibility of predators, and this applies to gulls as well (Burger 1977b).

Although many species use the same habitat and even the same substrate within that habitat, interspecific associations may still be one of the strongest factors affecting nest and colony site selection (Burger 1981a). Some species select colony sites based on the presence of other species. Interspecific associations can result in social parasitism (or social enhancement), whereby one species derives a benefit from nesting with another (Gochfeld 1980a; Pius and Leberg 1998; Coulson 2002). For example, Black Skimmers derive advantages from nesting with Common Terns and gulls that mob predators to drive them from colonies, thereby protecting the nests, chicks, and eggs of Skimmers from predation (Burger and Gochfeld 1990b). Black Skimmers also nest with Gull-billed Terns in Louisiana, similarly deriving antipredator benefits (Pius and Leberg 1998). The association of Black Skimmers with other species occurs in many different places (Gochfeld and Burger 1994; Leberg et al. 1995). Usually, Skimmers have little effect on the species they nest with, although in some cases Skimmers prey on the eggs or chicks of species nesting nearby (Gochfeld and Burger 1994). Black-crowned Night-Herons and Glossy Ibis also derive benefits from nesting with egrets, and often fly away rapidly and silently when a predator or person approaches the colony, letting the larger, white egrets soar around, vulnerable to the intruder.

Usually, the advantages of nesting with another species, especially in the case of social parasitism, can result in disastrous effects for the species that stays to defend the nests and colony. In the 1880s up until the early 1900s, populations of most waterbird species plummeted as a result of market hunting for food, sport, and for plumes for ladies hats (Ehrlich et al. 1988). Egret plumes are white, feathery, and delicate, and they were highly prized by the millinery trade (Figure 4.9). By 1900, nesting egrets had disappeared from their nesting areas. This stimulated the formation of the National Audubon Society, which adopted the Great Egret as its icon (see Chapter 3; Erhlich et al. 1988). Egrets were not the only species hounded to local extinction—women wore whole terns on their hats. The market for fashionable ladies hats was not limited to Europe or the metropolitan areas, as the hats with plumes were advertised in the Sears catalogs

Figure 4.9 Fashionable ladies hats in the late 1800s featured egret plumes as well as whole birds. (Photo from https://commons.wikimedia.org/wiki/File:3813GrandChpeau1911.png.)

delivered to rural America (Moore-Colyer 2000). Even nontarget species without pretty plumes were eliminated as collateral damage during the devastation visited on their more "attractive" co-occupants.

Colonial-nesting species are not the only ones to derive benefits from the actions of other birds. Several species that generally do not nest close to conspecifics and are therefore considered solitary nesters, nest near or within colonies of birds to derive benefits. In the Northeast estuaries and bays, Oystercatchers and Willets nest in Common Tern colonies, and Piping Plover nest in Least Tern colonies (Lauro and Burger 1989; Burger and Gochfeld 1991b,c). Ducks, such as Mallard and Gadwall, often nest in tern or gull colonies, hiding their nests in dense vegetation, but deriving some benefits from the protective behavior of the gulls and terns. Like Willets, these solitary nesting species tend to "sit tight" and not flush from their cryptic nest sites until almost stepped on.

Competition

Generally, the greatest competition for food or nesting sites is with conspecifics that have the same requirements. Therefore, nesting with other species reduces the competition because each species has slightly different nests, colony sites, and food requirements (Burger 1981a, 1984a; chapters in the work of Schreiber and Burger 2001). Such a reduction is particularly possible when some species nest on the ground, whereas others nest in low vegetation, or in shrubs. More birds can be packed into a nesting colony when there are both many species, and high physiognomy of the land and vegetation. Similar mechanisms also apply to seabirds nesting on offshore islands. Slope, presence of ledges or grassy knolls, and shrubs with different growth forms all contribute to providing more niches for birds to nest. This results in competition, but there are mechanisms to mediate competition, including arrival times and body size (Burger 1981a, 1983). Larger-sized species usually win in direct competition, and therefore, they do not have to engage in as many total aggressive interactions (Burger et al. 1977a; Burger 1978c, 1983; Figure 4.10).

The species of intruder matters as does the species of defenders. Figure 4.10 presents data from New Jersey for a complex heronry in shrubs, where the vegetation was less than 3 m high. The data show that although larger species generally win aggressive encounters, the amount of aggression they exhibit is partly a result of their "win" rate. That is, the largest species (Great Egret) and the smallest species (Glossy Ibis) engage in the least aggressive activity. For Great Egrets, it is because they always win and other species just avoid them. For Glossy Ibis, it is because they nearly always lose, and so they just avoid aggression and give way to the larger species. For similarly sized species, the win rate is similar. This results in Great Egrets nesting in the highest (and safest) nesting sites, and Glossy Ibis nesting in the lowest places. The overall pattern holds for many other places, including Argentina, Texas, New York, and Mexico, regardless of the vegetation height (Burger and Trout 1979). Cattle Egrets are the exception, probably because they are a nonnative species that has become more aggressive in North America than in their native Africa, and can thus outcompete the Great Egrets for nest sites. Cattle Egrets nest higher in the colony than would be predicted from their body size (Figure 4.10; Burger 1978b).

There are semicolonial species, such as Willet, that nest within hearing distance of one another and produce very loud alarm calls, thus deriving some antipredator advantages, as well as early warnings of predator or human approaches. Their nesting associations were not noticed for years, because the nests are quite spaced out and difficult to find (Burger and Shisler 1978b).

Finally, some species may seem to seek other species, but this apparent association is largely a matter of timing. For example, American Oystercatchers prefer to nest on sand, but will also nest on wrack or in higher *Spartina* places (Lauro and Burger 1989). Their association with Common Terns and Black Skimmers arises when these species start to nest on wrack already occupied by the Oystercatchers, which arrived earlier. The Oystercatchers are choosing the wrack habitat as do the later arriving terns and skimmers (Burger and Gochfeld 1991b). Thus, their association with

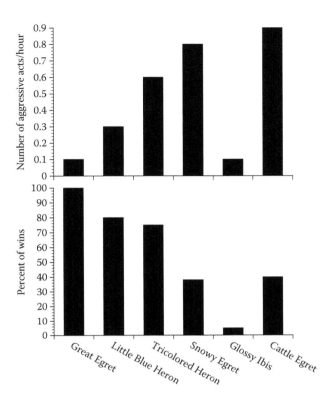

Figure 4.10 Aggression of herons and egrets in nesting colonies. The frequency of aggressive interactions per hour and the percentage of wins are shown. (After Burger, J., *Wilson Bull.*, 90, 304–305, 1978a; Burger, J., *Annu. Rev. Biol.*, 56, 1443–1467, 1981a; Burger, J., *Behav. Neurosci.*, 97, 492–501, 1983.)

tern colonies may be passive, and indeed there are many islands in Barnegat Bay with nesting Oystercatchers that have no tern colonies.

Predators

Avoiding predators, along with being near feeding opportunities, is often given as one of the primary factors that affect where and how birds nest. In general, seabirds nest on remote islands far from predators. Because many of these islands are small, or the suitable habitat there is limited, sea-birds often nest in large aggregations or colonies of thousands of pairs (Schreiber and Burger 2001; Coulson 2002). Seabirds and some other waterbirds that nest in coastal or inland areas often nest in dense colonies, partly as an antipredator strategy as other birds provide early warning, predator swamping, and active defense. Predators can account for variations in breeding success, as well as habitat choice (Nisbet and Welton 1984).

One of the disadvantages of nesting in colonies is that they are often noisy and visible, with a predictable location from year to year. This makes them easy for predators to find. Colonial species usually nest in places inaccessible to ground predators (such as islands) or in trees where they cannot be reached. There are three difficulties with these strategies: (1) if there are bridges or ice bridges (in the winter), mammalian predators may reach these small islands; (2) avian predators can reach such places; and (3) there are now predators on barrier islands that were once safe because they were not attached to the mainland until the construction of bridges.

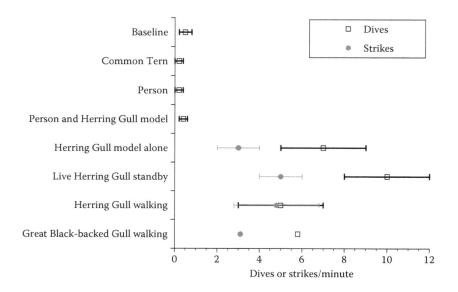

Figure 4.11 Defensive behavior of Black Skimmers toward Herring Gulls and model Herring Gulls when Skimmers nest in monospecific colonies (e.g., no other species to defend their eggs or chicks). Dives are when they dive down toward the intruder, and strikes means they made contact with the person (rarely), gull, or model. Shown are means ± standard error. (After Burger, J., Gochfeld, M., *The Black Skimmer: Social Dynamics of a Colonial Species*, Columbia University Press, New York, 1990b; Burger, J., Gochfeld, M., *Aggress. Behav.*, 18, 241–248, 1992b; Unpublished data.)

Another of the disadvantages of nesting in colonies, in addition to competition for space, is that colony mates may be predators as well. Black-crowned Night-Herons nesting in colonies may eat the eggs and chicks of neighboring gulls (Burger 1974), egrets (Raye and Burger 1979), or terns (Hall and Kress 2008). Ground-nesting species also have predators that nest among them. Herring and Great Black-backed Gulls eat the chicks not only of conspecific neighbors, but of other species as well (Burger 1979a). In some colonies where individual gulls were voracious predators, culling predatory gulls has been used as a successful management technique, especially to protect Common Terns (Magella and Brousseau 2001), although this can sometimes backfire (Guillemette and Brousseau 2001).

Black Skimmers, known for their behavior of fleeing a colony and letting terns or other species defend the colony against predators, become quite aggressive against Herring Gulls (Burger and Gochfeld 1992b). Skimmers respond more strongly to Herring Gulls, which clearly take their eggs and chicks, than they do to people (Figure 4.11). In these experiments, models were moved through the colony on runways, and the Skimmers responded strongly to a Herring Gull model (a stuffed gull). In some cases, terns are aggressive to Skimmers that nest nearby (Figure 4.12).

Flooding/Severe Storms

For many ground-nesting species nesting in coastal habitats, flooding and severe storms are a threat to themselves and their offspring. Although tides are predictable, the severity of high tides during storms, and the consequences of nor'easters, hurricanes, and superstorms are not predictable. Floods can range from a few washed-out nests and eggs or drowned chicks, to complete washouts (Burger and Gochfeld 1990b, 1991b; Burger et al. 2001b). This topic will be described more fully in Chapter 5 as it is a major factor for reproductive losses for many species, particularly those nesting on low-lying islands. For colonies on higher ground, prolonged heavy rains and chilling are a more

Figure 4.12 Common Tern attacking a Black Skimmer whose nest is too close to the Tern's nest.

serious factor than flooding. In some years, severe and prolonged rains resulted in zero reproduction for several tern and skimmer colonies in Barnegat Bay, for example (Burger and Gochfeld 1990b).

There are at least three interesting aspects of birds' responses to flooding: (1) the loss of nests, eggs, or chicks to flood tides; (2) whether birds have immediate responses to tidal flooding; and (3) whether birds have more long-term responses to flooding. There are many studies that report losses of nests, eggs, and chicks to flooding (Burger and Shisler 1980a; Schreiber 2001). In many cases, nests wash away, eggs and chicks are flooded out of nests, and chicks die of hypothermia (Burger and Gochfeld 1990b, 1991b; Schreiber 2001). However, there are short-term responses to flooding: birds can move chicks from flooded areas, nests can be built higher (and more substantial) in areas prone to flooding, they can be built higher as flood waters rise, or nests can even float up (Burger 1978d, 1980a). Most birds do not move eggs, although we have observed Clapper Rails carry both eggs and chicks in their bills to higher ground. When they nest in wetter areas of *Spartina alterniflora* (near creeks or low areas), Laughing Gulls build significantly higher and wider nests, and they repair them more often (Burger 1978d, 2015a). Ultimately, however, birds must build nests in areas that are less prone to flooding (which may be more difficult in the future as sea level rises).

The ability of birds to respond to tidal flooding was tested experimentally for several species of marsh-nesting birds (Burger 1979b, unpublished data; Figure 4.13). These experiments were conducted 3 days before the normal spring high tide, early in incubation. Nest height was measured a day before and a day after high tide (control or experimental condition administered in between), and again 24 h later. Conditions shown in Figure 4.13 were "control" (no change or manipulation), "cup" (plastic cup placed in the nest with water and eggs in it), "tide" (tidal inundation of nests), "top" (top of nest rim removed), and "side" (only one side of the nest removed). For all species except Mallard, the responses to "tide," "top," and "side" were significantly greater than the "control" or "cup" conditions (Figure 4.13). These experiments indicate a targeted response to specific exposure conditions. That is, all of the species responded somewhat to the immediate effects of either their nest and eggs getting wet, or part of their nest being destroyed (which happens with violent waves

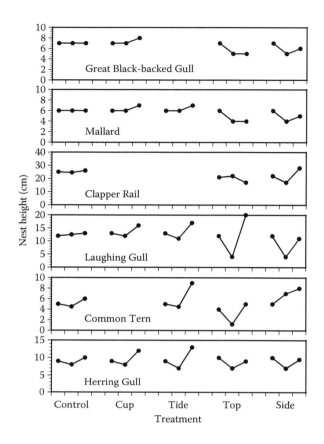

Figure 4.13 Nest behavior in marsh-nesting birds. Shown are responses of six species of marsh-nesting birds to tidal inundation, placement of a cup with water in their nest, removal of the top of their nest, and removal of the side of their nest. All birds, except Mallards (*Anas platyrhynchos*), responded to tidal inundation and experimental conditions. (After Burger, J., *J. Comp. Physiol. Psychol.*, 11, 189–199, 1979b; Unpublished data.)

and high tides). But the degree they can respond is limited by where the nest is located and how high they can build it. If their nest is too low, very high tides will simply wash over it, and on higher ground, they can only elevate the nest about 30 cm. At least we have never seen one built higher than that (Figure 4.14).

Some species of birds have eggs that can withstand some saltwater inundation without dying; for example, ducks (Welty 1975), Clapper Rails (Mangold 1974), and Laughing Gulls (Burger 1979b). Herring Gulls, however, did not evolve in marshes or under flooding conditions. Rather, until the middle of this century, they nested largely in Maine and farther north on rocky islands and rock ledges (Pierotti and Good 1994). However, when they moved into salt marshes, they became vulnerable to flood tides. In a series of experiments, Herring Gull eggs were found to survive submersion in saltwater for a period of 120 min (about 70%), and they survived better in saltwater than in freshwater (about 55%) during the first week of incubation (7°C and 26°C). However, during the third week of incubation, the percent surviving remained the same for the 7°C treatment, but at 26°C the rate of survival dropped to 22% in saltwater and 3% in freshwater (Ward and Burger 1980). These data suggest that both timing of inundation relative to egg development, and water temperature affect survival of eggs, and that with increased inundation later in the season, hatching drops drastically.

(a)

(b)   (c)

Figure 4.14 Laughing Gull nests under normal conditions are fairly flat (a); when exposed to flood tides, the gulls add material to raise the eggs above the water (b and c).

Human Activities

Human activities that cause disturbance during breeding is a severe problem for ground-nesting species, particularly those that still nest on beaches, such as Piping Plover and Least Tern. Most of the other species (Common Tern, Black Skimmer, Herring Gulls) have already fled to the salt marshes where human disturbance is much less. Many studies have examined the effects of people on nesting birds, particularly colonial species (Burger 1991a, 1994a; Boersma et al. 2001), whereas others have examined the effects of human activity on foraging birds (Stillman and Goss-Custard 2002; Burger and Niles 2013a,b). In many cases, the investigators were the experimental disturbance, and in others, the investigators simply waited until other people appeared and recorded how birds responded (Shealer and Haverland 2000; Burger et al. 2007a). These studies generally show that birds respond to people by flying away, and can even abandon their nests. Nisbet (2000) argued that some colonies of waterbirds should be managed for multiple human uses (scientific study, recreation, photography, education) to promote habituation and reduce responsiveness.

For birds nesting on islands, boats and their occupants pose a problem. Jet skis (personal watercraft) are often more of a problem than boats because they can get closer to nesting islands that are surrounded by shallow water. As might be expected, there were greater effects on nesting birds when watercraft came closer to nesting islands, than when they stayed away. A reasonable buffer zone around nesting colonies seems to be 100 m (Burger 1998a; Rodgers and Schwikert 2002; Blumsetin et al. 2003; Rodgers et al. 2005; Burger et al. 2010a). Common Tern colonies that experienced repeated and continuous exposure to watercraft eventually abandoned the colony site (Burger 1998a). Similarly, herons and egrets respond to human presence near their colonies by flying away from the colony and landing at a distance, but they rarely leave their nests when a boat goes by.

Reproductive Success

Birds manifest varying degrees of philopatry, returning to nest year after year at colony or nest sites where they have been successful in raising young—that is, as long as the site is still otherwise suitable and available. Many factors affect reproductive success, including proximate (affecting current reproductive success) and ultimate factors (affecting evolution of reproductive behavior) (Burger 1982a,b; Hamer et al. 2001). Proximate factors include human disturbance, weather, toxic chemicals, predators, competitors, and food scarcity (Hamer et al. 2001; Visser 2001). Many of the proximate factors can be ultimate factors as well, if they operate over long periods. Most of these are not going to be discussed here, except for how habitat affects success.

Birds generally do not continue to return to breeding areas where they are not successful for any of these reasons. Abandonment of nest or colony sites can occur slowly over time (weeks or years), or rather suddenly. Slow abandonments of a colony site could occur if a few birds are unsuccessful each year, or if some of their nesting habitat is destroyed, or after 1 or more years of reproductive failure, whereas rapid abandonment occurs when the entire colony fails or the entire site or nesting area disappears or is flooded out, as can easily happen in coastal areas. Wide sandy beaches for nesting Least Terns and Piping Plover can wash away with severe winter storms. Salt marsh islands or their wrack can be eroded away by sliding ice sheets in the winter, and severe hurricanes can drastically change the shoreline. Long-term studies, however, are required to understand the mechanisms of abandonment of colonies. In Black Skimmers studied over a 5-year period ($n = 19$ colonies), abandonment of colony sites was greater in colonies destroyed by predators than in those destroyed by flood tides (Burger 1982a). Flood tides that result in low productivity are not as predictable as predators; mammalian predators that arrive at a colony may stay, and avian predators can find the colony the following year.

For some species, however, the choice of colony site is dependent on nearby foraging opportunities, temporally and spatially. This is discussed more fully later in this chapter. The factors mentioned earlier all combine to produce temporal, horizontal, and vertical stratification patterns of nesting. Although these patterns are flexible, and no doubt change somewhat from estuary to estuary, the principles are the same. These patterns will be explored below for Barnegat Bay, but they apply to New York–New Jersey Harbor Estuary, Cape Cod/Buzzards Bay, and Chesapeake Bay (when the species are the same). We have documented similar nesting patterns in Mexico, South America, Africa, and Madagascar, where we have examined nesting patterns of herons and egrets (Burger 1978b,c; Burger and Gochfeld 1990d).

TEMPORAL, HORIZONTAL, AND VERTICAL STRATIFICATION

The most important habitat gradient along the Atlantic coast for birds is from open water through coastal environments to upland terrestrial habitats. Because birds are highly mobile, many species can be found anywhere along the gradient, "normal distributions" change during the year, and distributions are altered during hurricanes or other inclement weather events. Coastal landscapes are longitudinal habitats in that to maintain the same conditions, animals may have to move up and down the coast. To change habitats they move perpendicular to the coast, either toward the sea or toward the land (Burger 1991b). In some cases, the habitats are too narrow for successful nesting or foraging. In other cases, this is an advantage because birds can quickly access other habitats. Because the landscapes are narrow, birds living there can quickly access the ocean or the upland for feeding. The very narrow nature of the coastal habitat, however, also means that people, predators, and competitors can quickly access their habitats as well. Furthermore, coastal environments are more ephemeral than land-based habitats. When habitat shifts rapidly, there is less philopatry (fidelity) to colony or nesting sites (McNicholl 1975; Haymes and Blokpoel 1978; Burger and Gochfeld 1990b, 1991b,c; Hamer et al. 2001). Thus, there are both positive and negative consequences of site philopatry for colonial species nesting along the coast (Matthiopoulos et al. 2005).

Spatial stratification refers to the slight differences in habitat preferences that nesting birds exhibit along different coastal gradients: from the ocean to the mainland, from beach to the barrier island dunes, and from bay water to the *Phragmites* and bushes higher on the marshes (Burger 1985a). Spatial complexity, even in relatively small areas (e.g., a patch of *Phragmites*, cherry, and Poison Ivy in the middle of a salt marsh) provides variation that animals can use in habitat selection (Burger 1981a). Although ecologists mainly examine complexity and habitat stratification in terrestrial environments, there is also stratification in the aquatic environments (e.g., in eelgrass beds, *Zostera marina*), which in turn affects distribution of small prey fish (Bologna 2002; Fertig et al. 2013). The complexity of habitat use by forage (or prey) fish has been extensively studied in New York–New Jersey Harbor Estuary, Barnegat Bay, and Delaware Bay (Nemerson and Able 2003; Able 2005; Able et al. 2007; Able and Fahay 2014). In all areas, juvenile fish use a range of habitats and provide linkages between estuarine and oceanic habitats.

Temporal Stratification

There is also temporal stratification in that birds initiate breeding at different times during the breeding season. Breeding can be highly synchronized, either within or among species (Burger 1979d). Synchrony within a species reflects their annual cycle of sex hormones, but is subject to local conditions. Synchrony among species may occur with the sudden appearance in the neighborhood of a favored food source such as the arrival of schools of *Ammodytes* (Austin and Austin 1956). Breeding synchrony enhances the advantages of nesting in colonies because of predator swamping (Burger 1981a; Coulson 2002) (Figure 4.15). If many chicks hatch on the same day, predators can take only a small portion, whereas if the same number of chicks hatched over a period of weeks, predators might take them all (Darling 1938). Synchrony reduces the total time prey are available to predators (Figures 4.15 and 4.16).

Temporal stratification in nesting phenology depends on whether birds are residents (remain all year), when birds arrive on the breeding grounds, when they can obtain enough food resources to lay eggs, when food resources are optimal for feeding young, and other factors (Lack 1968; Weimerskirch 2001; Hamer et al. 2002; Carey 2009). These factors affect the overall general timing of birds nesting in Northeastern bays and estuaries, resulting in a wide possible range in timing of breeding. By contrast, Arctic breeders such as the Red Knot have only a narrow window of opportunity in which to arrive, mate, lay, incubate, and hatch their young.

One endpoint that is possible to examine across sites and years is the date of egg-laying because it can be easily and objectively determined. There is a great deal of variation in this measure, both within and between species. Figure 4.17 illustrates the dates of egg-laying for several species in Barnegat Bay and nearby New York–New Jersey Harbor Estuary (Burger 1985a; Burger, Gochfeld, Elbin, Palestis, Unpublished data). Thus, Herring and Great Black-backed Gulls lay before Laughing Gull, which puts them in direct competition (Burger and Shisler 1978c; Burger 1979a). Great Black-backed Gulls, being larger than Herring Gulls, have the competitive advantage (Burger 1983). In aggressive interactions between the two, Great Black-backed Gulls usually win, and they obtain the best nest sites. In both species, aggressive interactions increased from less than 0.2/h during incubation to more than 2/h during the chick phase. Similarly, Herring Gulls and Laughing Gulls compete for the higher places in *Spartina alterniflora* or at the edge of *Spartina patens*. Because Herring Gulls are nearly three times heavier than Laughing Gulls, there is no contest in who wins. Herring Gulls can force Laughing Gulls

Figure 4.15   Skimmers are synchronous when they nest in large colonies. Here, Skimmers nesting on Tow
Island in Barnegat Bay are laying synchronously—they all have two to three eggs in nests that are
evenly spaced. The complete clutches will mostly have four eggs.

from their preferred habitat (Burger and Shisler 1978c). Similarly, Oystercatchers lay before Common
Terns, and when the terns arrive, Oystercatcher may already be nesting on the wrack the terns also use.
Although they do not compete for food, Oystercatcher predation on tern eggs is common.

Although egg-laying requires that females have sufficient resources to lay eggs, additional
resources beyond what the parents need are required during the chick phase to provision them
(Lack 1968; Visser 2002). Some authors have suggested that the timing of breeding (or of the chick
phase) is ultimately determined by food availability (Møller 1980; Becker et al. 1985; Weimerskirch
2001). Examining these concepts requires examining the prey base, which is often difficult to do
for birds that forage over the bays or open ocean. However, using a sonar fishfinder, Carl Safina
found that prey fish peak abundance coincided with the period when Common Terns were feeding
their chicks, and declined sharply later in the season (Safina and Burger 1985, 1989a,b). Late nesting terns are at a disadvantage. The prey fish had not been abundant when the terns started laying,
indicating that the tern phenology is timed to optimize food availability for growing chicks. This is
a gamble that does not always pay off. Some years chicks grow slowly and may starve, leading to an
average productivity well below one chick per pair, whereas other years terns may be able to raise
two or three chicks per pair. Low food availability during egg-laying may result in smaller eggs or in
smaller clutches (one or two eggs rather than three eggs). As a proximate factor, the fish availability

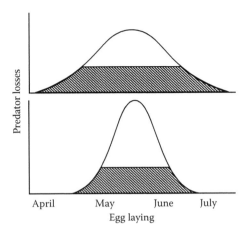

Figure 4.16   The effect of having synchronous egg-laying (bottom)—predators can take a smaller proportion of the chicks than if laying is spread out over a longer period (top). This advantage continues into the chick phase; when most chicks are available in a short period, predators can take only so many. The horizontal line indicates the number of eggs or chicks that predators can take each day.

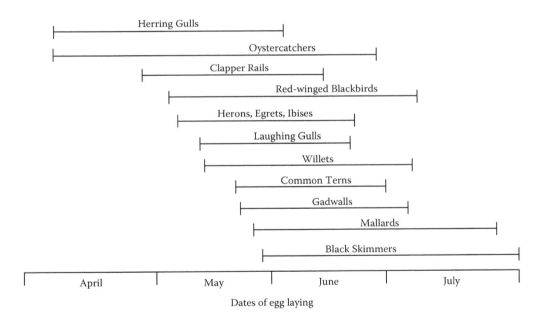

Figure 4.17   Temporal pattern of egg-laying dates in birds in Northeast colonies. The periods are long owing to seasonal and yearly differences, and to renesting.

at egg-laying may be correlated with and predictive of fish availability 3 weeks later at hatching. If fish arrive late, terns may delay egg-laying for days, as a proximate adaptation.

Black-crowned Night-Herons arrive earlier than the other egrets and herons, and often nest in *Phragmites* or low bushes. Parenthetically, *Phragmites* are denigrated as an invasive plant species of little use by birds. Native Americans used *Phragmites* extensively (Kiviat and Hamilton 2001), and some birds do so as well. Black-crowned Night-Herons and Northern Harrier (*Circus cyaneus*) sometimes nest on the ground in dense *Phragmites*, which provides cover, shade, and nest material.

Table 4.1   The Effect of Herring Gulls on Abandonment and Formation of New Colonies in New Jersey and DelMarVa Peninsula

Species	Total Number	Abandoned Colonies in New Jersey	New Colonies in New Jersey	Abandoned Colonies in DelMarVa	New Colonies in DelMarVa
Common Tern	48%	40%	0%	66%	0%
Forster's Tern	35%	33%	0%	33%	0%
Black Skimmer	17%	66%	0%	100%	0%
Laughing Gull	33%	60%	0%	0%	14%

Source: After Erwin, R.M. et al., *Auk*, 98, 550–561, 1981.
Note: The percentage of colonies with Herring Gulls is presented.

There are disadvantages to nesting with other species that can outcompete them for nest sites. This is partly a matter of arrival time and size (see below), but it also involves aggressive interactions and predation. The interactions between the larger gulls and smaller species often involve colony abandonment. The larger, aggressive species renders the habitat unsuitable. The smaller species may abandon, and either join other existing colonies or find new suitable and available sites. Although this topic is discussed further in Chapter 7, it should be mentioned that the arrival of breeding Herring Gulls in Barnegat Bay in the 1950s and Great Black-backed Gulls in the 1970s forced many of the smaller marsh nesting species out of prime colony sites. As is clear from Table 4.1, many of the colonies that the smaller species abandoned were occupied by an expanding population of Herring Gulls, and the smaller species rarely colonized a new site where Herring Gulls already nested.

Horizontal Stratification

Birds have generalized breeding requirements that relate to providing sufficient space to incubate and raise young, while avoiding predators (Hamer et al. 2001). Birds require land to breed (nests, eggs, and chicks are space-based). The open waters of the Atlantic coast are occupied by pelagic species such as seabirds and some diving ducks, which must come to land to breed. Tidal marshes are found in relatively narrow pockets along coastlines, with the main vegetation being *Spartina* (Teal and Teal 1969; Warren et al. 2002; Teal and Weishar 2005). The combination of salinity gradients, low floristic and structural complexity, regular tidal fluctuations, occasional catastrophic flooding, and high winds in tidal marshes creates an unpredictable environment (Greenberg et al. 2006).

Although birds exhibit flexibility in their choice of nesting sites, they prefer particular types of habitats (Wilson and Vermillion 2006). Gulls, terns, skimmers, and shorebirds nest on the ground, usually on bare sand or in places with sparse vegetation, or they build nests in marshes. Herons, egrets, and ibises prefer to nest in trees or shrubs, but will sometimes build nests on the ground. Ducks, Willet, and Clapper Rail build nests low on the ground, usually in marshes. Clapper Rails pull the vegetation over the nest so that it forms a canopy, protecting them from the view of predators flying overhead. Willet nests are usually partly covered by vegetation, whereas American Oystercatchers build nests in the open, unvegetated sand or wrack, relying on crypsis to camouflage their eggs. Willets sit tight, whereas Oystercatchers sneak away from their nests while an intruder is still distant.

The generalized habitat pattern for nesting is shown in Figure 4.18. Although this varies by habitat type, it indicates the rather wide habitat selection of marsh-nesting species (Burger 1985a). Within this wide range, species have preferences; Common Terns prefer to nest on wrack, but the limited amount of wrack means that some birds nest directly on *Spartina alterniflora* or *Spartina patens*. This type of diagram illustrates the normal wide range of habitats that can be used, as well as the problems that results when sea level rise reduces the available habitat.

There is also horizontal stratification in winter (Figure 4.19) (Goss-Custard et al. 1995). During the winter off Barnegat Light, Northern Gannets are typically seen more than a kilometer offshore;

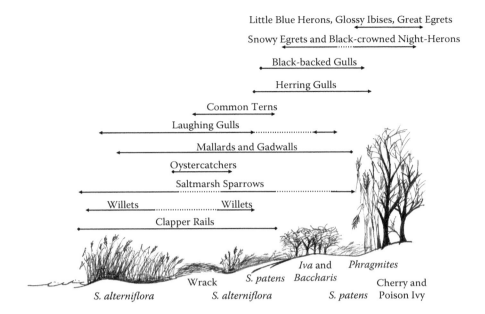

Figure 4.18   Horizontal stratification of nesting birds on salt marsh islands. The dotted line indicates they may occur in these habitats to forage, but do not nest there normally. (After Burger, J., in *Habitat Selection in Birds,* ed. M. Cody, 253–281, Academic Press, Waltham, MA, 1985a; Unpublished data.)

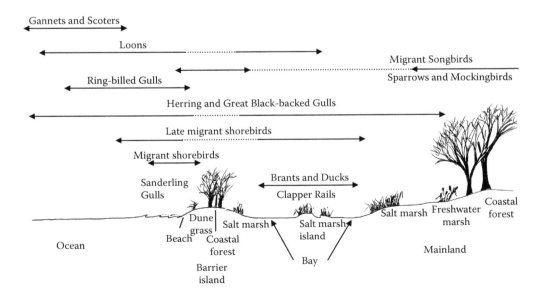

Figure 4.19   Horizontal stratification of wintering birds. The dotted line indicates a gap in their distribution.

many Common Loons and sea ducks (scoters and eiders) are found within 1 km of shore or in the inlet. Brant frequent eelgrass beds in the estuaries, and Snow Geese feed mainly on the salt meadows. Dabbling ducks, such as Mallards and Black Ducks, can survive in saltwater habitats but are more likely to occur in inland ponds and lakes.

Vertical Stratification

Generally, the greatest competition for food or nesting sites is with conspecifics that have the same physical, nutritional, and energetic requirements. Therefore, nesting with other species reduces the competition because each species has slightly different nest, colony site, and food requirements (Burger 1981a, 1984a; Schreiber and Burger 2001). Such reduction is particularly possible when some species nest on the ground, whereas others nest in low vegetation, or in shrubs. More individual birds can be packed into a nesting colony with high vegetation (e.g., trees), when there are many species. Similar mechanisms also apply to seabirds nesting on offshore islands; slopes, ledges, flats, and shrubs with different growth forms provide varied nest sites, including underground burrows.

Whereas coastal marshes and wetlands have a horizontal aspect, bushes and shrubs within salt marshes or wetlands and coastal forests provide vertical stratification. This stratification provides additional opportunities for nesting, which is particularly important for herons, egrets, and ibises. Although larger coastal forests may provide trees for these species to use for nest sites, such forest patches have largely been destroyed in the Northeastern bays and estuaries, although some maritime forests remain in the New York–New Jersey Harbor Estuary (Burger and Gochfeld 1993b; Craig 2013).

Among herons, egrets, and ibises, competition for nest sites is strongly affected by body size, as well as vegetation structure able to support nests (Burger 1978b,c, 1981a). Birds nest closer together when there is less visibility (Burger 1977b), and this is even true in heron/egret/ibis colonies in dense vegetation (Burger and Miller 1977). The more dense the vegetation, the closer birds can nest, regardless of body size (Burger 1977b). Even in colonies on the ground, visibility plays a role, and the internest distance is inversely related to visibility (vegetation density around the nest; Burger 1977b). Generally, larger species of birds are able to obtain the highest nest sites that will support their weight, thereby removing them from mammalian predators that enter a colony, as demonstrated by studies in the United States, Mexico, Argentina, and Africa (Table 4.2; Burger 1977a, 1978b,c, 1982b; Burger and Trout 1979; Burger and Gochfeld 1990b).

Cattle Egrets are the exception, as they are more aggressive than their size predicts, and usually obtain a higher nest site than their size would suggest. The vertical stratification represents behavior and body size (Table 4.2). Because Cattle Egrets are nonnative to North America, their behavioral responses are different from those of other similarly sized species, and they are different from those of Cattle Egrets in their native Africa (Burger 1978b, 1982c). Cattle Egrets are no longer as successful in the Northeast as they were in the 1970s and 1980s—although the reasons for this recent decline are unclear. Because Cattle Egrets forage on insects in fields and along roads, they are not competing with native herons and egrets for prey or foraging space, and the decline seems unlikely to be due to food shortage.

Double-crested Cormorants are another exception. Over the past half century, the Cormorant has been extending its breeding range southward, from New England, and it is a recent colonist of New Jersey (Dorr et al. 2014). In several of the Northeastern bays and estuary islands, Cormorants nest on abandoned dock pilings and other structures that remain in the water. They also nest on trees within heron and egret colonies. Cormorants nest in the New York–New Jersey Harbor Estuary, Delaware Bay, and Chesapeake Bay, but not in Barnegat Bay, although nonbreeding individuals are numerous. Barnegat Bay does not have many areas with trees or abandoned piers, without people. On Huckleberry Island (New York), for example, Cormorants mainly nested in dead trees (62%), whereas the egrets and Night-Herons nested primarily in live trees (67–90%). Herons and

Table 4.2 Bird Length and Mean Nest Height (cm) above Ground in Argentina (2 Colonies), Mexico (2 Colonies), Texas (3 Colonies), New York (3 Colonies), New Jersey (5 Colonies), and Massachusetts (1 Colony)

Species	Size Length (cm)	Argentina		Mexico		Texas			New York			New Jersey					Mass.
Great Blue Heron (Ardea herodias)	95					27	37	23									130
Great Egret (Casmerodius alba)	80	21	7		23	25	26				40	127	11		37	42	
Roseate Spoonbill (Ajaia ajaja)	70	18	8	25	24	22	20										
Little Blue Heron (Florida caerulea)	55										32		28				
Tricolored Heron (Hydranassa tricolor)	55			20	19	13						13	29				
Yellow-crowned Night-Heron (Nyctanassa violacea)	52														120		
Snowy Egret (Egretta thula)	50	9	9	18		17	24	10	14	29	29	101	6	22	120	35	102
Black-crowned Night-Heron (Nycticorax nycticorax)	50									20	23	95	12			28	90
White-faced Ibis (Plegadis chihi)	47	3	6					8.5									
Glossy Ibis (Plegadis falcinellus)	47								8	7	20	5	98		0	6	
Green Heron (Butorides striatus)				6	8												
No. of species in colony		4		5	3	4	4	4	2	3	5	3	4	3	6	4	3

Sources: Data from Burger, J., *Wilson Bull.*, 90, 304–305, 1978a; Burger, J., *Condor*, 80, 15–23, 1978b; Burger, J., *Wading Birds*, National Audubon Society, New York, 45–58, 1978c; Burger, J., *J. Comp. Physiol. Psychol.*, 11, 189–199, 1979b; Burger, J., *Am. Midl. Nat.*, 101, 191–210, 1979c; Burger, J., *Annu. Rev. Biol.*, 56:1443–1467, 1981a; Burger, J., Trout, R., *Condor*, 81, 305–307, 1979; Burger, J., Gochfeld, M., *J. Coastal Res.*, 9, 221–228, 1993b; Unpublished data. Body length from Robbins, C.S. et al., *Birds of North America*, Golden Press, New York, 1966. With permission.

Note: In general, the largest species nest the highest. Size denotes body length in centimeters.

Cormorants kill trees with their droppings. In live trees, Great Egrets nested highest, followed by Black-crowned Night-Herons, and then Snowy Egret (Burger and Gochfeld 1993b). The Cormorants nested at similar heights to the Great Egrets. In this case, the competition was avoided by the cormorants nesting in the dead trees (trees that they probably killed with their guano), and the other species nesting in the live trees.

FORAGING

Habitat use for nonbreeding birds is a function not only of water types and distribution, habitat structure, and vegetation types, but also of prey types, prey abundance and availability, and foraging methods (Shealer 2001). It is a matter of striving toward optimal foraging (Pyke et al. 1977). Human activities, such as fisheries, can affect prey availability, and thus foraging success and, ultimately, populations (Atkinson et al. 2007). Foraging thus has two important aspects: habitat suitability and prey (or food) availability. Prey can be depleted by birds (Rowcliffe et al. 2001), and it is essential to consider potential prey depletion when determining the amount of foraging habitat necessary for coastal birds.

In general, seabirds capture prey by a variety of methods, including pursuit swimming, plunge-diving for fish or invertebrates, surface-plunging from the ocean surface, hop-plunging, hover-dipping, picking food items off the surface of water, and kleptoparasitism (piracy) from other birds. Gulls and some other seabirds pick up fruit or insects from the ground, follow boats, scavenge on offal along the shore, pirate food from other seabirds, and forage at landfills (Ashmole 1971; Sealy 1973; Burger and Gochfeld 1981a, 1984b, 1991d,e; Hackl and Burger 1988; Shealer et al. 1997; Shealer 2001). Gulls, terns, and skimmers forage in shallow tidal creeks; gulls and terns feed behind boats or near other human activities; gulls feed at landfills (garbage dumps) and inland lakes and impoundments (Burger and Gochfeld 1981a; Burger 1987b, 1988c; Patton 1988). Gulls and terns also forage pelagically under certain conditions. Common Terns forage farther out at sea than Forster's Terns, and most of the gulls stay in the narrow coastal zone, although Laughing Gulls will go 40 km inland to forage (Dosch 2003). Some ducks form large rafts and forage on the open sea (diving ducks), whereas others feed at the marine–land interface in bays, estuaries, marshes, fields, and even inland aquatic habitats (dabbling ducks). These patterns are depicted in Figure 4.20.

Shorebirds feed along the shoreline on the mainland, along barrier islands, or around offshore islands (Burger et al. 1997a; Withers 2002; Colwell 2010). They feed by picking up items from the sand, mud, or from shallow water or along wrack lines, while standing and running—all methods that tie them to the narrow band of shoreline. Obviously, there are exceptions. In coastal areas, Piping Plovers forage along the high tide line or in back bays and shallow pools behind dunes, or in inlets and marshes (Burger 1991a; Elliot-Smith and Haig 2004; Elliot-Smith et al. 2009). Foraging and nest-site behavior can be modeled to evaluate habitat suitability and restoration potential (Maslo et al. 2011). Satellite imagery and Geographic Information System have greatly improved our ability to model habitat relationships (Gottschalk et al. 2005).

Other shorebirds, such as American Oystercatchers, specialize on bivalve and mollusks, and are limited to saltwater (Nol and Humphrey 1994). Species diversity and abundance of many species, including shorebirds (Burger et al. 1982a, 2007a; Withers 2002), vary by habitat. Tidal stage also influences foraging location; the highest number of shorebirds feed at low and rising tide compared with other times (Burger et al. 1977b, 1997a, 2004b; Fleischer 1983; Burger and Niles 2014). Shorebird species diversity varies within habitats that are close together, partly as a function of time of day, tide stage, tide height, and seasonality (Burger and Gochfeld 1983b; Burger 1988c; Warnock et al. 2001). Interference competition is another factor affecting habitat choices, but it is often difficult to document (Triplet et al. 1999; Vahl et al. 2005a,b; Vahl and Kingma 2007).

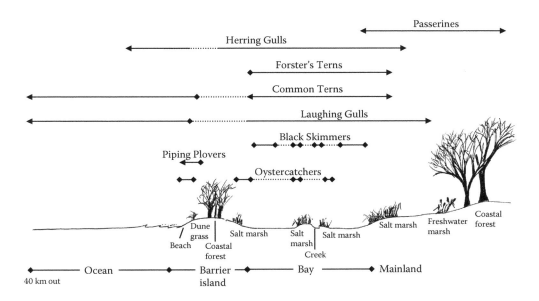

Figure 4.20 Foraging patterns of birds that nest along the shore. The dotted line indicates a hiatus in their foraging distribution. They may occur in these habitats to forage, but do not nest there normally.

In Delaware Bay, for example, shorebirds use a wide range of habitats, from tidal to freshwater streams flowing into the bay. The largest number of shorebirds, however, is along the bays and at creek mouths (e.g., West Creek; Figure 4.21). The shorebirds foraged from 40% to 60% of the time, regardless of tide cycle or time of day (Burger et al. 1997a). Red Knots spent the most time alert (more than 10% of the time).

More than most other coastal species on Delaware Bay, shorebirds experience frequent interruptions of their foraging because they mainly feed where people engage in a range of activities. Several authors have examined the effect of human activities on foraging shorebirds (Thomas et al. 2003; Goss-Custard et al. 2006; Tarr et al. 2010; Burger and Niles 2013a,b, 2014). For example, West et al. (2002) found that many small disturbances of foraging shorebirds can have a greater effect than a few large ones. In addition, on beaches without people, shorebirds may spend 90% of their time foraging, but this value drops when people are present. In Piping Plover, it can drop to about 50% when people are present (Burger 1994a; Figure 4.22). Piping Plovers do not seem to habituate easily to people.

Herons, egrets, and ibises feed in lagoons, marsh pools, and edges of canals. They are not likely to feed in the open water as most forage while standing in water, along the shore, on mudflats, or in wetlands as illustrated by a Snowy Egret (Figure 4.23). Wading birds forage at different depths of water, mainly related to body size and leg length (Powell 1987). Larger birds with longer bills can handle larger prey items (Le V Dit Durell 2000). As might be expected, long-legged waders forage in a greater diversity of water depths than can shorter-legged birds. The smallest species (e.g., Little Blue Heron, Snowy Egret, White Ibis) had a maximum foraging depth of 18 cm; medium-sized species (e.g., Great Egret) had a maximum foraging depth of 28 cm, and the large Great Blue Heron had a maximum foraging depth of 39 cm (Powell 1987). Furthermore, Snowy Egrets and other species foraged in manipulated pools with enhanced prey (double the fish density) more often than in unmanaged or depleted pools, and they preferred shallow pools (Master et al. 2005) (Figure 4.23).

Clapper Rails mainly hunt for Fiddler Crabs along tidal creeks and the edge of marshes (Rush et al. 2012). Passerines usually forage near their nesting sites, searching for food in salt marsh vegetation, shrubs, and *Phragmites.*

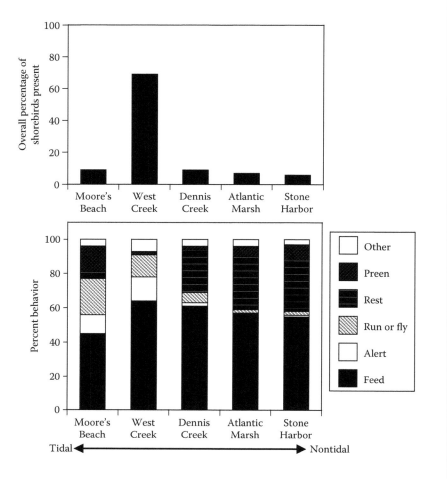

Figure 4.21 Habitat selection and activity patterns of shorebirds at five locations in Delaware Bay. Data are based on observations of individual birds in each habitat. (After Burger, J. et al., *Oil Spills*, Rutgers University Press, New Brunswick, 1997a.)

Food resources and foraging methods differ among species as a function of species size as well as age within species (Brown 1980; Burger 1987b, 1988d; Shealer 2001). Younger birds are often less successful than older ones, and compensate for it by feeding in different habitats or by different feeding methods. Young Laughing Gulls selected different habitats to forage where there were fewer adults feeding, and where their success was higher than elsewhere, although not as high as adult success (Burger and Gochfeld 1981a, 1983c).

Similarly, foraging within sight of one another can result in social enhancement, whereby birds coalesce at a spot where another bird appears to be successful. For example, foraging Sandwich Terns along the Mexican shore of Yucatan coalesced from distances of 0.5 km or more to join birds diving for fish (Gochfeld and Burger 1982a). Common Terns flying along the shore will fly rapidly to a flock of terns they see diving over a school of fish driven to the surface by Bluefish (Safina and Burger 1985, 1988a–c; Safina 1990). This behavior is facilitated by the flashing white pattern of gulls and terns, which is visible to conspecifics in the air at a great distance (Simmons 1972). In salt marshes, Herring Gulls and egrets gather at small salt marsh pools where people are gathering nets and traps used to catch bait fish for anglers. Great Egrets will even come to fishing docks to snatch fish from people that are cleaning their catch (Figure 4.23).

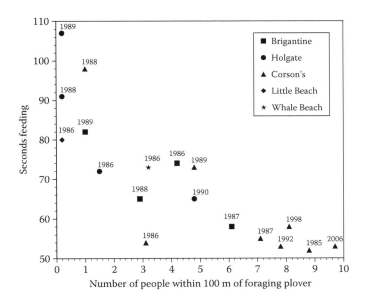

Figure 4.22 The effect of the presence of people (and their numbers) on foraging in Piping Plovers. When the number of people increases, Piping Plovers spend less time foraging. The dates refer to sampling data in that year at the given location. Each point is the mean of multiple 2-min focal bird observations. (After Burger, J., *Estuaries*, 17, 695–701, 1994a; Unpublished data.)

(a)   (b)

Figure 4.23 Great Egret waiting at a fishing dock for someone to return with food (a). Snowy Egret foraging (b).

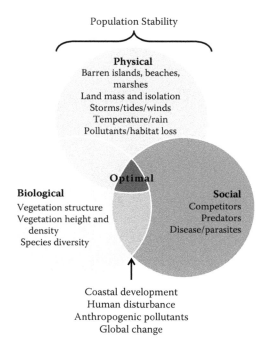

Figure 4.24   Model showing the relationship of physical, biological, and social factors to reproductive success in birds nesting in coastal regions. Physical, biological, and social factors have been considered by biologists for some time, but human dimensions, pollutants, and global change act on all three spheres.

SUMMARY AND CONCLUSIONS

Birds breeding, foraging, and roosting in coastal environmental face a range of physical, biological, and social pressures that determine where and when they nest, species composition of nesting assemblages and foraging groups, resting assemblages, and ultimately, reproductive success, survival, longevity, and population stability (Figure 4.24). Optimally, birds nest in those places that allow them to produce more young than if they nested in other places, live the longest, and contribute the most to the next generation. Over time, the proximate factors of tides, storms, habitat availability predators, and competitors will not only determine reproductive success in any one breeding season, but will morph into ultimate factors that shape their survival and evolution. The difficulty, discussed more fully in Chapter 7, is that global change (habitat loss or modification, temperature changes, sea level rise) may be happening at a pace that is too fast to allow the physical and social environment to change as well. Birds can fly, but if they have no habitat left to fly to, then they will be unable to breed, or they must change or adapt to new habitats. Climate change and sea level rise may be the greatest threats to seabirds and other coastal-nesting birds in the long run (Schreiber 2001), by rendering habitat unsuitable or unavailable. This chapter provided the background to assess information on population levels and changes, and metal levels in birds nesting in the Northeastern estuaries. This information will help shape future biomonitoring plans, management initiative, conservation, and the wise use of our estuarine resources. We endorse the goal of sustaining functioning estuarine ecosystems that provide the goods, services, and ecocultural values important to the health and well-being of people and the organisms that use these systems.

CHAPTER 5

Population Trends of Colonial Waterbirds in Barnegat Bay

INTRODUCTION

Population studies ideally include the size, structure (age distributions), and replacement of individuals over time. These measures aid in determining whether a given population is stable, increasing, or decreasing for the purposes of management and conservation. In the early 1900s, population data were developed for determining hunting quotas, particularly for ducks and waterfowl, and later for big game, and still later for conservation of endangered and threatened species. It is impossible to manage, conserve, or sustainably exploit populations without some idea of their numbers, or at least an index of population size and temporal trends. Except for endangered species, where all individuals may be counted, it is impossible to count every individual, so population estimates are based on sampling.

The rate at which a bird population increases or decreases depends on the number of eggs laid (fecundity), the survival of nestlings to fledging (productivity), and survivorship of individuals to breeding age (recruitment). Population numbers may be augmented by immigration or depleted by emigration (Weimerskirch 2001). In addition, age and phenotype can affect recruitment (Becker and Bradley 2007; Ezard et al. 2007). For birds living in coastal and marine environments, population size and trends including emigration, immigration, and recruitment, are influenced by interactions of food availability and location, oceanography, and climate, as conceptualized in Figure 5.1. Although traditional examinations of population regulation dealt mainly with intrinsic and extrinsic environmental factors, longer-term patterns that result from oceanographic features, weather, human disturbance, and climate change are usually not incorporated.

Reproductive success in colonial birds is often reported in terms of the number of young birds raised to fledging age per nest or per pair. In most breeding seasons, reproductive success is not an all-or-none phenomenon. Some pairs raise chicks, whereas others fail. The timing of severe weather events, storms resulting in extreme high tides and flooding, and days of chilling rainfall, can devastate colonies. These occur every few years, and may result in zero productivity for some species in some years. The timing of the severe weather events, or of food availability, or even diseases, with respect to the stage of the breeding season can be important. This is illustrated with Figure 5.2 and Table 5.1.

All of the effects mentioned earlier are exacerbated if the storms interfere with parental feeding. Wind and waves impair visibility of fish below the surface. Parents may be absent for longer periods, which will compromise brooding of young chicks. Even well-feathered chicks can suffer hypothermia if food is not readily available and adults are unavailable to brood them. Chicks that are too large to be brooded can die from prolonged chilling rain. This is a frequent problem for Black Skimmers, and has resulted occasionally in total failure of the first brood and synchronous relaying.

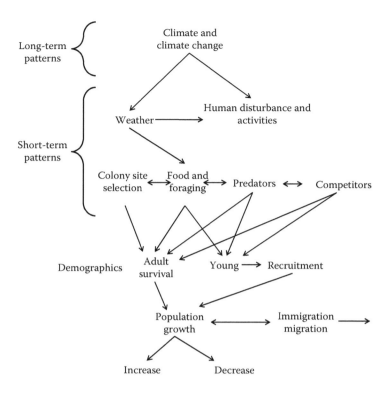

Figure 5.1   Schematic layout of the relationship of ultimate (long-term) and proximate (short-term) factors affecting population growth or decline influenced by climate and weather, human activities, ecological interactions, and demographics, acting over the long and short term.

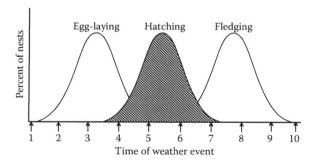

Figure 5.2   Hypothetical chronology of a breeding season showing temporal distribution of egg-laying, hatching, and fledging. Interpret this with Table 5.1, which discusses the impact of storm events at different stages of the breeding cycle.

We have monitored population levels of several coastal-nesting species in Barnegat Bay, documenting spatial patterns and temporal trends. In Barnegat Bay, most colonial species nest on salt marsh islands (as they do in the Chesapeake, the New York–New Jersey Harbor Estuary, and Long Island). The salt marsh islands in other bays are also vulnerable to sea level rise, storm surges, and coastal flooding. Indeed, numerous papers and reports from Massachusetts to Virginia mention sea level rise, erosion, and shrinking of island nesting areas as critical issues. In the other bays and

Table 5.1 Effects of Severe Weather Events Such as Flooding and Prolonged Rain on Avian Populations during Different Reproductive Phases

Number in Figure 5.2	Reproductive Phase	Effects of Extreme High Tides and Flooding or Prolonged Cold Rain Storms Assuming No Effect on Food Availability			
		Eggs	Chicks	Fledglings	Overall
1	Before egg-laying	Birds may relocate nest sites.			Minimal impact.
2	Egg-laying	Washed out eggs. Abandoned nest sites.			Low impact. Birds relocate and renest.
3	Incubation	Loss of all eggs and nests. Birds may relay or abandon and move to an alternative colony site.			Variable impact. Relaying may be successful.
4	Egg-laying Hatching	Remaining eggs washed out.	The few young chicks are brooded or succumb to flooding and cold.		Variable impact. Relaying may be successful.
5	Hatching and brooding		Many young chicks succumb to chilling if not brooded.		High impact. Relaying occurs, but late hatching chicks disadvantaged.
6	Brooding/feeding		Chicks too old to brood may thermoregulate or perish.		High impact, may entirely eliminate Skimmers.
7	Hatching Fledging		Older chicks well feathered. Late hatching chicks vulnerable.	Fledglings can move and thermoregulate if parental feeding continues.	Moderate impact.
8	Fledging			Fledglings moderately tolerant, but this is a period of high mortality.	Moderate impact.
9	Fledging			Fledglings becoming independent. Still require parental guidance.	Small impact because of interruption of feeding.
10	Fledging After fledging			Fledglings independent, still have to perfect feeding.	Negligible impact once the majority have fledged.

Note: Shown are how timing of a weather event can affect breeding biology and success. The number refers to Figure 5.2.

estuaries there are larger islands, some of which are relatively free from predators and people, ideal conditions for ground-nesting colonial species. Further north there are higher islands, with diverse habitats ranging from open sandy or rocky shores to woodlands (Seitz and Miller 1996). Higher habitats are not as vulnerable to sea level rise and severe floods, but even rocky islands may be subject to erosion (USFWS 2011b).

We focus on populations of colonial birds in Barnegat Bay in this chapter, and in Chapter 6 we examine populations in the other Northeast bays. The discussion at the end of this chapter focuses on the birds in Barnegat Bay, and the discussion in Chapter 6 compares populations in all the bays, including Barnegat Bay.

BARNEGAT BAY COLONIES

Barnegat Bay is designated as an Important Bird Area (IBA; Frank 2010). The IBA process in New Jersey was open and facilitated by a committee, coordinated by C. Frank. Many individuals and organizations wrote sections of the IBA book, which provides a good description of the Atlantic coast of New Jersey from Barnegat Bay south to Cape May.

For 40 years, we have followed population fluctuations of several species nesting on the islands in Barnegat Bay. Although most of these birds have been nesting on salt marsh islands, there are three sandy islands that have been used by nesting birds, and these were created by sand deposition from dredging of the inlet channels. Some sandy islands (e.g., Big Mike's) were initially used by nesting Least Terns and Black Skimmers, but these quickly became unsuitable because of *Phragmites* incursions. There were a number of higher sandy places on some of the Cedar Bonnet islands where Least Terns initially nested, but *Phragmites* invaded there as well. A pair of Northern Harriers (Marsh Hawk) nested on the ground there for a few years, but eventually the *Phragmites* closed in. Black-crowned Night-Herons often nest in these reeds. However, after a few seasons the Night-Herons abandoned as well, mainly because of intense human activities and nearby development.

SANDY BEACH HABITATS

There are beach habitats in the Barnegat Bay ecosystem that formerly had colonially nesting birds. Least Terns nested on the outer beach of Long Beach Island. Before the 1990s, there were active colonies of Least Terns, Common Terns, and Black Skimmers on several barrier island beaches, but these species have largely been extirpated from these habitats because of excessive human activities and predators that have increased because of human activities (Kotliar and Burger 1984, 1986; Burger 1989). Foxes and Raccoons are now common on barrier islands, attracted to the noisy bird colonies. Least Terns sometimes nest at Holgate (part of Forsythe National Wildlife Refuge) on the southern tip of Long Beach Island.

The decline of Least Terns was rather sudden in Barnegat Bay (Figure 5.3). There were usually between 500 and 850 adults counted on ground surveys on Barnegat Bay beaches, with a slight increase from 1977 to 1991, but then the number plummeted to zero by 1998. Figure 5.3 shows the statistically significant increase ($P = 0.03$) and decrease ($P = 0.001$). Least Terns nested in as many as seven colonies, but the number of suitable places decreased quickly. Efforts to restore them to previously used sites were successful in the short term, and colonies remained for several years, but eventually terns deserted because of daily human disturbance (Kotliar and Burger 1984; Burger 1989).

Colony failures for Least Terns occurred all along New Jersey's Atlantic coast. For many years, under the Endangered and Nongame Species Program of the New Jersey Department of

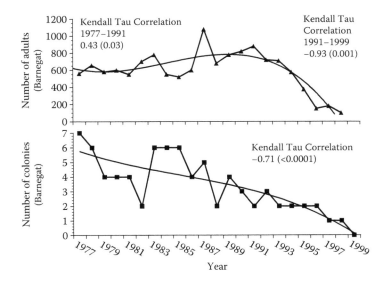

Figure 5.3 Number of adult Least Terns nesting only in colonies in Barnegat Bay on barrier island sandy beaches and on sandy islands. Numbers increased slightly (*P* = 0.03) from 1977 to the peak in 1987, then declined rapidly from 1987 to 1999 (*P* < 0.0001). In the 2000s, they sometimes nested successfully at Holgate.

Environmental Protection, there was an extensive monitoring effort that included assessing reproductive success for Least Terns (Burger 1989). Once Least Terns were declared endangered in New Jersey, significant management actions were undertaken, including wardening, fencing, or posting colonies to prevent human disturbance, coupled with predator control. It was essential to examine the effect of the management and conservation practices. At the time, Burger was in charge of the Least Tern and Piping Plover monitoring for New Jersey.

Using a criterion of raising 0.25 young/pair as "success" (which is low), reproductive success of Least Terns in the state generally increased from about 20% of the colonies fledging more than 0.25 young/nest, to more than 60% of the colonies exceeding this level in 1996 (Figure 5.4, top graph; Burger 1989, unpublished data). It is clear from Figure 5.4 that of the colony failures, human disturbance was the leading cause, followed by predators (Burger 1989; Burger et al. 1994b, Unpublished data). Predators were a major cause of colony failure in the early years, but their relative effect declined over time because of management; human activities, on the other hand, caused more failures (Figure 5.4, bottom graph). As might be expected, failure due to flooding remained relatively the same during this period. These data provide strong support for continued management of Least Tern colonies in the state, along with protecting Piping Plovers.

A few pair of Piping Plovers and a few American Oystercatchers nest solitarily on the barrier beaches of Barnegat, relying on crypsis for protection. Even these species have trouble (Burger 1987a, 1991a; Maslo et al. 2012). Piping Plovers, a federally listed species, often lose nests and chicks to predators. Successful breeding of Oystercatchers occurs mainly on salt marshes (Lauro and Burger 1989). Intense wardening, fencing, and education efforts are required to even maintain Piping Plover populations in New Jersey (Burger 1989; Burger et al. 2010c). The recent Superstorm *Sandy* (in 2012) removed many sandy areas on beaches used by Piping Plover, requiring sand restoration efforts.

The Least Terns mainly moved from Barnegat Bay to colonies further south, in Corson's Inlet, Stone Harbor, and Hereford Inlet, or north to Sandy Hook (Figure 5.5). Some of these sites had isolated sandy islands that provided suitable habitat for Least Terns and Black Skimmers, as well as

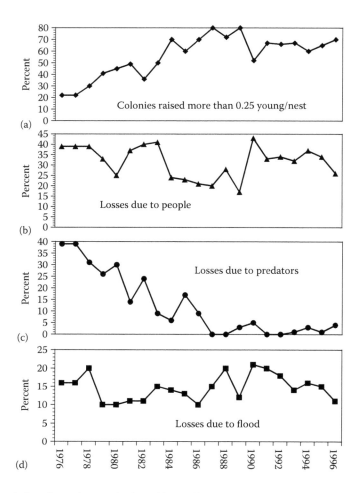

Figure 5.4 Burger led an intensive monitoring of Least Terns from 1976 to 1996 along all of New Jersey's coastline, documenting number of birds in colonies, productivity, and causes of colony failure. (a) The percent of colonies achieving relatively low success of 0.25 young fledged per pair. The number of successful colonies increased because of management activities. The percent of failures attributed to people (human development or direct disturbance) (b), predation (c), and flooding (d). (After Burger, J., *Condor*, 86, 61–67, 1984b; Burger, J., *J. Coastal Res.*, 5, 801–811, 1989. With permission.)

Piping Plover. Predator-free islands are ideal habitat, although Stone Harbor Point and Sandy Hook are connected to land areas with such predators.

SPATIAL VARIATION

Over the 40-year period, we concentrated mainly on a series of islands, largely or entirely salt marshes that were used over a long period by a range of species. The terns and Skimmers nested on wrack in *Spartina alterniflora*; terns and Laughing Gulls nested in *Spartina alterniflora*; and Herring and Great Black-backed Gulls nested on the highest places on the marsh, often close to or under bushes. The egrets and herons nested where there were bushes and small trees (Cherry, Poison Ivy), although some species nested on the ground because of limited bush/tree nesting sites. Until 1989, Least Terns nested out on the open beach, although into the early 1990s there were still available sandy islands (e.g., Big Mike's, one of the Cedar Bonnets) where they could nest on unvegetated sand.

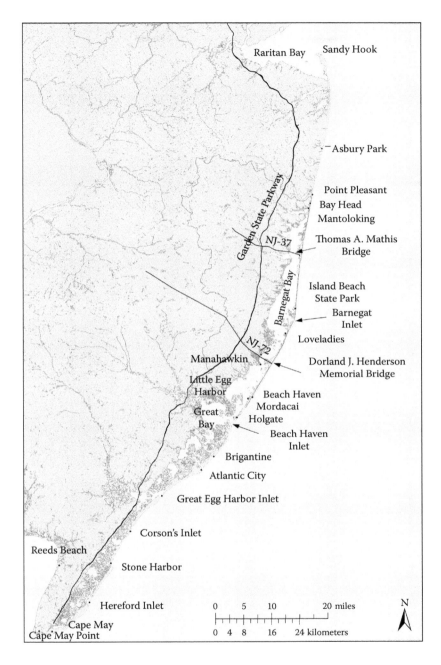

Figure 5.5 Map of the Jersey Shore, showing Barnegat Bay and other sites where Least Terns likely moved, including Sandy Hook to the north, and Corson's Inlet and Hereford Inlet to the south.

The main islands studied over the 40-year period are listed in Table 5.2 (their locations can be found in Chapter 7; Figure 7.10). We also report the number of nesting pairs or adults, which varied over the years as suitable habitat varied, succession occurred (*Phragmites* invaded), or human activities and development increased. Least Terns are not included in Table 5.2 because suitable, open sandy island habitat soon disappeared. Also, in the early years of our studies, Common Terns

Table 5.2 Range in Numbers of Adults of Several Species Present on Salt Marsh Islands in Barnegat Bay during the Years of Our Studies

Colony Name	Black Skimmer	Common Tern	Forster's Tern	Laughing Gull	Great Black-backed Gull	Herring Gull	Great Egret, (Snowy Egret Always in Same Colony)	Black-crowned Night-Heron
Number of marsh Colonies	25	37	4	22	18	25	9	12
				Marsh Islands				
NW Lavallette		39–750			2–4	2–38		
SW Lavallette		25–200			2–36	2–450		
N Lavallette		15–795			3–65	3–186	25–55	2–18
S. Lavallette		10–600			2–6	3–54	12–35	1–8
Little Mike's		2–600						
Buster	2–3	5–826		14–20		2–4		
East Point		18–325		2–16				
Clam	1–4	45–980	20–355	45–467	18–60	600–800	2–75	3–34
High Bar		9–28		41–123		2–25	1–6	2–5
East Vol	2	5–40		35–142	1–6	5–46		
West Vol	15–36	10–400		2–40	2–34	40–157	6–46	3–18
Gulf Point	1–2	42–110				3		
W. Sloop Sedge	13	87		2–121	5–67	100–204		
E. Sloop Sedge		1a		1–14	2–7	2–50		
Sandy Island		1a			2–54	50–61		
Flat Creek	2–4	5–33		35–36				
West Carvel	2–10	19–58		21–40	2–46	100–150	3–16	1–14
East Carvel	1–45	50–265		19–124	2–10	12–35		
West Log Creek	8–20	5–20		46–197				
Log Creek	3–14	1–63		23–85				
Pettit	1–20	12–600		2–6	1a	1–3		

(Continued)

Table 5.2 (Continued) Range in Numbers of Adults of Several Species Present on Salt Marsh Islands in Barnegat Bay during the Years of Our Studies

Colony Name	Black Skimmer	Common Tern	Forster's Tern	Laughing Gull	Great Black-backed Gull	Herring Gull	Great Egret, (Snowy Egret Always in Same Colony)	Black-crowned Night-Heron
				Marsh Islands				
Cedar Creek	1–8	10–114		26–42				
East Cedar Bonnet		1–4						1–3
NW/SW Cedar Bonnet	1–13	2–143		2–3		1–4		
Thorofare	2	20–58						1–5
Egg	1–8	11–145		16–65	3–56	3–360		
East Ham	3–55	10–42		10–25	4–45	20–100		
West Ham	4–28	20–331			3–65	10–250		
Marshelder	82	23–475			3–5	20–86	3–22	4–8
GoodLuck	1–2	1–212		6–180				
East Sedge	8–35	10–83				11–250		
East Long Point	1–2	2–32		18–400	2–3	2–6		
West Long Point		43		16–312	2–3	2–8		2–6
Little	2	21–235						
Mordecai	7–271	10–106		2–6		2–12		
Hester		6–121	25–121					
Goosebar		21–34	25–64				13–35	1–4
Barrel	6–14	48–125	25–223				12–82	2–13
Middle Sedge		12–90						
				Sandy Islands				
Big Mike's		5–112					2–16	1–6
Dredge Spoil	12–20	5–64			3–120	15–680	25–85	8–26
Tow	16–375	5–123		2–121	6–40	2–118		

Note: The colonies at the end are dredge/sandy islands. The marsh islands are listed from north to south in the bay.

[a] Not considered a colony because only one pair nested, and no other species nested with them.

tried to nest on the sandy islands, but they were soon eliminated by the expanding Herring and Great Black-backed Gull population.

There has been a great deal of variability in the number of nesting pairs on islands over the 40-year period, from none to several hundred pairs. The largest Common Tern colonies have generally been in the north (Lavallettes), Laughing Gulls were most abundant in the middle of the bay (e.g., Clam Island), whereas most Black Skimmers have been most abundant in the south (e.g., Mordecai). Herring Gulls and Great Black-backed Gulls, which moved into New Jersey as nesting birds only in the late 1940s and late 1970s, respectively, nested first in the north and moved south in Barnegat Bay, taking over the highest places on many of the marshes. Table 5.2 lists the major colony sites and the composition of colonial birds nesting on these islands. The prefixes for the four Lavallette Islands are our own designations. The range in the number of pairs includes only those years when the colony site was occupied; obviously, many have been abandoned because of changes in habitat (see Chapter 7).

Several inferences can be drawn from these data: (1) Black Skimmers only nested on islands with Common Terns. (2) Black Skimmers, Forster's Terns, and Laughing Gulls did not nest at the northern colony sites (e.g., Lavallettes). (3) Common Terns nested on the greatest number of colony sites (mainly because of they are able to use *Spartina alterniflora*, *Spartina patens*, and wrack). (4) Common Tern numbers were higher on the north end of Barnegat Bay. (5) Although Herring Gulls nested on a few more islands than Great Black-backed Gulls, eventually both species nested on most islands that were suitable and (6) Herons and egrets had far fewer suitable colony sites. The latter is attributable to there being fewer islands that are high enough to support *Phragmites*, bushes, and small trees, which these long-legged waders prefer.

TEMPORAL TRENDS

The number of colonial birds nesting in Barnegat Bay has changed dramatically over the 40-year period. The overall populations trends, which will be described and discussed in detail in the following subsections, are the following: (1) terns and Skimmers decreased; (2) Laughing Gulls decreased; (3) Herring Gulls and Great Black-backed Gulls increased, and the former are now decreasing; (4) Glossy Ibis and Cattle Egrets (*Bubulcus ibis*) decreased; (5) Black-crowned Night-Herons decreased and then stabilized; and (6) Great Egrets increased whereas Snowy Egrets decreased. Thus, the population patterns of different species varied and shifted, particularly as habitat changed and the large gulls have moved southward in their nesting distribution (Burger 1978a, 1979a; Pierotti and Good 1994; Good 1998).

Black Skimmers

Black Skimmers are endangered in New Jersey (see Table 3.2). Over most of their range, Skimmers nest mainly on sandy beaches, sandy islands, oyster bars, or sand spits, but in Barnegat Bay they have been forced to move to narrow sandy strips on salt marsh islands (Figure 5.6). In southern New Jersey, there are Skimmer colonies on sandy beaches and islands (e.g., Stone Harbor Inlet), but even there they face tidal floods, human disturbance, and predators. On salt marsh islands they prefer to nest on sandy patches or on wrack. As wrack ages, it settles and in some cases it is too close to the edge of an island where it is subject to flooding (Figure 5.6b). In the late 2000s, when these habitats were not available, a few Skimmers started to nest directly on *Spartina patens*, which occurs slightly higher on the marsh than *Spartina alterniflora* (Figure 5.6c). For the past several years, the only Skimmers in the bay that raised young did so on *Spartina patens*. Black Skimmers usually nest with terns or other species, and in Barnegat Bay they always nest with Common Terns (Burger and Gochfeld 1990b).

Figure 5.6 Black Skimmer colony on a sandy beach (a), Skimmer nest on flooded wrack (water in the nest) (b), and Skimmer clutch on *Spartina patens* (c) (on an unusual recent substrate).

The number of breeding adult Black Skimmers nesting in Barnegat Bay has decreased rather steadily since 1976 (Figure 5.7; Kendall nonparametric correlation tau = −0.58, P < 0.001). In the preincubation and incubation period, Black Skimmers normally are present as pairs during the day (Burger and Gochfeld 1990b). Thus, the data in Figure 5.7a can be divided in two to estimate the number of breeding pairs. The population of Skimmers in Barnegat Bay decreased slowly in the 1970s and early 1980s, and varied wildly in the late 1980s and 1990s, with an isolated peak of more than 2500 birds in 2003. The population declined precipitously the following year and has continued to decline thereafter, with an increase in 2015.

The variation in the late 1980s and early 1990s was attributable to severe high tides in winter that washed away wrack from some islands, and to active management that restored it on other islands. During this period, we went out in March with several boats and collected fragments of wrack from low-lying islands, and ferried it to a few islands that had slightly higher ground. There, we created artificial wracks for Black Skimmers and Common Terns. We made large wooden frames out of driftwood we found on the islands. We placed the vegetation in these frames, and left it to dry. Two months later, the returning Skimmers and terns were quick to colonize these wrack areas—a management success (Burger and Gochfeld 1990b). These structures were not permanent, and storms usually washed them away the following winter.

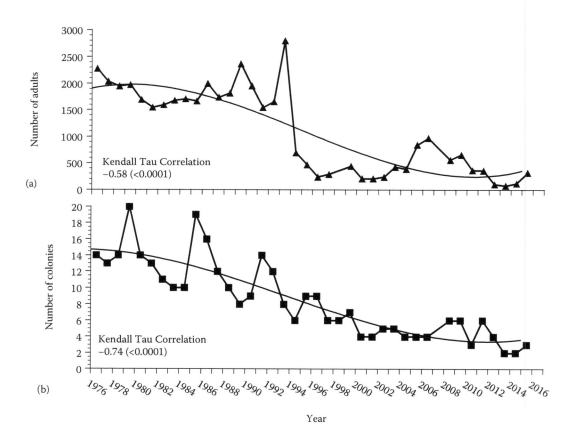

Figure 5.7 Number of adult Black Skimmers nesting on salt marsh islands in Barnegat Bay, 1976 to the present. The population declined significantly from the unique 1993 peak to very low numbers (a) in very few colonies (b). Skimmers normally stay on the colony site in pairs during the day as they are crepuscular and nocturnal foragers. Thus, the counts can be divided in half to estimate the number of breeding pairs.

In some years (1986, 1988, 1993), the wracks from the previous year survived the winter and were bleached by the sun, making them particularly attractive to the Skimmers. During many years in the mid 1980s and 1990s, most Skimmers breeding in Barnegat Bay nested on these artificial wracks, as no other suitable nesting habitat was available. This wrack construction work was done mainly by volunteers. Personnel and costs prevented us or the state from implementing this experiment as a sustainable management program, to the detriment of the Black Skimmers. As with the Least Terns, the number of occupied Skimmer colonies also declined (Figure 5.7b; tau = −0.74, $P < 0.0001$).

After the early 1990s, Skimmers still attempted to nest in Barnegat Bay, using the few wracks and sandy places left. An increase in the number of Skimmers in the early 2000s was largely attributable to the natural accumulation of wrack on some islands that had previously been too high to have any natural wrack. Some Skimmers nested on Mordecai Island on a small patch of sand, but after 2012, this sandy area was flooded out in most years. The Skimmers that tried to nest on *Spartina patens* on Mordecai were usually not successful, unlike in some higher colonies. Reproductive success for Black Skimmers was very variable, and strongly impacted by heavy rain storms and flooding (Figure 5.8b). High tides can flood out nests at any time, leaving eggs or dead chicks scattered at the edge of the *Spartina*. Heavy rains also resulted in lowered reproductive success because Skimmers often fail to brood chicks. If heavy rains occur when the chicks are small

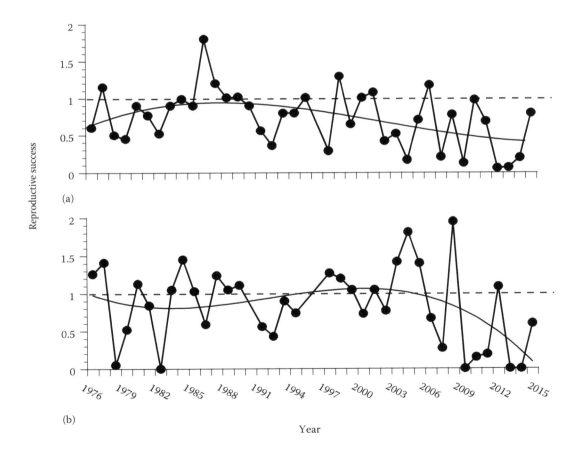

(a)

(b)

Year

Figure 5.8 Average annual reproductive success (young fledged per pair) for Common Terns (a) and Black Skimmers (b) nesting on salt marsh islands in Barnegat Bay. Prolonged summer storms completely wiped out Skimmer chicks in some years.

(a)   (b)

Figure 5.9   Black Skimmer fledgling trying to take flight from *Spartina alterniflora* marsh adjacent to its nest site (a). Black Skimmer chicks forming a crèche on a sandy beach colony (b).

or too large to fit under their parent's wings, then they die from thermal stress. Skimmer chicks will gather in a crèche if the colony is disturbed (Figure 5.9b). Skimmers seem to be unique in that 2-week-old, mostly feathered chicks are particularly vulnerable to chilling rain and hypothermia (Gochfeld and Burger 1994). Complete colony failures were usually attributable to very high flood tides or to 2–3 days of heavy rain—for example, in 1979, 1982, and 2009. Since 2010, few Skimmers have nested in Barnegat Bay, and they raised few young, except in 2012. As the Black Skimmer population in Barnegat Bay declined, there was an increase in the breeding Skimmers colonies to the south of Barnegat Bay (e.g., Stone Harbor Inlet islands). There was a slight upturn in breeding Skimmers in the bay in 2015.

A major drop in the number of pairs of Skimmers largely occurred in 1989 to 1994. This corresponds to a period of rapid change in habitat use. Figure 5.8b provides reproductive success data for all the Skimmer colonies in Barnegat Bay, regardless of whether they were nesting on wrack in salt marshes, sandy patches on salt marshes (Mordecai), or sandy beaches on barrier islands (e.g., Holgate, Barnegat Inlet, or Barnegat Lighthouse). During the first years of our study, Black Skimmers nested on sandy barrier beaches, as well as on wrack in salt marshes. Skimmers continued to nest on the barrier island beaches into the mid 1990s. During this period (1992–1994), reproductive success was higher on sandy barrier beaches (overall mean = 1.10 fledged young/pair) than on salt marsh colonies (mean = 0.79). Colonies on sandy islands (e.g., Hereford Inlet) had much higher success (mean = 1.52 fledged young/pair) than those nesting either on wrack or sand on salt marshes.

Black Skimmers nesting on barrier island sandy beaches, however, were subjected to intense and increasing human activity from sunbathers, swimmers, fishermen, and joggers. By 1997, there were no Black Skimmers nesting on the barrier island beaches bordering Barnegat Bay (this corresponds to the earlier loss of nesting Least Terns in these same barrier beaches, except for Holgate). Black Skimmers continued to nest on sandy areas on salt marshes (e.g., Mordecai). We attribute the movement away from the barrier beaches to excessive human disturbance, including the taking and removal of eggs. In the 1980s and 1990s, we observed people removing eggs of colonial species; in one case, they used eggs to write "I love you" in the sand, and in other cases, we saw people putting them in baskets for consumption and carrying them to their car.

In 1997 to 2000, reproductive success was still higher on these sandy patches, but began to decline (Table 5.3). In this period, reproductive success averaged 1.19 young fledged/pair on sandy patches, and only 0.81 young fledged/pair on salt marshes (mainly wrack). During this transition period, there was clearly a trade-off between nesting on sandy patches on salt marsh islands, and nesting on wrack.

From 2001 to 2005, reproductive success for Black Skimmers on both sand and wrack on salt marshes was relatively high (mean of 1.61 young fledged/pair), largely because of the large, usually successful

Table 5.3   Mean Reproductive Success (Young Fledged/Pair) of Black Skimmers during the Transition Years from Nesting on Sandy Barrier Island Beaches to Salt Marshes in Barnegat Bay, Compared to Sandy Beach Colonies in Southern New Jersey

Year	Barnegat— All Sandy Habitats	Barnegat— Salt Marshes	South Jersey— Sandy Inlet Islands
1989	1.11	1.25	0.97
1992	1.14	0.34	1.12
1993	0.90	0.86	1.40
1994	1.24	0.74	2.60
1997	1.44	1.02	0.96
1998	1.47	0.17	NA
1999	1.04	1.05	NA
2000	0.80	1.01	NA

Source: Burger, Unpublished reports to New Jersey Endangered and Nongame Species Program.
Note: NA, not available.

colony on the sandy beach at Mordecai. However, by 2006, reproductive success began to decrease markedly because of the erosion of Mordecai's sandy beach. Except for 2008 and 2012, fewer than one young/pair fledged, and in 2013 and 2014 none fledged. In 2015, more Skimmers again nested on wrack on Egg Island and on *Spartina patens* on other islands, and they succeeded in raising some young.

Common Terns

Common Terns prefer to nest on dry land (sandy or rocky beaches), but have been forced to move to salt marsh colonies in Barnegat Bay and Chesapeake Bay (Burger and Gochfeld 1991b; Figure 5.10). Unless the sand is on islands with sufficient water barriers around the island, mammalian predators can walk or swim there. Common Terns nesting on sandy beaches are exposed to high levels of mammalian predation and human disturbance (Burger and Gochfeld 1991b). Although Common Terns initially nested on sand on Big Mike's Island, Tow, and Dredge Spoil Island, they were eventually forced out by expanding populations of Herring Gulls and Great Black-backed Gulls. The large gulls, however, also moved into the higher areas of salt marshes (with *Iva* and *Baccharis* bushes), forcing the terns on those islands to nest on lower habitats, which were more vulnerable to flooding. On salt marsh islands, the early-nesting terns choose to nest on wrack (both natural and artificial). If wrack is densely occupied by terns or unavailable, terns will nest in *Spartina* areas, which are vulnerable to flooding. Similar shifts from mainland areas to islands have happened with gulls in the Wadden Sea; the birds shifted colonies to islands starting in the early 1990s (Koffijberg et al. 2006).

Over the 40 years of our study, reproductive success in Common Terns has been very variable, depending on heavy rains and flood tides. Where terns nest in the same places as Skimmers, their reproductive success has been similar (Figure 5.8a). In the first 14 years of the study, reproductive success was 1 chick/nest or more. In the middle years, pairs raised at least 1 chick/nest only 3 years, and in the latter period, pairs fledged 1 chick/nest in only one year. Recently, reproductive success has been consistently poor, and many adults have probably emigrated out of Barnegat Bay (Palestis and Hines 2015). Dispersal is a fundamental process in population dynamics (Breton et al. 2014), and the ability to do so may result in major shifts in Common Tern nesting patterns along the Atlantic coast. Tims et al. (2004), for example, reported colonization of nesting colonies in Buzzards Bay by immigrants from outside the region, and similar intercolony movements occur in the Great Lakes and elsewhere (Haymes and Blokpoel 1978). Reproductive success, however, needs to be considered in light of the number of pairs that are present, and that number has been decreasing along with reproductive success (Figure 5.11). Common Tern numbers increased, however, in 2015.

(a)

(b)

Figure 5.10 Where available, Common Terns nest on sandy patches at the edge of salt marsh islands (a). The nests are often placed near sparse vegetation (b).

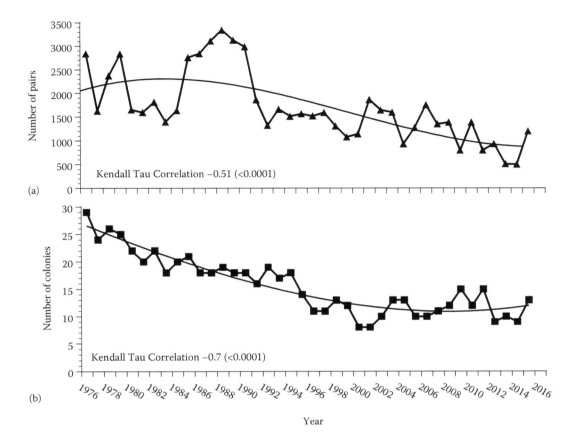

Figure 5.11 Number of pairs (a) of Common Terns nesting in Barnegat Bay colonies from 1976 to the present, with peaks in 1989 and 2007. Number of occupied colony sites (b), whether successful or failures. Both curves show significant declines to the present ($P < 0.0001$). (After Burger, J., Gochfeld, M., *The Common Tern: Its Breeding Biology and Behavior*, Columbia University Press, New York, 1991b; Burger, J. et al., *J. Coastal Res.*, 32, 197–211, 2001b; Unpublished data.)

Forster's Terns

For many years Forster's Terns bred primarily in one to three colonies in the southern part of Barnegat Bay on the salt marsh islands close to the inlet (Burger et al. 2001b). Later, Forster's Terns colonized Clam Island in the middle of the bay by Barnegat Inlet (Burger, Unpublished data). From 1976 to 1998, the number of Forster's Terms stayed fairly stable at 200 to 500 adults counted (150–225 pairs; Burger et al. 2001b), but once they colonized Clam Island in 2000, the number of nesting adults increased to as many as 800 (Figure 5.12). However, with sea level rise and increased tidal flooding of Clam Island, they deserted this island after it flooded several times each nesting season. Currently, they nest only on one or two islands in the south bay. Since a high in 2010, they have declined rapidly, and their reproductive success closely tracked that of Common Terns as they are vulnerable to the same tidal flooding, predators, and decreases in food supply. Forster's Tern also increased in 2015 (Figure 5.12).

Forster's Terns nesting on Clam Island failed because of flooding. It was not their fault. Laughing Gulls had usurped the more favorable habitat on Clam Island, relegating the Forster's Terns to lower areas than they might otherwise have chosen. Terns arrive later than the gulls, which succeed in

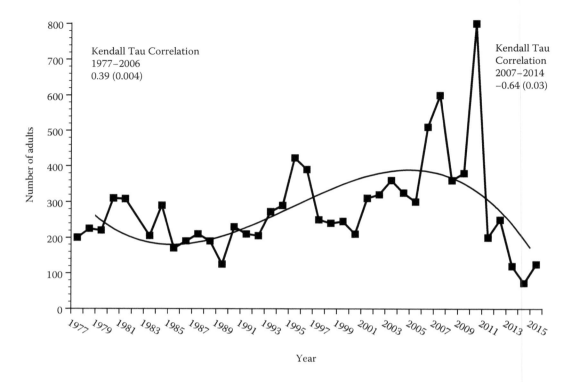

Figure 5.12   Number of adult Forster's Terns nesting in Barnegat Bay from 1977 to the present, showing peaks in 1995, 2006, and 2010. Typically, only one or two colony sites were occupied each year.

taking the highest places. Forster's Terns typically nest around marsh ponds, which are subject to flooding.

Great Black-Backed Gull and Herring Gulls

Herring Gulls invaded the salt marshes of New Jersey in the middle of the last century, having moved from rocky islands farther north (Figure 5.13). Initially, they nested on sandy islands, but they later moved to salt marshes. They increased rapidly until 1980–1981, and then they began to decline (Burger 1977a; Figure 5.14). The first Great Black-backed Gull nested in New Jersey salt marshes in 1976 (Burger 1978a). They immediately competed with Herring Gulls for the highest places in the marsh to nest, and forced the Herring Gulls to lower elevations on the marsh. The highest elevations in many of the islands where these large gulls nest are often in areas with *Iva* or *Baccharis* bushes, which provide both camouflage and protection from sun and rain. Great Black-backed Gulls are significantly larger than Herring Gulls, giving them an advantage in any nesting encounters. Furthermore, Great Black-backed Gulls maintain a larger territory around their nests, and exclude Herring Gulls, as well as conspecifics.

In addition, both Herring Gulls and Great Black-backed Gulls nest on Dredge Spoil Island just inside Barnegat Inlet. This island is entirely sand, and quite high in elevation; the gulls are always successful nesting there. Although Common Terns and Black Skimmers have attempted nesting on this island, the gulls preyed on their eggs and chicks, causing the terns and Skimmers to abandon the site. Although we tracked population numbers of both species (Figures 5.14 and 5.15), we did not follow reproductive success in all years.

(a)

(b)

(c)

Figure 5.13 Herring Gulls originally nested in rocky islands to the north (a), but moved south and in New Jersey first nested on dredge spoil dumped there many years ago (b). The colony is now heavily vegetated with little open sand (c).

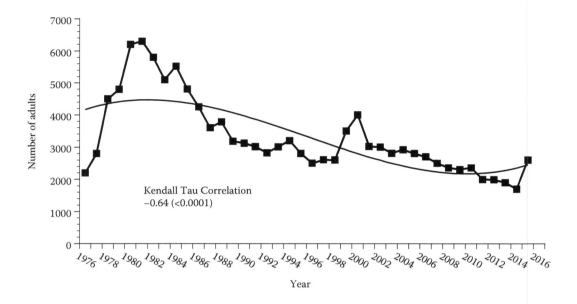

Figure 5.14   Number of Herring Gull adults nesting in Barnegat Bay from 1976 to the present. The graph shows a population increase in early 1980s, followed by a significant decline ($P < 0.0001$).

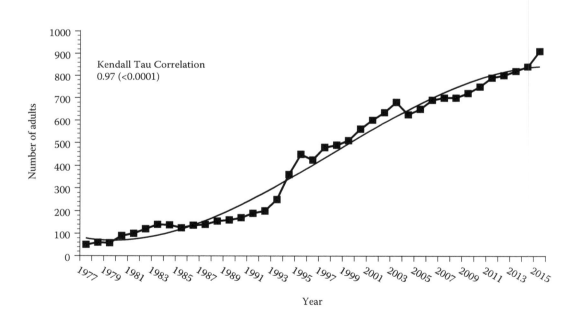

Figure 5.15   Number of adult Great Black-backed Gulls nesting in Barnegat Bay from 1976 to the present, showing a statistically significant, monotonic increase ($P < 0.0001$).

Laughing Gulls

New Jersey has traditionally been a stronghold for nesting Laughing Gulls (Burger 1996b, 2015a), with about 7000 to 8000 pairs nesting in Barnegat Bay, and similar numbers nesting at Stone Harbor and other colonies in southern New Jersey (Burger 1996b, 2015; Burger et al. 2001b). The main Laughing Gull colonies in Barnegat Bay were on Clam Island and Long Point Island, with smaller colonies on several other islands (see Table 5.2; Figure 5.16). Long Point Island was a relatively high *Spartina* island, but it was exposed to the full forces of wind and waves because it was far from the protection of both the mainland and the barrier island. Over the years, Long Point was washed away. Many of the islands with smaller colonies were also exposed to high tides, and eventually the gulls abandoned them because of tidal flooding.

Laughing Gulls usually nested in the *Spartina* itself (rather than on wrack), and selected the highest places, which were often near bushes (*Iva*, *Baccharis*). By selecting the highest places in the marsh, Laughing Gulls usually avoided egg and chick losses due to tidal flooding. In addition, they construct large, bulky nests, which sometimes float up, and settle once the waters recede (Burger 1978d, 1979b).

Nesting in the highest *Spartina* in salt marshes no doubt worked for Laughing Gulls until the invasion of Herring Gulls. However, the much larger Herring Gulls outcompete Laughing Gulls for prime nest sites because they arrive on the colony sites more than a month earlier, and they are three times heavier (Burger 1979a). Laughing Gulls either moved to other islands free of the larger gulls, or suffered much lower reproductive success where they did attempt to breed.

Finally, when the populations of nesting Laughing Gulls started to increase in Jamaica Bay in the 1980s (from 20 to about 7600 nests), they became a threat to aircraft at John F. Kennedy (JFK) International Airport (see Chapter 6). A gull-shooting program, initiated in 1991, resulted in shooting more than 50,000 adult Laughing Gulls (Brown et al. 2001a). Many of these gulls had been banded in New Jersey as young (Gochfeld et al. 1996). The population decline in the early 1990s corresponded to the shooting or "culling" program on Long Island. Ultimately, more than 50,000 gulls, mainly Laughing Gulls, were killed (Brown et al. 2001a). We believe that the

Figure 5.16 Laughing Gulls nesting in Barnegat Bay, showing small internest distances of three pairs.

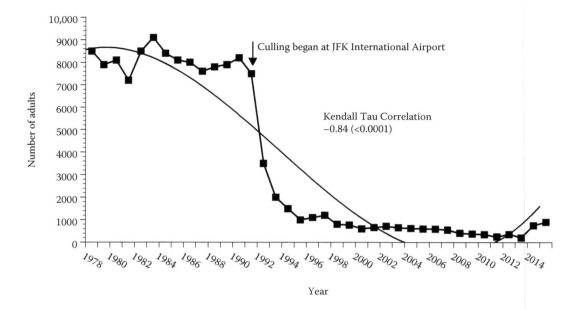

Figure 5.17   Number of adult Laughing Gulls nesting in Barnegat Bay from 1977 to present. The graph shows stable numbers through the 1980s, followed by a collapse related to the killing or "culling" of more than 60,000 Laughing Gulls at JFK International Airport. It is rare that a single factor can be so clearly associated with a population crash.

massive shooting program at JFK International Airport was the main cause of the decline of breeding Laughing Gulls in Barnegat Bay (Figure 5.17).

Brown et al. (2001a), noting that shooting had reduced the bird–aircraft strikes rate, voiced concern about the shooting contributing to a regional decline. Laughing Gulls disperse northward from New Jersey following the breeding season, joining the Laughing Gulls of Jamaica Bay (Burger, Unpublished wing-tag data). This resulted in their wholesale destruction. The combination of loss of habitat due to sea level rise, competition for space with the larger gulls, and the massive shooting of Laughing Gulls at JFK International Airport resulted in a decline in the number of Laughing Gulls breeding in Barnegat Bay (Figure 5.17).

Great Egrets, Snowy Egrets, and Black-Crowned Night-Herons

Great Egrets, Snowy Egrets, and Black-crowned Night-Herons nest in colonies on salt marsh islands in Barnegat Bay. However, they only nest on islands that are high enough to have bushes. The one exception is Black-crowned Night-Herons that nest on the ground in *Phragmites* (e.g., Log Creek). They once nested on Cedar Bonnet Island, but eventually these islands were developed, and the Night-Herons were forced to move. In general, the egrets and Night-Herons nest where there are *Iva*, *Baccharis*, Cherry and Poison Ivy bushes, as well as any other small shrubs (Figure 5.18). These are the highest areas on the marshes, and thus are least likely to flood, although over the last years of the study some Night-Heron and Snowy Egret nests on the ground were flooded out.

The number of islands with places high enough to support large bushes is limited, and the area of bushes on any one island is also limited. Thus, there is competition for the higher nesting sites (see Chapter 4), and Great Egrets succeed in getting the highest places (Burger 1978b, 1979c). The exception is Dredge Spoil Island inside Barnegat Inlet. This island is quite high because of the

Figure 5.18 Egret and Night-Heron colony in bushes in Barnegat Bay.

periodic deposition of dredge material and is not vulnerable to tidal flooding. It has a relatively large area of bushes and *Phragmites*.

Egrets and Night-Herons are sensitive to human disturbance, particularly when they have small chicks. Chicks may jump out of nests, climb around, or run when people move through the colonies. Thus, we normally entered colonies only during the incubation period (to collect eggs and count nests), and then again when chicks were nearly fledged and were able to find their way back to their nest sites (after we captured them to pull breast feathers). Since egg-laying is not as synchronous as some other colonial-nesting species, this limited our visits, our censusing, and our collection of eggs of the later-nesting species. It is difficult to count nests in heron colonies because they are often in dense vegetation, making it difficult to follow a path through the colonies. In addition, many colonies in Barnegat Bay are in dense stands of Poison Ivy. Census data for egrets and Night-Herons comes from the surveys conducted by the Endangered and Nongame program of the New Jersey Department of Environmental Protection based on periodic aerial censuses.

Great Egret populations have increased in Barnegat Bay, whereas those of other species have decreased (Burger et al. 2001b; Unpublished data, NJDEP). We present population levels only for the most common species: Great Egret, Snowy Egret, and Black-crowned Night-Heron (Figures 5.18 and 5.19). However, our previous analysis (Burger et al. 2001b) reported no significant population trends for Little Blue Heron, Tricolored Heron, Yellow-crowned Night-Heron, and Glossy Ibis. The numbers for all these species in Barnegat Bay are low and variable, making it difficult to ascertain trends. Furthermore, these darker species are more difficult to count from the air, tend to stay lower in the shrubbery, and are less likely to fly from the aircraft. It is much easier to count Great Egrets and Snowy Egrets because they sit on the top of the vegetation, or fly to the nearby marsh to await the departure of the aircraft.

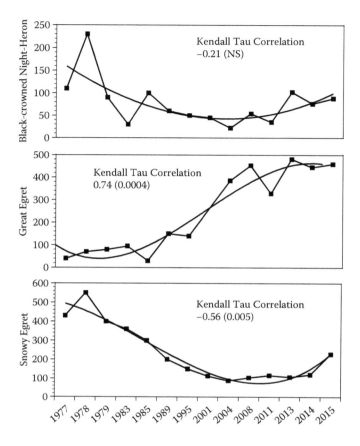

Figure 5.19 Population levels for Great Egrets, Snowy Egrets, and Black-crowned Night-Herons nesting in Barnegat Bay. Colony numbers are from the Endangered and Nongame Species Program and New Jersey Department of Environmental Protection. (Unpublished data, courtesy of C. Davis.)

SUMMARY AND CONCLUSIONS

The number of pairs of colonial nesting birds in Barnegat Bay has fluctuated considerably over the past four decades, both on individual islands and in the bay overall. There has been a decline in the number of breeding pairs of most species, and in the number of colonies for nearly all species, but especially for Least Terns (that no longer nest on the sand on salt marsh islands of Barnegat Bay), Common Terns, Black Skimmers, and Snowy Egrets. The causes for population changes are varied, but are summarized in Table 5.4.

The causes of declines include human development (barrier island beaches where terns and Skimmers once nested), human disturbance (barrier beaches), competition with larger species (gulls), sea level rise (increased flooding of nests and chicks), loss of salt marsh islands (sea level rise and erosion), and shooting (Laughing Gulls at JFK International Airport). It is possible that fish availability has declined, affecting some species (terns, Snowy Egrets) more than more omnivorous species (gulls, Night-Herons). Many of these threats are rather intractable, suggesting that active colony management and restoration (or recreation of habitat) is needed to sustain some populations in Barnegat Bay. The relative success of Great Egrets is partly attributable to their selection of the highest islands (some of which are dredge spoil). Dredge spoil islands could be augmented with additional dredge spoil material, and even salt marsh islands could be augmented by broadcasting

Table 5.4   Summary of Population Trends for Colonial Bird Breeding in Barnegat Bay, with Suggested Causes of Changes

Species	Trend	Causes of Change
Least Tern	Declined to extirpated on islands in Barnegat Bay	Human development of beaches, and human activities in and around colonies, combined with predators (human commensals and predator access over bridges) resulted in extirpation from the Barnegat Bay islands and most of the barrier beaches, although in some years they nest on Holgate (Forsythe National Wildlife Refuge).
Black Skimmer	Declined	Sea level rise led to increased flooding and lowered reproductive success for a series of years; loss of suitable wrack on islands and decreases in available sand on islands reduced habitat availability; arrival of nesting gulls on one sandy island led to abandonment.
Common Tern	Declined	Loss of habitat due to tidal flooding and sea level rise; loss of habitat because of lack of suitable wrack on salt marsh islands. Role of prey availability unknown.
Forster's Tern	Increased, now declining	Loss of habitat owing to tidal flooding and sea level rise.
Great Black-backed Gull	Increased	Expansion into high areas of the salt marshes allowed populations, augmented in the early 1970s and 1980s by garbage dumps, which increased survival of young gulls.
Herring Gull	Increased and then declined	Expansion into high areas of the salt marshes allowed populations to expand, augmented in the early 1970s and 1980s by garbage dumps which increased survival of young gulls; increased competition with Great Black-backed Gulls for the highest places and closure of landfills led to recent declines.
Laughing Gull	Declined	Nesting habitat competition with larger Herring and Great Black-backed Gulls forced them into lower elevation nesting sites, which were subject to tidal flooding; loss of some nesting islands owing to sea level rise; massive shootings at JFK International Airport in the 1990s killed many New Jersey birds and decreased populations.
Great Egret	Increased	There has been no loss of habitat yet for this species that nests on the highest bushes on the salt marshes.
Snowy Egret	Decreased	Snowy Egrets compete directly with the larger Great Egrets for nesting space, which is limited; some are forced to nest on the ground; loss of foraging intertidal habitat may affect this species more than larger, longer-legged species.
Black-crowned Night-Heron	Increased then declined	Nest with egrets; some loss of suitable nesting habitat; elimination of one colony by residential development.

soil on the islands. Furthermore, in some years, terns and Skimmers nested successfully on salt marsh islands because we constructed wrack (carried from other, lower islands) on the highest places of traditional colony sites (Burger and Gochfeld 1990b; Palestis 2009). Black Skimmers, Common Terns, and Least Terns can be drawn back to protected barrier island beaches by adding spoil to create flood-free zones and using decoys to attract them (Kotliar and Burger 1984; Burger 1988b, 1989). Predator control may be essential for terns and Skimmers being drawn to barrier beach sites since Red Foxes, Striped Skunks, and Raccoon populations can be quite high, even in settled areas (Burger and Gochfeld 1991b; N. Tsipoura, L. Niles, personal communication). Natural oceanic processes of normal sand movement along the coast, and salt marsh accretion are hindered by extensive human development and infrastructure (bulkheads, jetties) along the coast of New Jersey. Active management to maintain populations of colonially nesting species requires time, money, and commitment.

CHAPTER 6

Population Trends of Colonial Waterbirds in Other Northeast Bays

INTRODUCTION

Understanding the spatial and temporal differences in population levels of colonial species nesting in Barnegat Bay provides local data for a bay that is more than 60 km long. However, it is also useful to examine whether the trends evident in Barnegat Bay are local or regional, and the extent to which species differ in their population trends in other Northeast bays. Since populations, by definition, are circumscribed by scientists (e.g., birds of Barnegat Bay, birds of New Jersey, birds of the Northeast), it is often difficult to determine and define the deme (members of the species that are potentially interbreeding), and the extent of exchange among colonies or populations that constitute a metapopulation (Spendelow et al. 1995).

The birds in Barnegat Bay clearly move around within the bay based on environmental conditions. Over time, some colony sites are no longer usable because of habitat changes, flooding (sea level rise and subsidence), predators, or human activities (see Chapters 5 and 7). That is, even species that normally show high fidelity to specific colony sites can be forced to move because their colony sites are no longer available or suitable. Movement of specific individuals among the many colony sites in Barnegat Bay, however, can be determined only by banding. Our early banding studies indicated exchange among islands for Common Terns, Herring Gulls, and Laughing Gulls. Most of these changes, however, were attributable to their colony and nest sites no longer being available. Recent studies by Palestis in the bay also indicate movement among colonies by Common Terns (Palestis and Hines 2015).

Just as birds can move among colony sites in Barnegat Bay, there is movement among the different bays and estuaries. For example, if increases in colony numbers in a given location are greater than is possible by reproduction, then recruitment from other colonies is occurring. When new colonies form and grow rapidly, then the increases are attributable to recruitment from elsewhere. For example, Laughing Gulls did not breed on Long Island, New York, for well over a hundred years, but then a colony formed on a marsh in Jamaica Bay near the end of the runway at John F. Kennedy (JFK) International Airport (Buckley et al. 1978). The colony grew exceedingly rapidly, mainly from New Jersey recruits. Gulls nesting at the end of one of their major runways presented a risk to aircraft, particularly when young birds rested or roosted on the runway itself. An extensive gull control program demonstrated that most of the banded birds came from New Jersey (Dolbeer et al. 1993; Gochfeld et al. 1996).

Similarly, declines or increases in total birds in Barnegat Bay could be attributable to movement among colonies in the Northeast. Adult survival for some species (e.g., Roseate Tern; Spendelow et al. 2008) is similar when colonies are expanding or decreasing, suggesting that the declines in number of terns breeding in any one bay may not be attributable to different annual survival rates of adults.

Figure 6.1   A breeding colony of Laughing Gulls in Texas. The Gulf of Mexico coast is a stronghold for nesting Laughing Gulls. There they nest in mixed-species colonies with terns, egrets, herons, and spoonbills.

In this chapter, we review data on population levels of terns, gulls, herons, and egrets from other Northeast bays and estuaries, including the Massachusetts, Long Island, New York–New Jersey Harbor Estuary, Delaware Bay, and Chesapeake Bay. Complete census data are not available for all species for all bays. Monitoring requires resources, personnel time, and flight time. Methods (helicopter, fixed-wing, ground-based) and personnel time may change from year to year. Ideally, multiple censuses in a given year should coincide with the peak of birds incubating. Productivity measures, likewise, need to be timed to fledging periods. A shift in phenology (very early or late nesting) in 1 year may result in underestimates for that year.

Regional trends provide context, particularly if a decline in one area is offset by an increase in another. Species are not uniformly distributed from Maine to Virginia. For example, Laughing Gulls breed in southern Maine and Massachusetts, at Jamaica Bay on Long Island, and in New Jersey, but not in Delaware Bay, although they breed nearby in the Stone Harbor marshes on New Jersey's Atlantic coast. Laughing Gulls also breed in the Chesapeake, and are common and abundant in Florida and along the Gulf of Mexico coast (Figure 6.1), and in the Caribbean. Thus, their breeding populations are not evenly or continuously spaced all along the Atlantic coast (Burger 2015a).

MASSACHUSETTS BAYS AND THE REGION

This section includes the Boston Harbor, Cape Cod, and Buzzards Bay. Mostello (2014) has published statewide estimates of breeding birds (Figures 6.2 through 6.4). The importance of the islands in Boston Harbor for avian habitat has only recently been acknowledged, particularly for nesting wading birds (Paton et al. 2005; Parsons and Jedrey 2013). For example, 50% of all Black-crowned Night-Herons and Snowy Egrets in Massachusetts nest in the few islands in Boston Harbor.

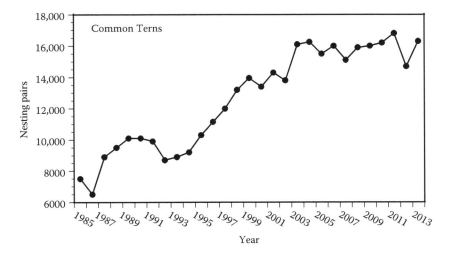

Figure 6.2 Population trends for Common Terns in Massachusetts (coastwide). (After Mostello 2014.)

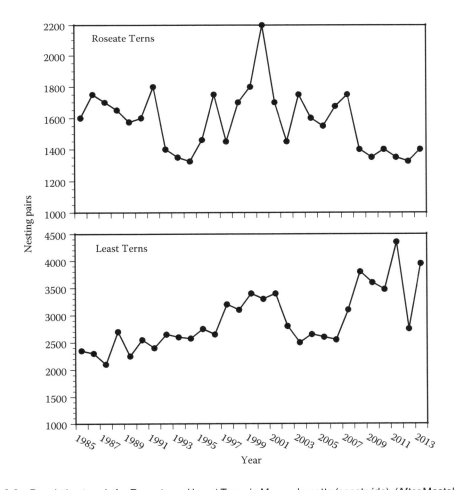

Figure 6.3 Population trends for Roseate and Least Terns in Massachusetts (coastwide). (After Mostello 2014.)

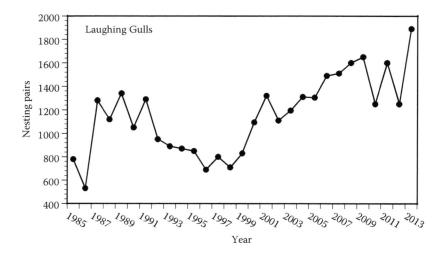

Figure 6.4 Population trends for Laughing Gulls in Massachusetts (coastwide). (After Mostello 2014.)

With the formation of the Boston Harbor Islands National Park, opportunities to study, monitor, and facilitate management are greatly improved, and a monitoring plan has been developed for the area (Trocki 2014).

The State of Massachusetts conducts an annual inventory of terns, Laughing Gulls, and Black Skimmers (Melvin 2010; Mostello 2014). Cooperators in Massachusetts surveyed more than 140 coastal sites for these species. In 2013, 79 sites were occupied, and the following number of pairs were counted: Common Tern (16,336), Least Tern (3977), Roseate Tern (1330), Laughing Gulls (1863). They found only three pairs of Black Skimmers at the northern limit of their range, and one pair of Arctic Terns at their southern limit. Overall population trends for some species and islands are available since 1970, and data from 1985 to 2013 are shown in Figures 6.2 through 6.4. Common Terns approximately doubled from about 7500 pairs to more than 16,000. Least Terns increased from under 2500 pairs to nearly 4000 pairs. Roseate Terns remained relatively stable at 1500 pairs. Laughing Gulls increased from about 800 pairs to nearly 1900 pairs (Mostello 2014). The Laughing Gull data (Figure 6.4) illustrate the challenges in interpreting trends. From 1991 to 1996, a sharp downward trend might predict that the species was on its way to local extinction but they then increased.

The Monomoy National Wildlife Refuge hosts one of the major assemblages of Common Terns on the Atlantic coast (Massachusetts 2015). Although the terns have increased in Massachusetts, not all species have fared as well. Several species have declined in Massachusetts, including Double-crested Cormorant (by 19.5%). This comes as a surprise to those who have witnessed its exploding population in New York and in New Jersey. Herring Gulls have declined by 42.2%, Great Black-backed Gulls by 40%, Black-crowned Night-Herons by 45.3%, and Snowy Egrets by 36% (Melvin 2010). The gull decline is misleading since it reflects a rebound from the nonsustainable explosion in the mid-twentieth century. Night-Herons and Snowy Egrets are declining in many areas.

Given the changes in populations of some species in the region, Parsons and Jedrey (2013) developed a colonial waterbird management plan for the Boston Harbor islands. Their strongest recommendations, which apply generally to island habitats throughout the Northeast, include: (1) seasonal closures of "keystone" islands through signage and patrol, (2) protection of specific species that are declining or vulnerable, (3) enhancement of nesting habitat on some islands, (4) targeted studies to identify factors affecting reproductive success, and (5) educational signage on islands accessed by the public. As Trocki (2014) notes, management and archiving of data are also critical for sustainability of any colonial waterbird populations.

BUZZARDS BAY

Buzzards Bay includes three islands of great historic importance for terns: the Bird, Ram, and Penikese Islands (Mackay 1899). Common Terns were abundant breeders in the bay into the 1880s, almost disappeared during the decades of exploitation, and recolonized again in the 1920s. Protections were put in place as early as 1895, and even at that period island erosion threatened the availability of tern habitat (Mackay 1899). By the 1950s, tern numbers in Buzzards Bay had increased to about 15,000 pairs nesting on four to five islands. However, gulls colonized the islands in the 1930s as part of their southward range extension, and by the mid-1950s they were seriously impacting tern numbers. Here, too, gulls displaced terns from favorable nesting habitat, causing tern populations to decline, until closure of garbage dumps in the 1970s–1980s resulted in a reversal of the gull population trajectory (Massachusetts 2015). As elsewhere in the 1980s, gull numbers were going down and tern numbers were going up. The total number of Common Terns on the Buzzards Bay islands increased from about 500 in 1970 to about 7000 in 2010. Terns were also potentially impacted by exposure to polychlorinated biphenyls (PCBs) from the New Bedford Harbor Superfund site; PCBs have been implicated as endocrine disruptors causing feminization of tern embryos—however, the relationship is uncertain (Nisbet et al. 1996). Both Ram Island and Bird Island are critical sites for the endangered Roseate Tern, and together they harbor about half the Northeast population, fluctuating widely between 1200 and 2000 pairs. Both islands suffer from sea level rise and erosion and require aggressive habitat management and restoration (Massachusetts 2015).

LONG ISLAND SOUND

The tern colonies at Great Gull Island and Falkner Island have been monitored and protected for many years. Great Gull Island (7 ha) lies between Orient Point and Fisher's Island, and is managed by Helen Hays and her colleagues from the American Museum of Natural History. In any given year, about half the Northeastern Roseate Tern population breeds there. Falkner Island, in the mid-Sound, about 3 mi south of Guilford, Connecticut (currently about 1.2 ha), is also severely impacted by sea level rise and erosion (down from about 1.8 ha). For many years, Jeffrey Spendelow monitored the Roseate and Common Terns nesting there, and enhanced nest sites for the former species. Roseate Tern numbers have substantially decreased in recent years.

The Long Island Sound Study (LISS 2010) published wading bird populations for Long Island Sound including most of Long Island. Although the areas overlap those covered by Brown et al. (2001b), the years 1998–2010 do not. Great Egrets increased from about 750 to 1250 pairs (inferred from the graph). Snowy Egrets show a slight decrease from about 800 to 550 pairs. Black-crowned Night-Herons were stable at about 1600 pairs (LISS 2010).

NEW YORK–NEW JERSEY HARBOR ESTUARY

The New York–New Jersey Harbor Estuary is large and complex, with several rivers entering into different bays. It has numerous islands, some natural, some made from dredge spoil, and others are augmented by dredge spoil. Its salt marshes are largely in Jamaica Bay and in the Hackensack Meadowlands (Quinn 1997; Raichel et al. 2003). Colonial nesting birds have been monitored by the New York City Audubon Society under the Harbor Herons Project, initially led by Katherine Parsons, among others, and later by Susan Elbin and Elizabeth Craig. This is a particularly important project because New York City and environs is one of the most densely populated areas in the world, and to showcase heronries for the public is an important ecological and societal goal. Furthermore, unlike many beach-nesting birds that are declining, small, cryptic, and difficult to see, herons and egrets are large and easy to observe, particularly when they are foraging in bays, tidal creeks, and marshes. Moreover, they are elegant, majestic, and often occur in groups. The value of the Harbor Herons Project is that it has been conducted yearly since 1982 using consistent methodology, albeit with a change in observers over time. The project provides invaluable data on population trends for conservation and management.

Approximately 20,000 acres (ca. 8100 ha) of tidal wetlands remain in the New York–New Jersey Harbor Estuary, representing only a fraction (20–25%) of the historical marshes in the area (Steinberg et al. 2004). Island colony sites total only about 249 ha, and not all of this area is suitable for nesting. Herons and egrets have nested on 18 islands (Craig 2009, 2013). Most of the islands used for heronries are uninhabited by people, and most are designated as Important Bird Areas by New York State (National Audubon characterization). Herons and egrets require structure for nesting, including shrubs and trees. The most abundant species in these heronries, accounting for 92% of the recent nesting assemblages, are Great Egret, Snowy Egret, Black-crowned Night-Heron, and Glossy Ibis (Figures 6.5 and 6.6). The four main species increased from the early 1980s until the mid-1990s. High counts for the four species were maintained during the 1990s. During the past three decades, Great Egret populations increased. Glossy Ibis populations rose, declined slightly, then rose to a high in the mid-2000s, but have now declined dramatically. Black-crowned Night-Heron populations declined by about 40% since the early 1990s (Craig 2013). Cattle Egrets are relatively recent additions to Northeast heronries. The first record for Cattle Egret in North America was in 1952 (Crosby 1972). The species increased dramatically through the next decades, until it began an inexplicable

(a)   (b)

Figure 6.5 Huckleberry Island (a), off Westchester County, was one of the New York–New Jersey Harbor traditional heronries. Birds nested mainly in trees. Double-crested Cormorants had the highest nests and their droppings have killed the trees (compare with Figure 13.1). At Prall's Island, herons and egrets also nested in trees. (b) Kathy Parsons uses a ladder to reach high nests. (Photo courtesy of Katherine Parsons.)

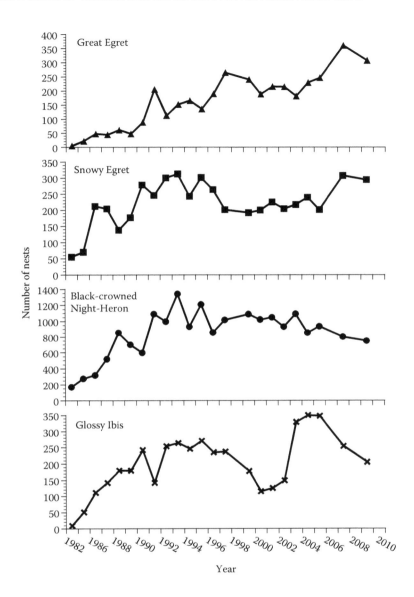

Figure 6.6 Population levels of Great Egrets, Snowy Egrets, Black-crowned Night-Herons, and Glossy Ibis from the heronries in the New York–New Jersey Harbor Estuary. (After Craig, E., New York City Audubon's Harbor Herons Project: 2013 Nesting Survey Report, *New York City Audubon*, New York, 2013.)

decline in the 1990s, and is now almost gone from the New York–New Jersey Harbor heronries (Craig 2013). The species has had a similar trajectory in Barnegat Bay (Burger/Gochfeld data).

It is worth noting that the number of nesting wading birds monitored by the Harbor Herons Project has varied greatly. There was a large increase in total nests from 1982 to about 1991–1992, and then numbers decreased (Craig 2013). In the early 1990s, the largest colonies were in the Arthur Kill and Kill van Kull near Staten Island. People may have forgotten, but during the night of January 1–2, 1990, 567,000 gal of No. 2 fuel oil leaked into the Arthur Kill from an underwater Exxon pipeline (Burger 1994b,d). In the immediate vicinity, oil destroyed 20% of the marsh vegetation, leaving the mudflats covered in oil (Burger 1994e,f). Unlike most oil spills, several scientists were already working in the Arthur Kill, and so pre-oil data were available. As with all coastal oil spills, there was acute avian mortality, but these were not the only effects (Parsons 1994). Parson's studies with a number of wading bird species showed lowered reproductive success and disrupted foraging ecology in the breeding season following the spill. Most losses were attributable to chick starvations because of degraded wetlands. Parsons (1994) found changes in colony use, species abundance, and reproductive success. Snowy Egrets showed the greatest adverse effects (Table 6.1). These data are invaluable because Parsons (1994) had prespill data from the same colony and used the same methods to evaluate the impact of the oil spill that happened close to the colonies.

These data show a decline in reproductive success following the oil spill. In the case of Black-crowned Night-Herons, a direct causal relationship can be seen. This species preys heavily on Fiddler Crabs, which were severely affected by the spill (Burger et al. 1994c). The oil spill occurred in winter and forced some crabs to the surface, while oil plugged holes sealing in other crabs. Fiddler Crabs kept coming to the surface in the days after the spill, and then perished during the winter nights. Their mortality was high.

At the same time, the long-term censuses show that the total number of heron nests in the region declined (Burger et al. 1994c). These data suggest a link with oil exposure. Although it is not possible to demonstrate cause and effect, the herons slowly deserted these colonies in the Arthur Kill and Kill van Kull after the spill. Thus, in the early census years, there were large active heronries in the Kill van Kull and Arthur Kill, but these colonies were abandoned by 2001 (Craig 2013). These events illustrate major shifts in the location of heronries, although some heronries have been occupied since the 1970s (South Brother Island, Goose Island). Some heronries, such as Huckleberry Island, increased from 0 to more than 300 pairs, and then declined rapidly (Figure 6.7).

The most remarkable difference in nesting assemblages in the New York–New Jersey Harbor Estuary has been the expansion of Double-crested Cormorants into these colonies. Over the past three decades Cormorants have undergone rapid expansion in the New York area (Parsons 1987; Hatch and Weseloh 1999; Elbin and Craig 2007; Figure 6.8) as well as in Delaware Bay. There is concern that Cormorants are disrupting heronry dynamics and habitat with their excessive guano killing trees and rendering sites unsuitable after a number of years. Craig et al. (2012, 2014) investigated the potential impacts of Cormorants and other waterbirds on plant and arthropod communities on two islands in the New York–New Jersey Harbor Estuary. On the colony where Cormorants were more recent settlers, they found reduced understory richness and total plant cover underneath cormorant nests, compared to places where there were no nests, but there were no differences between cormorant and other waterbird nests. The effects of Cormorants on colony dynamics and habitat of egrets and Night-Herons may involve competition for nest sites or nest material, or aggression during the nesting season (Cuthbert et al. 2002; Weseloh and Moore 2006; Somers et al. 2007). In the Great Lakes, Double-crested Cormorants interfere with reproduction in Herring Gulls (Somers et al. 2007).

The Harbor Herons Project has largely focused on the long-legged waders (herons, egrets, ibises, Night-Herons), and not on gulls, terns, and cormorants. However, Brown et al. (2001b) examined

Table 6.1   Changes in the Number of Chicks Fledged per Nest Before and After the Exxon Oil Spill on Arthur Kill (Mean ± Standard Deviation)

Species	Before the Spill (1986–1989)	After the Spill (1990)	After the Spill (1991)	P Value
Black-crowned Night-Heron	0.64 ± 0.43	0.57 ± 0.37	0.43 ± 0.39	0.03
Snowy Egret	0.47 ± 0.43	0.22 ± 0.28	0.09 ± 0.21	0.03
Cattle Egret	0.51 ± 0.42	0.47 ± 0.42	0.32 ± 0.36	Not significant
Glossy Ibis	0.48 ± 0.43	0.32 ± 0.36	0.13 ± 0.26	0.06

Source: Parsons, K.C., In: *Before and After an Oil Spill: The Arthur Kill*, ed. J. Burger, Rutgers University Press, New Brunswick, 215–236, 1994.

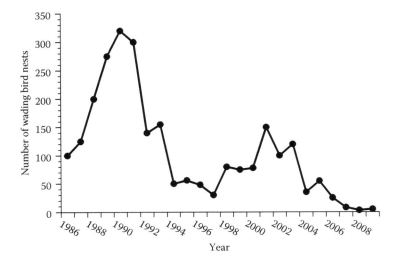

Figure 6.7 Change in the number of wading bird nests on Huckleberry Island (see Figure 6.5 for colony photo). (After Craig, E., New York City Audubon's Harbor Herons Project: 2013 Nesting Survey Report, *New York City Audubon*, New York, 2013.)

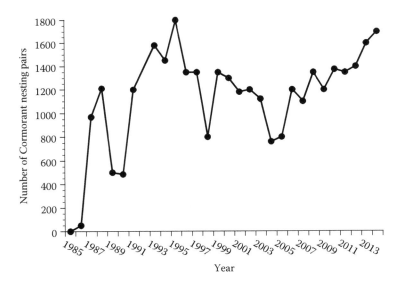

Figure 6.8 Changes in Cormorant populations in the New York–New Jersey Harbor Estuary. (S. Elbin, personal communication.)

changes in nesting populations of colonial waterbirds in Jamaica Bay Wildlife Refuge (JBWF), part of the New York–New Jersey Harbor Estuary, but not part of the Harbor Herons Project. During this period (late 1990s), some species increased (Laughing Gull, Great Black-backed Gull, Forster's Tern), some decreased (Herring Gull, Snowy Egret, Common Tern), and some disappeared (Cattle Egret). Herring Gulls declined from about 4000 pairs in the mid-1970s to about 2500 pairs by the late 1990s (Figures 6.9 and 6.10). The declines were partly a result of decreases in open-faced garbage dumps. Similarly, the populations of Common Terns declined, whereas Least Terns numbers

(a)   (b)

Figure 6.9 Herring Gull colonies in New York–New Jersey Harbor Estuary are sometimes on sand under trees (a) or among bushes on salt marsh islands (b).

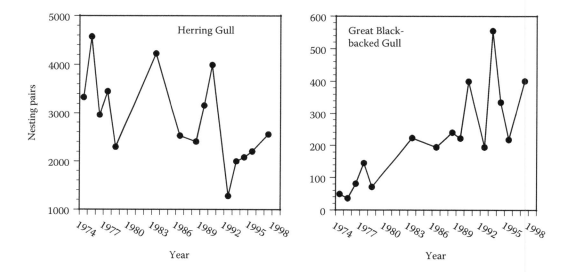

Figure 6.10 Trends in population levels of Herring and Great Black-backed Gulls in Jamaica Bay. (After Brown, K.M. et al., *Northeast. Nat.*, 8, 275–292, 2001b.)

varied (Figure 6.11). Populations of wading birds were variable, but in general trends show population declines, even in Great Egret (Figure 6.12).

Laughing Gulls colonized JoCo Marsh on the JBWF in 1979, after an absence from Long Island of 100 years (Post and Riepe 1980). Over 3 years, the colony grew exponentially from 2 to 20 to 200 pairs, eventually increasing to a count of 7629 pairs by 1990, spread over 16 ha of marsh at the end of JFK International Airport, creating a potential hazard to aircraft (Brown et al. 2001a,b). At this point, the Port Authority began a gull shooting program in 1991 to reduce the potential for gull–aircraft collisions. The population declined more than 50% as a result (Figure 6.13; Dolbeer et al. 1993). Many of the adult Laughing Gulls that were killed had been banded in New Jersey as hatchlings (Burger 1985b; Dolbeer et al. 1993; Gochfeld et al. 1996), and subsequently there was a steep decline in Laughing Gulls in Barnegat Bay. Laughing Gulls often disperse northward after the breeding season (Figure 6.14), and many obviously ended up at JFK Airport where they were shot.

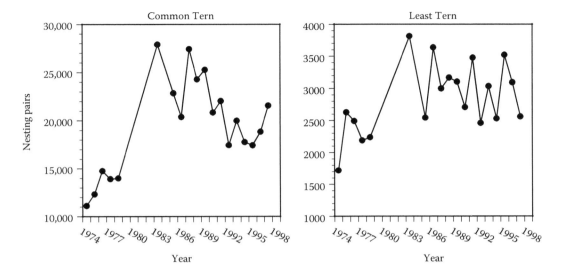

Figure 6.11   Trends in population levels for Common Terns and Least Terns at Jamaica Bay. (After Brown, K.M. et al., *Northeast. Nat.*, 8, 275–292, 2001b.)

Laughing Gulls may continue to increase on Long Island generally, as Washburn et al. (2012) conducted an evaluation of suitable salt marshes for the species and identified 66 suitable salt marsh islands without evidence of nesting. As salt marshes in New Jersey become unsuitable, Laughing Gulls may move to Long Island.

The first Forster's Tern nest on Long Island was discovered in Hewlett Bay in 1981 (Zarudsky 1981). The population grew slowly during the 1980s, and there has been a stable population of Forster's Terns on JoCo marsh for many years, with 100–200 pairs nesting each year since the early 1990s (D. Riepe, personal communication). The terns breed in the marshes, and forage over the small ponds and creeks in the marsh, and do not pose a hazard on the runways.

It is also important to place the population status and trends information for colonial water-birds within the context of Long Island generally. Information on population levels for Long Island was compiled by Brown et al. (2001b) through 1996 (see Table 6.2). Additional information was provided by Ian Nisbet. During the 1990s, the following trends were noted: (1) Great Egrets and Black-crowned Night-Herons increased. (2) Snowy Egrets, Glossy Ibis, and Herring and Great Black-backed Gulls decreased from peaks in the mid-1980s. (3) Laughing Gulls only nested at Jamaica Bay; they increased exponentially in the 1980s, then extensive culling reduced both Long Island and New Jersey population by about 50% in the 1990s. (4) Common, Forster's, Roseate, and Least Terns increased and then stabilized over the past decade. (5) Black Skimmers slowly increased. These trends differ somewhat from those found only at Jamaica Bay (Brown et al. 2001b), illustrating the importance of examining colony dynamics over wider geographical areas (see Table 6.2).

DELAWARE BAY

Delaware Bay is open to the sea, making it a good shipping port. However, unlike the bays discussed earlier, it has very few islands and relatively few salt marsh islands, hence it has few nesting colonies of birds. Pea Patch Island was once the largest heronry north of Florida (Parsons et al. 1998). There are a couple of other, low-lying sandy islands, but they flood during high tide.

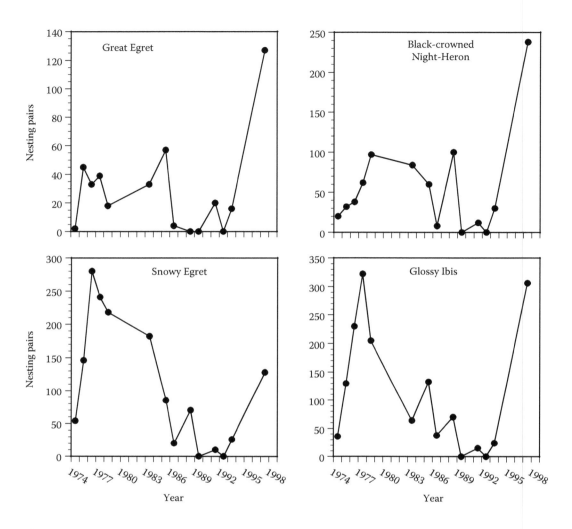

Figure 6.12   Trends in population levels of Great Egrets, Snowy Egrets, Black-crowned Night-Herons, and Glossy Ibis in Jamaica Bay. (After Brown, K.M. et al., *Northeast. Nat.*, 8, 275–292, 2001b.)

The State of Delaware has conducted two statewide surveys in the periods 1983–1987 and 2008–2012 (BBAE 2015a,b). The data are mainly recorded as the number of colonies where the species occurred. The reports noted that (1) the number of colonies remained the same for Great Egret and Snowy Egret; (2) the number of Common Tern colonies decreased from 4 to 1; (3) the number of colonies decreased to zero for Little Blue Heron, Cattle Egret, Herring Gull and Great Black-backed Gulls, Forster's Tern, Least Tern, and Black Skimmer; (4) the number of colonies increased for Tricolored Heron (1 to 2), Black-crowned Night-Heron, Glossy Ibis (1 to 2), and Laughing Gull (1 to 4). Thus, overall, there were fewer nesting colonies for most species. The trend in the number of breeding pairs is not available, except for those species that no longer breed in the Estuary.

Heronries

The most notable nesting area in Delaware Bay is the 125-ha Pea Patch Island, home to Fort Delaware, in the mid-river off Delaware City. With 12,000 pairs of herons and egrets in 1993, it

Figure 6.13 Population trends in Laughing Gulls breeding in JoCo marsh at the end of JFK International Airport. (After Brown, K.M. et al., *Environ. Manag.*, 28, 207–224, 2001a; *Northeast. Nat.*, 8, 275–292, 2001b.)

was considered the largest mixed-species heronry in the eastern United States (Parsons et al. 1998). Although the current population is estimated at 1000 to 2000 breeding pairs (C. Bennett, personal communication), it is still an important nesting area for Great Blue Herons (*Ardea herodias*), egrets, and ibises (Parsons et al. 1998; Delaware Birding Trail 2015). Developmental abnormalities and low survival of birds in the 1990s, along with declining populations, raised concerns about the role of contaminants in these declines. However, Rattner et al. (2000) reported no differences in contaminant levels or biomarkers between Pea Patch Black-crowned Night-Herons and those from a seacoast colony. They concluded that contaminant levels were below those likely to impair reproductive success.

Population levels are available for Pea Patch Island only for the period 1993–1997 (Table 6.3). Thereafter, the colony was reduced to much lower numbers, but in the early 1990s it was a thriving colony (Figure 6.15).

The Heislerville impoundments in southern New Jersey are bordered by an extensive salt marsh, mudflats at low tide, a narrow rim of beach, and upland forest. The impoundments were originally salt hay farms that were diked. When the farms were abandoned, the dikes gradually broke, letting in water. There remains a small island with tall (now dead) trees occupied by a mixed-species heronry. This heronry has grown considerably over the years, and the following counts were made in 2012, 2014, and 2015 (Table 6.4). Cormorants were first documented nesting at Heislerville in 2010 and became the most numerous species, nesting in the tree tops. It is likely that this accelerated the death of trees and reduced vegetation cover. Within a few years, the Night-Herons declined and the Glossy Ibis disappeared.

A traditional and long-used heronry (at least 75 years) was located at Stone Harbor on the Atlantic coast, but was suddenly abandoned in 1991 for unknown causes (Jenkins and Gelvin-Innvaer 1995). The colony was discovered the following year near Avalon, about 10 km to the north. It is likely that some of these wayward herons colonized Heislerville (27 km from Stone Harbor). Gulls and shorebirds regularly move back and forth between Stone Harbor and the Delaware Bay shore.

Figure 6.14 Burger holding a wing-tagged Laughing Gull used to document dispersal from Barnegat Bay following the breeding season. This gull and many other gulls were ultimately shot at JFK International Airport. (From Brown, K.M. et al., *Northeast. Nat.*, 8, 275–292, 2001b.)

Shorebirds

Delaware Bay is most noted for its shorebirds—thousands stop over at Delaware Bay during spring migration (May and early June). Delaware Bay is also home to a growing population of Osprey and Bald Eagles, and many other marsh birds (e.g., Willets, Clapper Rails, Saltmarsh Sparrows). However, habitat is limited for terns, gulls, and Skimmers. Many of the marshes are relatively low in elevation, and without the strong ocean tides and storm winds, there are no extensive wracks that could serve as nesting substrate.

Instead, the special draw of Delaware Bay is the migration stopover of Red Knot and other shorebirds in May, foraging on Horseshoe Crab eggs. The Red Knot was added to the United States Fish and Wildlife Service threatened list in 2015. Red Knots are clearly not colonial waterbirds, but they are iconic and indicative of the plight of birds in Delaware Bay. Even as Delaware Bay's Red Knots have received considerable worldwide attention because of their dependence on the eggs of Horseshoe Crabs and their precipitous population declines (Figure 6.16), other species have also declined, including Semipalmated Sandpipers (*Calidris pusilla*) and Ruddy Turnstones (*Arenaria interpres*) (Mizrahi et al. 2012). The Semipalmated Sandpiper is the most numerous of the shorebirds using the Delaware Bay shore (Figure 6.17).

Table 6.2 Trends in Colonial Waterbird Populations in the Northeast (from Previous Sections of This Chapter and Chapter 5)

	Massachusetts Coast[a]	Long Island	Jamaica Bay	New York–New Jersey Harbor Estuary	Barnegat Bay	Delaware Bay[b]	Chesapeake Bay
Sample period	1985–present	1974–1998	1974–1998	1982–present	1976–2015	1983–1987 to 2008–2012	1977–2003
Great Egret	NA	Increased	Mainly Stable	Increased to peak 2011	Increased	Stable[c]	Increased
Snowy Egret	Decreased	Very variable, slight decrease	Decreased	Increased then decreased	Decreased	Decreased[c]	Decreased
Black-crowned Night-Heron	Decreased	Increased	Stable	Peak 1993 then stable	Stable	Stable[c]	Decreased
Glossy Ibis	NA	Peak 1985, decreased	Decreased	Increased to peak 2005 decreased	Decreased	Decreased[c]	Increased
Great Black-backed Gull	Decreased	Peak in 1988 decreasing slowly	Increased	NA	Increased	Decreased	Increased
Herring Gull	Decreased	Peak in 1985, Decreased rapidly	Decreased	NA	Increased, then decreased	Decreased	Decreased
Laughing Gull	Increased	Increase[d] to mid-1990s then culled	Increased	None present	Decreased	None	Decreased
Black Skimmer	Marginal in Massachusetts	Increased	None present	None present	Decreased	NA	Decreased

(Continued)

Table 6.2 (Continued) Trends in Colonial Waterbird Populations in the Northeast (from Previous Sections of This Chapter)

	Massachusetts Coast[a]	Long Island	Jamaica Bay	New York–New Jersey Harbor Estuary	Barnegat Bay	Delaware Bay[b]	Chesapeake Bay
Common Tern	Peak 2004 Increased, now stable	Increased at GGI	Decreased	None present	Decreased	NA	Decreased
Forster's Tern	NA	Increased after 1981[e] then decreased	Increased	None present	Increased, then decreased	Decreased	Stable
Roseate Tern	Peak 2000 Stable	Increased on GGI Decreased Falkner	None	None present	Not present	Not present	Not present
Least Tern	Increased	Increased to peak late 1980s	Variable	None present	Decreased	Decreased	Increased, variable
Sources	Melvin 2010; Mostello 2014	Brown et al. 2001b	Brown et al. 2001b	Craig 2013 Winston 2014	Burger et al. 2001b; Burger and Gochfeld data	BBAE 2015a,b; Love and Carter 2001	Williams et al. 2007; Brinker et al. 2007; Prosser 2014

Note: Data are not available for all species; shown are whether in that area populations increased, decreased, or were stable. Not all species were examined in all locations. GGI, Great Gull Island; NA, not available.

[a] Mostello (2014) only reports on four species, but Melvin (2010) includes others. Includes Boston Harbor, Cape Cod and Buzzards Bay.
[b] These data from the Breeding Bird Atlas are for the State of Delaware. They report the number of colonies (not number of individuals). Thus, we could only note those species where there was a colony in 1983–1987, but none in 2008–2012.
[c] Data only 1993–2001 (Love and Carter 2001).
[d] First Long Island nesting, 1979.
[e] First Long Island nesting, 1981.

Table 6.3 Population Numbers (Estimated Breeding Pairs) of Colonial Waterbirds Nesting on Pea Patch Island (Delaware Bay) from 1993 to 1997

Species	1993	1994	1995	1996	1997
Great Blue Heron	389	395	311	323	394
Great Egret	603	688	781	721	745
Snowy Egret	2456	1583	1079	887	800
Little Blue Heron	1327	655	379	255	412
Tricolored Heron	30	15	2	1	3
Cattle Egret	4642	3887	1557	2299	1528
Black-crowned Night-Heron	519	649	511	585	525
Yellow-crowned Night-Heron (*Nyctanassa violacea*)	43	76	30	41	19
Glossy Ibis	2100	547	900	2000	1668
Unidentified nests	82	17	12	14	25

Sources: Parsons, K.C., *Colon. Waterbirds*, 18, 69–78, 1995; Parsons, K.C. et al., Wading birds and cholinesterase-inhibiting insecticides: An examination of exposure and effects in free-living populations. Interim Report to Delaware Department of Natural Resources & Environmental Control, 1998.

Figure 6.15 Heronry at Pea Patch Island when thousands of egrets nested in the trees and bushes. Note the Cattle Egret in the lower right. (Photo courtesy of Katherine Parsons.)

Red Knot populations illustrate the overall plight of the shorebirds (Niles et al. 2012; Dey et al. 2015). Delaware Bay is the major stopover area for Red Knots in the spring (Niles et al. 2008). Although some Knots also stop on Virginia beaches, the numbers using these beaches are small and have declined over the past three decades (Watts and Truitt 2014). During the spring stopover at Delaware Bay, Red Knots feed almost exclusively on Horseshoe Crab eggs (Tsipoura and Burger 1999; Karpanty et al. 2006). Some of the declines are attributable to decreases in Horseshoe Crab egg availability and habitat degradation (Niles et al. 2008, 2012), and the relative role of

Table 6.4　Counts of Nests at Heislerville Heronry, Close to Delaware Bay

	Number of Nests May 19, 2012	Number of Nests May 15, 2014	Number of Nests May 23, 2014	Number of Nests May 17, 2015
Double-crested Cormorant	70	180	250	190
Great Egret	30	40	50	25
Snowy Egret	450	10	30	26
Black-crowned Night-Heron	100 +	19	25	51
Glossy Ibis	25	0	1	0

Source: Gochfeld, M., and Burger, J., Unpublished data.
Note: For photo of heronry, see Figures 1.2b and 13.1b.

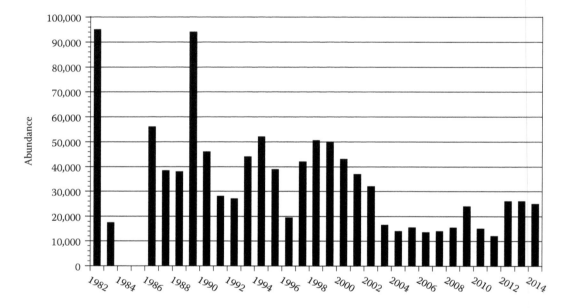

Figure 6.16　Peak counts of Red Knots on Delaware Bay. (After Dey, A.D. et al., Draft update to Status of the Red Knot [*Calidris canutus rufa*] in the Western Hemisphere, New Jersey Environmental Protections, April 2014, Unpublished Report, 2015.)

contaminants will be discussed later. Wardening has greatly reduced direct human disturbance on many Delaware Bay beaches during May when the migrant shorebirds are present. Foraging conditions on Delaware Bay, as measured by Horseshoe Crab egg densities, are statistically related to the proportion of Red Knots that achieve the 180-g body weight necessary for successful migration to the Arctic and subsequent breeding. An extensive program, involving scientists from several countries, is conducted each spring to capture, mark, weigh, measure, and color mark Red Knots during each year (Figures 6.18 and 6.19). The color mark is a small plastic tag with a three-character code that can be read with binoculars. The color of the tag indicates where it was captured.

New Jersey has instituted a moratorium on the take of Horseshoe Crabs for bait, which has the potential to increase the populations of Horseshoe Crabs, which in turn should increase the number of egg masses on the spawning beaches. The moratorium was instituted in 2008, and the reported commercial Horseshoe Crab harvest in the mid-Atlantic states has declined (Figure 6.20).

Figure 6.17   Semipalmated Sandpipers are the most numerous spring migrant at the Delaware Bay stopover.

Although there have been surveys of spawning female and male Horseshoe Crabs, and of egg densities, there have been no significant trends in Horseshoe Crab populations since 2004 (Zimmerman et al. 2014; Dey et al. 2015). It is difficult to conduct these surveys because of the timing and spatial extent of Horseshoe Crab spawning. However, there has been a trend toward an increase in the proportion of Red Knots that meet the goal of 180 g body weight before departure at the end of May (Figure 6.21; Dey et al. 2015).

These data suggest that (1) the abundance of Red Knots has remained low, but stable over the last decade, although far below their peak numbers in the early 1980s; (2) foraging difficulties are related to Horseshoe Crab egg density; (3) the proportion of Knots reaching the weight criterion of 180 g decreased, but is now showing a slight increase; and (4) long-term measures of Horseshoe Crab spawning and egg densities have not shown a significant improvement (Dey et al. 2015). This suggests continued concern for Red Knots, as well as for several other species of shorebirds that have shown similar declines.

Since the spectacle of thousands of Red Knots and shorebirds migrating through in the spring is largely limited to Delaware Bay, we present comparative information for other New Jersey bays to demonstrate the importance of Delaware Bay. The data were collected in the late 1970s and early 1980s, before the large decline in spring migrant shorebirds along the Atlantic coast (Table 6.5). As is clear, Delaware Bay clearly played a much more significant role than the other bays (Niles et al. 2008). However, in the 1980s, late summer and fall migrant shorebirds also concentrated at Jamaica Bay (Burger and Gochfeld 1983b) and at Cape Cod (Burger et al. 2012b).

(a)

(b)   (c)

Figure 6.18 The shorebird project at Delaware Bay requires intense effort of many people to complete the research and conservation work in the 3-week period that the birds are present. Here Joanna Burger and Larry Niles are banding a Red Knot with green flags (a), and Humphrey Sitters (b) and Mandy Dey (c) are measuring bill lengths and processing the birds.

In the 1980s, 8000–10,000 Red Knots arrived annually on the Massachusetts shorelines on their southward migration, but this has dropped to a few thousand in recent years (Fraser 2015). Thus, during spring migration, Red Knots concentrated mainly at Delaware Bay, whereas in fall migration they once concentrated in the brackish ponds at Jamaica Bay and at Cape Cod. The massive spring migration stopover at Delaware Bay is only about 2–3 weeks in duration, whereas the southbound migrants stay about a month or more at Cape Cod (Niles et al. 2008, 2010; Burger et al. 2012b).

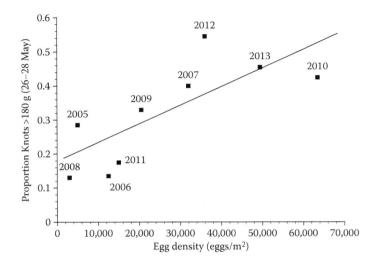

Figure 6.19   Relationship of Horseshoe Crab egg density and proportion of Red Knots reaching a body weight of 180 g (the threshold for successful travel to Arctic breeding grounds). (Data courtesy of Amanda Dey, NJDEP Endangered and Nongame Species Program.)

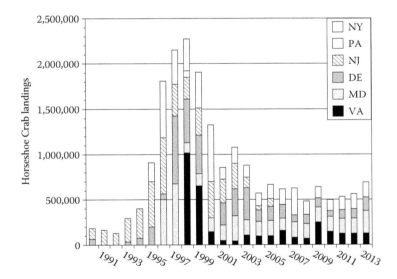

Figure 6.20   Reported commercial Horseshoe Crab harvest for the different states. The New Jersey ban on harvesting became effective in 2006. (Data courtesy of Amanda Dey, NJDEP Endangered and Nongame Species Program.)

Since the early 1980s, many species of shorebirds have declined, including the Red Knots. Considering the counts in the late 1970s and early 1980s is instructive, although it is difficult to compare high counts among bays. The data in Table 6.5 were from aerial counts by helicopter or plane (except Jamaica Bay) during the same general period, using the same protocols, and thus are comparable. Aerial counts have continued on Delaware Bay, but have not been routinely conducted on other bays since then (Figure 6.22). The first column shows a far greater number of birds recorded on the beaches compared with inland ponds and mudflats.

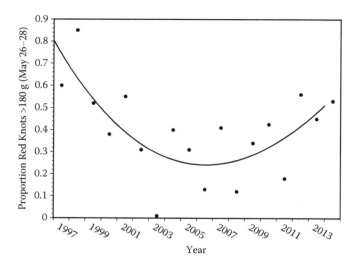

Figure 6.21 Proportion of Red Knots reaching the criterion of 180 g near their usual departure date (May 26–28). The line shows a significant quadratic trend over 1997–2013 (with a 95% confidence interval; Dey et al. 2015). It is significant even without the low value from 2003. Increases after 2006 correspond to the New Jersey ban on Horseshoe Crab take. (Data courtesy of Amanda Dey, NJDEP Endangered and Nongame Species Program.)

Table 6.5　Highest Counts of Shorebirds at Several Bays

Species	NJ Delaware Bayshore (Inland Counts)	Barnegat Bay	Raritan Bay	Jamaica Bay
Semipalmated Sandpiper	56,250 (1300)	1450	2650	16,200
Ruddy Turnstone	46,260 (20)	2	2463	0
Red Knot	29,980 (0)	255	715	3200
Sanderling (*Calidris alba*)	17,190 (10)	375	1144	6200
Approximate total	150,000 (1300)	2000	7000	25,500
Date of high count for knots	May 27, 1982	May 29, 1979	June 2, 1982	August 1979
Reference	Burger 1986 and Unpublished data	Burger, Unpublished data	Burger 1986 and Unpublished data	Burger and Gochfeld 1983b

Source: Data from aerial counts for Delaware and Raritan Bay, from helicopter counts in Barnegat Bay, and from ground surveys in Jamaica Bay, 1979–1982. (Burger, Unpublished data.)

CHESAPEAKE BAY

There are literally hundreds of refereed papers and reports on the physical features, ecology, condition, and health of Chesapeake Bay (Schubel 1986; Leatherman et al. 1995; Cronin 2005; Viverette et al. 2007), but we did not find consistent annual data on nesting colonial waterbirds (but see Watts and Byrd 2006; Erwin et al. 2007a and below). Much of the work in the bay has involved studies of their declining native oysters (Rothschild et al. 1994). Chesapeake Bay is one of the few places in the world where native oysters are still harvested from the wild (CBF 2010) (in New Jersey they are "farmed"). In the past three decades, Maryland and Virginia have suffered more than

Figure 6.22 Laughing Gulls feed on Horseshoe Crab eggs, on a receding tide as the crabs finish spawning and move out to the bay (Reed's Beach, New Jersey).

$4 billion in cumulative annual losses because of the decline in businesses related to oyster harvesting. Harvests are less than 1% of historic levels (CBF 2010).

In 2007, Erwin et al. (2007a,b) edited a volume of the journal *Waterbirds* on Chesapeake Bay and vicinity. There are accounts of diving ducks, bay-breeding waterfowl, Bald Eagles, waterbirds, marsh birds, Piping Plover, and seabirds. Many of the nesting colonies look similar to those in Barnegat Bay—salt marsh islands and sandy islands (Figure 6.23). In that volume, Williams and colleagues (2007) reported on the status of nesting wading birds within the Chesapeake Bay and Atlantic Barrier Island–Lagoon System. Tables 6.6 and 6.7 show the number of breeding pairs of waders common to the bay. Over the period sampled, Great Egrets and Glossy Ibis increased, and the other species decreased. In the same year, Brinker et al. (2007) published a paper that included gulls and terns as well as waders. Significant declines were found in the period 1985–2003 for Black Skimmers, Common Terns, Laughing Gulls, and Herring Gulls. As previously reported for Chesapeake Bay, Brown Pelican, Double-crested Cormorant, and Great Black-backed Gull increased.

A technical advisory workshop held in 2014 provided some trends data on the major species in Chesapeake Bay (Table 6.8). Data may vary slightly from the Williams et al. (2007) report, because of the inclusion of some barrier beach areas in the Williams study. As with all such studies, numbers can vary depending upon which colonies were surveyed, which were included in the study, what time in the breeding season they were surveyed, and the inevitable differences due to

Figure 6.23 Colony sites in Chesapeake Bay are often flat sandy islands or shell substrates. (Photo courtesy of Michael Erwin.)

Table 6.6 Trend Data for the Number of Breeding Pairs of Some Colonial Birds in the Maryland/Virginia Chesapeake Bay Region

Species	1977	1985/1986	1993	2003
Great Egret	1421	1481	3291	3601
Snowy Egret	5508	3069	4633	2336
Cattle Egret	1811	3468	3799	657
Black-crowned Night-Heron	2703	945	668	935
Glossy Ibis	673	1488	2415	2052
Great Black-backed Gull		50	75	675
Herring Gull		45,200	2000	2800
Laughing Gull		790	0	0
Common Tern		2100	1600	1300
Forster's Tern		1200	700	1000
Least Tern		370	320	620
Black Skimmer		150 (320 in 1986)	250	75

Sources: For wading birds: After Watts, B.D., Byrd, M.A., *Raven*, 77, 3–22, 2006; Williams, B. et al., *Waterbirds*, 30, 82–92, 2007. For gulls and terns: After Brinker, D.F. et al., *Waterbirds*, 30, 93–104, 2007.
Note: Numbers are estimated from graphs.

personnel conducting the studies. In all cases, the counts are indices of abundance because of these methodological issues.

As with the other bays and estuaries examined, the number of colonial birds in Chesapeake Bay are changing. Some species are decreasing (terns, Great Blue Heron), whereas others are increasing (Bald Eagle, Cormorant, Brown Pelican). Population sizes of some species have varied (e.g., Least Tern, Forster's Tern). Black Skimmers have declined (Brinker et al. 2007). Common Terns and Royal Terns are coastal species, feeding on marine prey fish. Declines in the latter species may truly reflect declines.

Table 6.7   Trend Data for Breeding Pairs in Some Colonial Waterbirds Nesting in and around Chesapeake Bay

Species	1977	1993	2003	2013
Double-crested Cormorant	0	500	2500	4800
Brown Pelican	0	300	1600	2000–2500
Osprey	1200 (in 1973)	3000 (in 1985) 3500 (in 1995)		More than 6000
Bald Eagle[a]		3000	700	
Great Blue Heron[a]	2500	12,000	15,000	
Common Tern[b]	5100	8130	3236	
Royal Tern (*Thalasseus maximus*)	4800	6586	3332	
Forster's Tern	1600	3692	3236	

Source: Erwin, R.M. et al., *J. Coastal Conserv.*, 15, 51–60, 2010.
Note: In some cases, numbers are extrapolated from their graphs.
[a] Bald Eagles have started displacing Great Blue Herons on their nesting colonies (Watts 2013).
[b] Continuous data for Poplar Island shows Common Terns declined from more than 1000 pairs in the 1990s, to more than 500 pairs in the early 2000s, and then declining to 205 pairs in 2014 (Prosser 2014).

DISCUSSION

It is always tempting to examine population levels in small regions where individual scientists can track the numbers of nesting birds consistently over time. Often, people follow one or two colonies, and may do so for many years (e.g., Hays on Great Gull, Nisbet on Bird Island, Spendelow on Faulkner Island). Because the aforementioned colonies include Common Tern and Roseate Tern, and are some of the major colonies of each, these three studies provide key and important information on these two species. Furthermore, these colonies are monitored and studied throughout the breeding season, rather than being monitored on only two or three visits. It is difficult to monitor many colonies in a large geographical region because of personnel and travel costs, and the difficulty of episodic weather or storm events in one or two locations. Reproductive differences in 1 year may reflect chance events rather than real trends or differences among regions (Burger et al. 1994b). It was possible to monitor the colonies in Barnegat Bay because the islands were small and could be visited in a single day or a 2-day period, and we had a fast boat and a faithful companion (Fred Lesser).

Birds of North America Accounts

One source of population trends data is the *Birds of North America* (BNA) accounts for each species, which are available online and are updated periodically (http://bna.birds.cornell.edu/bna/species). In Table 6.8, we provide several variables that can be used to understand the intrinsic potential for recovery following a perturbation, as well as population trends reported in the accounts.

Data on the age of first breeding, clutch size, and longevity provide an indication of the potential for a population to recover following stressors, or other perturbations. These reproductive parameters have presumably evolved over millions of years to optimal levels for the species survival, even in the face of fluctuating environments. What does not kill a species, makes it strong. However, anthropogenic factors produce rapid and extreme fluctuations that potentially cause local population explosions and extinctions. Under these conditions, the egrets and herons have a greater potential for rapid recovery because they have about the same longevity as gulls and terns, but lay larger clutches. This allows them to take advantage of good conditions of nesting habitat, foraging habitat, and prey abundance to raise more young. Furthermore, the herons and egrets have a wider foraging niche than the terns and skimmers. The terns and skimmers forage by only a few methods (plunge-diving

Table 6.8   Demographic and Population Information for the Primary Species Discussed in the Book

Species	Age First Breed (Years)	Clutch Size Mean (Range)	Longevity	Population Trends Along Atlantic Coast	Reference
Great Egret	3	3 (1–6)	22 years, 10 months	Steady increase; NA population now about 180,000 pairs	McCrimmon et al. 2011
Snowy Egret	1–2	4 (3–5)	22 years, 10 months	Declines since 1980s	Parsons and Master 2000
Black-crowned Night-Heron	1–2	4 (3–5)	21 years	General decline; 12,944 breeding pairs along Atlantic coast	Hothem et al. 2010
Great Black-backed Gull	4–5	3 (1–4)	>19 years	Increased, but then decreased	Good 1998 Anderson and Garthe 2013
Herring Gull	4–5	3 (1–4)	15–20 (36 years)	Now declining; about 106,000 pairs along Atlantic coast	Pierotti and Good 1994
Laughing Gull	2–3	3 (2–4)	19 years	Variable, now decreasing along Atlantic coast	Burger 1996b, 2015
Common Tern	2–3	2–3	26 years	Increased from early 1970s to early 1990s, then steady; NA population about 150,000 pairs	Nisbet 2002
Forster's Tern	2	3	12 years	No long-term trend; about 5766 pairs along Atlantic coast	McNicholl et al. 2001
Roseate Tern	2	2 (1–2)	25 years, 6 months	Increased from 1988 to 2000, then decreased to 2007, then stable; Northeast total about 3000 pairs	Nisbet et al. 2014
Black Skimmer	3–4	4 (1–5)	20 years	Decreasing[a]; about 4200 pairs from Chesapeake Bay north (about 30,000 from Carolinas to Texas)	Gochfeld and Burger 1994
Red Knot	2	4 (3–4)	19 years	Decreasing[b]; North American migrant population about 15,000 to 26,000	Baker et al. 2013

Source: These data come from the *Birds of North America* (BNA) accounts.

[a] We are revising the skimmer account, and current information shows it declining in the Northeast.
[b] Red Knot was recently (2015) placed on the U.S. Federal Endangered and Threatened Species list.

and skimming, respectively), on one type of prey (small fish). Herons and egrets, on the other hand, forage on a wider range of foods, including invertebrates, frogs, snakes, and fish (Kushlan 1978, 2000a,b; Hancock and Kushlan 1984). In this sense, they are more resilient in the face of changing environments.

The BNA accounts are useful, but one drawback is that not all have been updated within the past 15 years, and thus do not reflect recent changes. The following conclusions can be drawn from the BNA accounts: (1) Great Egrets are increasing; (2) Snowy Egrets, Black-crowned Night-Herons, Herring Gulls, Black Skimmers, and Red Knots are decreasing; and (3) the other species are variable. Other than the Great Egret, no species seems to be increasing nationwide. The accounts clearly note, however, that the trends are variable among years and locations, making a continentwide—or even a coastwide—analysis of population size difficult.

Other Analyses

Other analyses of population trends also provide information. In a symposium on the rise and fall of gull populations, some authors noted that Herring and Great Black-backed Gulls probably peaked in the mid-1970s to the early 1990s in the Northeastern United States (Anderson and Garthe 2013; Anderson et al. 2013), followed a similar pattern in Canada, but may have stabilized (Cotter et al. 2013), and have declined in the North and Baltic Sea coasts (Garthe 2013). Ring-billed Gull (*Larus delawarensis*) populations also rose and fell during the same time interval in eastern North America (Giroux 2013). The causes are similar: abundant food (offal, landfills) allowed expansion southward, but closure of landfills and reduced fisheries by-catch removed superfavorable conditions for the gulls.

Population trends data are also available for the Canadian Great Lakes, based on four sampling periods (1976–1980, 1989–1992, 1996–2002, and 2007–2009; Weseloh et al. 2011). Some species increased (Double-crested Cormorant, Great Egret), Black-crowned Night-Heron were stable, and some species declined (Great Black-backed Gull, Herring Gull, Common Tern). Common Terns have increased in some areas of Canada (the four Atlantic provinces), even though they declined around the Great Lakes (Morris et al. 2012). Common Terns have also declined in the German Wadden Sea (Szostek and Becker 2012), and in Italy where they nested on salt marshes (Scarton 2010). Forster's Terns have decreased in San Francisco Bay, California (1982–2003; Strong et al. 2004).

In a recent volume on marine birds in the eastern United States, Nisbet et al. (2013) provided short species accounts of several of the species considered in our book. Their evaluations were as follows:

Great Black-backed Gull: Their populations spread to Maine in the 1920s and continued southward, increasing into the 1990s, when they peaked. Thereafter, they decreased.

Herring Gull: The population increased in the United States in the twentieth century, and peaked in the 1980s; their populations have declined by 50% since then.

Laughing Gull: Variable patterns with increases and decreases owing to the culling program at JFK Airport.

Common Tern: The population has been variable; it increased in the 1930s then declined to the early 1970s, when it again increased in the 1980s and thereafter from Long Island north, but declined south of Long Island.

Forster's Tern: Reported numbers have been increasing rapidly, but increases may be an artifact of earlier undercounting (particularly in salt marshes).

Least Tern: Variable trends, with no recent data.

Roseate Tern: Their populations peaked in the 1930s, declined to a low in 1978–1979, increased in 1994–1995, and is now in decline.

Black Skimmer: The population is decreasing.

In their synoptic summary assessments, the authors report the large gulls increasing and then decreasing, Laughing Gulls increasing, Roseate Terns decreasing, and the other terns increasing (Nisbet et al. 2013, pp. 92, 93). One problem they note is the paucity of data for most species since the late 1990s, the period when we found the greatest changes in populations in Barnegat Bay. It is since the 1990s that there has been an increased rate of sea level rise in the Northeast region (Sallenger et al. 2012). Nisbet et al. (2014) noted that many of these species nest in salt marshes, which we suggest are the most vulnerable to sea level rise. Their data set, and the others described in this chapter, clearly point to the need for continuous monitoring to derive a clear picture of abundance and distribution of coastal and estuarine birds. The great declines noted by BNA accounts, Erwin et al. (2010) for Chesapeake Bay, Nisbet et al. (2013) for the Northeast, and our book clearly indicate changes since the 1990s in population numbers.

Role of Reproductive Success

In this chapter, we discuss population estimates. At any time and place, population numbers are influenced by adult survival, immigration/emigration, annual reproductive success, and recruitment. The count or census result is an estimate of the population influenced by timing, spatial extent, methodology, and experience of the counters. Few places conduct annual statewide counts of any of the waterbird species. Ideally, to be useful in understanding changes in populations levels among the bays, reproductive success data should be available among the different estuaries for the full range of years (that data on populations are available), using the same endpoints and methods. Reproductive success generally declines over the breeding season in many species. For example, in Common Terns this decline is usually attributable to decreased food availability, inexperience of late nesting birds, and sometimes to predation and kleptoparasites (Arnold et al. 2004).

In the absence of ideal data sets, we present the best available data, from a study conducted by Parsons et al. (2001). They collected data on hatching success and nestling survival to 15 days of age, from 1986 to 1998, which is remarkable (Table 6.9). They found significant differences among colonies. These illustrate that there are differences in reproductive success regionally and locally that clearly can affect population trends, immigration, and recruitment.

These data, averaged over several years, show differences in two measures of reproductive success for both species, especially in nestling survival from hatching to day 7. Even if there were no

Table 6.9   Hatching Success (Percent of Eggs Hatched) and Nestling Survival (Percentage Alive at 7 Days) for Two Species Nesting in Four of the Northeast Estuaries

Location	Black-Crowned Night-Heron (% Hatching Success)	Black-Crowned Night-Heron (% Nestling Survival)	Snowy Egret (% Hatching Success)	Snowy Egret (% Nestling Survival)
Delaware Bay (Pea Patch Island; 1993–1998)	84.8 ± 21	38.2 ± 37	79.3 ± 23	21.6 ± 31
New York–New Jersey Harbor (Prall's, Shooters, Isle of Meadows; 1986–1998)	85.2 ± 21	78.7 ± 29	85.0 ± 21	60.4 ± 35
Cape Cod (Sampson's Island; 1990–1994)	82.2 ± 22	57.8 ± 30	84.6 ± 20	70.8 ± 32
Boston Harbor (Sarah Island; 1993–1994)	87.0 ± 17	87.3 ± 21	88.1 ± 18	76.3 ± 27
F (P)	6.04 ($P < 0.001$)	49.7 ($P < 0.001$)	2.11 (not significant)	64.9 ($P < 0.001$)

Sources: Parsons, K.C., *Colon. Waterbirds*, 18, 69–78, 1995; Parsons, K.C. et al., *Waterbirds*, 24, 323–330, 2001.
Note: Data are presented as means ± standard deviation.

(a)   (b)

Figure 6.24 Black-crowned Night-Heron chicks (a) and Snowy Egret chicks (b) made excellent subjects for Katherine Parsons et al.'s (2001) study of hatching rates in Ardeids (see Table 6.9).

differences in hatching rate (as they found for Snowy Egret), the significant differences in nestling survival results in differences in reproductive success. Reproductive success was lowest for Pea Patch Island in Delaware Bay, which appeared to be due to predation (Parsons 1995). Low cholinesterase levels suggested a possible contribution of pesticides to abnormal nest attentiveness. During 5 years of study, wading birds nests on Pea Patch Island decreased from 8318 to 3139 (Parsons et al. 2001). Some of the differences among studies may have been due to the duration of studies (only 2 years for Boston Harbor). For example, with data for only a few years, episodic weather events or differences in predation rates could account for some of the differences between the species (Figure 6.24).

Comparisons among Species in the Northeast Bays and Estuaries

Many states have regular monitoring programs for colonial birds (e.g., Massachusetts, New Jersey), but these may not occur every year (e.g., New Jersey counts every 5 years, at best). Colonies may be censused only once or twice a year. Methods may differ from year to year depending on personnel and resources. Not all colonies may be monitored each year, and only one or two species may be targeted (e.g., Least Terns). Species differ considerably in their countability—egrets are easy to see whereas Night-Herons and Ibis will be undercounted unless they are flushed. Even with these limitations, any data (well documented) will be useful for conservation and management purposes, and for state planning. Furthermore, involving volunteers in a Citizen Science framework is done in several states. Despite the skepticism expressed in some quarters, we view Citizen Science as providing eager, trainable personnel that serves an educational role, while building capacity and constituency. People protect the things they know and appreciate.

Recognizing the limitations and unevenness of the data, Table 6.2 compares trends in waterbird populations for the Northeast. Some of the data are current, whereas others ended a decade ago. Still, they present a picture of regional trends.

Another evaluation, conducted by the Colonial Waterbird Working Group (CWWG 2007), reported that Black-crowned Night-Heron declined by 45% across the Northeastern United States since the 1970s, and Snowy Egrets declined by 20–30%, whereas Great Egret populations increased by approximately 100%. Taken altogether, the above sources indicate the following trends for the Northeast colonies: (1) Great Egrets are increasing; (2) Snowy Egrets, Herring Gull, and Black Skimmer are decreasing; (3) Black-crowned Night-Herons show a variable trend; (4) Common Terns, Forster's Tern and Laughing Gull are increasing in the north and decreasing in the south; and (5) Red Knots are decreasing, but may have stabilized. Knot populations, however, are so low that they have been placed on the U.S. threatened list.

Overall, most species are decreasing in part or all of their northeastern range—only Great Egrets and Double-crested Cormorants (not on our census list) are increasing everywhere. Different methods of examining the data, however, result in different conclusions. In a recent analysis, Washburn et al. (2016) concluded that Herring Gull populations fluctuated in the Long Island area, but were stable in New Jersey; however, they reported that Great Black-backed Gulls have increased overall. In contrast, we found Herring Gulls were declining in New Jersey. In the following discussion, we summarize the findings and suggest possible reasons for the trends. We do not focus on global change and sea level rise, which is the subject of Chapter 7.

Increasing Trends

The overall increase in Great Egrets, despite the declining or stable populations of the other wader species, suggests that something different is affecting Great Egret populations compared to those of the other species. In part, the success of the Great Egrets may result from their larger size; with sea level rise and a decrease in available mudflat or shallow water foraging habitat (see Chapter 7), Great Egrets can forage in a wider range of habitats than the smaller Snowy Egret (Powell 1987; Lantz et al. 2010, 2011). Being able to forage in a wider range of water depths provides access to both a wider selection and abundance of prey, and being larger than Snowy Egrets also allows Great Egrets to forage on a wider range of prey species. The answer may lie underwater with declines in predatory fish populations, allowing in increases in the medium-sized fish that Great Egrets eat, at the expense of the smaller prey fish that Snowy Egrets and terns require.

Decreasing Trends

Snowy Egrets, Herring Gulls, and Black Skimmers declined overall. The cause of the decline in Snowy Egrets is unclear, but may be attributable to reproductive disruptions due to pesticides and metals (Parsons and Master 2000), and loss of shallow foraging areas in the bays because of sea level rise or subsidence (see Chapter 7). It is also possible that Snowy Egrets are suffering from competition with Great Egret for nesting space (Burger 1981a), and that Double-crested Cormorants are altering the vegetation, making it no longer suitable for nesting egrets (Elbin and Craig 2007; Craig et al. 2007, 2012). The Cormorants degrade the habitat through the destruction of vegetation and the alteration of soil conditions. In some areas, particularly the New York–New Jersey Harbor Estuary, Cormorants have increased more than any other species, and they nest in established heronries (Elbin and Craig 2007). In many of these colonies, the primary species are Great Egrets, Snowy Egrets, and Cormorants. The Great Egrets are much larger and can win the competition for nest sites with Snowy Egrets (Burger 1978c, 1981a), but Snowy Egrets may not win against Cormorants.

Herring Gulls—and perhaps Laughing Gulls in the southern part of our study area—declined partly because of the closing of open landfills (Belant et al. 1993). Landfills provided abundant food for young of the year that otherwise have difficulty locating and capturing prey (Burger 1981c; Burger and Gochfeld 1981a). Pesticides and heavy metals (Breton et al. 2008) and dioxin (Ryckman et al. 2005) have been linked to lowered reproductive success in Herring Gulls. However, a more important cause of Herring Gull declines, we believe, is the more recent and rapid expansion of Great Black-backed Gulls. The latter are much bigger than Herring Gulls, and easily win any encounters for nest sites high enough to avoid flooding (Burger 1978a).

Black Skimmers have declined in many parts of their northeast range because of lack of stable habitats that do not flood, and are free from predators (Burger and Gochfeld 1990b; Gochfeld and Burger 1994). Although they once nested on the open sandy beaches of barrier islands, these habitats are largely unavailable because of sea level rise (see Chapter 7; Brinker et al. 2007), human use and disturbance, and the increase in predators (e.g., cats, dogs, Red Foxes, Raccoons; Kadlec 1971; Burger and Gochfeld 1991b). Unlike Common Terns, Black Skimmers do not normally nest on *Spartina*, but instead nest on wracks of vegetation thrown up on the marsh by high winter tides. The excessively strong winter tides and storms of the past few years have washed right over islands, leaving no wracks, especially in New Jersey.

Variable Trends

Black-crowned Night-Herons showed a quite variable trend, but were decreasing in most places. Various pesticides and metals have been linked to decreases in reproductive success (Davis 1993). It is also possible that oil spills have resulted in declines in some of their prey (e.g., Fiddler Crabs; Burger et al. 1994c) or affected them directly (Parsons 1994). In addition, competition with other species (the larger Great Egret, and aggressive Cormorants) may have reduced their options.

Common Terns, Forster's Terns, and Laughing Gulls showed an increasing trend in the northern part of our Northeastern United States study area, and a decreasing trend from New Jersey to Chesapeake Bay (Breton et al. 2014). Common Terns nesting in salt marshes in New York and New Jersey are increasingly subjected to tidal flooding, especially given the increase in severe storms and sea level rise (Burger and Lesser 1979; Sallenger et al. 2012; Palestis and Hines 2015). It is unclear why Forster's Terns have remained stable in some parts of their northeastern range. It may be that the less marine-foraging Forster's Tern do not rely on some marine forage fish that have declined. In contrast, Common Terns rely on Sand Lance (*Ammodytes* spp.), which have greatly declined (Robards et al. 1999), and the terns also rely on Bluefish schools forcing prey fish to the surface (Safina and Burger 1985, 1988c). Recreational fishermen report that Bluefish are arriving at the bays later than usual, and this mistiming may mean that they are not available to force prey fish to the surface when Common Terns have growing chicks—a critical window of need and susceptibility.

Some of the declines are attributable to predation (Nisbet 1975; Burger and Gochfeld 1991b; O'Connell and Beck 2003). Historically, rat predation heavily impacted terns, but their relative importance has declined. Today, humans deliberately and inadvertently feed wildlife and have become more tolerant of wildlife, to the point where it is difficult to "control" Red Foxes, Raccoon and feral cats, which are serious predators on eggs, chicks, and even adult birds. Humans themselves are important predators on the wintering grounds for both shorebirds and terns.

The geographical difference in trends is partly a result of loss of suitable habitat on salt marshes in the south (sea level rise; Brinker et al. 2007), coupled with more suitable sandy and rocky *terra firme* habitat farther north (Breton et al. 2014). The increase in Common Terns in the north is dramatic. Breton et al. (2014) reported a 7-fold population increase from 1983 to 2004 in Buzzards Bay (Massachusetts). From the mid-1980s to 2015, Common Terns in Barnegat Bay decreased from a high of 3000 nesting pairs to less than 500 pairs (Burger, Gochfeld data). Foraging may be better in northern waters (Goyert 2014), and young birds prospecting for colony sites (Dittmann et al. 2005) may shift from New Jersey north if prey and suitable habitat are less available in their natal colony. Decreasing trends in Common Tern populations have also been found in Venice, Italy (1989–2008; Scarton 2010), which was attributed to habitat problems due to sea level rise.

New Jersey had been one of the centers of Laughing Gull populations (Burger 1996b). Laughing Gulls declined in New Jersey largely because of the culling of gulls near JFK Airport (Burger 1985b; Dolbeer et al. 1993), and because of competition for nest sites with the much larger Herring Gulls (Burger 1979a). Laughing Gulls always lose because they arrive at the breeding colonies 2 months later than the larger gulls, and their small size ensures that they lose any aggressive encounters (Burger 1979a). Laughing Gulls are doing better farther north, where the larger gulls nest on sandy beaches (e.g., Captree, Long Island), leaving the marshes to Laughing Gulls.

SUMMARY AND CONCLUSIONS

Many states and organizations conduct surveys of colonial waterbirds, although these surveys are not always conducted every year, on every colony, for every species. States may monitor colonies every year (Massachusetts) or every 5 years (New Jersey) for only some species (e.g., Massachusetts, Chesapeake Bay) or for species that may provide a problem (e.g., Double-crested Cormorants). The data presented in this chapter indicate that very few colonial waterbird populations have remained stable even in a given state, much less the region.

Over the past 40 years, some species have generally increased (e.g., Great Egret, Double-crested Cormorants), whereas others have declined (e.g., Black Skimmer, Snowy Egret). Some have increased and then declined. Herring Gulls, once a northern nesting species, moved south into Massachusetts, New York, and New Jersey, and—with the closing of open-faced garbage dumps—declined. Some species have decreased in the southernmost bays and estuaries, and increased farther north (Common Terns, Least Terns, Forster's Terns). The causes of declines appear to be competition with other species, loss of nesting or foraging habitat, increased predation, and increased flooding because of the increasing frequency and severity of storms coupled with rising sea level.

The population dynamics of the colonial birds nesting in the Northeast bays and estuaries have been influenced over the past 70 years by three major factors: (1) human development and activities reducing or changing available habitat; (2) range expansion south of Herring Gull, Great Black-backed Gull, and Double-crested Cormorant; and (3) climate change and sea level rise that have rapidly degraded some habitat and rendered others more vulnerable to flooding. The rapid changes in population levels that have occurred over the 40 years of our studies indicate the importance of continued monitoring over a regional level, along with determinations of reproductive success and the causes of colony abandonment and colony failures. Conservation and management can only proceed when all of the necessary information can be integrated into a comprehensive, long-term plan. The natural resiliency of coastal colonial species (large clutches, long lives, colony and nest site adaptability) can only operate if sufficient habitat is available and free from disturbance by people and predators.

CHAPTER 7

Global Warming, Sea Level Rise, and Suitable Nesting and Foraging Habitat

INTRODUCTION

The Earth's climate has been fluctuating wildly over the nearly 4 billion years since the first "life" appeared. Time and again, species have escaped or adapted to these changes or faced extinction. Massive extinctions of most life on Earth have occurred in the past and could occur in the future—perhaps the near future (Kolbert 2014). Great gaps in the fossil record provide evidence of five great extinctions over a billion-year period. Elizabeth Kolbert (2014) suggests that *The Sixth Extinction*, because of climate change and mankind's global mismanagement, is imminent.

Fast forward to the Pleistocene beginning an estimated 2.58 million years ago. By that time, most modern families of vertebrates including early hominids, precursors of our species, were in existence. During the 2.5 million years of the Pleistocene, great ice sheets formed, advanced, and shrank multiple times, changing landscapes, coastlines, and biodiversity. Since the Pleistocene, human exploitation has exterminated numerous species, particularly those on islands (Pimm et al. 1995; Vitousek et al. 1997). Examples of victims include all nine species of New Zealand Moas (Dinornithiformes) exterminated in the mid-1400s by the Maori, the iconic Dodo (*Raphus cucullatus*) eliminated from Mauritius by European colonists in the mid-1600s, and biggest of all, the Elephant Birds (Aepyornithidae) eliminated from Madagascar before 1800. The pace of exploitation and habitat destruction has accelerated in the past 500 years and can only worsen as the human population explosion increases, but their combined impact is likely to be exceeded by the indirect effects of human alteration of many aspects of global ecology—none so pervasive as climate.

The massive volcanos and meteorite(s) associated with massive extinctions of most species on Earth did not kill by fire or physical impact, but by "climate change." Today, climate change is again proceeding rapidly (IPCC 2014), and is challenging the adaptive mechanisms of all species, including humans. Coastal nesting waterbirds are particularly vulnerable. Climate change has been shown to affect many biological systems. Increasing concentrations of carbon dioxide and other gases are expected to warm the Earth several degrees in the next century by a mechanism known as the greenhouse effect, and the excess carbon dioxide is predicted to increase the acidity of the ocean greatly impacting marine life (Titus 1990; IPCC 2007, 2014). These changes will have a profound impact on environments around the world, including in North America.

Could these changes have been foreseen a half-century ago? The 1960s was a boom period for environment and conservation awareness. Rachel Carson's *Silent Spring* (1962) galvanized attention around the unintended consequences of pesticides. Major leaps forward in environmental regulation included the Clean Air Act (CAA) (1963), the birth of the Environmental Protection Agency (EPA) (1970), the Clean Water Act (CWA) (1972), and the banning of some of the most persistent pesticides. People talked seriously about energy, conservation, and recycling. Ecological modeling and predictions were an exciting frontier. In 1965, Frank Fraser Darling and John P. Milton convened a conference of leading ecologists tasked with predicting the Future Environments of North America. The impressive proceedings volume (Darling and Milton 1966) addressed the unintended consequence of population growth and technology. In those days, the only climate modification involved seeding rain clouds. The theme of the volume was human manipulation, exploitation, and devastation, and what could be done to intervene (Darling and Milton 1966). That volume played an important role in focusing the environmental movement for more than a decade, and was considered "one of the most important documents of our current era" (Kartman 1968). But it shows us that 50 years ago, although sediment was accreting and salt marshes were spreading, climate change and sea level rise were not on their radar screen. Yet, sea level had been rising steadily for the preceding 50 years. Of course, there were no web pages to make such data searches instantaneous, but apparently none of the conference participants happened to look.

Sea level rise has had a major impact on the ecosystems we study. But coastal wetlands are vulnerable to macroclimatic changes involving temperature and precipitation. These can impact the vegetation structure and ranges of invertebrates and vertebrates independent of sea level rise (Osland et al. 2015). Although there is a range of proposed mechanisms for global change and sea level rise, including the expansion of seawater because of warming, the fact of global warming and sea level rise remains despite political efforts to paint this as controversial (Broccoli et al. 1998). Changes and responses to climate change are occurring at all ecosystem levels, from zooplankton (Roemmich and McGowan 1995) and forests (Robichaud and Begin 1997), to top predators (Hazen et al. 2013). Evidence comes from far-flung places, from Arctic regions (Gaston et al. 2005) and Alaska (Hinzman et al. 2005), to birds nesting in the Antarctic (Croxall et al. 2002; Barbraud and Weimerskirch 2006). Sea level rise is occurring everywhere, but the rate of rise is not uniform worldwide. These geographic variations are caused by dynamic processes arising from ocean circulation patterns, variations in temperature and salinity, and by equilibrium processes, including changes in the Earth's own gravity patterns (Sallenger et al. 2012). In the Northeastern United States, sea level rise is about 3–4 times greater than the global average (0.2–0.6 m by 2100; IPCC 2007), creating a hot spot of accelerated sea level rise (Sallenger et al. 2012). The consequences of such global change are great (Boesch et al. 2000), especially for coastal-nesting birds whose habitat is shrinking (Burger 1990a; Erwin et al. 2007a,b, 2010; Nisbet et al. 2013) as are habitats for coastal communities (Miller et al. 2014) (Figure 7.1).

Climate change is not universally unwelcome. Global warming is proceeding faster at northern latitudes, expanding growing seasons, and alleviating winter cold stress (McCarty 2001; Bradshaw and Holzapfel 2006). However, we see such climate change as an environmental justice issue for many high-latitude communities and tribes, whose homes and subsistence hunting are jeopardized by rising water, shrinking ice, and declining species.

Loss of animal species is happening at an alarming rate in the global ocean (McCauley et al. 2015). Although climates have changed without our help, climate scientists generally agree that

Figure 7.1   A view of the New Jersey coastal community with wall-to-wall bulkheads.

human activities are driving the relatively rapid climate change that is evident over the past half-century (AAAS 2014; IPCC 2014). Changing climate brings changes in vegetation, habitats, foraging and nesting space, food availability, and storm and flood frequency. The ability of natural and human ecosystems to adapt to these changes is important for scientists, public policy, and the general public. Changes in species presence, numbers, and geographic distribution influenced by global climate change will have major implications for conservation and the management of endangered species (Wells et al. 2010).

In this chapter, we review the effects of climate change on coastal environments, other (non-avian) species, and trophic nodes, on seabirds, on shorebirds, and on coastal nesting species. We then present 40 years of data on climate change and sea level rise effects on birds nesting on salt marsh islands in Barnegat Bay as a case study. Whereas there are many studies of climate change effects on seabirds and shorebirds, there are relatively few on coastal-nesting species, which are extremely vulnerable to sea level rise, particularly because the narrow coastal band of beach, dune, and marsh has nowhere to go if it is inundated (Burger 1990a; Erwin et al. 2010). It will be decades to centuries before new habitats establish themselves at the frontier of a migrating coast driven by the rising sea.

ENVIRONMENTAL EFFECTS OF GLOBAL WARMING AND SEA LEVEL RISE ON COASTAL HABITATS

Global climate change is expected to affect temperature, precipitation patterns, ocean and atmospheric circulation, rate of sea level rise, acidification of the sea, and the frequency, intensity, timing, and distribution of hurricanes and other severe storms (Michener et al. 1997). Although global warming has many effects, the overarching physical impacts on coastal systems stem from sea level rise and the inundation of low-lying islands and mainland areas. Severe storms can alter wetland hydrology, geomorphology, biotic structure, energetics, and nutrient cycles (Michener et al. 1997). These, in turn, impact biotic communities, including the organisms that sustain food webs.

Periodic, but noncyclical, changes in plankton abundance have been recognized for a century, and are related to "hydroclimatic influences" (currents, temperature, and primary productivity) (McManus et al. 2015). With productivity varying from below, and predator pressure altered by fisheries above, the rise and fall of many fish stocks has challenged both fisheries science and ornithologists, both of which have a vested interest in predicting where and when particular fish species will increase or crash. Planktonivorous species such as Sand Eels (*Ammodytes* spp.) play a key role in transferring energy from planktonic to predator trophic levels, supporting populations of large fish, mammals, and birds. Temperature changes can cause shifts in where, when, and which plankton will be available, thereby driving many aspects of the food web for periods varying from weeks, to seasons, to decades (Safina and Burger 1985).

Differences in wetland form and function result in different outcomes from global warming and sea level rise (Leatherman et al. 1995; Cahoon et al. 2006; USCCSP 2009). Local geomorphology, climate regime, hydrology, and human development (including ditching and diking, groins, and jetties) result in differences in sediment supply, primary productivity, decomposition, subsidence, and autocompaction. These, in turn, lead to different vulnerabilities to sea level rise. Both direct and indirect biotic processes will have a major influence on beach and marsh vulnerability to sea level rise (Cahoon et al. 2006). One of the major difficulties in predicting impacts is the large uncertainties in estimating effect of each of these factors and the complexity of their interactions. The magnitude and even the direction of hydrological and sedimentology effects on each of the receptors is unclear (see Sutherland 2004). Managers and environmental planners must take into account these physiognomic and biotic differences when determining vulnerabilities of coastal areas, and when planning for the future (Gornitz et al. 2001; Figure 7.2).

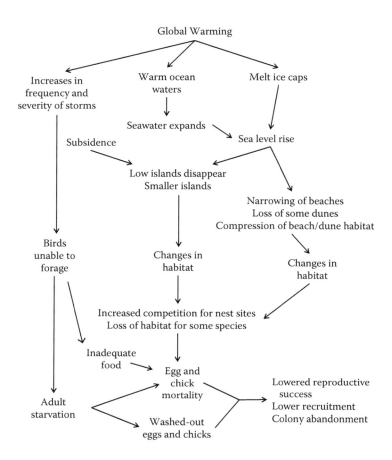

Figure 7.2 Schematic layout of the effects of climate change and global warming on birds nesting along coasts.

Worldwide, erosion impacts about 70% of the sandy beaches. The loss of sandy beaches is ultimately impacting fixed human structures and development along coasts, with dire consequences. Estimates of the rate of long-term sandy beach erosion are 2 orders-of-magnitude greater than the rate of sea level rise (Zhang et al. 2004). Thus, the effect is multiplicative, and has consequences for human health and well-being (WHO 2013) as well as for ecosystem health. Because of the extensive human development along coasts, the effect of sea level rise is severe for both humans and ecosystems. Recent storms, such as Superstorm *Sandy* (2012), illustrate the interdependence of both types of communities (Burger 2015b). Intact marshes and dunes buffered the residential communities behind them, somewhat mitigating the impact of the storm surge. Unimpeded by wall-to-wall development, a rising sea would gradually push ecological communities farther inland, creating new beaches and marshes as has happened over centuries after the Ice Ages. Fixed barriers, whether coastal cliffs or our built environment, block these processes. Only severe storms that destroy buildings, marinas, and other infrastructure can erase the coastal communities, and then only when people do not immediately rebuild. This effect will strengthen as cities build taller and more durable sea walls for protection. Natural and residential communities are on an inexorable collision course.

Coastal erosion, combined with sea level rise, also leads quickly to a narrowing of the beach and sand dune habitat with resulting biotic effects. A narrower beach provides less physical space for sensitive invertebrates and vertebrates, including birds. On a microscale, there is likely to be less elevation differences, fewer undulations, and fewer wrack lines. Fewer undulations on a sand beach means fewer higher elevation places for birds such as Piping Plover and Least Tern to nest—even on undeveloped beaches. Furthermore, dunes may also be narrower from the ocean side to the landward side (Feagin et al. 2005). When dune vegetation is restricted to a narrow band, the normal succession of plant communities breaks down, resulting in a potential loss of late-successional plants and their associated fauna (Feagin et al. 2005). Coastal shorelines will shift landward (Zhang et al. 2004). If there is no space to shift landward, each coastal zone will be truncated, providing less space for the diversity of coastal species, and the possible squeezing out of some plant communities, as well as some animal species showing northward shifts (Figure 7.3).

These changes are not theoretical or uncertain at all. They have been observed annually for decades, including on our local Long Island and New Jersey shoreline. People have responded to shrinking beaches by expensive and only temporarily successful beach nourishment projects. The newly widened beaches are better at supporting beach blankets than beach ecosystems.

It is not only the ocean beaches and dunes that are vulnerable to sea level rise, but low-lying sandbars, salt marshes, and even mainland marshes. Sandbars and sand spits are the first to go, and unless new ones are formed, these loafing, foraging, and occasional nesting sites are gone from the system. Ocean currents create dynamic patterns as they move sand from one area to another, forming and then erasing temporary islands or sand spits. Cartwright Island, off eastern Long Island, represents a constantly shifting archipelago shaped by tidal currents and storm tides. In some years, Cartwright Island is indeed an island, separated from Gardiner's Island by more than a kilometer of ocean. In other years, it is the end of a long sand spit connected to the larger island. Post and Raynor (1964) describe its breakup and coalescence, and in most years part of the island supports nesting Oystercatchers, Skimmers, and Common Terns.

The impacts of climate change are multiple. Salt marshes are vulnerable to both sea level rise and subsidence occurring over a period of decades. Meanwhile, more frequent and more severe storms erode islands over a period of years. The combined effect reduces area above water, particularly at flood tides, and alters the amount and type of vegetation. At the same time, changing temperature affects the structure and function of plant communities, whether aquatic or terrestrial, and these in turn affect animals dependent on these communities.

Figure 7.3   There are northward shifts in the winter ranges of many species, such as this American Robin feeding on holly berries.

CLIMATE CHANGE AND NONAVIAN SPECIES

Climate change has the potential to affect every species group in nearly every biome and habitat (Parmesan 1996, 2006). Effects on birds may be more obvious because the birds themselves are abundant, diurnal, and easy to see. We mention several nonbird examples to illustrate the breadth of changes and how they relate to birds. Climate warming in the California current is leading to declines in the zooplankton in the region (Roemmich and McGowan 1995). Since 1951, the biomass of macrozooplankton in the waters off southern California has decreased by 80%, whereas the surface water warmed by more than 1.5°C. Shifts in water temperature also resulted in a drop in inorganic nutrients, which resulted in lower productivity. Because zooplankton is a significant component of the base of the food web, such differences have consequences for a wide range of marine organisms, including top trophic level birds and mammals.

Invertebrates on land have fared no better. There have been climate-driven changes in butterfly communities in the Northeastern United States (Breed et al. 2013). Using data from butterfly surveys in Massachusetts, Breed et al. (2010) demonstrated increases in many species at the northern limit of their range, and declines in nearly all species at the southern limits of their range. They suggested that there is a northward expansion of warm-adapted species, and a retreat of cold-adapted species (Breed et al. 2013). We have noted similar phenomena in New Jersey, as well as changes in the phenology of emergence and brood periods in butterflies (Gochfeld and Burger 1997), and this seems to have intensified in the past several years. New Jersey has lost the Arctic Skipper (*Carterocephalus palaemon*, a northern species), and some southern species (e.g., Common Checkered Skipper [*Pyrgus communis*]) (Figure 7.4) are now overwintering. In the western United States, there are altitudinal shifts, with species distributions shifting higher up mountains because of shifts in global temperatures. Similar altitudinal shifts have been noted in butterflies from studies in Europe (Konvicka et al. 2003; Roth et al. 2014).

On the basis of models of species distributions, chlorophyll *a*, and bathymetry, Hazen et al. (2013) predicted up to a 35% change in core habitat for some marine species in the Pacific Ocean, with a significant northward displacement of biodiversity across the North Pacific. For already

Figure 7.4   Common Checkered Skipper, a southern species, is now overwintering in New Jersey in most winters, and is therefore more established. (Burger, J., Gochfeld, M., Unpublished data.)

stressed species, loss of habitat and increased migration times could lead to population declines or could inhibit recovery. The general northward movement, noted for both butterflies and pelagic predators, is mirrored by studies with birds.

In North America, plants and plant communities are also on the move northward, abetted by warming climates (Wolkovich et al. 2013). Alternatively, plants are blooming much earlier than in the past. The date of first flowering for spring flower plants documented in Concord, Massachusetts, by Henry Thoreau in the 1850s has advanced by about 2 weeks, concurrent with a 3°C average temperature rise (Ellwood et al. 2013). Growing seasons and plant hardiness zones have shifted in response.

CLIMATE CHANGE EFFECTS ON BIRDS

Although birds can fly, they are not immune to the effects of climate change, and they are especially vulnerable to sea level rise. Although the changes may be more obvious for coastal, colonially nesting species, climate change and sea level rise also affect coastal foraging space as well as species nesting inland. The phenology or seasonal timing of nesting is one indicator of warming. Birds can respond to increasing temperatures by arriving earlier and laying eggs earlier, or by shifting their breeding farther north (Huntley 2006). A note of caution is warranted; some of the phenological changes in bird distributions can be attributable to age-related differences in populations—older birds may breed earlier than younger or first-time breeders (Ezard et al. 2007). Moreover, the proximate factor affecting timing for fish-eating birds is probably the timing of fish availability (Marples and Marples 1934). In the sea, cold water currents and upwellings promote productivity, whereas warming oceans foretell dwindling food resources (Safina 2014a,b).

The effect of climate change on birds has reached the popular press, and is a priority for conservation organizations, as evidenced by the production of a special issue of *Audubon* magazine on climate change and birds. One article notes that 314 North American birds are at risk from climate change. However, the models are ambiguous, for warming may allow expansion of breeding ranges into habitats that may end up being under water (Nijhuis 2014; Yarnold 2014).

Studies of the effect of climate change on birds seem to focus around three groups: wintering migrants (songbirds and others), seabirds, and shorebirds. Changes in these three groups clearly demonstrate a general phenomenon that is affecting all birds, from coastal marshes to high latitudes. There are relatively few studies on the effects of global warming on coastal waterbirds. Some of these topics will be discussed in the following sections.

Effects on Noncoastal Species

In response to warming climate change, many migratory bird species in the United States and Britain are arriving and nesting about 10 days earlier than decades ago (Butler 2003), about the same as the advance in flower phenology. The ranges of birds have shifted northward and upward (Roth et al. 2014). There are poleward shifts in the distribution of winter ranges for 254 North American species that were examined using Christmas Bird Count data (La Sorte and Thompson 2007; NAS 2010; La Sorte and Jetz 2012). These geographical shifts northward will have major implications for avian distribution and abundance throughout North America (Figure 7.3). Models also predict changes in avian arrival and distribution, and that the nesting boundaries of many birds may shift more than 1000 km northward (Huntley et al. 2006). This shift may lead to a decline in species diversity in some geographical regions with enrichment in others. Over the course of our own lives and experience, we have seen several "southern" bird species such as the Cardinal (*Cardinalis cardinalis*) become established around our childhood homes in southern and central New York. Warming temperatures are associated with earlier arrival times of migrants such as documented for six species in the U.S. northern Great Plains for the period 1910–1950 (Travers et al. 2015).

There are also changes in the phenology of winter migrants in Europe. Sparks and Mason (2004) found that some migrants in the United Kingdom arrived and departed earlier, whereas others did not. Maclean (2014) reported that climate change has caused or allowed changes in the distribution and site abundance of birds in the winter in Western Europe. Models predicting the effect on migratory birds of climate change suggest a variety of possibilities, including that increasing winter temperatures may lead to declines in the proportion of migratory species, and increasing spring temperatures may lead to increases in the number of migratory species (Schaefer et al. 2007). At Albany, New York, a winter bird feeder in the 1950s would have attracted six species in a typical day. Today, the number expected has doubled (personal observations).

The proximate mechanism for early egg-laying may be food supplies (Carey 2009), especially for new recruits to a population (Becker and Bradley 2007). The aquatic food web will be responsive to temperature so that climate change may result in a mismatch between food and usual breeding times, and species survival may be partly a result of whether the species survived previous climate changes in past centuries or millennia (Carey 2009). Ellwood et al. (2013) provide evidence that mismatching may already be occurring on land as the first arrival dates of migrant birds at Concord, Massachusetts, have not kept pace with the earlier flowering of spring plants. There may be an increased risk of phenological miscuing (responding inappropriately to climate change), whereby birds may respond by breeding earlier or later than is optimum (Crick 2004). There may also be mismatches because different species may not shift at the same time or rate, and they may not shift fast enough to keep up with the plant species they depend on (Visser 2001; Visser and Both 2005). Rapid climate change may select for species adaptability, and species that are specialists or inflexible in habitat choice or foraging strategies may not survive in their present numbers. Invasive species are characterized by adaptability, and climate warming is abetting their spread (Wolkovich and Cleland 2011).

Effects on Seabirds

Climate variability has a strong effect on marine ecosystems (Schreiber 2001; Hoegh-Guldberg and Bruno 2010), largely because much of the increased heat will be absorbed by the oceans (IPCC 2007). Ocean warming will affect seabirds through foraging habitat and food availability (Sydeman et al. 2012), as well as during the breeding season (e.g., heat stress; Oswald and Arnold 2012). Seabirds are particularly vulnerable to changes in the ocean because most species live their whole life on, above, or in the ocean, except when they breed on offshore islands (Schreiber and Burger 2001; Figure 7.5). They forage over or in the ocean, exploiting prey that are directly affected by ocean conditions. Thus, any change in ocean currents, upwellings, temperature, or salinity directly affects their prey base, including prey numbers, prey types, and the availability of prey. Any of these changes affect seabird foraging locations, and possibly foraging methods, which in turn can affect where seabirds can nest. That is, they have to nest on remote islands and be able to feed within a reasonable travel distance of their nests.

Studies have included effects of climate change and fisheries bycatch on seabirds (Barbraud et al. 2012), food quantity or availability (Montevecchi and Myers 1997; Thompson and Ollason 2001; Viverette et al. 2007; Safina 2014a,b), population changes because of food availability (Wolf et al. 2010; Bustnes et al. 2013a), the spatiotemporal mismatch of seabirds and their prey (Grémillet and Boulinier 2009), effect of shrinking sea ice on seabird populations (Croxall et al. 2002; Jenouvrier et al. 2005), delayed breeding (Barbraud and Weimerskirch 2006), and climate variability and population dynamics (Jenouvrier et al. 2003). One interesting effect of climate change suggested by Møller et al. (2006) is that dispersal is influenced by temperatures, both during the natal dispersal year and in the first breeding year.

Whereas individual studies demonstrate a particular effect, several species in the same environment may exhibit different patterns, dependent on the temporal scales examined (Hyrenbach and Veit 2003). Such studies often result in inconsistencies among species and locations, opposite effects, and paradoxical effects (Croxall et al. 2002). This makes finding a common response difficult—in, fact there may not be a common response. A meta-analysis of more than 208 publications on

Figure 7.5 Colony of Cape Gannets (*Morus capensis*) nesting on Marcus Island, off South Africa, observed by Mike Gochfeld. Nesting density is high in this colony managed for guano production. In North Africa, Northern Gannets breed on offshore islands.

seabirds and climate indicated that there is a lack of information on low-latitude seabird–climate phenomena, modeling studies, and the integration of oceanographic, food web, and population dynamic models (Sydeman et al. 2012). Such studies are necessary to predict future changes and provide information for possible management options.

Effects on Shorebirds

Shorebirds (sandpipers and plovers) and long-legged wading birds (egrets, herons, ibis) are the third group that has been examined for possible effects from climate change and sea level rise. Although it might seem that there is little similarity in these groups, the common theme is that they feed in coastal beaches, marshes, mudflats, and shallow water. Any loss of mudflats and shallows reduces the amount of available foraging habitat (Erwin et al. 2006).

Most shorebird species that breed in North Temperate and Arctic habitats migrate along coasts, gathering at staging areas to feed and fatten before continuing on to breeding or wintering grounds (Clark et al. 1993) (Figure 7.6). Many of these species breed at high latitudes. The Willet, American Oystercatcher, and Piping Plover breed in the Northeast. These species face the same habitat changes as the gulls, terns, and egrets considered in this book.

Because of their widespread distribution, shorebirds are heralded as integrative sentinels of global environmental change (Piersma and Lindström 2004). Many shorebirds travel over large segments of the globe during their annual cycle. Some species (e.g., Red Knots) migrate more than 13,000 km, one way (Niles et al. 2010). Morrison et al. (2001) found that for 80% of shorebird species where there were sufficient data to analyze, there were consistent declining trends. Declines of many North American shorebird populations are a cause for concern (Morrison et al. 2001) and a reason to examine how global climate change is affecting them.

Climate change is most rapid and evident near the poles. Antarctic ice sheets crack apart. Arctic sea ice shrinks. The iconic Polar Bear (*Ursus maritima*), seemingly stranded on a tiny ice island, became the poster child for climate change. These changes impact food availability and hunting,

Figure 7.6 Shorebirds (mainly Red Knots) foraging along the shore, with two Laughing Gulls trying to get to the shoreline with the Horseshoe Crab eggs.

and impact reproduction physiology and behavior for both male and female Polar Bears. Earlier ice breakup and poorer hunting means the bears are in poor nutritional condition (Stirling et al. 1999). Bear reproduction is only the "tip of the iceberg" when it comes to global population ecology.

Most of the studies dealing with climate change and shorebirds involve either studies of potential habitat shifts, individual shifts in distribution, or both. Habitat shifts are likely to occur in nesting habitats because one potential effect of global warming is that the Arctic climate will become warmer, rainier, and windier in summer (Lindström and Agrell 1999). These habitat changes may affect both north Temperate and Arctic stopover areas (places shorebirds stop to refuel before migrating farther) as well as breeding areas. Much of their preferred breeding areas may disappear as "habitat types" move farther north (eventually into the Arctic Sea when no land is available) (Lindstrom and Agrell 1999). There will most likely be a compression of breeding habitat (Rehfisch and Crick 2003), much like what is expected in beach/dune coastal areas in the Northeastern United States.

The advances in phenology of nesting birds have been greater in northern latitudes (50° to 72°N) (Root et al. 2003). The timing and arrival of shorebirds is often responsive to ambient temperature. The first arrival of several species of shorebirds is earlier than it was a decade ago (Rehfisch and Crick 2003). Early arrival and a longer breeding season could provide additional opportunities for replacement clutches or double clutching, but this has not been explored. Obtaining sufficient data to show shifts in breeding locations, however, is difficult because shorebirds generally nest solitarily and are spaced out across the tundra, making it difficult to track sufficient numbers over a wide geographical scale.

Tracking numbers of shorebirds on wintering grounds, however, is easier because the important shorebird wintering areas are known (and limited in number), and large numbers can be monitored. Maclean et al. (2008) analyzed counts of waders from about 3500 sites over 30 years in Western Europe, and found that seven species of waders have undergone substantial shifts of up to 115 km to the northeast. Similarly, in the United Kingdom, the distribution of nonbreeding shorebirds has shifted in relation to climate change (Austin and Rehfisch 2005). The distributions for eight of nine species of shorebirds changed, and these birds normally breed as far apart as Greenland to the high Arctic of Russia. Shorebirds are shifting their wintering distribution to the north and east, where wetter and muddier shores support a higher biomass of their invertebrate prey (Austin and Rehfisch 2005). These shifts, which can be positive if prey are abundant or negative if there is a decrease in available mudflats for foraging, should be carefully monitored. A common demographic method of monitoring would greatly increase the utility of such data, especially for managers, conservationists, and public policy makers (Robinson et al. 2005).

The studies cited earlier focused on surveys of shorebirds at established, recognized stopover or wintering locations. For our purposes, changes in coastal shorebirds are particularly relevant. Impacts on Eurasian Oystercatchers (*Haematopus ostralegus*) were associated with mean temperature rise, rather than with the extremes (Pol et al. 2010). Norris et al. (2004) examined changes in the abundance of Common Redshank (*Tringa totanus*) breeding on UK salt marshes; they reported a 23% decline between the mid-1980s and the mid-1990s. The decline, however, seemed related to grazing pressure, suggesting the difficulty of isolating changes due to climate from those of other stressors. These studies illustrate the difficulty of examining single species responses. The answer, perhaps, is that stressors act together; isolating one factor is difficult.

Another approach to determining the potential effect of climate change and sea level rise on shorebirds is to examine potential habitat losses, as these will affect all coastal birds. Galbraith et al. (2002) developed models to examine potential changes in foraging habitat for migrating and wintering shorebirds at five sites in the United States that support important numbers of shorebirds. Assuming a conservative global warming scenario of 2°C within the next century, they projected losses between 20% and 70% of current intertidal habitat on the Pacific, Atlantic, and Gulf coasts of the United States (Figure 7.7). The most severe losses were likely to occur in places where the coastline is unable to move landward because of development (seawalls, bulkheads). Such coastal development is particularly severe in parts of the Northeastern estuaries, making ecological accommodation difficult.

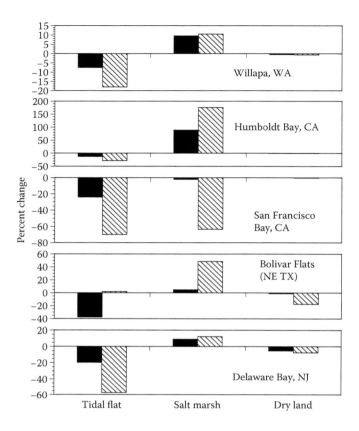

Figure 7.7 Predicted changes in habitat in several U.S. bays and estuaries that would result from a 2°C rise in global temperature modeled for 2050 (shown as black bar) and 2100 (shown as hatched bar). (After Galbraith, H. et al., *Colon. Waterbirds*, 25, 173–183, 2002.)

A similar study in Florida examined all coastal landforms, and found a 16% loss of habitat from the present until 2100, largely because of sea level rise (Convertino et al. 2012). The authors examined distribution of nesting, breeding, and wintering occurrences for a number of coastal-nesting and wintering shorebirds. Although they found that habitat loss, fragmentation, and connectivity were separate factors affecting their analysis, the end result was a loss of breeding and wintering habitat for shorebirds (Convertino et al. 2012).

Populations of many shorebird species are declining, partly because of loss of wintering and migrating habitat, partly because of exploitation on the wintering grounds, and perhaps because of changes in their high-latitude breeding sites. Shorebirds appear to be particularly vulnerable because: (1) they are dependent on beaches, mudflats, and other tidally affected habitats; and (2) mudflats or tidal flats are declining in quantity in a number of bays. When Galbraith et al. (2014) developed a model to examine the vulnerability of 49 species of North American shorebirds, and they predicted that 90% are at an increased risk of extinction. This does not bode well for shorebirds or for other coastal birds that rely on tidal mudflats.

Given the increased risk that global warming imposes on shorebirds, and other species that rely on intertidal habitats, the question becomes—"What is to be done?" Some scientists resort to habitat modeling as a method of predicting how much habitat will be lost, for Piping Plovers, for example (Sims et al. 2013). Although these exercises provide insights into habitat selection and habitat needs of a species, and the potential for habitat loss under different sea level rise scenarios, they require testing and implementation on a large scale. Habitat restoration may mitigate the changes. Creating mudflats can be successful, particularly if sediment supply is sufficient (and constant), especially because invertebrates and shorebirds colonize them rather quickly (Atkinson 2003). Unfortunately, many restoration sites are small, and thus do not meet the needs for tidal mudflats eliminated by sea level rise. That is, the massive loss of mudflat habitat to sea level rise is not being replaced by the formation of new mudflats or the creation/restoration of mudflats.

AVAILABLE HABITAT AND CHANGES IN BARNEGAT BAY

The preceding sections provided an overview of some of the factors involved in examining the effect of global warming and sea level rise on a broad taxonomic and geographic range of species. Studies have demonstrated the effects of global warming on bird distribution, phenology, and diversity. In the northern hemisphere, birds are moving north for breeding and wintering. The studies generally note that species using coastal environments, particularly tidal mudflats, are vulnerable to projected decrease in availability of this habitat. However, the studies mainly deal with landbirds (migrant passerines), seabirds, or with shorebirds. There have been few studies of gulls and terns and wading birds. The few coastal species modeled by Nijhuis (2014) will have major northward shifts of range along the Atlantic coast. Similarly, Hu et al. (2010) modeled the northward shift in suitable habitat for the endangered Black-faced Spoonbill (*Platalea minor*) in Southeast and East Asia. This will greatly complicate conservation efforts, because its projected future habitats are not currently protected.

While these studies demonstrate distribution and phenology changes in response to global warming, and provide models demonstrating what habitat loss will be by 2050 or 2100, we can look at existing long-term data sets to document such changes. We provide such data for birds nesting in Barnegat Bay over the past 40 years.

Policies and Perceptions of Sea Level Rise

Global warming and sea level rise are clearly occurring in the world, and along the Atlantic coast (NPCC2 2013; IPCC 2014). Figure 7.8 shows a century of sea level rise at Atlantic City, New Jersey.

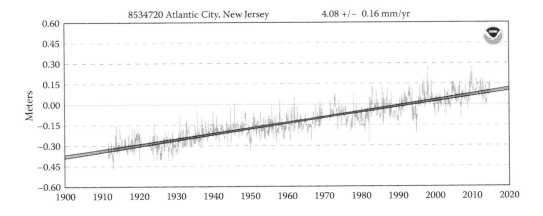

Figure 7.8   Sea level rise in Atlantic City, New Jersey, from the early 1900s to the present. (After NOAA, Mean Sea Level Trend, http://tidesandcurrents.noaa.gov/sltrends/sltrends_station.shtml?stnid=8534720, 2014.)

Sallenger et al. (2012) note that in the Northeastern United States, sea level rise was about 3–4 times greater than the global average, creating a hot spot of accelerated sea level rise. Miller et al. (2014) reported that in New Jersey: (1) global sea level rise in 2010 was 3.3 mm/year; (2) global sea level is predicted to rise 80 cm by 2100 (compared with an average of 40 cm; IPCC 2014); (3) sea level rise will be greater in New Jersey (exceeding 1 m by 2100) than in other states; (4) sea level rise will combine with severe storms to increase beach erosion, and flooding of coastal communities (and presumably erosion of salt marshes). Whereas environmentalists urge major social and economic changes to slow and moderate climate change, states and local governments are focusing on "adaptation." Municipal governments, especially in New Jersey, are proposing the pumping of sand into beaches and building dunes; these, however, are short-term solutions. Although some communities may eventually accept managed retreat (moving from low-lying places; Alexander et al. 2012), others are raising houses on stilts, and insisting that the government subsidize risky development (Titus 1990, Mayer 2013).

Much of the environmental response to climate change is driven by public opinion about climate change—whether it exists, whether it will impact the United States, and whether it will impact them personally. Perceptions have the ability to compel or constrain political, social, and economic action (Leiserowitz 2005). Several surveys around the world have claimed that the issue of climate change is of declining interest (Ratter et al. 2012). Furthermore, many people believe that climate change will not affect them, but will affect other people in distant lands (Leiserowitz et al. 2010)—Polar Bears and Penguins, but not people. Although people often respond on surveys that climate change is considerable or severe, their perception of personal liability is low (Lieske et al. 2014). Clearly, there is a disconnect between the scientific data showing increasing global warming and increasing sea level rise (IPCC 2014), and the public perception of little cause for worry (Pidgeon 2012). There are public groups that champion the effects of global warming and sea level rise. A recent People's Climate March in New York City is a case in point (HR 2014).

There is another disconnect between perceptions of climate change and global warming with respect to ecosystems, particularly coastal ones. The recent devastation that Superstorm *Sandy* wrought on the Northeast is evidence of both climate change and sea level rise. Yet, there is still a lack of awareness of the role of coastal ecosystems in protecting both human and ecological communities (Burger 2015b; Burger and Gochfeld 2014b, 2015). Following *Sandy*, people were generally more concerned about personal property and safety than about the environment or future storms (Burger 2015b).

Environmental changes are greatly affecting avian communities. The key factors of interest for birds are storms and sea level rise, tides, vegetation (not enough or too much), and presence of wrack. Birds must balance the benefits of living closer to the waterline where food is abundant against the benefits of living higher and dryer. When considering the implications of global warming and sea level rise on birds, it is useful to consider conditions in the past (20 years ago or so), current conditions, and future conditions (Table 7.1). Table 7.1 describes the condition of each of these factors for the winter, spring, summer, and fall for three periods (past, present, and future). These observations are based on our experiences when we first started (1970s, past), presently (current), and what we expect in the future. A brief description of birds that might be present is also provided.

Nesting Birds in Barnegat Bay as a Case Study

Barnegat Bay has a wide diversity of birds nesting, foraging, and wintering in the bay, associated marshes, and ocean front (Burger et al. 2001b; Chapter 4). Given the expected changes in sea level, vegetation, and wrack on salt marsh islands, we predicted how nesting habitat, frequency of storms, severity of storms, and sea level rise will affect the different species (Table 7.2). These conclusions are based on our 40 years of experience with birds nesting in Barnegat Bay. The overall effects include decreased availability of suitable habitat and particularly loss of high quality nesting habitat, which means elevations high enough to minimize flooding losses and presence of appropriate substrate and vegetation (depending on species). Loss of available and suitable habitat results

Table 7.1   Seasonal Effects from Global Warming

	Winter	Spring	Summer	Fall
		Past		
Sea level rise	Cyclic/slow rise	Cyclic/slow rise	Cyclic/slow rise	Cyclic/slow rise
Tides	Normal overwash of low-lying islands	Overwash of extremely low-lying islands	Infrequent overwash of islands	Some overwash of low-lying islands with severe storms
Vegetation	Dormant, remains stable	Remains stable, begins growth	Full growth and productivity	Seed production (harvest by farmers)
Wrack	Deposited on shores of intermediate and high islands by high winds/tides	Wrack remains, may be moved either higher or lower on the islands	Wrack weathers and turns white/grayish; terns nest on it	Weathering, compaction, and sinking; decays after year or two
Biological effect on birds	Winter visitors and migrants return from north; residents remain	Departure of some winter visitors; arrival of summer breeders; initiation of breeding	Completion of breeding; departure of adults and young	Departure of summer breeding birds; arrival of migrants and winter visitors
		Current		
Storms/Sea level rise and subsidence	Slow increasing of severity and frequency of storms; increase in sea level rise	Slow increasing of severity and frequency; increase in sea level rise	Slow increasing of severity and frequency; increase in sea level rise	Slow increasing of severity and frequency; increase in sea level rise
Tides	Frequent high tides	Sporadic higher tides than usual	Similar high tides	Sporadic higher tides than usual
Vegetation	Dormant, stable	New growth, some shift in *Spartina patens* and *Spartina alterniflora*	Continued growth; some loss of *Iva* and *Baccharis* because of more frequent salt inundation	Normal seed production
Wrack	Storm tides put wrack higher on some islands, remove all from others	Storm tides remove some wrack that would otherwise have remained in the summer	Fairly stable	Storm tides remove some wrack that would otherwise have remained in the summer
Biological effect on birds	Loss and erosion of low-lying islands; severe storms remove predators from some islands; changes may result in bays freezing allowing predators access to some islands	Changes in vegetation provide more competition between birds for higher nesting places; loss of high places in marsh removes safe nesting sites for Clapper Rails	Continued competition among birds for safe nesting places high enough to avoid high tides; risk of predation by larger birds on smaller species	Fewer foraging and nesting places for migrant birds and winter residents because of loss of some low-lying islands
		Future		
More frequent storms/Sea level rise and subsidence	Long-duration storms; rising sea level may inundate marshes, erode islands, and change suitability and availability	Storms alter vegetation structure; wash away or deposit wrack; increase in sea level rise due to climate change	Long-duration storms and extreme tidal flooding, wipe out nesting colonies (see Table 5.1); longer duration; increase in sea level rise	Loss of marsh area

(Continued)

Table 7.1 (Continued)   Seasonal Effects from Global Warming

	Winter	Spring	Summer	Fall
Tides	More frequent extreme high tides that overwash more of the lower islands, and go higher on intermediate-elevation islands; will destroy or erode low-lying islands	More frequent high tides will overwash a greater number of islands	More frequent high tides will overwash more low-lying islands; more breeding failures	More frequent high tides erode banks of islands and along salt marsh creeks
Vegetation	Dormant; some vegetation eroded from island edges	Vegetation shifts; reduction in horizontal extent of vegetation types (*S. alterniflora* and *S. patens*); may cause death of low-lying shrubs because of salt exposure during growing season; may break branches of shrubs	Vegetation shifts; may cause death of low-lying shrubs because of salt exposure during growing season; full growth	Seed production; some vegetation at the edges of islands may be eroded away by severe storms
Wrack	Severe high tides wash away wrack from many low-lying islands that usually have wrack: may remove all dead vegetation from nearby islands	Severe high tides wash away wrack from many low-lying islands that had wrack remaining from the winter	Severe high tides wash away wrack from many low-lying islands that had wrack remaining from the winter; greater inundation speeds decay (reducing wrack for nest year)	Continued washing away of wrack from islands; greater inundation speeds decay (reducing wrack for following years)
Biological effect on birds	Continued loss of nesting islands; loss of wrack for resting and roosting sites wintering birds; less foraging space for ducks and geese	Less wrack available for nesting species (Skimmers, Oystercatchers, Terns); shift use of vegetation by salt marsh species, creating more competition for higher sites and for safe sites (herons, egrets)	Continued competition for both wrack and safe (no flooding) nesting sites for all birds	Increased loss of wrack, narrowing of horizontal *Spartina* zones, loss of shrub zone, increased breakage of shrubs

Note: We provide a summary of the potential effects of sea level rise (and subsidence) on birds in Barnegat Bay and other salt marshes in the Northeastern coastal estuaries comparing today with about 20 years ago, and with the future.

in birds shifting to other islands or emigrating out of the Bay. Birds that remain and try to nest will experience increased nest density and the consequences of crowding, including competition, and loss of time for breeding activities because of increased neighbor aggression. Habitat is lost to water (sea level rise, flooding, erosion) and vegetation changes from storm damage and salt intrusion. Increased nesting density increases potential for predation by colony neighbors. The impact of these factors varies by species, habitat, and interspecific nest associations. Salt intrusions have the potential to kill small trees and shrubs used for nesting by herons (Figure 7.9). Because of the size relationships, most nesting associations will result in greater competition, more aggression, more predation, and lowered reproductive success.

Table 7.2   Predicted Impacts and Species' Responses to Climate Change and Sea Level Rise Based on Our Experience with Waterbirds of Barnegat Bay

	Nesting Habitat	Increased Storm Frequency	Increased Severity of Storms/Wind	Sea Level Rise
General for birds		Greater costs for incubation and brooding; potential for more washouts; less foraging time; potential for thermal stress on chicks	Longer flooding of marsh; greater costs for incubation and brooding; more washovers; break up of nests on the ground; frequent hypothermic mortality of chicks	Loss of entire islands or low-lying sections once available for nesting; nest space compression; changes in vegetation
Clapper Rail/ Willet	High places in *S. alterniflora*, usually with dome of grass over nest	Loss of some nesting places and foraging space; greater costs for incubation; difficulties of precocial chicks to have enough time for foraging on their own	Loss of nesting and foraging space; loss of foraging time; longer flooding resulting in some nests and eggs under water for longer, suffering deceased hatching rate	Loss of nesting space; Rails being compressed in a narrower potential nesting and foraging space; inter- and intraspecific competition for high nesting spaces
American Oystercatcher/ Black Skimmer	Wrack, edges of wrack near *Spartina*	Loss of courtship and foraging time; more washouts on wrack inundated by tides	Greater likelihood of flooding out, with extended inundation of nests; fewer safe nesting sites; greater likelihood of chicks drowning or dying from thermal stress	Fewer nesting sites safe from flooding; less horizontal space for nesting; less wrack for optimum nesting; more crowding and competition with terns for space
Common/ Forster's Tern	Wrack, or in *S. alterniflora* and *S. patens*	Loss of time for courtship and foraging; more washouts on wrack inundated by tides; fewer wracks; greater competition with Skimmers for nest space	Loss of time for breeding activities; more washouts higher on the marsh; possible loss of wrack and decreased nesting space	Loss of low-lying nesting islands; less nesting space on what were once intermediate-elevation islands; possible spring washovers; greater competition with Skimmers for space
Great Black-back Gull/ Herring Gull	*S. patens*, or under shrubs or *Phragmites*	Loss of foraging time; competition with Herring Gulls for nest space, and Black-crowned Night-Herons for nesting space under shrubs	Loss of foraging and courtship time; loss of nesting space under shrubs because of salt inundation of shrubs	Loss of nesting space because of compression due to less horizontal space on nesting islands; compete with Night-Herons, and among gulls for nesting space
Laughing Gull	*S. alterniflora* and *S. patens*	Loss of nesting spaces high enough to avoid flooding; less time for courtship, nesting, and foraging; greater likelihood floods; greater competition with Herring Gulls	Less nesting space free from flooding; possibility of nests floating away; greater likelihood of chick deaths due to drowning	Loss of low-lying nesting islands; loss suitable nesting habitat on intermediate-lying islands; intense competition with larger gulls for nesting space

(Continued)

Table 7.2 (Continued)   Predicted Impacts and Species' Responses to Climate Change and Sea Level Rise Based on Our Experience with Waterbirds of Barnegat Bay

	Nesting Habitat	Increased Storm Frequency	Increased Severity of Storms/Wind	Sea Level Rise
Great Egret	High in shrubs of *Phragmites*; rarely on ground	Loss of foraging time and nesting time; increased competition with Night-Herons and Snowy Egrets for nest space in shrubs, partly destroyed by rain/winds	Loss of foraging time and nesting space; increased thermal stress on nests, eggs and chicks; compression because of habitat loss for nests in shrubs; destruction of nests in shrubs due to wind; thermal stress on chicks	Loss of some horizontal space on low-lying and on intermediate-elevation islands; some loss of shrubs used for nesting because of salt inundation: competition with Night-Herons and Snowy Egrets for nest space in shrubs
Snowy Egret	Intermediate level in shrubs or *Phragmites*; rarely on ground	Loss of foraging time and nesting time; increased competition with Night-Herons and Great Egrets for nest space in shrubs, partly destroyed by rain/winds	Loss of foraging time and nesting space; increased flooding of nests, eggs and chicks; compression due to habitat loss for ground nests; destruction of nests in shrubs due to wind; deaths due to thermal stress of chicks	Loss of some horizontal space on low-lying islands, and on intermediate-elevation islands; some loss of shrubs used for nesting because of salt inundation: competition with Night-Herons and Great Egrets for nest space
Black-crowned Night-Heron/ Glossy Ibis	On ground under shrubs and *Phragmites*; low in shrubs or *Phragmites*	Loss of foraging time and incubation costs; increased flooding of ground nests; increased competition for nest sites with other ground/shrub nesting species	Loss of foraging time and nesting space; increased flooding of nests, eggs and chicks; compression of habitat for ground nests; possible destruction of nests in shrubs (wind)	Loss of horizontal nesting space; loss of bushes that provide nesting space: increased competition with gulls, and other species for ground sites

Note: More detail on breeding habitat use can be found in Chapter 4. Birds with similar requirements are lumped together for brevity.

The preceding observations are based on our combined experience, and those of other biologists in the region, and are likely to be applicable to other low-lying colonies in the bays and estuaries of the Northeastern Atlantic coast. We also examined changes in colony site suitability (after Burger et al. 2001b, continued data collection). Several species of colonial birds were monitored every year, and we were able to examine changes that occurred during this period (Figure 7.10).

Colony Site Selection in Common Terns

Early in our Barnegat Bay studies, we recognized the importance of colony and nest site selection, and the role that availability of habitat plays in their choices (see Chapter 4). In the mid-1970s, we examined colony site selection in Common Terns (Burger and Lesser 1978). Because this study served as a basis for much of our later work, our methods are described briefly. Using boats and helicopters, we surveyed the barrier beaches, salt marsh islands, and mainland beaches from Mantoloking at the north end of Barnegat Bay to Holgate and Beach Haven Inlet in the south (about 75 km). No Common or Forster's Terns nested anywhere except on salt marsh islands. In those presatellite/pre-GPS days we used nautical maps, aerial photographs, ground data, and tax maps to

Figure 7.9 Saltwater inundation is killing some of the trees, which now support only Snowy Egret nests.

collect the following data for all islands in the bay: length, width, and area of each island, distance to closest mainland and barrier island, distance to nearest other island, presence and length of mosquito ditching, distance to open water, vegetation types, and presence of open bay water on at least one side. The *Spartina* cover percent was determined in the field. At the same time, we regularly monitored the islands for nesting birds, recording the number of eggs, nests, and adults. We later compared the characteristics of the islands used by terns, with all other islands.

Common Terns nested on 34 of a possible 259 islands (Burger and Lesser 1978). Using the data collected on the islands that were used by Common Terns, we computed a "nesting space" model. In other words, we used the description of habitat used by Common Terns to examine whether there were other islands in Barnegat Bay that fit within this "space" (Table 7.3). Of the 225 islands that were not used by terns for nesting, only 4 fit within the "space" selected by the terns, and these were all used over the next 5 years. Only two other islands were used as colony sites, and they fit the criterion except for having less *Spartina* than the other colonies sites that were used. Moreover, two islands—East Cedar Bonnet and Sandy Island—occasionally had a one pair or two pairs of Common Terns, perhaps holdovers from colonies that existed there before our study, or failed pioneers at sites that never caught on.

We interpret these data as indicating that virtually all the suitable colony sites in Barnegat Bay were already occupied by Common Tern colonies in the mid-1970s. The question remains, however—"Why nest on so many different islands, rather than in fewer large colonies?" Elsewhere

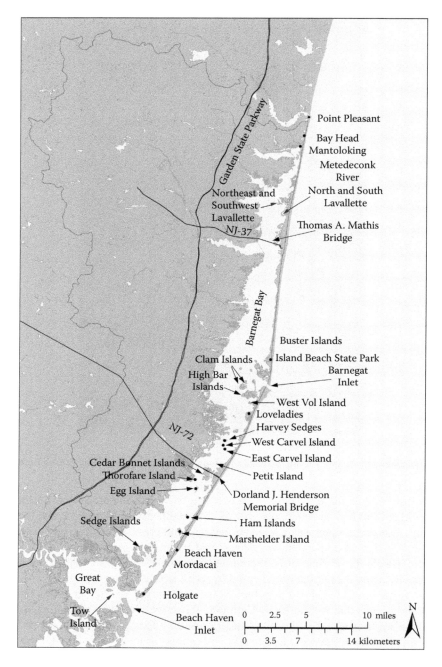

Figure 7.10   Map of Barnegat Bay showing the major tern-nesting islands studied. These islands are also used by other colonial species, as well as Oystercatchers and Willets.

Table 7.3 Colony Site Selection by Common Terns in Barnegat Bay, New Jersey

Characteristic	Terns' Habitat Choice	Islands That Do Not Meet Terns' Criteria	Remaining Suitable Islands
Total number of islands	34 of 259 islands used by terns		225
Island size	0.6–108 acres (0.24–44 ha)	131	94
Width of island	More than 20 m wide	3	91
Length of island	Less than 1050 m long	2	89
Distance from nearest island	3–580 m	14	75
Distance from mainland	More than 23 m	6	69
Distance from barrier island	Less than 3250 m	10	59
Exposed open water on one side	Open exposed water on at least one side	24	35
Shallow water depth by island	At least 0.2 m deep on shallowest side	23	12
Percent *Spartina*	More than 50% *Spartina*	3	9
	At least 19% *S. alterniflora*	4	5
	No more than 60% patents	1	4
Suitable islands that are unused[a]		4	0

Source: Burger, J., Lesser, F., *Ibis*, 120, 443–449, 1978.
Note: The table shows the relationship between the characteristics of islands selected by Common Terns, and the characteristics of all salt marsh islands in Barnegat Bay in the mid-1970s. Terns nested on 34 of the 259 islands. Given are the number of islands excluded by each variable lying outside of the Terns' range for each characteristic.
[a] All three of these islands were used by nesting Common Terns over the next 5 years, and only two others were ever used as a colony site.

(Cedar Beach, Great Gull Island), there were thousands of pairs of Common Terns in a colony. The largest Barnegat colonies had hundreds of pairs. There are a number of possible reasons related to salt marsh as a habitat, including: (1) Common Terns in Barnegat Bay prefer to nest on wrack, which is always limited in amount (Burger and Lesser 1978; Burger and Gochfeld 1991b); (2) much of the available *Spartina* is sufficiently low that nests would encounter tidal flooding; (3) predators are drawn to larger colonies; and (4) spacing out of colonies reduces foraging competition among Common Terns in the bay and in offshore waters.

Our surveys of these islands, along with the data collected on the physical characteristics of salt marsh islands in Barnegat Bay, provided a basis for examining colony abandonments, mainly because of sea level rise and flooding (Table 5.2 lists the colonies in Barnegat Bay).

SEA LEVEL RISE AND AVIAN RESPONSES IN BARNEGAT BAY

Sea level rise in Barnegat Bay is a function of both subsidence and actual sea level rise. For the birds, however, the effects are the same. Global climate change and sea level rise have had several impacts on birds nesting in the bay, none of them favorable (Table 7.2). Sea level rise has the potential to flood islands and render some habitats unsuitable or unavailable. Furthermore, sea level rise in combination with severe storms has the potential to erode away islands, particularly those with open water around them. Table 7.4 (a version of Table 5.2), illustrates the major effects for the primary species considered in this book. Excluding cases of single birds nesting in 1 year (marked by an asterisk, *), there were 131 colonies sites for the tern/skimmer/gull species. Some were occupied continually, some for only a few years. Colonies that are gone because the island is submerged or eroded leaving no available habitat are indicated with dark shading. Light shading indicates sites rendered unsuitable mainly because of flooding. Unshaded cells marked with "(G)" indicate that

Table 7.4   Status of Barnegat Bay Tern, Gull, and Skimmer Colonies

Colony Name	Black Skimmer	Common Tern	Forster's Tern	Laughing Gull	Great Black-Backed Gull	Herring Gull
Salt Marsh Colonies (n)	25	37	4	22	18	25
NW Lavallette		39–750**			2–4**	2–38**
SW Lavallette		25–200			2–36	2–450
N. Lavallette		15–795**			3–65**	3–186**
S. Lavallette		10–600			2–6	3–54
Little Mike's		2–600**				
Buster	2–3	5–826		14–20		2–4
East Point		18–325		2–16		
Clam	1–4	45–980**	20–355	45–467**	18–60	600–800
High Bar		9–28		41–123		2–25
East Vol	2	5–40		35–142	1–6	5–46
West Vol	15–36**	10–400**		2–40 (G)	2–34**	40–157**
Gulf Point	1–2	42–110				3
W. Sloop Sedge	13	87		2–121	5–67**	100–204**
E. Sloop Sedge		1*		1–14	2–7	2–50
Sandy Island		1*			2–54**	50–61**
Flat Creek	2–4	5–33		35–36**		
West Carvel	2–10 (G)	19–58 (G)		21–40	2–46**	100–150**
East Carvel	1–45 (G)	50–265 (G)		19–124 (G)	2–10**	12–35**
West Log Creek	8–20	5–20		46–197		
Log Creek	3–14	1–63		23–85		
Pettit	1–20	12–600**		2–6	1*	1–3
Cedar Creek	1–8	10–114		26–42		
E. Cedar Bonnet		1–4				
NW/SW Cedar Bonnet	1–13**	6–143**		2–3		1–4
Thorofare	2	20–58				
Egg	1–8**	11–145**		16–65**	3–56**	3–360**
East Ham	3–55	10–42		10–25	4–45	20–100
West Ham	4–28 (G)	20–331 (G)			3–65**	10–250**
Marshelder	82	23–475			3–5	20–86
Good Luck	1–2	1–212**		6–180**		
East Sedge	8–35 **	10–83 **				11–250**
East Long Point	1–2	2–32		18–400	2–3	2–6
West Long Point		43		16–312	2–3	2–8
Little	2	21–235				
Mordecai	7–271	10–106 **		2–6		2–12**
Hester		6–121**	25–121**			
Goosebar		21–34	25–64			
Barrel	6–14	48–125**	25–223**			
Middle Sedge		12–90				
Sandy Islands	N = 1	N = 3		N = 1	N = 2	N = 2
Big Mike's		5–112				
Dredge Spoil		5–64 (G)			3–120	15–680
Tow	16–375 (G)	5–123 (G)		2–121	6–40	2–118

Note: This table shows the range in numbers of pairs present on islands over all the years of occupancy. Dark shading represents islands that are gone (washed away, underwater, or broken apart), and light shading represents colonies where suitable habitats are no longer available or they flood in most years. No shading indicates islands still suitable. Cells marked with (G) indicate areas displaced by large gulls; double asterisks (**) indicate colonies active in the past 4 years (2012–2015).

* Not counted as a colony because only one pair nested without conspecifics or other species.

** Currently viable colonies with nesting at some time in the past four seasons (2012–2015).

the birds abandoned the site because of the presence, aggression, and/or predation by Herring or Great Black-backed Gulls. Unshaded cells marked with a double asterisk (**) are colonies active in the past four season (2012–2015). Other unshaded cells indicate sites that have available, suitable habitat but have not been occupied recently.

Of the total of 131 colonies (all species combined), 59 (45%) are in the lightly shaded suboptimal category, and 15 (11%) are unusable or gone. Almost two-thirds of all the colony sites occupied by terns, Skimmers, and gulls have become unsuitable or unavailable. In the early part of our study, expanding gull populations were the main culprit rendering former tern nesting islands unsuitable (frankly dangerous). Common and Forster's Terns, Black Skimmers, and Laughing Gulls, four species vulnerable to larger gulls, accounted for 88 colonies. Of these, 46 (52%) were lost to rising water and 8 (9%) were lost to gulls. As gull populations have stabilized and even shrunk, the main proximate factor reducing habitat availability is the shrinkage of island space above flood tides. The ultimate factors are sea level rise and subsidence.

Black Skimmers, endangered in the State of New Jersey nested either on sand or on wrack on salt marsh islands. In most of their range, including the Cape May area of New Jersey, Black Skimmers nest on sand islands or beaches. Along the central New Jersey coast, beach-nesting Skimmers were crowded off the beaches by people (Burger and Lesser 1979; Burger and Gochfeld 1990b). In the 1970s to the late 1990s, there still remained some fairly large sandy patches on several salt marsh islands (East Carvel, West Carvel, Mordecai, and Tow Island). In addition, there were relatively high wrack mats on other islands where the Skimmers nested (Buster Islands, East Point, Gulf Point, East Ham, Marshelder, Pettit). However, these habitats have shrunk. Small patches of sand remaining on Tow and Mordecai are washed over by monthly high tide, and the rest of Tow Island was invaded by gulls. In response, Black Skimmers have attempted to nest on higher places on the marsh on *Spartina patens*, but they are not usually successful. By 2010, there were few Skimmers left in the Bay, most having joined large colonies in southern New Jersey. However, in 2015 a slight population buildup occurred.

Common Terns are widespread in the Bay, but only 13 of 37 colonies (35%) remain active today. Some islands become too low to have successful nesting Common Terns because normal tides wash over the nests and eggs, there is no wrack left, and some colony sites are completely under water. In the 40 years of this study, all four fates occurred. Lost colony sites included disappearance (4 of 37 colony sites = 11%), loss of suitable habitat (13 of 37 = 35%) and gull invasion (3 of 37 = 8%). These data indicate that the islands in Barnegat Bay are in transition with respect to nesting Common Terns. As sea level rises, islands are washed and eroded away by severe storms and hurricane; others lose wrack or other suitably high nesting sites. Others fail consistently because monthly high tides flood out most nests. Of those with suitable habitat, a quarter have lost nesting terns after the invasion of large gulls.

Forster's Terns are an interesting case. New Jersey is near the northern limit of its regular breeding range on the Atlantic coast. It nests on Long Island, where the first nest was found in 1981 (Zarudsky 1981), and it has nested in Massachusetts (McNicholl et al. 2001). In the early 1970s, Forster's Terns nested only on one island in the southern part of Barnegat Bay. These terns nest mainly in *Spartina alterniflora*, rather than on wrack, and nests are usually adjacent to small permanent ponds that have some prey fish. Their nesting habitat tends to be wetter and their nests tend to be taller than for Common Terns. Forster's Terns colonized Clam Island (middle of the Bay), but with sea level rise, the places they nested were routinely flooded out, and they no longer nest on Clam Island. Their nesting is now confined to islands in the south.

Laughing Gulls usually nest on the higher places in *Spartina alterniflora*, and build substantial nests (Burger 1996b, 2015a). Some nests are built up much higher than usual (30–50 cm) to elevate the eggs above the tide. Sometimes nests simply float up, and settle once the tides recede (Figure 7.11). It would seem on the face of it that such places would remain because sea level rise should simply shift the location of the *Spartina alterniflora* on the islands. However, this is not the case. More than half of former colony sites were simply eroded away by severe storms and high tides (East

(a)   (b)

Figure 7.11   Nest of a Laughing Gull built up to withstand the recent tidal inundations—note the clam shell being incubated (a); and (b) a Great Egret nest is on the ground because the shrubs have died from tidal inundations and no longer support heavy nests.

Long Point, West Long Point), or suitable habitats disappeared because of regular flooding. Other islands became unsuitable when Herring Gulls forced the smaller Laughing Gulls out (Burger and Shisler 1978c, 1979, 1980a; Burger 1979a). Faced with large gulls and rising water, Laughing Gull numbers have not rebounded after the slaughter conducted over a several-year period in the 1990s at John F. Kennedy (JFK) International Airport (see Chapter 6; Brown et al. 2001a).

At present, Herring Gulls and Great Black-backed Gulls have not suffered as much from sea level rise as the Laughing Gulls because they nest on the highest places on the islands, and they have a wide range of places they will nest (in *Spartina patens*, under bushes, on spoil). Of the 25 colony sites used by Herring Gulls, 11 (44%) are currently or recently active. However, Herring Gulls have abandoned Clam Island, Northwest, and South Lavallette because of flooding (Burger and Shisler 1979; Burger 1980a, 2015a). Herring Gull numbers have also declined elsewhere in the Bay and regionally. For Great Black-backed Gulls, 50% of colony sites are recently active.

Herons, egrets, and ibises nest on sandy islands (dredge spoil) and on the highest parts of salt marshes with *Iva*, *Baccharis*, *Phragmites*, Poison Ivy, and Cherry bushes (Burger 1978c, 1979c). Although heronries do not usually get flooded out during the breeding season, the habitat gets destroyed during the nonbreeding season. That is, strong storms with high winds and very high tides wash over salt marsh islands, even with low shrubs. Continual overwash results in high salt inundation, and eventually the *Iva* and *Baccharis* die back or are broken down so often they do not survive. Similarly, some areas of *Phragmites* and Poison Ivy have disappeared. These islands are no longer suitable because there are no longer strong bushes to serve as the nucleus (Clam Island, NW Lavallette, North Lavallette, East Point, and West Carvel). Species that nest on the ground (ibises, Night-Herons) have had difficulty because some very high storm tides have washed out ground nests in some years. Black-crowned Night-Herons also nest low in bushes, and such nests so far have been relatively free from flood tides (see Table 7.9).

In Table 7.5, we summarize the fate of colonies. Islands washed away correspond to the dark shading in Table 7.4. Islands that have lost most or all suitable habitats correspond to light shading in Table 7.4. Islands with suitable habitat but subject to flooding and currently unoccupied are unshaded in Table 7.4 and those considered "Lost" because of gulls' presence correspond to (G) in Table 7.4. The Viable Habitat/Active Colony indicates those that have had nesting birds in the past 4 years.

Three sandy islands in Barnegat Bay are used by nesting birds, including Big Mike's (next to Little Mike's), Dredge (in Barnegat Inlet), and Tow Island (by Beach Haven Inlet). These are not included

Table 7.5 Reasons for Colony Abandonment for Colonial Waterbirds Nesting in Barnegat Bay

Species—Total Islands Used	Islands Washed Away	Islands Where Suitable Habitat Disappeared	Islands Too Low for Successful Nesting	Islands Lost Because of Gulls Competition	Viable Habitat
Black Skimmer—25	2	12	4	3	4
Common Tern—37	4	13	4	3	13
Forster's Tern—4		2			2
Laughing Gull—22	3	10	3	2	4
Herring Gull—25	4	7	3	0	11
Great Black-backed Gull—18	2	5	2	0	9
Great Egret, Snowy Egret—10	0	5	1	0	4
Glossy Ibis, Black-crowned Night-Heron—11	0	5	3	0	3

Note: The number of colony sites fitting each category is shown. Islands given as too low for successful nesting are those that were completely gone or in the process of disappearing.

in Table 7.5; the first two are much higher than the salt marsh islands, and were constructed mainly of sand deposited by the Army Corps of Engineers. Only Dredge and Tow have active Herring and Great Black-backed Gull colonies, and all three had heronries at one time. Only the Dredge Island heronry is currently (2012–2015) active. All three had Common Terns nesting there, and Dredge and Tow had Black Skimmers, but these colonies were forced out by gulls. Tow had a small salt marsh once used by Common Terns, but this disappeared with sea level rise. Herring Gulls also outcompeted nesting terns, Skimmers, and Laughing Gulls for higher nest sites on the DelMarVa Peninsula (Erwin et al. 1981).

Sea level rise and the loss of islands and suitable habitat are a major contributor to the declines in colonial-nesting birds in Barnegat Bay (Figure 7.12). Initially, competition with the larger Herring

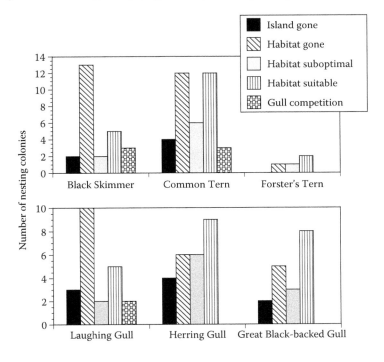

Figure 7.12 Effect of global warming and sea level rise on island suitability for nesting. Fates shown are from the early 1970s to 2015. Suitable islands are shown as vertical lines.

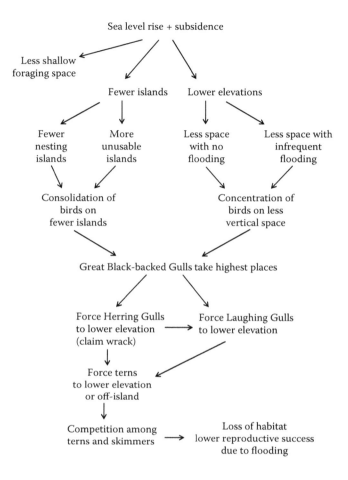

Figure 7.13   Schematic of cascading effects with increased competition at the top (Great Black-backed Gulls) and loss of suitable habitat free from floods (because of global warming) at the bottom.

and Great Black-backed Gulls forced the smaller species from some islands. Today, it is thus a matter of the cascading effects of sea level rise, subsidence, reduced suitable habitat, and competition among species for safe nesting sites (Figure 7.13). Displaced birds have had to move elsewhere.

FORAGING BIRDS IN BAYS AND ESTUARIES

In addition to nesting space, coastal nesting birds and migratory species rely on coastal environments for foraging opportunities. Foraging spaces in coastal environments include the open ocean, open bay waters, mudflats, salt marsh creeks, coastal marshes, costal shrubs and forests, and mixed-species habitats (such as *Iva, Baccharis, Prunus, Phragmites*). Decreases in these habitats, whether from human-induced habitat loss (e.g., development, disturbance) or sea level rise, results in loss of foraging space. This is particularly true in coastal environments where foraging time is often limited by tidal phase. Mudflats are only exposed for a limited time during each tidal cycle, and when optimal foraging time according to tidal cycle is added to diurnal cycles, foraging time may be even more limited (Burger et al. 1977b; Figure 7.14).

Figure 7.14   Snowy Egrets foraging on small fish along a creek bank.

Most studies examining mudflat "space" have modeled predicted sea level rise (Galbraith et al. 2002, 2014). Although loss of tidal mudflats affects all the bays and estuaries along the Atlantic coast, its effect will be most severe in places where concentrations of migratory species are highest. For example, Delaware Bay is one of the most important gathering places (stopover) for spring migrating shorebirds (Niles et al. 2008, 2009), and for a 3-week period in May over a million shorebirds once passed through. Today, the shorebird numbers are a 20th as great as they were 30–40 years ago. During their migratory stopover, Red Knots, Ruddy Turnstones, and Semipalmated Sandpipers feed mainly on Horseshoe Crab eggs (Tsipoura and Burger 1999), and gulls also exploit this temporary abundance of food.

Changes in the ability of shorebirds to obtain enough food during stopovers could affect the timing of annual cycles (Carey 2009). They may arrive on Arctic or Subarctic breeding grounds with insufficient fat reserves to take advantage of the short breeding season. Shifts in the geographical distribution of birds may occur (Huntley et al. 2006), which will affect population dynamics, as well as habitat use and annual cycles (Crick 2004). And these are the direct effects on the birds. Effects on their prey and food supply provide additional feedback loops (Montevecchi and Myers 1997; Barbraud et al. 2012). If birds breed later, as has been shown for Antarctic species (Barbraud and Weimerskirch 2006), then there may be a mismatch between chick hatching and food supplies resulting in starvation. Mismatches may occur on migration, for example, if Horseshoe Crabs finished breeding in April, before the migratory shorebirds arrived.

IMPLICATIONS FOR FUTURE POPULATIONS

One of the interesting aspects of global change is whether species can adapt to these changes, and if so, how? Although in this chapter we have dealt mainly with climate change and sea level rise, global change also includes habitat loss, exotic species, and human harvesting or use (discussed in earlier chapters). Many species are imperiled by these changes, including habitat loss (Erwin et al. 2007a,b; Chaine and Clobert 2012), exotic and invasive species (Sih et al. 2010), pollution (Burger and Gochfeld 2001b; Butchart et al. 2010), and harvesting (Nettleship et al. 1994). From our vantage point, in a small boat looking out over Barnegat Bay, it is the loss of nesting habitat that is most devastating. When we study flocks of migratory Red Knots, it is the excessive harvesting of Horseshoe Crabs that threatens the shorebirds. There is a new potential threat—the expansion of intertidal oyster culture on the intertidal area off the very beaches where shorebirds feed. The harvesters, fishermen, and medical technology will not care about sea level rise, but a small change in water temperature may have a significant impact on Horseshoe Crab breeding behavior and cycles.

These negative impacts are amplified by the intense human development along coasts that prevents natural succession and erosional processes from occurring. There is simply no room in most of the coastal areas from Boston Harbor to Chesapeake Bay for the land to retreat, building salt marshes as it moves. In some places, such as Barnegat Bay (this book) and Chesapeake Bay (Erwin et al. 2007a, 2010), nesting habitat for birds is disappearing at an alarming rate. Massive restoration efforts may reverse some of the habitat losses (see Chapter 15), but they will require collaborations and governmental support.

In Barnegat Bay, it is likely that some Common and Forster's Terns—species that live for 20+ years—will remain and nest in the future, moving higher on available marshes. Their populations may continue to decrease because of loss of suitable and available nesting sites, until their once familiar calls may fade from memory. They have seriously declined in Barnegat Bay, and without management to build wracks on salt marsh islands (Burger 1989; Burger and Gochfeld 1991b; Palestis 2009) or massive broadcasting of spoil to raise elevations, they may well disappear as a nesting species in the near future. Laughing Gulls may continue to decline as less *Spartina alterniflora* is sufficiently high to support nests that do not flood out. Herring and Great Black-backed Gulls will continue to flourish until the high ground disappears. Heronries will survive as long as there are bushes and reeds.

The data from Barnegat Bay suggest that there are no clear winners with sea level rise. Many of the nesting colonial species that were previously forced from nesting on barrier island beaches by people, and were then competing with large gulls for the higher places on salt marshes, are now being forced from low-lying salt marsh islands. The overall population numbers have declined in Barnegat Bay for terns, Skimmers, Laughing Gulls, and Snowy Egrets. Whereas Black-backed Gulls are increasing, Herring Gulls are declining. Humans, too—the "baymen" of the past century—are losing habitat, as some of their old hunting and fishing lodges and shacks fall victim to flood damage or fall into the bay (Figure 7.15).

The question remains—which species will adapt and how quickly, and which will not (Sih 2013)—and how can managers make use of their behavioral responses to human-induced, rapid environmental change to increase or protect bird populations? The plasticity of behavior, perhaps in the ability to enlarge the nesting or foraging space that a species requires, will influence which species adapt and which do not. It may not be an age for specialist species.

The plight of some colonial nesting birds in the Northeast bays and estuaries considered in this book is made more severe by the fact that this region is a hot spot of accelerated sea level rise; sea level rise is 3–4 times greater in this region than the global average (Sallenger et al. 2012). Furthermore, increases in the maximum high tide results in more frequent and catastrophic flooding of nests (van de Pol et al. 2010). Regional studies, such as this one in Barnegat Bay, in

Figure 7.15   Salt marsh houses and old shacks in Northeast bays, used for hunting, were once common on salt marsh islands in Barnegat Bay. Here, the old shack on Clam Island (a) was partly burned down and swept away by high tides. Burger lived on Clam Island for 3 years in the 1970s when the island supported a thriving colony of Laughing and Herring Gulls, some others remain (b).

Chesapeake Bay (Erwin et al. 2011), and on the salt marshes at Jamaica Bay (New York City; Hartig et al. 2002) indicate an overall decline or degradation of salt marshes.

SUMMARY AND CONCLUSIONS

Birds that nest and forage in coastal areas face habitat loss, invasive species, competitors, predators, and human disturbance. The threats from sea levels rise and subsidence resulted in loss of many salt marsh islands in Barnegat Bay, as well as a loss of habitats on low-lying islands. These changes have significantly affected species nesting in lower elevations, especially Black Skimmers, Common Terns, Forster's Terns, and Laughing Gulls. Some Heronries, as well Herring and Great Black-backed Gulls' nesting habitats have been altered by salt intrusion. On the barrier beaches, habitat was lost for beach-nesting species such as Least Tern and Piping Plover, and traditional beach-nesting species (Common Tern, Black Skimmer) were forced from beaches by increased human use and disturbance in the 1960s and 1970s, as well as intrusions of human commensals (dogs, cats).

The changes in Barnegat Bay are similar to those that have occurred in other estuaries and bays in the Northeast. Islands that remain suitable for nesting are those that are higher, like many in the New York–New Jersey Harbor Estuary. However, there is erosion of islands and loss of suitable habitat is general, even with vigorous management (e.g., Falkner Island, Massachusetts islands). A rise of a meter or more over the next century will have major effects on nesting birds. Colonies that are very high in elevation, or are rocky and able to resist erosion (e.g., Great Gull Island) will survive and flourish. It is likely that Common Terns unable to nest successfully in Barnegat Bay and salt marsh islands in Long Island will move to these higher colonies. Terns from various breeding areas gather together in large, postmigratory assemblages before heading to wintering grounds in South America (Nisbet et al. 2014). This will facilitate birds returning to large successful colony sites the following spring, and will have the effect of putting more eggs in fewer "baskets" or colonies, making the population more vulnerable.

Similarly, the rise in sea level experienced along the Northeast from Chesapeake to Cape Cod will continue to have major effects on mudflats and foraging habitat for shorebirds and wading birds. Herons, egrets, and ibises that forage in shallow water will experience a decrease in foraging space, with a decrease in prey availability. Specialists, such as Black-crowned Night-Herons and Clapper Rails that eat Fiddler Crabs (*Uca pugnax*), will have added foraging problems because of decreased space. Common and Roseate Terns that forage off the coast will not experience foraging difficulties as directly, but Forster's Terns that forage on small prey fish in salt marsh pools may face difficulties if they have frequent connections to the sea, preventing fish from concentrating. The loss of tidal mudflats and other coastal foraging areas suggests that migratory pathways for some species, such as shorebirds, may be in jeopardy. With less horizontal space to forage in during low tide, the suitable places will be more in demand, increasing competition, and concentrating shorebirds in places that are more visible to predators. The loss of tidal mudflat habitat to global warming will make threats (e.g., human activities, contaminants) even more important.

CHAPTER 8

Overview of Ecotoxicology for Birds

INTRODUCTION

Birds and other organisms face a wide range of stressors and threats, operating at different time scales and at different points in the life cycle. These include habitat loss and human disturbance, predators and competitors, weather variables and climate change, and contaminants (Figure 8.1). In most cases, these stressors interact; some mitigate the effects of others, whereas others enhance the effects. For example, severe storms that occur when food is already in short supply leads to increased reproductive losses. On the other hand, increased predation can reduce competition with neighbors because of decreases in density of nesting or foraging birds. Examining how stressors act alone or interact together is critical for conserving or managing populations. Understanding the role of metals and other contaminants in the environment, and in the health and well-being of organisms requires placing ecotoxicology and toxicodynamics within a framework of the biology and ecology of organisms.

The different stressors that affect survival and reproduction in birds vary for different species, and at different times (Figure 8.1). Sometimes, habitat quality or loss is the greatest threat, whereas at other times it is adverse weather, predators, competitors, or human disturbance. For example, loss of habitat is severely affecting populations of terns and Skimmers in Barnegat Bay (see Chapter 5). Predators contributed the most to loss of the Common and Roseate Tern colony at Cedar Beach (Long Island). Competitors (Herring and Great Black-backed Gulls) have rendered many former tern colony islands unsuitable, and are contributing to declines in the reproductive success of Laughing Gulls in Barnegat Bay. Human disturbance coupled with increases in predators contributed to the decline of Least Terns and Black Skimmers breeding on coastal barrier island beaches. Similarly, contaminants, such as dichlorodiphenyltrichloroethane (DDT) caused the decline of fish-eating birds such as pelicans in the 1960s (Anderson and Hickey 1970). The effect of contaminants on populations is likely when there is excessive exposure, such as happened with DDT in the 1960s, selenium at Kesterson in the 1980s (Ohlendorf et al. 1986a,b; Ohlendorf 2011), lead poisoning of albatrosses on Midway (Sileo and Feffer 1987; Burger and Gochfeld 2000b,c), and mercury currently (Evers et al. 2010, 2011). In the first case, the impact was apparent across the Northern Hemisphere, whereas in the other cases, the impact is local. Although Figure 8.1 shows a schematic layout of potential stressors on colonial birds, the actual percentage each contributes to reproductive losses or decreases in survival depends on the species and ecological conditions. Synergisms between different stressors can amplify the effects, and this is perhaps the greatest current worry with contaminants.

In this chapter, we provide an introduction to contaminants and examine ecotoxicology in birds. Many books have been devoted to ecotoxicology (Moriarty 1983; Hutchinson and Meema 1987; Eisner and Meinwald 1995; Linthurst et al. 1995; Walker et al. 1996; Dell'Omo 2002; Hoffman et al. 2002) and ecological risk assessment (Kolluru et al. 1995). This chapter provides a background

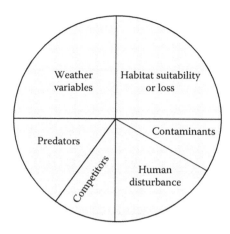

Figure 8.1   Pie chart of potential stressors affecting reproductive success and survival in birds. The relative relationship among stressors varies by species and environmental conditions (based on our observations across many species, colonies, and years).

relevant to the discussion of metals in the chapters that follow. We introduce organic and inorganic compounds that contaminate aquatic systems, and describe environmental fate and transport, toxicokinetics and toxicodynamics, and the challenges of establishing causal relationships in ecotoxicology. In the next chapter, we examine the toxicity and effects of individual metals on birds, with particular reference to biomonitoring using bird eggs and fledgling feathers. This information is provided in one place so that it is not repeated in each chapter devoted to metal levels in terns and Skimmers (Chapter 11), gulls (Chapter 12), and egrets and herons (Chapter 13).

Every environment is composed of chemicals, although we think of pristine, preindustrial environments as chemical-free. All living organisms contain the elements carbon, hydrogen, oxygen, sulfur, phosphorus, calcium, iron, and nitrogen organized into lipids, proteins, carbohydrates, and other key ingredients of cells. Living systems require some additional elements to function smoothly—sodium, potassium, calcium, iron, copper, chromium, manganese, magnesium, selenium, and zinc are among the essential elements for most forms of life. Too little of these elements is unhealthy and is referred to as deficiency states. Too much of any one of these elements can become toxic. Because these elements are essential for life, they are present in varying degrees in plants and animals that are eaten by other species, including humans, so they are obtained in normal or physiological quantities from diets. These naturally occurring sources of elements (including metals) are enhanced in areas where the soil has high quantities of an element owing to natural or anthropogenic sources. Likewise, airborne contamination from natural, industrial, agricultural, combustion, or urban emissions adds elements to the air or water, often in forms that do not occur naturally.

In addition to lipids, proteins, and carbohydrates, there are many combinations of compounds containing carbon and hydrogen, which are not necessary for life and which may be harmful in varying degrees. Some of these occur naturally; many do not. The "nots" are produced deliberately, or are unwanted by-products or waste products in industrial processes, and are referred to as anthropogenic in the environment and xenobiotic (foreign compound) when they enter the body. For many of these chemicals, the toxic properties have been harnessed for their beneficial uses as biocides (also called pesticides). These include insecticides, algaecides, mitocides, nematocides, and fungicides. These chemicals are considered beneficial when used correctly in time, place, and amount, and are harmful when they get into ecosystems and impact nontarget organisms. There are tens of thousands of organic chemicals, of which only a few have had detailed studies on their toxicologic or ecotoxicologic effects. Xenobiotics may be present in the air, water, soil or food, and may enter our bodies through inhalation, ingestion, or rarely through the skin.

Chemical compounds have many different properties that influence how they move through the environment, enter the body, and move through the body to target or excretory organs. Chief among these is water solubility. Most organic compounds and many inorganic compounds have low solubility in water, but are soluble in nonpolar solvents (organic solvents) or in lipids. The acidity of the environment changes the physical properties of chemicals, and many chemicals are more readily taken up from an acid environment than from a neutral environment.

BACKGROUND ON ORGANIC AND INORGANIC POLLUTANTS

Organic Compounds

This book is about the inorganic chemical pollutants, mainly heavy metals, so we make only brief reference to organic contaminants that are relevant to our discussions of exposure and toxicity of metals in aquatic birds. Many metals form organic compounds that are of ecologic importance, but by organic contaminants, toxicologists typically refer to chlorinated pesticides such as DDT, polychlorinated biphenyls (PCBs), dioxins, polyaromatic hydrocarbons (PAHs), and a host of recently reported commercial organic compounds that are ubiquitous in the environment. Metals can affect many organ systems, including the reproductive system, so it is important to mention that organic compounds can also affect reproduction. For example, chlorinated hydrocarbon insecticides such as DDT, dieldrin, and chlordane are persistent in the environment and in the body, where they are stored in lipid component of cells or in fatty tissues, and can exert severe endocrine disruptive properties. The discovery that DDT caused eggshell thinning and reproductive failure in birds was a pivotal discovery in environmental toxicology, although it took years for definitive evidence and a persuasive mechanism to be explained. David Peakall conducted seminal research on how pesticides induced breakdown of reproductive hormones, altered vitamin D metabolism and calcium uptake, and interfered with calcium deposition on the egg shell (Peakall 1993). When birds incubated the thin-shelled eggs, the eggs broke under their weight, leading to complete failures in colonies of pelicans and nests of peregrines (Figure 8.2). Other commonly encountered organics include the PCBs produced deliberately for their unique properties, as well as dioxins, furans, and PAHs produced as unwanted by-products in pesticide synthesis or combustion (Eisler 1986a,b). The Brown Pelican (Figure 8.2) was one of the species showing severe population declines and local extinctions because of DDT.

Figure 8.2 Colony of Brown Pelicans, a species whose populations plummeted following exposure to DDT, which resulted in eggshell thinning.

In the past 20 years, attention has focused on chemicals used as fire retardants (polybrominated diphenyl ethers) and on perfluoro compounds perfluorooctane sulfonate (PFOS) and perfluorooctanoic acid (PFOA) used in the manufacture of coatings such as Teflon™ (EPA 2014b). PFOA and PFOS are ubiquitous in organisms, including humans, and their toxicity (e.g., on the endocrine system) is being worked out (Post et al. 2012). For most of these chemicals, there is very little systematic data on effects or levels in birds or any other organisms. All organisms excrete in their urine or feces natural bioactive hormones or hormone metabolites from our naturally occurring estrogens and androgens. Most recently, attention has focused on humans as sources of the natural hormones and metabolites that they excrete into sewage systems (Carvalho et al. 2015). Likewise, most of the drugs we consume are metabolized and excreted. Thus, many active metabolic breakdown products of drugs end up in the sewage system. However, sewer systems designed to remove infectious organisms do not remove hormones, drug metabolites, or nanoparticles, so these are discharged to rivers, bays, and estuaries. This is recognized as an international problem (Houtman et al. 2014; Kostich et al. 2014; Sun et al. 2015a).

Inorganic Compounds and Metals

The numerous compounds that do not contain carbon are referred to generally as "inorganics." Among these are the so-called "heavy metals," many of which (e.g., chromium, manganese, copper, iron, selenium, zinc) are essential in minute quantities, and toxic at high levels. Among the heavy metals with no known essential functions in any organisms are mercury, lead, and cadmium. Metals may occur in pure metallic form, or as various kinds of ores including sulfides, oxides, and other more complex compounds in the Earth's crust, where they have lain undisturbed for billions of years until the past two millennia (mainly until the last two centuries). Mining and extraction industries extract, smelt, process, and purify these metals for economic use. The mining leaves behind slags and tailings with varying degrees of metal contaminants of varying ecological availability, whereas smelting releases metals and various gases to the atmosphere.

Although the metals occurring naturally in the earth are products of the early Earth environment and radioactive decay, the main ecotoxicological harm to ecosystems or people occurs when they are released, either by natural processes (volcanoes, erosion) or anthropogenic processes. Two elements, arsenic and selenium, are referred to as metalloids because of their chemical properties and their position in the periodic table. However, for the purposes of our discussions, they will be included under the rubric of "heavy metals." Our research focused on heavy metals in aquatic systems, and this topic occupies the remainder of this chapter.

GENERAL PRINCIPLES AFFECTING METALS IN THE ENVIRONMENT

Each metal is an element, and with the exception of a few radioactive isotopes, metals are not created or destroyed in the environment or in the body. Each metal may occur in several forms that chemists refer to as chemical "species." The process of analyzing metals to identify the species is referred to as "speciation." This term usually applies to broad categories of a metal, such as "organic species" or "inorganic species." These terms are somewhat confusing to biologists who use "species" to refer to particular kinds of plants or animals, and "speciation" as the process of evolving changes in the kind of plant or animal.

Metals may form inorganic salts such as chlorides, oxides, and sulfides, which have different properties such as solubility, bioavailability, and toxicity. For example, most metal chloride compounds are soluble in water, whereas many metal sulfides have low solubility. Metals may also form organic species, for example, through the process of methylation. These organic species behave very differently in the environment and in the body, both in terms of their fate and transport, and in terms

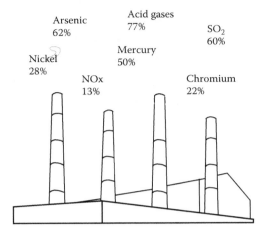

**Portion of U.S. Air Pollution That
Comes from Power Plants**

Figure 8.3   Contribution of coal-fired power plants to U.S. air pollution. (From the EPA: Cleaner Power Plants; http://www.epa.gov/mats/powerplants.html, 2014c.)

of their toxicity. Examples of organometallic compounds are methylmercury, which is more toxic than either elemental mercury or inorganic mercury compounds. On the other hand, most organo-arsenic compounds have much lower toxicity than inorganic arsenicals, such as arsenic trioxide.

Metals may form positive ions through the loss of electrons from their outer orbit. Most commonly encountered metals form only positive ions referred to as cations. Metals may have a +1, +2, +3, +4, +5, or +6 charge. In nature, metals occur mainly in "ores," which must be mined, crushed, and smelted so that the metals can be extracted. Smelting is one of the main sources of atmospheric pollution by arsenic. The 600 coal-fired power plants in the United States remain the major source of arsenic and mercury air pollution in the country (EPA 2014c; see Figure 8.3).

Metals enter the environment either through natural processes (volcanic, erosion, sandstorms) or anthropogenic releases from mining and milling, utilities, cities and waste sites, farms, or factories. The metals may be emitted from stacks and transported in the air and fallout on the land, or may be discharged in effluents into waterbodies. Since the 1970s, environmental regulations in the United States have significantly reduced both air emissions and water effluents, resulting in substantial improvement in air quality and water quality in many parts of the country (Lioy and Georgopolus 2011). However, some contamination continues and may even increase, such as the emissions of mercury from coal-fired power plants. Regulations require enforcement, and some industries have sued the Environmental Protection Agency to keep it from enforcing all the regulations required under the Clean Air Act and other laws that were designed to protect human health and the environment (EPA 2014c).

Sources of Metals

Metals occur in the earth's crust and in seawater in highly variable concentrations. Volcanic activity spews huge amounts of metal compounds into the atmosphere in the form of gases and dusts; the latter may remain airborne for years. More subtle erosion also moves surface soil downhill and downstream. Metals in seawater can also become airborne through dynamic interactions at the sea surface in a process called evasion. However, the anthropogenic release of metals from the earth's crust by mining and processing far outweighs these natural processes. Mining and quarrying have greatly increased the transfer of metals from the earth to the surface, air, water, and ecosystems. Smelting releases many chemicals into the atmosphere, whereas the wastewater or effluent contaminates surface water and, ultimately, groundwater. Metals are also released during many uses, during combustion of various fuels, and during disposal.

Once in the environment, metals and their compounds undergo a variety of complex geochemical processes including weathering, oxidation–reduction, dissolution in waters, and complexing with organic compounds. The biogeochemical processes also involve organisms. Even in prehistoric premining periods, organisms participated in the dynamic movement of metals, referred to as biogeochemical cycling. Organisms consume metals, distribute them through the body, modify them, and eliminate them as waste.

Mixtures

Organisms are not exposed to single chemicals, but to complex mixtures of chemicals with varying degrees of solubility and bioavailability. Mixtures of contaminants can act independently on the same or different endpoints, or can interact in various ways. One chemical might enhance the synthesis or retard the degradation or transport of another. Or chemicals may interact directly, forming new compounds or complexes with special properties, as in the case of the reported synergism between cadmium and chlorpyrifos (Chen et al. 2013). Chemical mixtures may have additive effects (the impact of the mixture is equal to the sum of the impacts of each individual component), or may be subadditive if one chemical reduces the toxicity of another. Some interactions are superadditive or multiplicative, resulting in synergism in which the resulting toxicity is greater than either chemical acting alone (Faria et al. 2010).

Beneficial Uses of Metal Toxicity

Early mining was conducted to obtain minerals for jewelry and pigments, and later for metalworking. Toxicity was well recognized so that slaves were used to work the rock face (Hunter 1955). In recent centuries, society has harnessed the toxicity of many metals for beneficial uses. The generic term biocide is used to encompass agents that kill any form of life, including microorganisms, plants, or animals. The pantoxicity of mercury has made it valuable as antiseptics (mercurochrome), fungicides, stabilizers in paint, preservatives in vaccines, and other uses. Organic mercury and organic tin compounds have been widely used as antifouling additives in marine paints to prevent the growth of barnacles and other organisms on boat hulls. This preserves streamlining and paint integrity, but "works" through the continual release of the toxic metal into the environment (Omae 2006). Lead arsenate was one of the most widely used insecticides prior to the availability of DDT, and residues of this pesticide still contaminate many orchard landscapes (Delistraty and Yokel 2014), rendering them uninhabitable by people (Hood 2006). Arsenic, familiar as a poison, has been used to defoliate crops in war and peace (Eisler 1988b). Agent Blue was an organic arsenical defoliate used in Vietnam, although Agent Orange was the primary one used (60% of use). During the extraction, production, distribution, application, and inappropriate disposal of these biocides, there is widespread exposure of nontarget species (Tardiff 1992). Organomercurials were used as fungicides to preserve seeds and prevent mold. This use has been banned for 50 years.

EXPOSURE ASSESSMENT AND TOXICOKINETICS

In the environment, metals can be present in the air, in soil or dust, in water or sediment, and in foods that are consumed deliberately or inadvertently. Once a metal or metal compound enters the environment, it may change form. It may be oxidized or reduced, or methylated or demethylated. It may dissolve in water. It may adhere to fine particulates (dust in air, or fine particles in water).

It may form complexes with organic matter. It may also change compartments or media. Volatile substances may evaporate from soil to air, or may escape from the water surface to air (evasion). Airborne substances may be deposited as particulates (dry deposition) or may be washed out of the atmosphere by rain events (wet deposition). These may then fall onto land and be reentrained into the air as dust or may fall onto water, and be converted by microorganisms and enter the food chain. Over most of the United States, wet deposition greatly exceeds dry deposition (NJ Mercury Task Force 2001), but over high-altitude arid regions the converse occurs (Sather et al. 2014). Eventually, some portion of the compound may settle in the sediment, form an insoluble compound, lay dormant, and will remain out of the ecological food web for centuries. Other portions are consumed by organisms and enter the food web.

Fate and Transport

The umbrella term for the events described above (source to organism) is "environmental fate and transport." This covers the movement from source, through environmental medium, food chain bioamplification, to contact with the organism. The term "toxicokinetics" refers to the sequence of events in the animal or plant, including contact, absorption into the bloodstream, and distribution to various organs or elimination by various excretory routes (typically urine and feces). The term "bioavailability" (or bioaccessability) refers to the ability of a metal to be released from its substrate, or conversely how tenaciously it is bound to the substrate. This is closely related to absorption, which depends on host properties such as the acidity of digestive processes or the regulation of metal transporters in the gut or respiratory tract. A substance with low bioavailability may pass through the organism with negligible absorption. An organism with specialized enzymes may be able to absorb a xenobiotic even when its bioavailability is low (Peakall and Burger 2003). The term "toxicodynamics" refers to the interaction of the xenobiotic with various components of cells and organs, to produce toxic effects (Table 8.1).

Toxicokinetics refers to how a xenobiotic is handled by the body from the time it is absorbed in the lungs, the gastrointestinal tract, or through the skin until it is excreted in urine or feces, exhaled by breath, or by sequestration in growing structures such as hair and nails, and in eggs. During this process, the xenobiotic is absorbed through membranes into the bloodstream, attaches to various transporter molecules such as albumin, and travels to various organs. The organs can be excretory, such as the kidney, or can be the target organs where the chemical does damage. Soluble species are

Table 8.1   Terms Used in Ecotoxicology with the Process They Describe

Terminology	Process
Bioconcentration	The process of increasing concentration in an organism compared to its environment. The ratio is called a bioconcentration factor.
Bioaccumulation	The process of accumulating chemicals into an organism's tissues. This occurs when intake is higher than excretion rate.
Bioamplification	The process by which a chemical reaches an increasingly higher concentration at every level of the food chain.
Biological half-life	Many chemicals that accumulate in the body are also excreted. When exposure has ceased, the chemical continues to be excreted, and the rate of excretion varies by chemical and species. The time required to eliminate half of the body burden is termed biological half-life.
Bioavailability or bioaccessibility	The ease with which a substance is released from its environmental matrix when it comes in contact with the body.
Absorption	The process by which a contaminant in the gut or lung is transferred into the bloodstream.
Toxicokinetics	The total processes of absorption, distribution, metabolism, storage, and elimination of a toxic chemical from the body.

readily absorbed from water, particularly at low pH (acid waters). Organic complexes retard metal absorption. Bioavailability of metals is generally lower from saltwater than from freshwater (e.g., chromium, zinc; Hunt and Hedgecott 1992).

An important step in toxicokinetics is the elimination of the xenobiotic from the body. Depending on its form or solubility, it may be excreted in urine or feces, or transferred to hair, nails, or feathers, or eliminated in eggs. Many of our studies use metal levels in eggs or feathers as bioindicators of body burden and exposure. For example, the amount of a xenobiotic that enters the egg depends on the properties of the compound and on the concentration in the female's blood circulating through the ovary. The formation of each egg therefore reduces the female's body burden of the chemical, and it is a reasonable prediction that each successive egg should have a lower concentration of the chemical, unless of course, the female is replenishing her supply of xenobiotic through her diet in the egg-laying period. Thus, the first-laid eggs may have higher levels of the contaminant (see below), and in the case of eggshell thinning, egg-laying sequence played a key role in egg breaking and failure to hatch (Figure 8.4).

Figure 8.4   A very famous *National Geographic* photograph shows a clutch of Mallard (*Anas platyrhynchos*) eggs, beginning with a large, but thin-shelled and collapsed first egg, to progressively smaller, but thicker-shelled eggs, with finally one live duckling emerging from the last, least contaminated egg. (From http://www.scienceclarified.com/Co-Di/DDT-dichlorodiphenyltrichloroethane.html.)

The image in Figure 8.4 is persuasive, but the results of actual studies were not always consistent. The earliest-laid eggs of Arctic Terns (*Sterna paradisaea*), Common Eiders (*Somateria mollisima*), and Long-tailed Ducks (*Clangula hyemalis*) had higher mercury levels than the latest-laid eggs, with an almost 2-fold difference in the ducks (Akearok et al. 2010). This suggests that it is important to incorporate egg-laying sequence into a sampling regime. We suggest that the first egg would give a better approximation of high exposures. However, in some species the first egg is the largest and most likely to lead to a successful fledgling—selecting the last egg might have the least impact on the species. Custer et al. (2010) reviewed several studies indicating that the relationship between laying order and contaminants is inconsistent or absent. Tree Swallows (*Tachycineta bicolor*) lay five to six eggs, and Custer et al. (2010) found that clutches with a higher total PCB content showed a decline across egg order, whereas clutches with lower PCB content showed an increase, perhaps because the female was taking in more PCBs on a daily basis than was eliminated in the eggs. There was a difference between years as well, with 2005 clutches showing an increasing trend. Brasso et al. (2010) found no relationship between Tree Swallow egg sequence and mercury concentration, but found a high correlation between female blood mercury and egg mercury.

Feather proteins are rich in sulfhydryl groups that form disulfide bridges, allowing proteins to fold into a functional configuration. Metals that bind to sulfur can break the bridges and denature the protein, thus inactivating its enzyme function. This accounts for the excessive metal toxicity. The same sulfur bridges of the keratin protein account for the incredible strength of the feather structure. All metals have an affinity for sulfur, and as a feather is forming, the blood supply delivers metals to the cells that are laying down the feather protein. Thus, metals are distributed throughout the feather, and once the feather is fully formed, its blood supply atrophies, leaving behind the metal concentration as a reflection of the blood level at the time the feather was actively growing.

TOXIC EFFECTS OF METALS AND TOXICODYNAMICS

In human health assessment, individuals matter. In ecological health, populations matter, although both are oversimplifications. For human health, endpoints include carcinogenic effects and noncarcinogenic effects on organs including the nervous system, kidney, liver, digestive, endocrine, and reproduction. Some of the effects are mediated through the endocrine system and are referred to as endocrine disruptive effects, including those caused by chlorinated hydrocarbons. Reproductive toxics affect different aspects of reproduction, from libido to fertilization, embryogenesis, and gestation.

In ecotoxicology, the focus is on populations, and with the exception of endangered and threatened species, individuals are important insofar as they are the sampling units that we study. Behavioral toxicity may interfere with normal development, with the social interactions between chick and parent, feeding behavior, predator avoidance, and then in adulthood, its reproductive behavior. Reproductive effects are often the key to a population decline, including hatchability, developmental defects, survival to fledging, and recruitment. Carcinogenic effects related to contaminants have been identified in the wild, but rarely do cancers affect avian populations. Indeed, we found only one example of cancer affecting populations. The population of Tasmanian Devil (*Sarcophilus harrisii*) has declined more than 70% in 20 years and is threatened with extinction by a unique cancer spread by biting (Hamede et al. 2015).

Although the overall assessment goals may differ (individual health for human risk assessment, populations for ecological risk assessment), some of the same indicators and endpoints provide information for health of both humans and ecosystems (Burger and Gochfeld 1996d).

Metals cause toxic effects via a variety of mechanisms: (1) many metals bind to sulfur that is critical for the action of many enzymes and for the synthesis of microtubules; (2) metals may interfere with genetic, hormonal, or metabolic functions; (3) metals bind to selenium, and some selenium enzymes protect the body against oxidative stress; (4) oxidative stress causes damage to many cells and tissues. Toxic effects can occur at different levels of biological organization, from the organ down to enzymes, molecules, and genes. Chemicals may inactivate enzymes, alter gene expression, or cause genetic damage. Although the bioaccumulation of xenobiotics in liver, kidney, brain, and muscle can be measured, the significance of such levels is not always apparent. The most relevant effects for populations are those with behavioral impacts affecting survival (antipredator behavior) or reproduction, as well as direct effects on the reproductive system and the underlying endocrine system. These include: (1) reproductive behavior (courtship feeding, mating and mate guarding, care of young), (2) gamete formation, (3) fertilization, (4) embryogenesis, (5) neurobehavioral development, (6) parental recognition, and (7) antipredator behavior (Figure 8.5).

Mechanisms of Toxicity

Toxic chemicals can exert impacts on different cell components (e.g., nucleus, mitochondria, membranes), on cells (cell death or necrosis), and on different organs (morphology, function). Effects will vary depending on the developmental stage (from fertilized egg to mature adult), and seasonally with stages in the reproductive cycle. Examples of mechanisms include:

- Cell death and necrosis
- Poisoning of mitochondria or other organelles reduces cell and organ function
- DNA damage and repair (adducts, chromosomal abnormalities)
- Inhibition of synapse formation in developing nervous system
- Enzyme inhibition by xenobiotic binding to sulfur or selenium
- Microtubules disassembly
- Oxidative stress (Hoffman et al. 2011)
- Acute versus chronic toxicity

A major distinction in ecotoxicology is whether exposure is a sudden, high-level event, such as an industrial effluent causing a major fish kill over a period of hours, or a longer-term, lower-level exposure resulting in cumulative toxicity (chronic toxicity). For example, the chronic exposure of adult Peregrines, Osprey, Eagles, and Pelicans to DDT did not kill them directly as it accumulated in their tissues over time, but the chronic exposure interfered with egg formation and thereby impaired reproduction and led to extinctions of local populations. Laboratory toxicologists have developed protocols that mimic acute exposure (e.g., 96-hour lethality) or chronic exposure (e.g., 30-day to 2-year exposures), which vary depending on the size and longevity of a species.

No Effect Levels, Effect Levels, and Lethal Doses

Toxicologists working in the laboratory have devised many measurement endpoints. Death of an organism is easily detected, so lethality is a common endpoint, particularly in early studies (before the 1980s) before more sophisticated techniques were developed to detect cellular damage. Two of the most widely published lethality metrics are the LD_{50} (lethal dose that kills 50% of the test subjects, which can be invertebrates, fish, rodents) and LC_{50} (lethal concentration that kills 50% of the subjects). Dose refers to the amount of xeniobiotic that reaches critical organs and is therefore difficult to measure. Concentration of the xeniobiotic in water or air is under the control of the experimenter and is more easily determined. For toxics in food, the term "dose" is usually used, whereas for toxics in water, the term "concentration" is preferred. Although the LD_{50} and LC_{50} measures have been supplanted by more sophisticated biochemical measures or quantitative techniques, there are still many published values that can be used to compare various chemical compounds. For example, because LD_{50} values are standard measurements, it is possible to compare

Breeding activities

Figure 8.5 Reproductive phases of birds. Laughing Gulls courting (a) and a Great Egret displaying to attract a mate (b). Herring Gulls court by giving "head tosses" to each other (c), which leads to copulation (d). Following copulation, birds start to build nests (e). Incubation follows, along with the brooding of small chicks (Black Skimmers) (f).

relative toxicities of different pesticides within a species. The lower the LD_{50} dose, the more toxic the pesticide (Figure 8.6).

Studies of birds rarely use lethality as an endpoint, but may focus instead on behavior or on some biomarker affected in the blood. Some studies measure the concentration of a substance in organs (e.g., liver or kidney), in eggs, or in feathers. Toxicologists may use a more general term "effective dose 50%"—the ED_{50} or EC_{50}—for the dose or concentration that produces a certain

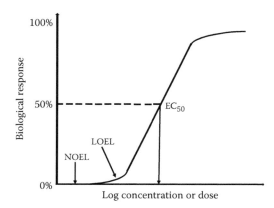

Figure 8.6   Dose–response curve. Note that the *X* axis is logarithmic and indicates the doses used in an experiment.

effect or endpoint in 50% of the test animals. Toxicologists use other measures such as the no effect concentration (NOEC) or the no observed adverse effects level (NOAEL), the lowest observed adverse effect level (LOAEL), and other values, which depend on the concentrations chosen for the study. A zero dose or concentration is usually used as a control. If the lowest concentration above the control produces no effect, it is considered a NOEC. If it does cause an effect, it is a LOEC. These values are then used in various ways by various agencies to set guidelines or standards.

In many cases, toxicity is measured experimentally, and in other cases, particularly for birds, it is inferred from observations and measurements on wild animals. There are different kinds of toxicity testing for different types of organisms. Water or effluent quality is usually measured by a series of standardized bioassays using aquatic plants or invertebrates (EPA 2002), whereas many chemicals are tested experimentally using rodents or fish such as Fathead Minnow (*Pimephales promelas*) or Rainbow Trout (*Oncorhyncus mykiss*).

Vulnerability and Susceptibility

We distinguish vulnerability from susceptibility. Vulnerability refers to an individual or species whose habitat, movements, or behavior subjects them to higher levels of exposure. Susceptibility refers to attributes (genetic or dietary) that may influence how strongly individuals or species respond to xenobiotic toxicity, or conversely, how high a concentration or dose is needed to achieve a particular toxicological response. For example, Heinz et al. (2009) compared the LC_{50} of methylmercury for embryonic development of 23 bird species with some species having LC_{50} >1000 ng/g (lower susceptibility), whereas others had LC_{50} below 250 ng/g (higher susceptibility). His work is examined in more detail in Chapter 9. Benthic organisms living in the sediment are most exposed to contaminants in aquatic systems, and studies of their species diversity have led to rough divisions of pollution-sensitive and pollution-tolerant species (Boesch 1973). This has been applied to the Chesapeake (Weisberg et al. 1997), New York–New Jersey Harbor (Weisberg et al. 1998), and Delaware Bay (Llanso et al. 2002).

Sensitivity or tolerance may arise through differences in ecology (habitat selection, diet, behavior) or in toxicokinetics (uptake and transporters, enzyme induction, or the metabolic processes that activate or detoxify xenobiotics). For example, Lucia et al. (2012), studying metal concentrations in migrating shorebirds along the French coast, showed that Red Knots accumulated arsenic and selenium more efficiently than Black-tailed Godwit (*Limosa limosa*), which accumulated more silver, copper, iron, and zinc, resulting in higher metallothionein production. This led to the conclusion that detoxification pathways differ among species and differ for different elements.

Tolerance: Adaptation and Evolution

For some groups of organisms, Species Sensitivity Distributions have been worked out (e.g., response of aquatic invertebrates to cadmium in water; EPA 2012b). The converse of susceptibility is tolerance. Tolerance usually arises when organisms can avoid, reduce uptake, sequester, or enhance elimination of a xenobiotic. If such changes occur in the short run in an individual, it is called adaptation. On the other hand, toxicity may eliminate individuals that are susceptible, leaving less susceptible (more tolerant) individuals as survivors, to give rise to the next generation of more tolerant individuals. Thus, tolerance can arise in just a few generations. The important analogy is the development of tolerance in bacteria exposed to antibiotics or in insects exposed to insecticides. If less than 100% of target organisms perish, the survivors have the opportunity to perpetuate the genes for resistance or tolerance, which allowed them to survive (Falfushynska et al. 2015). Transplantation experiments have demonstrated tolerance in Mummichogs (Prince and Cooper 2009). Fish from pristine waters were transferred to cages in a polluted estuary, whereas fish from the dioxin-polluted estuary were transferred to adjacent cages. The fish with prior exposure to dioxin had evolved tolerance. A graph can be drawn of species tolerance, for example, to cadmium (Figure 8.7).

ESTABLISHING CAUSATION OF TOXIC EFFECTS

Our knowledge of toxicology, in general, environmental toxicology, and ecotoxicology continues to grow with new compounds discovered, new toxic effects attributed to well-known compounds, new mechanisms, and new information on susceptibility of individuals and species. For high-level exposure and acute toxicity, it is often easy to establish a causal relationship: agent A has caused subject B to develop a disease C. Because ascribing toxic effects to a compound may have economic

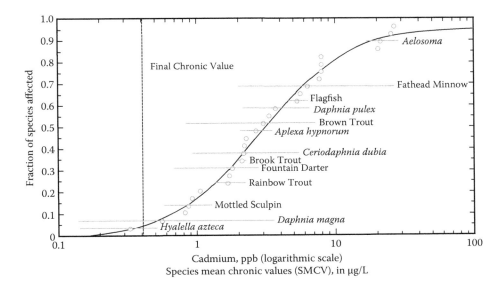

Figure 8.7   Mean chronic toxicity values for various organisms, showing the different levels of effects and different concentrations of cadmium in water. Used by the USGS to derive a Final Chronic Value for cadmium. (After USGS, Cadmium Risks to Freshwater Life: Derivation and Validation of Low-Effect Criteria Values Using Laboratory and Field Studies, http://pubs.usgs.gov/sir/2006/5245/pdf /sir20065245.pdf, 2010.)

consequences, assertions that a particular company's agent A caused illness may be challenged in the courts. High-level exposures that produce immediate effects in the laboratory are usually easy to detect and prove. If, however, the endpoint of interest is delayed, such as cancers with long latent periods, establishing causality may be very difficult. The overall approach to establishing causality in ecotoxicology involves three questions: (1) Is something happening (detection and definition)? (2) Can one or more causal factors be identified? (3) What are the impacts on individuals, populations, ecosystems, and society? An additional question of interest is whether stopping exposure will reverse the disease process and reduce the risk.

Detecting an Event: Is Something Happening Out of the Ordinary?

The first step is to recognize when an event of interest and concern is occurring, such as a die-off. A fish kill in a lake or river may be attributable to an anoxic event, a winter freeze, or industrial effluent, but the event itself is obvious, assaulting both the eyes and nose, and attracting attention from the media. The cause and effect are close in time (hours or days), and the causal linkage, often a single factor, may be obvious. Likewise, a massive die-off of birds following an oil spill, where birds are completely oiled, is obvious. A person poisoning a thousand grackles captures newspaper and regulatory attention (Stone et al. 1984). Botulism epidemics that kill numerous waterbirds tend to be subacute, occurring over a period of weeks as the birds frequent ponds or impoundments where the toxin has accumulated.

Increased mortality occurring over a period of months, or decreased reproduction, is more difficult to detect and assess, and the causes are often multifactorial. Raptor biologists were able to recognize widespread reproductive failure in Peregrines in only a few years. However, population declines may be occurring for years or decades before they become apparent. Then, it may take additional times to convince responsible agencies that the declines are real, to document the cause(s), and to take preventive actions.

The underlying approach to establishing causation of toxic events was articulated by Bradford Hill (1965–1966), and has survived with little modification (but also with little application in ecotoxicology). Sir Austin Bradford Hill, one of the leading biostatisticians of Britain in the mid-twentieth century, described an approach to establishing causation for workplace exposures to human disease. These have been widely used (and misused) in human and environmental toxicology. Although the application and interpretation of the Hill criteria remains controversial, they provide useful guidance for exploring a presumed causal relationship. The approach involves testing known information, observations, and toxicity data against the criteria (see Table 8.2). The more criteria that are met, the greater the confidence that causation has been established.

Fox et al. (1991) illustrated how ecotoxicologists should use this approach. For birds, however, it is often difficult to satisfy all or many of these in the wild because of the complexities of species differences, environmental variations, and mixtures. Hill himself was very clear that a causal relationship did NOT have to meet all of the criteria, and in many cases causation would have to be inferred in the absence of complete information. Confidence in a causal relationship may take years to evolve, and more years to achieve consensus. For example, the Surgeon General report (Surgeon General 1964) established that cigarette smoking caused lung cancer, but it took a half century for the industry and the courts to acknowledge this causal relationship.

The previously described impact of DDT on avian reproduction occurred over a period of years and decades. Peakall (1993) explained that eggshell thinning was detected in Peregrines in the 1940s, but it was almost 30 years before convincing evidence on mechanisms was gathered, leading to the banning of DDT. The population declines were noted in a number of ways. Among wintering Bald Eagles, for example, the ratio of juveniles to adults declined to almost zero. Egg breakage in Peregrine Falcon nests was first noted in Britain in 1948, and had become commonplace by early 1950s, associated with the disappearance of many nesting pairs (Ratcliffe 1970).

Table 8.2   Bradford Hill Postulates for Establishing a Causal Relationship between a Toxic Exposure and an Endpoint

Criterion	Description
Temporality	The effect has to occur after the cause. This is the only absolute criterion.
Strength of Association	A steep dose–response curve supports causation; however, a weak association does not negate a causal effect.
Biological Gradient	Greater exposure should lead to a greater incidence or intensity of the effect.
Specificity	Causation is likely if a very specific population at a specific site has a disease with no other likely explanation.
Consistency	Consistent findings observed by different persons in different places with different samples strengthens the likelihood of a causal effect.
Biological Plausibility	A known or causative mechanism consistent with other biological knowledge is valuable.
Coherence	Agreement between epidemiologic and laboratory findings increases the likelihood that an effect is causal. However, lack of laboratory support does not mean epidemiologic evidence can be ignored.
Experiment	Intervention by removing the putative cause and seeing a change in outcome is valuable.
Analogy	If similar substances are known to cause similar effects, this is useful evidence.

Source: After Hill, A.B., *Proc. R. Soc. Med.*, 58, 295–300, 1965.

Examination of shell thickness revealed a precipitous decline in 1946–1948. Although Rachel Carson's 1962 book, *Silent Spring*, had called attention to pesticide effects, Ratcliffe (1970) was cautious in pointing to pesticides as the primary cause of eggshell thinning and reproductive failure in Peregrines and several other species. The other global anthropogenic change he considered was radioactive fallout, but concluded on the basis of timing and spatial distribution that fallout was unlikely to have caused eggshell thinning (Ratcliff 1970). In the case of DDT and avian reproduction, the presumed cause was poisoning causing reproductive effects. Laboratory analyses revealed high concentrations of DDT and its metabolites (DDE and DDD) in eggs that failed to hatch and in the tissues of adult birds. In some cases, the adults had succumbed to injuries or had been shot, and in other cases they were collected for the purpose of analysis. The history of the detective work finally, and conclusively, linking the reproductive failure and population declines to pesticides has been told often.

Metal poisoning events are discussed under each metal in Chapter 9. These are typically chronic poisoning events. For example, lead poisoning is a significant cause of mortality among birds, particularly waterfowl that ingest spent lead shot while feeding, and among birds of prey, such as the California Condor (*Gymnogyps californianus*) that ingested lead shot while consuming wounded prey (Watson et al. 2009). Birds may manifest symptoms for weeks before succumbing. Likewise, an epidemic of selenium poisoning was manifest as developmental defects in embryos, resulting in population declines (Ohlendorf 2011).

SUMMARY AND CONCLUSIONS

This chapter reviews the basic principles and concepts of ecotoxicology with particular reference to metals and birds. It also provides examples of metal toxicity epidemics. Organisms are often used to monitor environmental quality, requiring an understanding of toxicokinetics on one hand, and ecology on the other. Understanding both is essential to choosing the most useful species for bioindicators. Metals occur naturally in the earth and the ocean, and they are liberated slowly to the general environment, or occasionally rapidly through volcanic activity. However, most metal contamination results directly from mining, milling, processing, fabrication, use, and disposal of metal objects. Metals have a complex environmental fate and transport. Although the elemental metal itself cannot be changed in nature or in the body, the metal compounds can undergo a variety of metabolic changes that influence transport, bioavailability, absorption, and toxicodynamics. Metal toxicity is often mediated by their affinity for the sulfhydryl groups and their ability to disrupt the disulfide bonds that confer the essential tertiary structure to enzymes and other proteins. Detecting effects on populations is challenging because there are many sources of uncertainty, including the intrinsic biological variability of food webs and target organisms. The chapter also covers the issue of causation, illustrating the difficulty of detecting outbreaks of metal poisoning, and the challenge of adducing that the outbreak is attributable to a particular metal or mixture.

CHAPTER 9

Effects of Metals in Birds

INTRODUCTION

Metals have been responsible for some large-scale avian mortality events. These include mercury poisoning of seed-eating birds and their predators from mercurial fungicide-treated grain, lead poisoning of waterfowl and vultures from ingested shot, and selenium poisoning of waterbirds from agricultural wastewater (references in Chapter 1). In this chapter, we focus on lead, mercury, cadmium, selenium, chromium, manganese, and arsenic, and examine their effects on birds, as well as explaining the rationale for the elements we chose to study. These are circled in the periodic table (see Appendix Figure 9A.1 at end of this chapter), while the parallelogram denotes the elements that are considered heavy metals. We examine laboratory and field experiments, and their relevance to metal levels in Barnegat Bay and the other Northeast estuaries. In our studies of waterbirds, many of which are listed as threatened or endangered, it was necessary to rely on sampling that did not require killing the birds. We collected eggs and fledgling feathers. For each of the elements we analyzed, we searched for published effect levels, toxic thresholds, or data from which effect levels in eggs or feathers could be estimated. Metals in eggs served both as biomarkers of maternal exposure and chick exposure, but also as a direct source of toxicity to the embryo. Metal levels in feathers served as a biomarker of exposure for the developing bird during the several weeks of its prefledging growth.

We found many published reports of contaminant concentrations in unhatched eggs or dead birds, but these are more confusing than edifying. The levels may have been irrelevant to the cause of hatching failure or death, or if fatal, provide no information on sublethal effects. We found many toxicologic studies that included information on dose, death rate, and associated concentrations in the kidney and/or liver. Relatively few studies, however, report the concentrations in feathers associated with effects or the levels in eggs associated with hatching failure or developmental defects. Finding usable effects levels has been challenging, particularly because many early studies did not indicate whether concentrations are reported as wet weight or dry weight.

There is a rich literature for lead, mercury, cadmium, and selenium in the environment and in birds. We reviewed the series of *Synoptic Reviews* by Ronald Eisler of the U.S. Fish and Wildlife Service published in the 1980s and the *Toxicological Profiles* developed by the Agency for Toxic Substances and Disease Research (ATSDR), a branch of the U.S. Centers for Disease Control and Prevention (CDC). The volume on *Environmental Contaminants in Biota* (Beyer and Meador 2011) provided valuable more recent reviews. In reviewing ecotoxicology in birds, we found an enormous variety of research paradigms, dosing regimens, forms of metals, exposure routes and durations, and endpoints studied, as well as variation in how analyses were performed and results reported.

We recognized that effect levels would be different for young birds versus adults, and that each endpoint would have a unique dose–response curve. By controlling the exposure levels to which experimental birds are exposed (in air, food, or water) or the doses given to them (by injection or gavage), a toxicologist can construct the dose–response curves and can identify or estimate no observed effect concentrations (NOECs), no or lowest adverse effect levels (NOAELs, LOAELs), thresholds (between NOAEL and LOAEL), LC_{50} or LD_{50} (the lethal concentration or dose for 50% of test subjects), and various benchmark doses. The latter can include, for example, ED_5 or ED_{10}, doses that cause an effect in 5% or 10% of the test subjects.

Early studies used lethality as a common endpoint to estimate the LD_{50} or LC_{50} value. Killing or maiming 50% of the individuals in an experiment or harming 50% of the species in an ecosystem, is pretty gross. Aquatic toxicologists (Aldenberg and Slob 1993) have derived HC_5 values as the hazard concentration sufficient to harm 5% of species in an ecosystem (but protective of 95% of the species present; Shore et al. 2011, p. 611). This is essentially an ED_5 for species rather than individuals, but it has not yet translated itself into bird populations. It would be ideal to be able to identify the highest NOAEL—the level at which no adverse effects occur.

Choice of Metals and Metalloids to Study

We chose to study lead, mercury, and cadmium, which are widely recognized as toxic aquatic pollutants and were recognized as significant contaminants in Raritan Bay (Greig and McGrath 1977) at the time we began our studies. Selenium is recognized as an important toxicant, causing severe developmental defects and high mortality in waterbirds. Moreover, selenium binds other metals and reduces their toxicity (Ohlendorf and Heinz 2011). Chromium contamination is localized in the New York–New Jersey Harbor Estuary from past industrial processing and profligate disposal practices in Hudson County. Until the 1970s, Jersey City was one of the two largest chromium ore processing centers (Stern et al. 2013). We included manganese, because despite much controversy and legal machinations, an organic manganese compound was approved as a gasoline additive to replace organic lead. We worried that it would become the next generation's toxic disaster. Fortunately, no company has been foolish enough to actually add manganese to gasoline in the United States. We also included arsenic in recent years because of unexplained high levels in some biota.

All our measurements are reported in parts per billion (ppb), which is equivalent to ng/g and µg/kg on a dry weight basis. All our analyses are for total metal; we did not speciate mercury to determine methylmercury, chromium to determine hexavalent chromium, nor arsenic to identify inorganic forms. To convert published wet weight levels to dry weight, we relied on moisture content analysis in our laboratory (Chapter 3). The conversion for organs (e.g., liver) is 3×, for gull and tern eggs 4×, and for heron/egret eggs 5×. We always report blood levels as nanograms per gram or micrograms per deciliter (wet weight) and feather and egg levels as ppb (dry weight).

LEAD

Lead poisoning in birds has been studied extensively, particularly with respect to lead shot, and many studies have been conducted by dosing birds with one or more lead pellets. Lead poisoning is particularly prevalent in waterfowl that ingest shot as grit while feeding (Bellrose 1959; Honda et al. 1990). Lead poisoning can be acute, subacute, or chronic. The neurotoxic effects are reflected in inability to fly, stand, and eventually to eat. Symptoms may occur over a period of a few days or weeks, leading ultimately to death. Top-of-the-food chain scavengers, such as vultures, show high levels of lead (Eisler 1988a). For example, an isotopic analysis of bone lead in Black Vulture and Turkey Vultures showed that they had been exposed to multiple sources of lead including lead shot, leaded gasoline, coal-fired power plants, and zinc smelting based on isotopic fingerprinting (Behmke et al. 2015).

For the California Condor, lead poisoning, both acute (shooting) and chronic (ingestion of wounded animals with lead shot) contributed to their virtual extinction in the wild. Lead poisoning is considered a major impediment to reestablishing viable wild populations, with an estimated 20% of the population requiring treatment for lead poisoning to prevent death (Finkelstein et al. 2012). When we visited the Condor area in southern California in 1985 just before the retrieval of the last wild Condors, a farmer told us that his friends shot Condors to get an endangered species off their land. More recently, three of the Condors released back into the wild, and later retrieved for treatment, had elevated blood lead and also had x-ray evidence of lead shot in their tissues, evidence that illegal shooting continues (Finkelstein et al. 2014). A study of Bald Eagle specimens found that 5% had died of lead poisoning (Eisler 1988a), probably from eating wounded ducks. Ingestion of lead shot raises blood lead to a much greater extent than a similar mass of lead shot embedded in tissues (Finkelstein et al. 2014). Finkelstein et al. (2010) demonstrated that feathers provided a more useful timeline of lead exposure than blood sampling.

Toxicity Effects Levels

Lead is unique among metals in not having an Environmental Protection Agency (EPA)–designated Reference Dose (RfD). EPA concluded that, for humans, behavioral development may be impaired "at blood levels so low as to be essentially without a threshold." EPA "considered it inappropriate to develop an RfD for inorganic lead for any effect" (IRIS 2004). An effect or endpoint can refer to death, disability, discomfort, or to any measurable change in some molecular, cellular, biochemical, physiologic, reproductive, anatomic, or behavioral variable. In early studies, mortality (or survival) was the main endpoint. Other readily observable endpoints were inability to stand (weakness), walk, or fly (locomotion), poor appetite and slow weight gain, and impaired or altered reproductive behavior or outcomes (Franson and Pain 2011). The enzyme delta aminolevulinic acid dehydratase (ALAD) is inhibited by lead, and it has been used for years as a biomarker of early lead poisoning (ATSDR 2007a). More recently, researchers have been able to measure the activity of other enzymes or gene expression both as modifiers of toxicity and as biomarkers themselves.

Because our work focused on metals in eggs and feathers, we were particularly interested in finding relevant published effect levels associated with these tissues (Table 9.1). There are many published accounts of organ concentrations of lead in birds that died of lead poisoning (listed in Franson and Pain 2011), but these do not indicate the organ levels at the time that subclinical or

Table 9.1 Published Effect Levels for Lead in Birds

Endpoint	Concentration	References
Human Reference Dose	None established. No evidence of a threshold	IRIS 2004
Effect levels in birds (frequently used in the literature)	4000 ppb in feathers	Scheuhammer 1987; Eisler 1988a; Burger 1995a; Burger and Gochfeld 2000a
Neurobehavioral effects in Herring Gull chicks	2300 ppb in feathers	This chapter (see below)
Effect levels in Night-Herons	2000 ppb in feathers	Golden et al. 2003a
Lower survival in Little Blue Herons	1000 ppb average level	Spahn and Sherry 1999
Organ tissue levels in sensitive species (American Kestrel)	1300 ppb (dry weight) in liver	Eisler 1988a
Liver levels in ducks and raptors	6000 ppb (dry weight) in liver	Franson and Pain 2011
Blood lead NOAEL based on enzyme inhibition	3–5 µg/dl	Work and Smith 1996
Blood lead levels (subclinical poisoning) ducks and raptors	20–50 µg/dl 200–500 ng/g (wet weight)	Franson and Pain 2011
Levels in food (prey) associated with adverse effects	100–500 ppb (wet weight)	Eisler 1988a

clinical effects first appeared. Even excluding birds known to be lead-poisoned, there is no consistent ratio between liver lead and feather lead (Burger 1993). Neither Eisler (1988a) nor Franson and Pain (2011) provided data on lead levels in eggs associated with embryotoxicity, reduced hatchability, or developmental defects.

The Mallard Duck has been a primary model for experimental studies of lead. Ducks given an oral dose of #6 lead shot experienced 9% mortality in 20 days after a single shot or 100% mortality after ingesting a #8 shot, indicating a strong dose–response relationship. A single oral dose of 200 mg lead shot raised blood lead to 1.0 mg/kg (approximately equal to 100 µg/dl). Many Mallard studies summarized by Eisler (1988a) reveal substantial variability in response to ingested lead, with some showing no effect on survival at 25 ppm in diet, whereas the target enzyme, ALAD, was depressed. Female Mallards accumulate lead to a higher level than males with the same exposure, probably because of a relative calcium deficiency (Scheuhammer 1996). Very few of the studies reported blood lead levels (still costly to analyze in the 1980s), and none reported on feather lead levels.

Some studies demonstrate an adverse endpoint in a contaminated waterbody compared to one less contaminated. For example, a study on Little Blue Herons (*Egretta caerulea*) from Louisiana reported lower chick survival in birds exposed to lead, although the lead level in feathers averaged only 1000 µg/kg (ppb), and many feathers did not have detectable lead levels (Spahn and Sherry 1999). Other studies compare levels of a metal in birds to a published effects level and draw inferences about risk or to background levels. For example, 5 of 20 Grey Heron and 4 of 18 Black-crowned Night-Heron chicks had lead in feathers "higher than the background level for lead" of 6000 ppb (dry weight; Kim and Oh 2014). On Midway Island, 15 moribund Laysan Albatross (*Diomedia immutabilis*) chicks with drooping wings, close to buildings with unremediated lead paint, had an average of 31,000 ppb (geometric mean [GM] = 21,300 ppb) in feathers compared with 1000 ppb in other healthy chicks on the island (Burger and Gochfeld 2000c). One lead-poisoned chick had only 16,000 ppb in its feathers. These levels would eventually have been associated with death.

Excluding birds with known or suspected lead poisoning, almost all of our published lead levels from Long Island, New Jersey, Puerto Rico, Egypt, and the Pacific are in the range of 800–4350 ppb, with the notable exception of Cattle Egrets from Egypt (mean, 9660 ppb; GM = 4290) (Burger et al. 1992b; Burger 1993).

Our Laboratory Experiments with Lead in Young Terns and Gulls

We chose to study developmental neurotoxicology in birds for several reasons. Most importantly, birds communicate primarily by vision and hearing, and therefore are better analogues for some human experiences than rodents that communicate primarily by touch, olfaction, and ultrasonics. Second, birds—or at least students of birds (i.e., Konrad Lorenz, Niko Tinbergen)—have been pioneers in developing the field of animal behavior or ethology. Third, the ecological correlates of bird behavior and their developmental milestones are well known. Fourth, birds were heavily exposed to environmental contaminants and we had already seen evidence of illness and reproductive failure, which we attributed, at least in part, to contaminants.

Our experimental protocol for the following series of experiments was to use intraperitoneal (IP) injection of a lead nitrate or lead acetate solution ("lead birds") or isotonic (0.9%) saline ("controls"). We believed this would emulate a high dose that a newly hatched chick might receive from its egg. We used doses of lead nitrate or lead acetate delivering from 0.05 to 2.0 mg of lead per gram body weight. Controls received an equal volume of saline solution (Burger and Gochfeld 1997b). We used Common Tern (*Sterna hirundo*) and Herring Gull (*Larus argentatus smithsonianus*) chicks collected either as pipping eggs or recently hatched young. We collected a single egg or chick from a clutch to minimize impact on the population. In the laboratory, chicks were fed by attendants and quickly became habituated to handling, which usually preceded a feeding. They were usually fed one fish at a time by a forceps (Figure 9.1, top). Terns were housed in plastic mouse cages and gulls

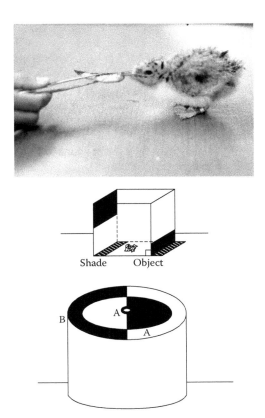

Figure 9.1 Our experiments with the effects of lead on Common Terns tested for speed of eating fish presented in a forceps (top). Thermoregulation testing apparatus (middle) provides shade without an object and an object that casts no shade. Visual cliff testing device (bottom) offers clear and opaque surfaces.

were kept in larger cat-sized cages. Chicks were injected by one of us (MG), who took no part in the behavioral observations, whereas the observers (JB and others) were blinded to the treatment status of the chicks they were testing, identifiable only by colored leg bands.

For our first foray into neurobehavioral testing, we tested 16 Common Tern chicks collected at age 1 day from different nests (Gochfeld and Burger 1988). We made pairs of chicks matched on weight ± 3 g. On day 5, one chick of each pair received a single IP injection of Pb (0.2 mg/g) of a 50 mg/ml solution. This was about 20% of the maximum tolerated dose, previously established. The other chick received an equal volume of saline IP. There was no mortality during the weeks of the study. Chicks were fed whole fish: Silversides (*Menidia*) or Sand Eels (*Ammodytes*) for 21 days until they reached the normal fledging weight of about 100–110 g.

At feeding time, the attendant picked up a single fish with a forceps, gripping it behind the gills, and presenting it directly to the chick, with the forceps coaxial with the beak (about 20% above horizontal) and the fish head protruding either to the chick's left or right. This approximates the typical feeding approach of an adult returning to the nest with a fish. A healthy chick immediately seizes the head of the fish and with a series of rapid head tosses, manipulates the fish so that it can be swallowed headfirst. During the rapid head tosses the beak opens slightly, allowing the fish to be turned so swallowing occurs headfirst (Gochfeld 1975b). Time to seize the fish, to swallow it completely, and the number of head tosses or dropped fish were recorded. Daily weight gain was plotted against grams of fish consumed.

Beginning on the day following injection, the weight of the lead birds always lagged, becoming significantly lower by 12 days postinjection. Overall, the lead birds ate less fish per day than controls on 16 of the 21 days, and significantly less on 2 days. The assimilation efficiency (weight gain/fish eaten) did not differ significantly until day 12. However, lead birds took longer than controls to swallow fish. These first 2 weeks are a time of rapid feather growth, which serves as a sink for metals including lead. Control birds adjusted the number of head tosses to the size of the fish consumed, more than did lead birds. Although headfirst swallowing is almost universal in nature, control birds were able to swallow small fish tail first. Lead birds that ended up with the tail in the beak were more likely to drop the fish. Lead birds dropped fish slightly more often than controls and were equally likely to retrieve a dropped fish (Gochfeld and Burger 1988).

Although weight gain lagged in lead birds, they did not appear ill, nor show typical signs of lead poisoning, which would include appetite loss, crouching, staggering, or unusual sleeping. At 24 days, all chicks were placed in an enclosure, and each of the authors attempted to identify the lead birds and the controls based on appearance and mobility. At the end of the test, the authors were correct in identifying six of eight lead birds and five of eight controls, not better than random. This suggests that the behavioral changes we observed were not attributable to birds suffering "malaise," locomotion impairment, or general illness, but were attributable to specific changes in the nervous system.

Our Neurobehavioral Test Battery

We developed a battery of neurobehavioral tests, essentially an obstacle course, to test perceptual, motor, and cognitive function (Burger and Gochfeld 1985b). We designed tests that would explore behaviors that are necessary for wild birds to survive. Compared to some standard neurotoxicology paradigms, our test battery was low technology, and sufficiently sensitive to detect effects on chicks exposed at 20% of the maximum tolerated dose. And they were behaviors that affect behavior and survival in the field.

Begging behavior is a primal feature of neonatal birds essential for survival. Depending on species, they may be noisy or silent, may simply open their beak to passively receive food, or extend their necks and climb on the backs of siblings to monopolize their parents. Survival favors the tall and lusty.

Fish swallowing is crucial for survival. Terns and gulls are semiprecocial and rely on their parents for all food for many weeks. The chicks were normally fed several times a day. We mimicked the natural feeding of parents by presenting a tern chick with an individual fish, held crossways in a forceps. Gull chicks are fed by regurgitation stimulated by the chick pecking at the red spot on the parent's beak (Tinbergen 1953). However, the gull chicks thrived on the food we provided by forceps, even though we did not regurgitate it. We presented the chick with a fish in a forceps, held just out of reach. We scored begging behavior for up to 1 min, from 0 = no begging to 6 = vigorous begging, jumping up and down, calling loudly and often, for at least 1 min.

Fish reversal feeding test. Chicks must be able to quickly seize a fish proffered by their parents, and swallow it to avert piracy by siblings or neighbors. We fed chicks the same way each time, with the head pointing to the chick's right. For the test, chicks were presented fish pointing to the left, and we timed how long it took the chick to seize, manipulate, and swallow multiple fish.

Righting is another lifesaving response. Chicks could easily find themselves upside down when rushing out to greet an arriving parent or rushing for cover at the approach of danger. A normal chick can right itself within a second or two. Before a feeding, we placed the chick on its back and held it for a second and then released it, timing the righting behavior with a stopwatch.

Locomotion also can save a life. We tested birds on a 4-cm-wide, elevated board held 1 m above the sand. We timed how long birds could stay on the board, up to 2 min. Although our beach colonies on Long Island were on flat, benign terrain, some Common Terns nest on steep sloping islands or even on rocks. John Coulson reported that chicks of the cliff-nesting Black-legged Kittiwake (*Rissa tridactyla*) behaved differently from other gull chicks. They always faced into the cliff, rather than looking outward, and they were essentially immobile, displaying none of the exploratory behavior of other species (Coulson 2011). Thus, adaptations to nesting habitat are important. We repeated the balance beam test by elevating one end of the board to a 20° angle. After scoring the chicks for 2 min, we raised the end of the board further, until the chick slid off, giving us the maximal incline angle.

Chicks must be able to recognize their parents. We examined *caretaker recognition*. The chicks were fed daily by a single caretaker. At the time of the test, the caretaker and another person sat at opposite sides of the table, each holding a spoon with food. Chicks were hidden under a bowl for about 30 s. Then, the bowl was lifted. Chicks were scored by how quickly they responded and by which person they approached. Control chicks responded more quickly (<6 versus >10 s) and were more likely to approach the caretaker they recognized.

Our *thermoregulation* test combined temperature regulation with perception. We used a 50-cm plexiglass cube (Figure 9.1, middle), placed on sand with a surface temperature of about 37–44°C in midafternoon. This is a natural condition encountered by tern chicks most days in the summer. We placed a wooden block outside, such that its shadow was outside the box and unavailable to provide shade, whereas its vertical surface offered an illusion of "cover." An opaque strip across the top of one side cast a shadow on the sand. The cube was oriented so that there were three choices: hot sand, a raised object and hot sand, and a shadowed area of cooler sand (about 10–15°C cooler). Chicks could choose between approaching the vertical surface, which might seem protective, or stepping into the actual shade. They were scored on how quickly they entered shade and how long they remained there during the 3-min test.

We designed a *cliff apparatus*, adapted from experiments with visual cliffs (Figure 9.1, bottom). Chicks were placed in the center of an elevated, clear plastic board that had both clear and opaque surfaces. Through the clear surface, they could see that it was "far down" to the ground, whereas the opaque surface looked more like sand. Chicks were scored on how often they looked down and whether they went to the clear (falling) or opaque (security) surface. The overall impact of lead on Common Tern and Herring Gull chicks in our first experiments is shown in Table 9.2. Table 9.3 provides more detail on the Common Tern responses.

Once we had established our testing paradigm, we became interested in identifying critical developmental windows when exposure affects some behaviors. We switched from terns to gulls because terns were appearing on state threatened or endangered lists, and as thoughtful stewards, we did not want to compromise tern populations. Herring Gulls offered both advantages and disadvantages. In the 1980s, the Herring Gull population in New York and New Jersey was rapidly expanding, whereas tern populations were shrinking. Herring Gulls were considered a pest species, and minimal justification was needed to obtain permits to collect large numbers of young birds— "the more the better" one regulator put it. Moreover, gull chicks adapt readily to human handling, eat a variety of readily obtained foods, and there was already a voluminous literature. We had studied Herring Gulls in the field for more than a decade, so the switch was easy. Unfortunately, the gulls were larger (adult size 2000 versus 120 g for the terns), took longer to reach fledging

Table 9.2 Summary of Lead Experiments with Common Tern and Herring Gull Chicks Injected Intraperitoneally with Lead as Lead Nitrate or Lead Acetate

Species	Sample	Injection Age	Dose	Growth	Significantly Different	No Difference	References
Common Tern	8/8	5	0.2 mg/g	↓↓	Righting Locomotion Visual cliff Thermoregulation Novel feeding	Begging Maximum incline	Burger and Gochfeld 1985b
Herring Gull	8–15 in each group	2 or 6 or 2-4-6	0.1 mg/g 0.2 mg/g	↓ at 0.1 ↓↓ at 0.2	Response time Distance moved Time to reach food Correct choice of caretaker		Burger and Gochfeld 1988b

Note: Sample size is given as number of lead/number of controls.

Table 9.3   Responses of Common Tern Chicks to Our Developmental Neurobehavioral Test Battery

Test	Lead versus Controls
Begging response	Score 0.34 versus 0.34, $P > 0.50$
Righting	0.96 versus 0.76 s, $P < 0.02$
Fish reversal feeding	First fish, 4.3 versus 0.9 s, $P < 0.04$
Locomotion—balance beam	Didn't walk and fell off, $P < 0.001$
20° incline	No difference
Maximum incline	Slight difference, $P < 0.04$
Visual cliff	No difference in any components or total score
Thermoregulation	Time to reach shade (trial 3), 64 versus 17 s, $P < 0.001$
	Total time in shade (trial 3), 38 versus 92 s, $P < 0.001$

Source: Burger, J., Gochfeld, M., *J. Toxicol. Environ. Health*, 16, 869–886, 1985b.

(6–7 weeks rather than 4 weeks), and they were much more expensive to house and feed. We used the same behavioral tests developed for the terns.

In our first set of gull experiments, we collected 24 one-day-old Herring Gull chicks and randomly assigned them to trios matched for weight ± 3 g (Burger 1990b). We randomly assigned one member of each trio to receive 0, 0.1, or 0.2 mg/g of body weight of lead as lead nitrate solution by IP injection. The control birds received an equal volume of saline. Behavioral tests were performed at varying intervals, and birds were followed for about 6 weeks. One of the first findings was that our lead injections on day 1 resulted in an 11% reduction in weight gain at 0.1 mg/g and 21% reduction at 0.2 mg/g (Burger and Gochfeld 1988b).

Over the 45 days, control birds performed best on more days than the lead birds. Balance was disturbed by the lead injection for the first 6 days postinjection. Caretaker recognition developed by day 5 in controls, by day 10 in the low dose group, and by day 14 in the high dose group. Depth perception on the cliff and thermoregulation were also impaired in the high dose group. Chicks injected on day 6 had more time to learn the caretaker before the injection and made more correct responses than chicks injected earlier. In subsequent studies, we explored the effect of different dosage regimes (Burger and Gochfeld 1995a,c), injecting some birds on day 2, dividing the dose on days 2, 4, and 6, or injecting birds on day 6 and day 12 (Table 9.4). Control chicks approached the

Table 9.4   Herring Gulls: Behavioral Consequences of Different Injection Schedules (Kruskal–Wallis Test for Difference among Treatments)

Herring Gulls	Control	Lead on Day 2	Lead on Days 2–4–6 1992	Lead on Day 6	Lead on Day 12 (Half-Dose)	Lead on Day 12 (Full Dose)[a]	P Value
Dose of lead (mg/g)	0	0.1	0.1 × 3	0.1	0.05	0.1	
Righting	2.4			2.5	2.5	2.9	0.001
Caretaker recognition on day 5 after dose	1.0	3.5	3.7	1	1	Slower	0.001
Locomotion—balance beam (30 s distance)	4.1 cm			2.7 cm	3.0 cm	2.1 cm	0.003
Maximum incline angle	46		46		47	49	NS
Visual cliff (peers)	5		5		4	4	NS
Actual cliff	5		6		4	5	NS
Thermoregulation: time to reach shade	17 sec	42	42	32	22	23	<0.001
Time in shade	80	39	45	43	78		<0.01

[a] The full dose on day 12 of 0.1 mg/g body weight is the same dose that was used on day 6.

caretaker 70% of the time compared to lead gulls. Gulls injected on day 6 chose more correctly but more slowly than gulls injected on day 2 (Burger and Gochfeld 1993c).

Sibling recognition plays a role in the development of gulls and Common Terns (Burger et al. 1988). In the laboratory, we created artificial family broods by raising three chicks together from 1 day of age. We then tested sibling recognition by confining a sibling at one end and a nonsibling at the other end of a runway. Sibling recognition can be demonstrated by age 4 days (Palestis and Burger 1999). By 15 days, control gulls recognized and quickly approached their "sibling" preferentially, whereas lead birds were slower and less correct (Burger 1998c).

Righting response and balance were affected immediately after injection, regardless of the timing of exposure, suggesting that this was an acute effect. Righting gradually improved in the days following injection, but never reached the level of the controls. The visual cliff and thermoregulatory responses were impaired in chicks injected on or before day 6, but injection at day 12, regardless of dose, had little effect. Flight practicing at 6–8 weeks was clearly reduced in the lead birds: 6 versus 10 jumps/min, 7 versus 14 cm, and 13 versus 19 flaps/min ($P < 0.01$). Some other tests showed no effect or were not sufficiently sensitive to detect an effect.

Neurobehavioral Development in the Field

Making birds sick in the laboratory is all well and good, but how do particular levels of exposure and impairment translate into the wild? Sick individuals may recover and lead normal productive lives or may perish quickly. We studied Herring Gulls at Captree State Park, Suffolk Co., New York, under required Federal, State, and Park Permits (Burger and Gochfeld 1994a). Herring Gulls typically lay three eggs unless the food supply is very poor. The 1000 pair Captree gull colony was flourishing at the time, benefitting from nearby garbage dumps, fishing boats, and picnickers. Nests were dense, and we selected a vantage point from which 20 nests could be readily observed, and we identified 12 control nests where chicks were not handled. A brood of three was well designed for our study. As soon as two chicks had hatched they were color marked, and we alternated assigning either the first or second chick to the lead group, the other to a control injection group. Lead birds were injected with lead at 0.1 mg/g and controls received an equal volume of 0.9% saline. The third chick, also color marked, was left as an uninjected control. All injections were performed by MG (Figure 9.2a), and all observations were by JB, who knew the birds only by their color band and the nest to which they returned to be fed.

Within a few days of injection, lead chicks had lower begging scores, poorer walking scores, and often stumbled. They were less accurate than controls in pecking at their parents' beaks to stimulate feeding. A chick that does not beg may not get fed. Moreover, vigorous begging stimulates parents to go out and get food (Figure 9.2b,c). This double jeopardy, less food and less feeding, led us to predict dire consequences for the lead chicks. At 16 days we caught and weighed the chicks, and lead birds were significantly lighter (131 versus 189 g, $P < 0.01$) than either controls. Begging score and walking scores were both significantly influenced by treatment (negative), by age (improvement), and by the treatment × age interaction. Lead treatment accounted for 63–78% of the variability (Burger and Gochfeld 1994a).

We divided the observation period into three equal periods of about 12–13 days each and compared the scores for begging, seeking cover, walking, standing and sitting. There were differences in all periods, but they were exaggerated in the middle period when chicks normally become very active, and lead birds lagged in exploratory behavior. This is also the period when chicks may wander into adjacent territories and be killed or injured (Table 9.5). We observed injuries and starvation among the lead chicks.

At 14 days postinjection, the survival of the lead chicks was slightly lower than for their control siblings or chicks in control nests. Closer to fledging (ca. 40 days), the survival was reduced ($\chi^2 = 7.39$; $P < 0.05$). At about 40 days of age, we surrounded the observational area with fencing and

(a)

(b)   (c)

Figure 9.2   (a) M. Gochfeld performed lead injections in the field and color-banded the chicks. Then J. Burger observed the subsequent behavior of chicks and their parents. (b) An adult with food stands over young chick, waiting for it to peck at the red spot on its bill. (c) Both parents trying to coax and feed a "slow" chick that later turned out to be lead-impaired. The cryptically colored 2-day-old chick is under the adult on the right.

Table 9.5   The Fate of Wild Herring Gull Chicks With and Without Lead Injection

	Lead-Treated (*n* = 20)	Control Chicks at Experimental Nests (*n* = 40)	Chicks at Control Nests (*n* = 36)
Nest completely lost or abandoned	2	1	0
Chicks Lost by 10 Days Due to Any Cause			
Due to starvation	4	0	0
Due to aggression	2	1	2
Unknown	1	0	0
Survivors at 14 days postinjection	13 (65%)	27 (68%)	28 (78%)
Chicks fledged	11 (55%)[a]	25 (63%)	26 (72%)

Source: Burger, J., Gochfeld, M., *Fundam. Appl. Toxicol.*, 23, 553–561, 1994a.

captured the chicks. We collected blood and breast feather samples. We were not really surprised to find that Suffolk County is not a pristine environment. The control birds blood lead averaged 100 ng/g wet weight—certainly not "normal," whereas the lead chicks averaged 260 ng/g ($P < 0.001$). Feathers averaged 1205 ± 344 (standard deviation) ppb for controls versus 4790 ± 1690 ppb for lead birds. Lead chicks that survived showed improvement in their begging, walking, and pecking. To our surprise, by day 40 the lead chicks had almost caught up in weight and were not significantly lighter than controls. Although we expected that the lead chicks, weaker, less vigorous beggers, might be shunned by their parents, in favor of more vigorous chicks, we were surprised to see that parents compensated, spending extra time caring for their exceptional young (Burger and Gochfeld 1996e). Thus, at least half of the lead birds survived to fledging age (Burger and Gochfeld 1994a). Lead was not good for them, but at least some of them survived to fledging.

We wondered why some of the effects in the field were less severe than in the laboratory. Spalding et al. (2000a) had provided a generalization that lethal doses were typically lower in the wild than in captive birds, and that sublethal effects were more readily apparent under controlled laboratory conditions. Our findings were in agreement. Even though we periodically took laboratory gulls out of their cages for exercise, gulls chicks in the wild get much more exercise than in laboratory cages.

Does exercise matter? We conducted an experiment with gull chicks injected with 0.1 mg/g of lead. Horrified by the high cost of animal treadmills, we modified a human treadmill for gulls, and allowed half of the lead and half of the control birds to exercise twice each day on the treadmill. The others, both lead and control birds, remained in their cages. We tested the conditioned and couch potato gulls on the treadmill. Birds had to learn to point forward and to keep pace with the belt (2 m/min) to avoid the visibly unpleasant experience of being carried to the rear and bumped. Endpoints were latency to orient forward, endurance, and avoiding being bumped. The lead birds that had no prior experience on the treadmill performed much worse on all measures ($P < 0.0001$). The experienced lead birds performed almost as well as the inexperienced controls (Burger and Gochfeld 2004b).

Lead in Feathers of Experimental Birds: Recalculating an Effect Level

In the studies discussed earlier, most of the lead-treated Herring Gulls in the laboratory, even at the 0.1 mg/g dose, had feather lead concentrations that exceeded 4000 ppb, which is often published as an effect level (Table 9.1). These birds did indeed show effects at varying stages between injection and fledging, leading us to assume that the effects level in feathers would be lower. The prefledging mortality of the lead birds was higher ($P < 0.05$) than for the controls. Laboratory-raised controls averaged 172 ppb in feathers, compared with 4840 ppb for the low lead birds and 9200 ppb for the high lead birds (Burger and Gochfeld 1990a).

Using the mean and standard deviation for the 11 surviving lead birds and 25 controls (see the preceding discussion), we calculated the 99% confidence limits. This yielded a value of 3175 as the lower confidence limit around the lead mean of 4790, and 1398 as the upper 99% confidence limit around the control mean of 1205 ppb. We rounded these to 3200 and 1400 ppb. The values of 4790 and 3400 ppb are within the range of effects levels in the literature (Table 9.1). By interpolation, we consider that a reasonable and conservative approximation of an effects level for gulls would be 2300 ppb for Herring Gull feathers.

Following a protocol similar to ours, nestling Black-crowned Night-Herons were injected with lead nitrate at doses up to 0.25 mg Pb/g, and were euthanized about 10 days later (Golden et al. 2003a). Lead in feathers was correlated with dose, and showed modest positive correlation with lead in tissues. Lead accumulated mainly in bone. As in our own studies (above), the birds showed only slight growth changes, even at the high dose, and did not show a predicted dose–response relationship between hematocrit or several other variables and dose. The delta-ALAD enzyme was depressed significantly in the medium dose (0.05 mg/g) group, but hematocrit showed no change because the controls had low hematocrit. The feather level corresponding to their medium dose at which effects were observed was 2000 ppb (Golden et al. 2003a). That is close to what we recorded in our study (2300 ppb), and we accept 2000 ppb as a reasonable effect level.

Mechanisms of Developmental Delay

We engaged our neurotoxicology colleagues at the Rutgers Environmental and Occupational Health Sciences Institute to explore a possible mechanism of lead toxicity and developmental delay. We studied gull chicks injected at age 1 day with 0.1 mg/g of lead nitrate and saline-injected controls. We sacrificed four birds in each group at 34, 44, and 55 days and analyzed the brain for the expression of neural cell adhesion molecules (NCAMs) and the NCAM-associated sialyltransferase (ST). Lead birds showed increase in the NCAM-ST and in golgi ST at 34 days, as well as reduction in N-cadherin at 34 and 45 days. L-1 expression was not affected by lead, and by day 55 there was no effect on any of the variables. We concluded that lead disrupts cell adhesion molecules at certain periods in brain development, leading to behavioral deficits, and also serving as potential biomarkers of neurotoxic exposures (Dey et al. 2000).

MERCURY

All forms of mercury are toxic to all forms of life. There is no evidence that mercury is an essential trace element for any organism. Levels of mercury in different bird populations have risen and fallen over the course of the past century as determined from museum specimens (Frederick et al. 2004; Vo et al. 2011). The widespread use of organomercurials as fungicides to prevent seeds from becoming moldy over the winter resulted in widespread poisoning of seed-eating birds and bird-eating hawks after the fungicides were introduced in the 1940s (Fimrite 1979). A museum study of feather levels in Swedish seed-eating birds documented this increase in the 1940s (Berg et al. 1966), and was instrumental in Sweden banning the mercurial fungicides. Mercury levels in birds and mercury poisoning declined rapidly after the ban in 1966 (Fimreite 1979). Museum feathers also showed a mercury chronology for Black-footed Albatross (*Phoebastria nigripes*), with significant increase occurring after 1940 and again at the end of the century (Vo et al. 2011). Other species have had similar museum-based chronologies as well, all documenting the rapid increase in the mid-twentieth century (Frederick et al. 2004).

Biomethylation and Bioamplification of Mercury

To the end of the past century, most industrial sources of mercury pollution, such as chloralkali plants, were closed or their effluent controlled. Many contaminated waste sites and waterbodies remain. However, in recent years, the main source of the methylmercury in estuaries has been from coal-fired power plant emissions and atmospheric transport and deposition (ATSDR 1999, 2013; NJ Mercury Task Force 2001). Mercury in the atmosphere has a complex chemistry, but the main importance is that the mercury is washed out of the atmosphere mostly in rain events known as wet deposition, which deposits the mercury on the surface of waterbodies or on the land (Fitzgerald and Mason 1996). Land-deposited mercury may be washed into waterbodies as well. The mercury in the water reaches the sediment in the bottom of streams, lakes, and bays where bacteria in sediment transform inorganic mercury into methylmercury through a process called biomethylation, a phenomenon first reported by Jensen and Jernelov (1969).

Methylmercury is taken up by bacteria and phytoplankton at the bottom of the aquatic food chain (Figure 9.3). These are consumed by zooplankton, which in turn are consumed by larval fish, small fish, larger fish, and eventually by predatory fish, birds, or humans. At each stage of the food

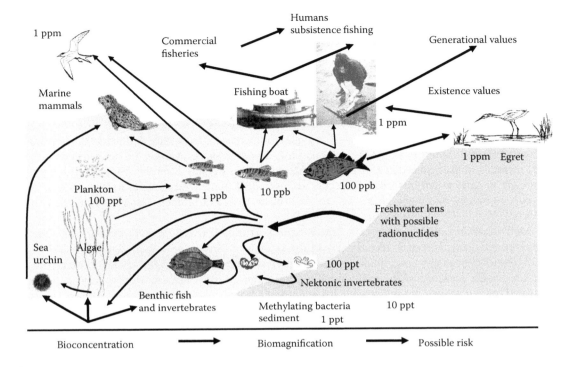

Figure 9.3 Schematic food web diagram illustrating the bioamplification of methylmercury from sediment to methylating bacteria to plankton, to small, medium, and large fish and their predators. (Modified from an Amchitka/Aleut food web, drawn by J. Burger.)

chain, more mercury is retained in the organism than is excreted. Thus, the concentration builds up through bioaccumulation, and at each stage there is a bioconcentration factor. The overall process is referred to as food chain bioamplification (or biomagnification). Mercury in aquatic systems also moves through the terrestrial food web, and not just through the aquatic one to birds (Cristol et al. 2008). At each stage, the concentration may be increased by about an order of magnitude (10 times), so that the concentration in the predators may be 6 orders of magnitude higher than the original concentration in the sediment. For example, the mercury concentration at the sediment surface may be 1 part per trillion (ppt), but the concentration in the predators may be a million times higher, 1 part per million (ppm). Most fish in the estuary have mercury levels in the range of 0.1–1.0 ppm (wet weight).

Methylation rates occur under various conditions depending on the availability of inorganic mercury ion (Hg^{2+}), and the temperature, pH, and organic substrate, conditions in which the sediment bacteria occur. MeHg contamination of aquatic ecosystems continues to be a major concern, spawning research, conferences, and attempts at regulations. The bioaccumulation of mercury in various organisms poses ecological risks that impact pollution prevention policies (Evers et al. 2011).

Methylmercury versus Total Mercury

Once methylated in the sediment, methylmercury is the main form of mercury present in the food chain and is the form most readily absorbed from the gastrointestinal (GI) tract. Once absorbed, there are complex methylation/demethylation reactions in different organs. Methylmercury is much more toxic than inorganic mercury, in part because of its ability to pass into cells and reach critical organs such as the brain. Most studies of mercury in fish have analyzed total mercury, reporting it as a surrogate for methylmercury. As an average across many studies that analyzed methylmercury, it is commonly accepted that about 90% of the total mercury in fish is MeHg, although this value varies from species to species and from study to study. Martín-Doimeadios et al. (2014) reported an average of 89% MeHg in six piscivorous fish species from two locations in Brazil, and an average of 80% for four herbivorous fish.

Species, and perhaps individuals, vary in their ability to demethylate methylmercury. In birds, the MeHg/total mercury ratio is very variable, as low as 10% in albatross liver (Kim et al. 1996), reflecting active demethylation. In young Forster's Terns, the percentage of total mercury in the liver that was methylmercury declined as the mercury concentration increased, indicating that demethylation was occurring (Eagles-Smith et al. 2009b). In adult birds, however, demethylation did not begin until the total mercury in liver exceeded a threshold of 8500 ppb (dry weight). Fimreite (1974) analyzed four species of Ontario waterbirds. Methylmercury averaged 36% of the total mercury in liver, 60% of the mercury in kidney, and ranged from 50% to 80% (mean 69%) in Common Tern eggs from different lakes.

Methylmercury readily passes the blood–brain barrier, whereas inorganic mercury does not. However, inorganic mercury accumulates in the brain after dosing with methylmercury (Friberg and Mottet 1989). In any case, the inorganic mercury is toxic in its own right, so that total mercury is a reasonable measure of the toxic burden. Our analytic results are reported as total mercury. Elevated mercury levels are responsible for many fish advisories to protect recreational and subsistence fisherfolk. However, commercially purchased fish can also have high mercury levels (Burger and Gochfeld 2005a).

Mercury in Feathers and Eggs

There is a very extensive literature on concentrations of mercury in bird tissues (particularly in liver), mainly from areas of known contamination. There are studies of mercury in unhatched eggs, but fewer studies of mercury in feathers, particularly in relation to effects. Mercury circulating in the bloodstream binds avidly with the sulfhydryl-rich keratin protein of the growing bird feather. Accordingly, growing feathers are a sink for mercury, and sequestration of mercury in feathers during molt can be an important detoxification mechanism (Burger 1993; Spalding et al. 2000a). This cannot be emphasized enough—feather growth has a strong protective effect against mercury because mercury is sequestered in feathers. The implication of this conclusion is that we should be looking for effects AFTER feather growth, especially in young birds.

There is no evidence that mercury in the feather affects the function or durability of feathers. Feathers are very suitable for monitoring mercury exposure because they reflect the circulating levels of mercury during the brief period when the feather is actively growing and still has a blood supply. Once growth is complete, the blood supply atrophies. Other metals also bind to keratin with varying affinities. External contamination of feathers is only a problem in environments with low levels of metal in food and high levels in air. Dauwe et al. (2003) found that external contamination was not a problem for mercury in feathers. Unfortunately, many otherwise valuable toxicology studies did not analyze mercury in feathers.

Mercury and some other metals are also excreted into eggs, thereby temporarily lowering levels in the female's body at the expense of her offspring. Only a small amount is in the shell (Burger 1994c). Metal levels in eggs have been reported in many studies, particularly with regard to hatchability of eggs, viability of offspring, and developmental defects (e.g., Henny et al. 2007; Heinz et al. 2009).

Feather mercury levels have been used to identify birds at risk. For example, Ackerman et al. (2008) reported that Forster's Terns were at risk of adverse effects based on feather mercury levels averaging 6440 ppb. However, the mercury levels in feathers from juveniles that died were not different from the average level in live tern chicks (Ackerman et al. 2008). Snowy Egrets breeding at a contaminated reservoir in Nevada had a GM feather mercury of 20,400 and 36,400 ppb in 2002 and 2003, respectively, with the highest value recorded for an individual bird at 58,300. Reproduction was impaired in this colony (Hoffman et al. 2009).

Mercury Toxicity to Birds

Although mercury affects many cellular, biochemical, and physiologic functions, the main biomarker of exposure is actual analysis of mercury (usually total mercury) in tissues and eggs. This is true for both experimental dosing studies and field observational sampling. Wolfe et al. (1998) reviewed 17 studies linking mercury levels in tissues to some outcome. More than half of the studies reported only the corresponding liver level, and seven reported levels in eggs. Feather and blood mercury levels were not reported. Unfortunately, there is no consistent ratio between mercury in feathers and mercury in liver that would allow extrapolation from the many published liver effect levels (Burger 1993).

Mercury toxicity to birds varies among species, and varies by the form of mercury (inorganic, organic), as well as by route of exposure or dosing (Eisler 1987a; Shore et al. 2011). Wolfe et al. (1998) reviewed the literature and identified many data gaps, most of which persist. Birds suffering from mercury poisoning manifest incoordination and impaired locomotion, poor appetite, weight loss, and hypoactivity. They manifest general, nonspecific signs of illness often sitting crouched, with fluffed plumage and drooping eyelids. They may be slow to respond to stimuli or hyperresponsive (White and Finley 1978). Methylmercury is more toxic than inorganic mercury, and young birds are more susceptible than adults (Eisler 1987a; ATSDR 2013). Lethal concentrations in the diet were usually >10,000 ppb, whereas reproductive effects were seen at concentrations of about 1000 ppb (Shore et al. 2011). Spalding et al. (2004) report an association between liver mercury levels >6 ppm (wet weight) (= 18,000 ppb dry weight) and a variety of fatal parasitic and metabolic conditions in Great White Herons (*Ardea herodias occidentalis*) in Florida.

Mercury in eggs may cause embryonic mortality, malposition, reduced hatching, reduced viability of chicks, or developmental defects (Eisler 1987a; Heinz et al. 2009; Herring et al. 2010). Reports of adverse effects (decreased weight, malformations, lower survival of egg or chick) cover a broad range of concentrations (see Table 9.6; Eisler 1987a; Burger and Gochfeld 1997a). Some studies have identified effects in populations. For example, Fimreite (1974) reported reduced hatching and fledging in Common Terns of Ontario at a mean total mercury in one lake of 6100 ppb (dry weight) with a lower 99% confidence limit of about 4400 ppb. A study of Night-Herons, egrets, and terns in Hong Kong found that mercury levels in eggs were sufficient to pose a population risk (hatchability, survival; Lam et al. 2005). In San Francisco Bay, 48% of Forster's Terns were deemed at high risk because of high mercury levels. Clapper Rails (*Rallus longirostris*) had an average of 9040 ppb mercury in breast feathers, and body condition (subcutaneous fat and muscle mass) was negatively related to mercury (Ackerman et al. 2012). Likewise, Black Rail (*Laterallus jamaicensis*) feathers in the San Francisco Bay estuary averaged 6940 ppb, and 32–78% of the birds exceeded various NOAELs (Tsao et al. 2009).

As a neurotoxic agent, mercury can affect orientation. Eared Grebes that crashed in a snowstorm on migration had higher mercury and lower selenium/mercury molar ratios in their brains than Eared Grebes captured alive. Assuming that mercury contributed to disorientation, an effect level would be at least 4200 ppb with a lower 95% confidence level of 3600 ppb (Burger et al. 2013g).

EPA derived an RfD for birds based on the Mallards fed methylmercury by Heinz (1979). The lowest dose (500 ppb in the diet) corresponding to 64 μg/kg (body weight per day), caused adverse reproductive and behavioral effects, and was designated as a chronic LOAEL. This is divided by 3 to find a corresponding NOAEL, which is 21 μg/kg/day in the diet (EPA 1997d).

Table 9.6   Effect Levels for Mercury in Birds

Endpoint	Concentration	References
Reference dose for methylmercury for humans	0.1 µg/kg/day	IRIS 2001
Chronic oral minimal risk level for humans	0.3 µg/kg/day	ATSDR 2013
Dietary level that would require risk evaluation	250 ppb (wet weight)	Shore et al. 2011
Avian reference dose	21 µg/kg/day in diet	EPA 1997d calculated from Heinz 1979 LOAEL
Levels in food or prey associated with adverse effects	1000 ppb (wet weight)	Eisler 1987a
Effects in fish	5000 ppb in muscle	Eisler 1987a; Wiener and Spry 1996
NOAEC for total mercury in eggs	Range 280–6400 ppb (dry weight) geometric mean = 1600	Shore et al. 2011
LOAEC for total mercury in eggs, reduced hatching and fledging	Range 3200–20,400 ppb (dry weight) geometric mean = 7600	Shore et al. 2011
HC_5 in eggs to impair reproduction in 5% of species	2400 ppb (dry weight) in eggs	Shore et al. 2011
Reduced egg weight, hatchability and chick survival in Ring-necked Pheasants (*Phasianus colchicus*)	2000 ppb (dry weight) in eggs 6000 ppb was LC_{50} for eggs	Fimreite 1971
Reduced hatching and fledging in Common Terns	Mean 8120 ppb dry weight Lower 99% CI = 4360 ppb	Fimreite 1974
Effects on egg hatchability	4800 (dry weight) led to 13% ↓ hatching and 28% ↓ nest success 2160 ppb dry weight led to 10% ↓ nest success	Eagles-Smith and Ackerman 2010
Egg hatchability for Snowy Egrets	4000 ppb dry weight ↓ hatch in wet years and 0 hatching in dry years	Hill et al. 2008
Embryo lethality	>1000 ppb (dry weight) in eggs	Eisler 1987a
LC_{50} for Snowy Egret eggs injected with mercury	750 ppb (dry weight)	Heinz et al. 2009
Egg level associated with chick hyperresponsiveness	4000 ppb (dry weight) in eggs	Heinz 1979
Egg level associated with total reproductive failure in Common Terns	14,600 ppb (dry weight) in eggs	Fimreite 1974
Feather levels associated with Toxicity effects in adult birds	4500–5000 to 9000 ppb	Eisler 1987a; Wolfe et al. 1998; Evers et al. 2008
Feather levels associated with male × male pairing in Ibis	4300–18,000 ppb	Frederick and Jayasenda 2011
Feather levels associated with developmental abnormalities in Common Tern chicks	Lowest is 800 ppb (dry weight) in feather Median level 1800 ppb	Gochfeld 1980b
Organ tissue levels in sensitive bird species	200 ppb	Eisler 1987a; Yeardley et al. 1998
Liver levels associated with lethality	Geometric mean = 180,000 (dry weight) HC_5 = 66,000	Shore et al. 2011
Adult liver levels impairing reproduction	6000 ppb (dry weight)	Shore et al. 2011

Note: All values converted to ppb or ng/g, dry weight.

Effects Levels for Mercury in Birds

Effects levels for various endpoints correspond to mercury concentrations in eggs of 4000–14,100 ppb (dry weight) for a "variety of waterbirds" (Wolfe et al. 1998). Levels of 790–2000 ppb (wet weight = 3200–8000 ppb dry weight) in eggs and 5000–40,000 ppb in feathers are associated with "impaired reproduction in various bird species" (Eisler 1987a, p. 63). In ducks, levels of about 10,000 ppb in feathers and 2 ppm (wet weight = 6000 ppb dry weight) in other tissues were associated with reproductive and behavioral abnormalities (Eisler 1987a, p. 62). Vo et al. (2011) reported that Black-footed Albatross feathers exceeded their "deleterious threshold" of 40,000 ppb. A National Research Council summary identified reduced hatchability of eggs laid by females with levels of 5000–11,000 ppb in feathers (NRC 1977). Zillioux et al. (1993) proposed that a liver concentration of 5000 ppb (wet weight) was a "conservative threshold for major toxic effects in waterbirds." Based on very few studies, reproductive effects were observed at brain levels below 6000 ppb.

Methylmercury affects various aspects of reproduction including egg size, clutch size, and laying behavior. "Floor eggs" as well as thin-shelled and shell-less eggs occur at increased frequency in female fowl fed methylmercury (Tejning 1967), and "floor eggs" were also increased in Mallards (Heinz 1979). The literature is replete with positive studies showing adverse effects at some dose, whereas there is a bias against negative studies, which are equally important in establishing effect levels (Fanelli 2012). "Floor eggs" are laid on the floor or ground rather than in nests.

Although contaminants can have profound impacts on birds, most impacts are subtle, hence difficult to document. Reproductive impairment may result in, contribute to, or be masked by, population declines, which are usually multifactorial. Many effects occurring in noniconic species will not be detected. Waterbirds, however, are large, conspicuous, and familiar. Many are top-of-the-food-chain predators. Their disappearance is likely to be noted and heralded. However, they are also long-lived, so it may be years or decades before population impacts are recognized, particularly if exposure levels are not high enough to affect adult survival. Table 9.7 indicates some effects levels, illustrating the great variability attributable to species differences and method differences.

Table 9.7 Levels of Total Mercury in Eggs (µg/kg = ppb) Associated with Adverse Effects in Birds

Species	Level (ppb) Dry Weight	Effect	Source
Pheasant	4600 (LC_{50})	↓ hatching	Spann et al. 1972[a]
Pheasant	6000 (LC_{50})	46% ↓ hatching	Fimreite 1971
Pheasant	5400	26% ↓ hatching	Borg et al. 1969
Mallards	8600 5000 (LC_{50})	↓ hatching embryolethality	Heinz 1979 Heinz et al. 2009
Forster's Tern	4800 2160	13% ↓ hatching 10% ↓ hatching	Eagles-Smith and Ackerman (2010)
Snowy Egrets	750	LC_{50} for eggs injected with mercury	Heinz et al. 2009
Night-Herons	200	More failures in wet years	Henny et al. 2002
Chickens	5000 and 10,000	72% and 83% reduced hatching	Tejning 1967

Sources: Burger, J., Gochfeld, M., *Environ. Res.*, 75, 160–172, 1997a; Eisler, R., Mercury hazards to fish, wildlife, and invertebrates: A synoptic review. U.S. Fish and Wildlife Service Report, No. 85(1.10). Washington, DC, 1987a. With permission.
Note: Published values converted to ppb dry weight.
[a] Used ethyl mercury *p*-toluene sulfonanilide.

Studies of Mercury in Eggs

Mercury and arsenic compounds are fed to chickens as growth promoters, and some studies found slight improvements in growth and hatchability at the lowest mercury doses (Borg et al. 1969; Fimreite 1971). Methylmercury, whether deposited in the egg from the maternal circulation or injected into the air sac, exhibited embryolethality. Laboratory exposure has greater effects than an equal dose deposited by the mother. Some species such as the Mallard are relatively resistant with an LC_{50} of 1800 µg/g (wet weight = 7200 dry weight) (Heinz et al. 2009). Unfortunately, risk assessments have used the Mallard as a reference species representing all birds (Heinz et al. 2009). For many of the species Heinz et al. (2009) studied, doses below 250 µg/kg (wet weight) were lethal including American Kestrel (*Falco sparverius*), Osprey, White Ibis (*Eudocimus albus*), Snowy Egret, and Tricolored Heron. Impaired reproduction in birds has been reported for the Common Loon (Burgess and Meyer 2008; Evers et al. 2008), Common Tern (Fimreite, 1974), California Clapper Rail (*Rallus longirostris obsoletus*) (Schwarzbach et al. 2006), and White Ibis (Frederick and Jayasena 2011). Feeding MeHg to female ducks resulted in ducklings that were hyperresponsive to fright stimuli and hyporesponsive to maternal calls (Heinz 1979).

Chickens fed methylmercury at 4.6 or 9.2 µg/g of diet accumulated about 5000 and 10,000 ppb dry weight) in their eggs and suffered 72% and 83% reduction in hatching (Tejning 1967). In Mallards, 16 µg/g methylmercury in the diet was associated with 75% hatching reduction (Heinz and Hoffman (1998). There was a 46% decrease in hatching success in pheasant eggs, averaging 1.5 µg/g wet weight (= 6000 ppb dry weight), and significant reduction in hatching success and fertility in eggs with 0.5 µg/g wet weight (= 2000 ppb dry weight) (Fimreite 1971). Similarly, Borg et al. 1969) reported a 26% decrease in hatching success of pheasant eggs, averaging 1.35 µg/g (= 5400 ppb dry weight). Pheasant eggs average 33 g (Giudice and Ratti 2001) so that these eggs contained between 17 and 50 µg/egg. These results are consistent in that 2000 ppb dry weight can be considered an effect level, and 6000 ppb is an LC_{50} for pheasant eggs.

Eagles-Smith and Ackerman (2010) describe an elaborate approach to determine mercury effect levels for egg hatchability and nest success. They report a mean total mercury concentration in Forster's Tern eggs of 1.2 ppm wet weight and a moisture content of 74%, corresponding to a 13% reduction in hatching and a 28% reduction in nest success. This converts to 4800 ppb (dry weight). They estimate that a 10% reduction in nest success would occur with an average egg content of 2160 ppb (dry weight). The mean egg mercury level in Black Skimmer eggs sampled in Barnegat Bay was 2130 ppb (dry weight) (Burger 2002c), indicating that some birds would have exceeded this effect level. Table 9.7 reports mercury levels in eggs associated with adverse effects in a variety of species.

The most comprehensive data on embryolethality and mercury come from studies by Gary Heinz and colleagues. They reported appropriate cautions on the generalizability of their data. Heinz et al. (2009) injected MeHg into the air sacs of eggs, and estimated the LC_{50} for 26 species of eight avian orders (Table 9.8). The species included egrets, gulls, terns, and a cormorant. They classified species as high, intermediate, and lower susceptibility. LC_{50}s for adults do not tell us much about reproductive success (other than dead birds reproduce poorly), but LC_{50}s for eggs would be directly related to survival. Waterfowl proved the most tolerant with the highest LC_{50} values. Mallard, the standard "bird" for many ecotoxicology studies, was second only to Double-crested Cormorant in its tolerance. LC_{50} values for the intermediate group ranged from 260 to 870 ng/g (wet weight). Table 9.8 also shows the approximate equivalent value on a dry weight basis in brackets, computed by multiplying the wet weight concentration by 4 for gulls, terns, and cormorants, and by 5 for Ardeids. These conversions are based on the moisture content determined in our laboratory (see Chapter 3). Unfortunately, the data for Great Egret were inadequate to extrapolate an

Table 9.8 Vulnerability of Bird Eggs of Different Species to Injected Mercury

Low Vulnerability or Sensitivity	Intermediate Vulnerability	High Vulnerability
Mallard (1790) [7200]	Clapper Rail (330) [1320]	American Kestrel (120)
Hooded Merganzer (1230) (*Lophodytes cucullatus*)	Sandhill Crane (*Grus canadensis*) (760)	Osprey (180)
Lesser Scaup (*Aythya affinis*) (1530)	Ring-necked Pheasant (440)	White Ibis (120) [600]
Canada Goose (970)	Chicken (440)	Snowy Egret (150) [750]
Double-crested Cormorant (2420)	Common Grackle (*Quiscalus quiscula*) (260)	Tricolored Heron (220) [1100]
Laughing Gull (1250) [5000]	Tree Swallow (320)	
	Herring Gull (280) [1120]	
	Common Tern (870) [3440]	
	Royal Tern (400)	
	Caspian Tern (*Sterna caspia*) (inadequate points)	
	Great Egret (inadequate points) [ca. 2000]	
	Brown Pelican (890)	
	Anhinga (*Anhinga anhinga*) (560)	

Source: Heinz, G.H. et al., *Arch. Environ. Contam. Toxicol.*, 56, 129–138, 2009.
Note: LC_{50} in ng/g (ppb) wet weight are given in parentheses. The comparable dry weight concentrations are in square brackets for the species in our studies.

LC_{50} value, but from the graph given we infer that it would not be below 400 ng/g (wet weight). Although White Ibis and Glossy Ibis are in different genera, we will use the calculated LC_{50} of the White Ibis as a surrogate for Glossy Ibis, which is not included in the Heinz et al. (2009) study. It is notable that the LC_{50} values for Snowy Egret and Tricolored Heron are close, and we will use the midpoint of 185 ng/g (wet weight) rounded to 1000 ppb dry weight for the congeneric Little Blue Heron. Shore et al. (2011) tabulated concentrations in eggs associated with adverse reproductive effects for 11 species, of which the Snowy Egret was also the most sensitive, with a LOEC of 3200 ppb (Hill et al. 2008).

Developmental Defects

Methylmercury has long been known to be mutagenic (Ramel 1969) and teratogenic (Weis and Weis 1977). It binds to the sulfhydryl (–SH) groups on the tubulin protein, leading to the catastrophic disassembly of microtubules, including those that form the cell spindle that guides chromosome separation during cell division. Over a period of several years in the early 1970s, Common Terns on Long Island, and elsewhere, experienced an epidemic of developmental defects (Gochfeld 1975a), presumably related to one or more environmental pollutants. Polychlorinated biphenyls (PCBs) and mercury were prime suspects (Hays and Risebrough 1972; Gochfeld 1975a, 1980b). Eye, beak, and skeletal defects were prominent. Some terns developed brittle feathers, causing feather loss and flightlessness (Figure 9.4a). Abnormal terns had higher blood mercury (6.44 versus 3.73 μg/dl, $P < 0.05$) and slightly higher feather mercury (1750 versus 1270 ppb, $P > 0.05$). Feather abnormalities occurred in a bird with only 800 ppb in its feathers (Gochfeld 1980b). However, there was no unique threshold above which birds were affected and below which they were not, suggesting that mercury was interacting with other factors (Gochfeld 1980b). PCB levels were elevated in some abnormal birds (Hays and Risebrough 1972). Dioxin was just appearing on the environmental toxicology radar screen and was not routinely analyzed at that time (Bowes et al. 1973). The defects were varied and nonspecific, and any teratogen might have been responsible. The importance of the epidemic was its sudden appearance in multiple areas and its equally sudden disappearance.

Figure 9.4   In the early 1970s, there were reports of developmental abnormalities in colonial waterbirds including feather loss in Common Terns (a), and crossed or crooked beaks and other skeletal, extremity, and abdominal abnormalities attributed to mercury and/or PCBs and related organic compounds. Double-crested Cormorant chick with an abnormal bill (b).

At the same time, colonial waterbirds on the Great Lakes were devastated by episodes of reproductive failure and congenital defects, which became known as the Great Lakes embryo mortality, edema, and deformities syndrome (GLEMEDS). Some of the defects (see Figure 9.4b) were similar to the bill defects seen in Long Island terns, but the syndromes were not quite the same. Many possible etiologies were explored. Eventually, Mike Gilbertson and colleagues (1991) applied the epidemiologic approach to causality articulated by Bradford Hill (1965) (see Chapter 3) and applied to birds by Fox (1991). They provided convincing evidence that GLEMEDS was caused by chlorinated organic compounds, particularly PCBs.

Mercury Effects in Songbirds of Terrestrial Ecosystems

Given the evidence of widespread mercury exposure among high trophic level fish-eating birds and birds of prey, it was surprising to learn that small songbirds, primarily insectivorous species such as wrens, also attain high levels of mercury (Tsipoura et al. 2008). Levels in some populations were high enough to affect behavior (Jackson et al. 2011). Marsh Wren (*Cistothorus platensis*) eggs from the contaminated New Jersey Meadowlands had higher mercury levels than expected, and mercury levels averaged higher (300 versus 180 ppb wet weight) in eggs that failed to hatch (Tsipoura et al. 2008). Song performance was altered in Carolina Wrens (*Thryothorus ludovicianus*), Song Sparrows (*Melospiza melodia*), and House Wrens (*Troglodytes aedon*) from the industrially contaminated area of the South River in Virginia. Birds with higher mercury levels in their blood sang shorter songs with fewer notes. Carolina Wrens in mercury-contaminated sites had a 34% reduction in nest survival, and female blood mercury was a predictor of nest survival (Jackson et al. 2011). They reported a very high blood level of 1.3 µg/g of mercury, associated with a 20% reduction in nest survival. This was a major breakthrough in linking individual female blood mercury levels with reproductive success and possible population impact. Jackson et al. (2015) led a broad-scale survey analyzing mercury in more than 8400 birds of 100+ species, reporting a GM blood mercury level of 0.25 µg/g wet weight equivalent to 25 µg/dl.

Mercury Studies in Florida Egrets and Ibis

Mercury pollution of the Everglades ecosystem has been a significant environmental issue, resulting in widespread "Don't Eat" fish advisories. Wading birds (egrets, herons, ibis), have been considered important bioindicators of this exposure (Frederick et al. 2002). Marilyn Spalding and Peter Frederick of the University of Florida conducted an extensive series of toxicity studies of methylmercury fed to Great Egrets and White Ibis. They provided an important generalization that lethal doses were lower in wild than in captive birds, presumably because of greater stressors in the wild, whereas sublethal effects were apparent at lower levels in laboratory birds under controlled conditions, allowing more biomarker measurements (Spalding et al. 2000a).

Spalding et al. (2000a) fed methylmercury to Great Egret chicks at concentrations in food of 0.0 mg/kg (control), 0.5 mg/kg (500 ppb), and 5 mg/kg (5000 ppb; wet weight in fish). At the higher concentration of 5 mg/kg, the egrets developed ataxia (staggering gait) and showed severe hematologic and histologic abnormalities and were euthanized before the end of the experiment. At the more "natural" concentration of 0.5 mg/kg (500 ppb), there were abnormalities in lymphocytes, bone marrow, thymus, lungs, and immunologic defenses (Spalding et al. 2000a). At this dose, birds showed significant reduction in overall activity, shade seeking, and feeding motivation. Although the sample size was small, there was still a weak association between blood mercury and number of feeding strikes and time to capture fish (Bouton et al. 1999).

From the same study, Hoffman et al. (2005) reported additional biomarkers of oxidative stress and organ toxicity affected by mercury (some examples in Table 9.9). Levels of several enzymes in the brain were not affected. The high dose group experienced substantial increases in blood mercury and mercury content of the liver and kidney, but there was little increase in the mercury content of the brain. This is consistent with the idea that the brain can demethylate mercury and eliminate it. One of the toxic mechanisms ascribed to mercury is oxidative stress, and this is reflected in the reduction of glutathione peroxidase, an antioxidant selenoenzyme directly inhibited by mercury.

Table 9.9   Mercury Concentrations in Food, Blood, and Organs, and Plasma Biomarker Results in Great Egrets

	Control	Low Dose	High Dose
MeHg in food	0	0.5 µg/g	5.0 µg/g
Mean blood Hg	0.17 µg/g [17 µg/dl]	10.3 µg/g [1030 µg/dl]	78.5 µg/g [7850 µg/dl]
Brain mercury	0.22 [6600]	3.4 [10,200]	3.5 [10,500]
Liver mercury	0.34 [1000]	15.1 [45,000]	138 [420,000]
Kidney mercury	0.28 [900]	8.1 [24,000]	120 [360,000]
GSH peroxidase	100%	85%	25%
Aspartate aminotransferase	100%	110%	75%
Lactate dehydrogenase	100%	70%	45%

Source: Hoffman, D.J. et al., *Environ. Toxicol. Chem.*, 24, 3078–3084, 2005.
Note: Original values given and values converted to ng/g (ppb) and from wet weight to dry weight are in square brackets using an approximate 3× conversion factor for organs based on moisture content in our laboratory. GSH, glutathione peroxidase activity.

Appetite and growth began to lag in the dosed birds only after they became fully feathered at about 9 weeks. This suggested that before that point, their growing feathers had served as a sink, offering excretory protection from their mercury exposure (Spalding et al. 2000b). This was considered a possible explanation for the lack of difference in health or survival that Sepúlveda et al. (1999a,b) found after dosing Great Egret nestlings orally with methylmercury over a 2-week period. The dosed birds received about 4× more mercury (estimated 1.54 mg/kg food intake compared with 0.4 mg/kg) than undosed control birds.

Whereas Great Egrets eat mainly large fish and are top-level predators, White Ibis eat mainly invertebrates and are exposed to lower levels of mercury in the same habitats. Frederick and Jayasena (2010) conducted a feeding experiment with four methylmercury concentrations (0, 50, 100, and 300 ppb) in the diet fed over a nearly 3-year period. These concentrations spanned the concentrations found in White Ibis foods in the Everglades habitat from which the nestlings were obtained. To assess the demographic significance of mercury exposure, the authors focused on nesting attempts and successes. Over the three nesting seasons, both males and females in the treated groups, particularly the high mercury group, had lower reproductive success than controls. Moreover, male–male pairings were significantly more common in the high and medium mercury dose groups, particularly in the first year (exceeding 50%). Table 9.10 shows the mercury levels in feathers for the 3-year period at the four dose levels, as well as the reproductive success.

The study also revealed a dose-related reduction in normal courtship behavior including head bobs, mutual bowing, and approaches to males by females. Heterosexual pairs also showed a dose-related reduction in the production of fledglings for the low dose ($P = 0.10$) and high dose ($P = 0.085$) groups, indicating a possible 50% reduction in reproductive success, at exposure levels as low as 0.3 ppm.

Females approached mercury-dosed males less often, reflecting impaired courtship behavior. Fledging was reduced 35%. These behaviors were only partly correlated with endocrine associations that they measured. In prebreeding female ibis, estradiol was negatively correlated with mercury, whereas in incubating males testosterone was positively correlated with mercury. However, there was no relationship between hormones and mercury in prebreeding males (Heath and Frederick 2005).

The high dose ibises reached a mean of 51,000 ppb in feathers and 3.95 μg/g (395 μg/dl) in whole blood, compared with 740 ppb in control feathers and less than 0.1 μg/g (10 μg/dl) in the blood of controls (Table 9.10). There was a decrease in male × male pairs with age at all doses and by year 3, the number was zero for control males. Based on Table 9.10, we infer an effects level between 4300 and 18,000 ppb. Fredericks and Jayasenda (2011) were impressed with the preponderance of

Table 9.10　Mercury Levels in Feathers (ppb) and Blood (μg/g) and Reproductive Success in Captive White Ibises Fed Three Methylmercury Doses over a 3-Year Period

	Year	Controls	Low Dose	Medium Dose	High Dose
Total mercury (ppb)[a]					
Feathers (ppb)	2006	740 ± 250	7150 ± 2600	15,240 ± 8650	23,769 ± 8770
Feathers (ppb)	2007	470 ± 110	8200 ± 1530	14,130 ± 5920	51,320 ± 12,330
Feathers (ppb)	2008	620 ± 210	4310 ± 1280	17,960 ± 9150	35,030 ± 1694
Blood (μg/g wet) converted to μg/dl	2008	0.07 ± 0.01 7 μg/dl	0.73 ± 0.09 73 μg/dl	1.60 ± 0.32 160 μg/dl	3.95 ± 0.68 395 μg/dl
Male × male pairs[b]	2006	20%	27%	43%	55%[c]
	2007	10%	36%	43%[c]	44%[c]
	2008	0%	18%	21%[c]	27%[c]

Source: Frederick, P., Jayasena, N., Proc. R. S. Biol. Sci., 282, 1851–1857, 2011.
Note: Data are presented as arithmetic mean ± SD.
[a] The original paper reported these data in ppm and they have been converted here to ppb for comparison purposes. Feather levels are dry weight and blood is wet weight.
[b] Inferred from a scatter plot.
[c] Significantly greater than controls.

male–male pairs occurring at all doses, despite an even sex ratio of available female partners (Table 9.10). Abnormal pair formation has been reported in other species where there is a skewed sex ratio because of differential mortality. However, the ibis observations suggest a more direct endocrine disruption effect on males, analogous to the dichlorodiphenyltrichloroethane (DDT)-induced feminization of embryos of California Gulls (Fry and Toone 1981).

At the end of the nearly 4-year study, the ibis were released to the wild, and regular attempts were made over the ensuing 3 months to resight them. Frederick et al. (2011) did not find a dose-related difference in resighting frequency, and had no further information on reproductive behavior or success. They noted that supplemental feeding was available, but that disease and predation risks would have been as great as or greater than for wild ibis.

Mercury in Herons in Nevada

Although the studies cited earlier documented many biomarker and behavioral responses to methylmercury exposure, it proved difficult to detect such gross effects as mortality (or survival) under natural conditions, even when behavior and reproduction were significantly impaired. This is generally the case, as Spalding et al. (2000a) indicated. Partly, a signal due to mercury (or other toxic substance) may be hidden in the noise induced by environmental variability (i.e., freezing temperature, rain, or drought) or by chance events.

A 10-year study (1997–2006) of Snowy Egrets and Black-crowned Night-Herons was conducted at the mercury-contaminated Lahontan Reservoir in the lower Carson River drainage (Nevada). It is estimated that millions of kilograms of mercury were released from mining activities in the late 1800s, and that large quantities of mercury, much of it bioavailable, remain in the reservoir and wetlands (Hill et al. 2008). The study analyzed 10 eggs/species per year for total mercury, and 5–43 blood samples/year were obtained as well. The study included years of severe drought and low water. In addition to water variability, mercury concentrations in eggs and blood varied significantly among years. In the aquatic feeding egrets, the mercury variability appears unrelated to water level, whereas in the more omnivorous herons, egg mercury levels were higher in wetter years, reflecting greater availability of contaminated aquatic prey (Hill et al. 2008).

Based on an earlier study showing reduced reproductive success in Snowy Egrets, but not Night-Herons (Henny et al. 2002), the authors established criteria levels of 0.8 µg/g (wet weight = 4000 ppb dry weight) in eggs and 2000 µg/kg in blood, although these values were often exceeded, without documented consequences. Unfortunately, the highest mercury levels were measured in the last year of the study (Hill et al. 2008). The first and last years of the study were high water years, which may have mobilized mercury from sediment, resulting in the higher levels in the aquatic-feeding egrets. Drought affected laying date, egg volume, and nest success differently in egrets and herons. There were three summary observations:

1. During drought years all measures of reproductive success in both species were depressed, regardless of total mercury concentrations. Snowy Egrets fared worse than Night-Herons (Hoffman et al. 2009).
2. There was no evidence of reduced reproductive success associated with egg levels of 0.80 µg/g of total mercury wet weight (= 4000 ppb dry weight) for Night-Herons.
3. For egrets, eggs exceeding 0.80 µg/g wet weight of mercury all failed during drought years. However, during wetter years when food was abundant, eggs with 0.80 µg/g often hatched (Hill et al. 2008).

There is an ongoing quest for biomarkers of susceptibility, exposure, or effect. As part of the Nevada study, Hoffman et al. (2009) examined a large suite of clinical chemistry, immunologic, and antioxidant biomarkers in the egrets and Night-Herons in relation to mercury, for the 2002–2006 period. Birds with higher mercury levels showed more damage to liver cells, had higher liver injury enzymes, and higher immune markers. Glutathione reductase, an important antioxidant, was reduced in birds with higher mercury levels, but birds showed adaptive responses to these changes. Although there were many significant differences among egrets and herons from the Superfund site and reference site, none occurred across all 4 years. However, many markers were significantly correlated with blood mercury in some—but not all—years, indicating that the relationships must be confounded by factors that were not measured in the study—perhaps other contaminants or changes in diet.

Across years, egg mercury ranged from 0.23 to 1.93 in Snowy Egrets, and from 0.26 to 1.01 in Black-crowned Night-Herons, with up to 90% of the former and 50% of the latter exceeding the 4000-ppb mark, and hatching severely reduced in some years. Blood mercury levels were very high, ranging up to 5.5 µg/g in Egrets and 7.4 µg/g in Herons (wet weight) (Hill et al. 2008, Hoffman et al. 2009).

CADMIUM

Cadmium is a universally toxic metal with no known essential function in any species. It is carcinogenic and teratogenic as well as highly toxic; however, there are relatively few studies in wildlife (Burger 2008c; Wayland and Scheuhammer 2011). One of the classic cases of aquatic contamination is the cadmium contamination of Foundry Cove at Cold Spring, New York, in the New York–New Jersey Estuary. From 1953 to 1979, a battery factory dumped industrial waste, primarily cadmium, into this tidal freshwater marsh along the Hudson River, about 80 km north of New York City. There were extensive studies of this contamination before and after the dredging in 1994–1995 (Mackie et al. 2007). The highest cadmium level was 1500 mg/kg (dry weight of sediment) in 1983, reduced to a maximum of 230 mg/kg (dw) by 2005. Although this was probably the most cadmium-contaminated site in the world, we do not find any studies of cadmium in birds there.

As with lead, cadmium levels have declined as demonstrated by industrial ecology studies (Boehme and Panero 2003). By the 1990s, cadmium concentrations had declined to about half the peak values in the 1960s (Wayland and Scheuhammer 2011). Airborne concentrations in Arctic Canada declined 3-fold to about 0.05 ng/m^3 from 1970 to 2000 (Li et al. 2003).

Cadmium Levels in Birds

Cadmium levels in birds vary widely taxonomically, geographically, and ecologically, with the highest levels found in seabirds and the lowest levels in landbirds (Wayland and Scheuhammer 2011). Modifying factors affecting uptake include species, age, diet, water hardness, and organic acids (Wren et al. 1995). Bioavailability is lower in the marine environment. Cadmium is readily absorbed from the intestinal tract, particularly in the presence of a calcium-poor diet (Scheuhammer 1996). Cadmium is effective in inducing metallothionein synthesis, which has been used as a biomarker of effect (Scheuhammer 1996). Much of the cadmium stored in tissues is bound to metallothionein, a low molecular weight protein rich in sulfhydryl groups, which transports heavy metals in the blood. There is inadequate information on susceptibility among species, and aside from liver and kidney levels, there is a dearth of information linking levels with effects. Young birds are more susceptible to cadmium than adults (Wayland and Scheuhammer 2011).

Cadmium toxicity has been assessed in a variety of laboratory studies (Eisler 1985a; Wayland and Scheuhammer 2011; ATSDR 2012a). Birds dosed orally with cadmium suffer from growth retardation; kidney, testes, and liver damage; heart enlargement; and anemia (Richardson et al. 1974; White et al. 1978; Eisler 1985a). Cadmium inhibits enzymes by binding to sulfur and breaking the sulfur–sulfur bonds, which confer the tertiary structure necessary for normal enzyme function (Kench and Gubb 1970). Cadmium alters lipid metabolism in ducks (Lucia et al. 2010). It is toxic to male reproduction, damaging Sertoli and Lydig cells and reducing testosterone secretion (Marettová et al. 2015). It also damages the ovary by altering antioxidant defense enzyme systems, resulting in enzyme disruption, cellular death (apoptosis), and lipid peroxidation of cell membranes (Yang et al. 2012).

Cadmium antagonizes zinc, an essential element, and conversely zinc confers protection against cadmium toxicity (Jacobs et al. 1978). Most metals, including cadmium, have an affinity for selenium and exert oxidative damage by inhibiting the antioxidant mechanism that requires selenium. Cadmium caused histopathologic damage to the liver and depressed antioxidant markers, and this was partially reversed by selenium (Li et al. 2013).

Effects Levels

Birds are considered relatively resistant to cadmium toxicity (Wayland and Scheuhammer 2011). This is based on the Mallard, which is also more resistant to mercury—and probably therefore to cadmium—than most birds (see above). Mallards fed 200 ppm for 90 days survived with no loss of weight, although females had lower egg production (White and Finley 1978).

Although cadmium is considered a highly toxic chemical in aquatic ecosystems, information on birds is limited. Adverse effects in fish and wildlife are probable when cadmium exceeds 3 ppb in freshwater, 4.5 ppb in saltwater, 1000 ppb in diet, 100 $\mu g/m^3$ in air (Eisler 1985a), or 470 ppb in feathers (Spahn and Sherry 1999). Eisler (1985a) did not report on cadmium levels in bird eggs or feathers, and 25 years later Wayland and Scheuhammer (2011) had little additional information on cadmium in eggs and feathers. By measuring various oxidative stress biomarkers, a blood cadmium level of 260 $\mu g/dl$ was associated with adverse effects (Wayland and Scheuhammer 2011). However, Espín et al. (2014) suggested that blood cadmium above 0.05 $\mu g/L$ (= 0.005 $\mu g/dl$), blood mercury above 3 $\mu g/L$, and blood lead above 15 $\mu g/L$ (= 1.5 $\mu g/dl$) would cause oxidative stress.

Wayland and Scheuhammer (2011) concluded that a threshold effect level for liver cadmium in adult birds was likely to be between 45,000 and 70,000 ppb (wet weight), and identified a kidney concentration of 65,000 ppb (wet weight) as an ED_{50} (effective dose 50%) for biochemical effects. Reproductive effects were linked to levels in liver, and most of the published reproductive effects occurred at liver concentrations >100,000 ppb (dry weight). The lowest published effect level we found was testicular damage occurring in chickens at a testicular concentration of about 11,000 ppb (Wayland and Scheuhammer 2011).

Embryotoxicity has been demonstrated at very low doses of cadmium injected into chicken eggs (Dżugan et al. 2011). The egg weight for the breed of chickens averaged of 64 g, and the hatchling weight was 70% of the original egg or 45 g (Alsobayel et al. 2013). Hatching was decreased at the 1 $\mu g/egg$ dose, corresponding to 14 ng/g (wet weight) or 56 ppb (dry weight). The LD_{50} was 3.9 $\mu g/egg$, corresponding to 220 ppb dry weight. Heinz et al. (1983) fed cadmium to ducklings and found that those dosed with 4 ppm cadmium in the diet ran significantly farther from a frightening stimulus than either control chicks or those fed 40 ppm.

Through the years, we have attempted to identify the effects levels for cadmium in feathers. The reviews by Eisler (1985a), Heinz (1996), and Ohlendorf and Heinz (2011) did not find data on effects levels, nor did they identify wildlife epidemics of cadmium poisoning in birds. Burger et al. (2009c), using data from Burger (1993), coupled with published correlation coefficients between kidney and feathers, estimated an effects level between 100 and 450 ppb, and possibly as high as 2000 ppb for lethal concentrations in some species. In a 4-week feeding study with chickens, 25 ppm of cadmium was associated with a significant reduction in liver weight, corresponding to a mean feather level of 211 ppb (Abduljaleel and Shuhaimi-Othman 2013). We will use the effects range 210–2000 ppb (see Chapter 14).

Biomonitoring for Cadmium with Feathers and Eggs

The elimination of cadmium into the egg is relatively low compared to other metals, and eggs are probably not very useful for biomonitoring cadmium (reviewed by Burger 1993). Wayland and Scheuhammer (2011) did not find or report data relevant to biomonitoring cadmium in eggs or feathers. Goutner et al. (2001) found relatively low levels of cadmium in eggs of waterbirds, unrelated to their trophic level, and agreed that eggs were not a sensitive biomonitoring tool for cadmium. We found that Common Terns eggs in Barnegat Bay had the highest egg cadmium levels of five species (mean = 28 ppb, dry weight). By comparison, Common Tern eggs from New York averaged 16 ppb (dry weight) (Burger and Gochfeld 1991a). Unlike mercury, we did not find a trophic level relationship for egg cadmium among five species in Barnegat Bay. In the New York–New Jersey Estuary, cadmium concentrations in eggs ranged from a mean of 3.5 ppb (dry weight) in Great Egrets to 7.2 ppb in Black-crowned Night-Herons (Burger and Elbin 2015a).

Burger (1993) reviewed 63 published studies of cadmium in feathers; only 6 of the mean values exceeded 1000 ppb. These included Red Knot and Bar-tailed Godwit from the Wadden Sea (12,000 and 7000 ppb, respectively) and Sooty Terns from Florida (4250 ppb). Adult Common Tern feathers from the New York Bight averaged 49 ppb with no significant difference between sexes.

Wayland and Scheuhammer (2011) were ambivalent about using feathers for biomonitoring cadmium, mentioning that Burger (1993) and Pilastro et al. (1993) encouraged such use, whereas Dauwe et al. (2003) and Jaspers et al. (2004) reported external contamination of feathers from metals including cadmium (but not mercury). However, the latter studies were close to industries emitting high levels of metal air pollution. It is a reasonable caution that external contamination may complicate interpretation of feather metals anywhere there is a high level of airborne contamination. However, in most places background cadmium levels in air are below 5 ng/m³ (ATSDR 2012a).

SELENIUM

Most reports on selenium in birds emphasize its toxicity (Eisler 1985b, 2000b; Wayland and Scheuhammer 2011). The relationship between selenium and other metals has been known for decades (Hill 1975). Selenium toxicity is ameliorated by other metals and, in turn, selenium confers protection against them (Ralston 2008). There is a popular tendency to think of selenium as primarily protecting against mercury toxicity, perhaps by binding and precipitating mercury selenide. Many metals have a high affinity for both sulfur and selenium. Sulfur is much more abundant in the body than selenium, hence more of the body burden of metals is bound to sulfur, which in many cases disrupts sulfur–sulfur bonds that are essential for normal enzyme function. Metals also bind selenium and inhibit selenoproteins, such as glutathione peroxidase, a key player in the body's antioxidant defense. Metals may also bind selenium and prevent synthesis of essential selenoproteins. Thus, protection is a two-edged sword: mercury detoxification by selenium is also the mechanism of mercury toxicity by interfering with selenium functions. The protective effect of selenium is not limited to mercury. For example, selenium protects against cadmium oxidative damage to the brain (Liu et al. 2014).

To achieve protection, while minimizing toxicity, there has to be some excess of selenium. Although Ralston (2008) has proposed that a Se/Hg molar ratio of >1 is sufficient to prevent mercury poisoning, this does not account for the binding of other metals to selenium, and ignores the fact that most mercury poisoning results from interaction with sulfur. No protective or optimum molar ratio has been identified, and in fish, there is such a substantial intraspecific variability in the molar ratio that we do not consider it useful for dietary risk communication (Burger and Gochfeld 2013a,b). Moreover, as the Se/Hg molar ratio increases, the balance shifts from selenium protection to selenium toxicity. Selenium can actually be more toxic than mercury under certain circumstances.

Early studies on metal interactions identified mercury and cadmium as protecting against selenium toxicity (Hill 1975). In a more recent example, Lam et al. (2005) computed the hazard quotients (HQs) for mercury and selenium. The HQ is the ratio between the exposures (usually the daily intake) divided by some threshold effect concentration. They analyzed 17 elements in eggs of Black-crowned Night-Herons, Little Egrets (*Egretta garzetta*), and Bridled Terns (*Onychoprion* [*Sterna*] *anaethetus*). They found that the HQs for mercury and selenium in the three species were similar—that is, that selenium posed as much of a risk for reproduction as mercury, particularly in species where the Se/Hg molar ratio was very high. In egrets, the molar ratio was 64:1, resulting in an HQ of 4.7 for selenium, compared with HQ of 1.6 for mercury (Lam et al. 2005).

Teratogenesis

Selenium and mercury are both teratogenic. Klimstra et al. (2012) injected methylmercury and selenium (selenomethionine) into bird eggs, both alone and together, and counted the deformed embryos or hatchlings. In Mallards, the lowest dose of selenium (100 ng/g) produced abnormalities in 63% of embryos, whereas the lowest dose of mercury (200 ng/g) caused one abnormality (2%). Some deformities (cranial and spinal defects) occurred only when the two were given together. Thus, the mercury enhanced the teratogenicity of selenium, but also seemed to protect against embryolethality (improved hatching). However, direct injection of selenium (or probably any compound) into an egg has more of an impact than the same amount of selenium deposited in the egg by the female (Spallholz and Hoffman 2002; Klimstra et al. 2012).

Selenium Poisoning of Birds in California

The main dietary form of selenium is the aminoacid selenomethionine, which is believed to interchange freely with methionine in protein synthesis, and to serve as a storage form of excess selenium. However, inorganic selenium exposure may occur under certain circumstances. In the 1980s, an epidemic of selenium toxicity affecting birds, fish, and other organisms at the Kesterson Reservoir in California resulted from a combination of naturally occurring selenium in rock, newly constructed water diversion systems, and agricultural water releases (Ohlendorf 2011). The concentrations of selenium in a series of evaporation pools increased, resulting in chronic toxicity affecting a variety of bird and other species. Excess irrigation water flowed over a bed of shale, leaching a soluble selenite $\left(SeO_4^{2+}\right)$ ion and making it bioavailable. Selenium bioamplified, such that ppb concentrations in the water became ppm concentrations in fish and birds. The chronic toxicity resulted in developmental deformities in both embryos and chicks of most bird species nesting in the Kesterson area. About two-thirds of the chicks showed deformities (missing or abnormal eyes, feet, beaks). Fish die-offs occurred as well (Ohlendorf et al. 1986b). After water influx was halted and clean soil was used to cover the contaminated sediment, selenium levels declined slowly (Wu et al. 1995).

Effect Levels for Selenium

Lethal concentrations for various aquatic species range from 60 to 600 ppb. Recommended criteria for the protection of aquatic life is 35 ppb for freshwater and 54 ppb for saltwater (Eisler 1985b). However, levels as low as 50 ppb in water inhibit the growth of algae and reduce hatching in trout eggs (Eisler 1985b). Avian data are mainly represented by experiments in Mallard ducks. In Mallards, a level of selenium in eggs above 3000 ppb (wet weight = 12,000 ppb dry weight) has been associated with reduced hatchability and congenital deformities (Spallholz and Hoffman 2002). In their review, Ohlendorf and Heinz (2011) identified a no-effect concentration in eggs of less than

3000 ppb (dry weight) and a toxicity threshold of 6000 ppb (dry weight), based on decreased hatchability and deformed embryos in Black-necked Stilts. This has been proposed as a toxicity threshold (Lam et al. 2005). The effects level may be somewhere below 6000 ppb. Eggs containing 3000 ppb are toxic to organisms that consume them (Eisler 2000a). Some of the average selenium levels in tern eggs exceed 3000 ppb, suggesting a cause for concern.

Mallards exposed for 12 weeks to 2200 µg/L of selenomethionine showed suppressed immune response, which did not occur in birds treated with 3500 µg/L of selenite (inorganic). However, liver function was impacted by both treatments reflected in increased alanine aminotransferase released by injured liver cells. The margin of safety for selenium appears low. In adult mallards exposed to selenite in diet, the LD_5 was 25 ppm and the LD_{100} was 100 ppm (Heinz et al. 1987).

Data on selenium in feathers are sparse. Background concentrations are given as 1000–2000 ppb, and occasionally 4000 ppb; 5000 ppb is given as a threshold level, warranting study (Ohlendorf and Heinz 2011). Burger (1993) reviewed 42 published studies giving mean feather selenium levels that ranged from 300 (Papua New Guinea; bird-of-paradise) to 54,000 ppb (a Red Knot from Wadden Sea). Seven of 11 Wadden Sea values exceeded 5000 ppb, compared with only 3 others out of 31 (Burger 1993). Most species we studied had mean feather levels below 2000 ppb. Common Terns and Herring Gulls, however, had 10% of individuals with levels above 5000 ppb.

MANGANESE

Manganese is an essential trace element that is neurotoxic (ATSDR 2012b). Liu et al. (2012) fed cocks diets containing 0, 600, 900, and 1800 mg/kg of manganese chloride for 30, 60 and 90 days. They reported that manganese-treated birds showed an increase in manganese content of tissues and malondialdehyde content in lymph and spleen tissue, with a reduction of antioxidants such as superoxide dismutase and glutathione peroxidase. There was a dose–response relationship with DNA–protein cross-linking in lymphocytes consistent with oxidative stress. Manganese also damaged the immune system and altered antioxidant defenses, resulting in lipid peroxidation and apoptosis. They reported similar toxicity in the testes, with reduction in zinc levels, and reduction in hormones likely to affect fertility (Liu et al. 2013).

Although manganese is a well-established neurotoxin, the evidence regarding developmental neurotoxicity is limited. Using the same behavioral paradigm as with our lead studies (see the preceding discussion), each of the 36 two-day-old Herring Gull chicks was randomly assigned to one of three treatment groups to receive either chromium nitrate (25 mg/kg), manganese acetate (50 mg/kg), or a control dose of sterile saline solution. We examined food-begging behavior, balance, locomotion, ability to right, perception, and thermoregulation. By 5 days postinjection, manganese-injected birds were slower to initiate pecking at food compared with controls (mean 1.2 versus 0.2 s; $P < 0.01$). The behavioral tests at day 18 showed significant differences between control and the exposed groups for time to right themselves; thermoregulation behavior; and performance on a balance beam, inclined plane, actual cliff, and visual cliff. Final weight at 50 days of age was 546 g for manganese, compared with 821 g for controls ($P < -0.002$). Adult weight is about 1500 g (Burger and Gochfeld 1995b). The results were very similar to our lead results: manganese caused developmental neurotoxic effects in birds. We estimated an effects level of 4000 ppb, similar to lead. Burger (1993) found 19 published mean values for feather manganese (15 from our laboratory) with a range of means from nondetectable to 17,000 ppb. All species we studied had mean feather manganese levels of less than 2000 ppb. Our results will provide background information in the event that organic manganese compound does become a widespread fuel additive.

CHROMIUM

Jersey City, New Jersey, was a center of chromium ore processing, and the processed slag, containing hexavalent chromium, was disposed of at nearly 200 sites (Stern et al. 2013). Many of the contaminated sites were remediated in the early 1990s. Hexavalent chromium is highly toxic and carcinogenic compared to trivalent chromium. There is redox cycling in the environment and in the body. We only measured total chromium in feathers and eggs. In the absence of clear evidence of toxic impacts on birds, it is difficult to ascribe dose–response relationships. Eisler (1986c) suggested that a tissue concentration >4000 µg/kg (ppb) indicated chromium contamination. In our experimental study with IP injection of chromium nitrate, exposed Herring Gull chicks showed impaired behavior and had a mean feather chromium level of 1200 ppb (see below). The utility of analyzing fur and feathers for chromium has been questioned because of possible airborne deposition (Outridge and Scheuhammer 1993). However, the ambient air concentration is generally below 0.1 ng/m^3 (ATSDR 2012c), making external contribution to feather chromium very unlikely.

There is no significant biomagnification of chromium in aquatic food webs (ATSDR 2012c). Indeed, the toxic effects of chromium are primarily found at the lower trophic levels (Eisler 1986c). Ecologically significant exposures are usually direct exposure of benthic invertebrates and vertebrate eggs and larvae to chromium. However, there is a wide range of adverse effects in aquatic organisms. These include reduced growth in aquatic plants, reduced survival, reduced fecundity, growth inhibition, and abnormal locomotion in invertebrates, and reduced growth, chromosomal aberrations, impaired disease resistance, and developmental abnormalities in vertebrates (Eisler 1986c). There is little information on organic chromium compounds in the environment, although the inorganic chromium is generally complexed to organic material that may affect its fate and transport in the environment and in the body.

Although chromium is not considered a potent neurotoxin or developmental toxin, we conducted behavioral studies on Herring Gull chicks injected with chromium (Burger and Gochfeld 1995b). Each of the 36 two-day-old Herring Gull chicks was randomly assigned to one of three treatment groups to receive chromium (III) nitrate (25 mg/kg), manganese acetate (50 mg/kg), or a control dose of sterile saline solution. Chromium nitrate is a very water soluble compound of trivalent chromium. By 5 days postinjection, chromium-injected birds were slower to initiate begging (mean 1.0 versus 0.2 s, $P < 0.01$) compared with controls. Behavioral tests at 18 days showed that chromium chicks had poorer balance, had weaker locomotion, were slower to right themselves, and were slower to find shade and remain there compared with controls. Final weight at 50 days of age was 821 g for controls and 528 g for chromium-injected birds ($P < 0.001$) (Burger and Gochfeld 1995b). At 50 days, the chromium-injected laboratory birds had an average feather chromium level of 1200 ppb. Some wild Herring Gull chicks have higher levels, indicating that the laboratory dose was in an environmentally realistic range and that birds in the wild may experience impact from chromium (Burger and Gochfeld 1995b). A review of 17 published studies found a mean of 8800 ppb, mainly driven by very high levels in some passerines (median = 2200 ppb; Burger 1993). Herring Gulls from the Great Lakes averaged 400 ppb (adults) and 700 ppb (young birds; Struger et al. 1987).

ARSENIC

Arsenic is well known as a highly toxic chemical. Unlike mercury, it is the inorganic arsenites and arsenates that are more toxic than most organic forms. In our studies, we measured only total arsenic. Arsenic contamination is widespread both from historic use of lead arsenate insecticides and from more recent use of arsenical defoliants. However, arsenic does not figure prominently in ecotoxicology studies of birds. The signs of arsenic poisoning in birds—weakness, slow and jerky motions, hyperactivity, fluffed feathers, huddling, and loss of righting reflex—are nonspecific and are characteristic of most poison victims on their way to coma, convulsions, and death (Eisler 1988b). Symptoms may occur within hours after an acute dose (Hudson et al. 1984). The LD_{50} for single oral doses ranged from 17,000 to 48,000 µg/kg in birds. Exposure data are usually presented on dose, with few reports on body burden in any tissue. Acute arsenic poisoning in Common Quail (*Coturnix coturnix*) was manifest mainly by liver damage (Nystrom 1984).

In humans, chronic exposure is associated with cancer (skin, bladder, lungs, kidney), vascular disease, and diabetes. However, chronic effects have not been well established in other organisms, leading Eisler (1988b) to conclude that the impact of particular concentrations of arsenic in fish, birds, eggs, and feathers, remains unclear. This continues to be true 25 years later (Hettick et al. 2015). Arsenic's main chronic effects are on reproduction, and it is a known teratogen (Eisler 1994; Boyle et al. 2008), mutagen and carcinogen (ATSDR 2007b). It is also an endocrine disruptor (Sun et al. 2015b). Several studies show that arsenic causes oxidative stress in invertebrates (Saha and Ray 2014) and vertebrates, and that it impairs immune responses (Saha et al. 2010). EPA has set a criterion value of 1300 µg/kg (ppb, wet weight) in fish muscle.

The most recent review of arsenic in birds (Sánchez-Virosta et al. 2015) found relatively few chronic exposure studies. Most of the studies had measured arsenic in internal organs (37%), followed by feathers (32%), eggs (32%), and blood (15%). However, tissue levels were generally not linked to effects. Birds dosed with inorganic arsenic develop symptoms similar to those of lead poisoning. As with mercury, the Mallard is relatively resistant, with a single dose LD_{50} of 323 mg/kg body weight, compared to LD_{50} of 48 mg/kg body weight for California Quail (*Callipepla californica*) (Eisler 1988b). Birds that eat mainly invertebrates have exposure to high amounts of inorganic arsenic, whereas birds that eat fish (or other birds) will be getting mostly organic arsenic of low toxicity.

We did not find satisfactory effects levels associated with arsenic in feathers. Embryotoxicity has been demonstrated by injecting a variety of arsenic compounds into eggs. The results are reported on a per egg basis. In Mallards, the threshold for congenital defects was 0.03 mg of arsenite per egg, and the LD_{50} was between 0.01 and 1 mg/egg. A Mallard egg averages about 50 g (Drilling et al. 2002), so that the embryotoxic concentration amounts to about 2400 ppb (dry weight) with an LD_{50} of probably about 8000 ppb dry weight by interpolation.

Arsenical Feed Additives

Organic arsenicals, particularly the pentavalent roxarsone™, have been used in animal feed to promote growth and combat infections. These pentavalent compounds are almost 100% absorbed from the GI tract, and almost 100% excreted. Although chickens rapidly excrete arsenic, they retain only a small percentage of the ingested dose that accumulates in their tissues (Eisler 1988b). This buildup proved sufficient over time to result in accumulation of significant carcinogenic concentrations of inorganic arsenic in chicken meat (Lasky et al. 2004). Although the Food and Drug Administration had not outright banned arsenic feeding to chickens, in 2011, Pfizer, producer of roxarsone, conducted a voluntary withdrawal of the drug, which was already banned in Europe. By 2013, EPA had banned the main drugs, roxarsone, carbarsone, and arsanilic acid, which had been incorporated into more than 100 veterinary products added to animal feed (Strom 2013). Unknown at the time, the drugs did produce subclinical toxicity in chickens (Xing et al. 2015), as genes related to inflammation, including NF-κB, iNOS, COX-2, PTGEs, and TNF-α, were upregulated by arsenic supplementation, up to 30 mg/kg As_2O_3. The data suggest a significant arsenic-induced inflammatory response after 3 months, which may be ecologically significant in birds that live longer than commercial fowl.

SUMMARY AND CONCLUSIONS

Lead, mercury, cadmium, and arsenic are toxic substances with no known essential function in any organism. Chromium, manganese, and selenium are essential to many species. Metals exert toxicity by binding to sulfur and selenium of enzymes, and thereby inhibiting their activity. Mercury binds avidly to sulfhydryl groups on tubulin, which is critical for cell division and nervous system formation. Metals also generate reactive oxygen species or enhance oxidative damage, partly by interfering with the antioxidant mechanisms involving selenium.

Lead poisoning in birds occurs commonly from ingestion of lead shot or lead pellets retained in tissue after a nonfatal shot. Although lead shot was banned for waterfowl hunting in 1991, it continues to be used for other game. Moreover, decades of shooting over water has left a legacy of spent shot waiting for a dabbling bird (USGS 2014). Our experiments with lead, using Common Terns and Herring Gulls as study species, showed behavioral impairment in several functions including locomotion, feeding, thermoregulation, visual cliff, and individual recognition. These effects were observed both in the laboratory and the field, and affected chick survival in nature. Studies dosing birds with methylmercury in food caused subacute toxicity. The youngest chicks could excrete mercury into their feathers, whereas older chicks rapidly accumulated mercury and showed impaired growth and histologic, immunologic, and neurological changes. Because small chicks sequester mercury in their developing feathers, chicks are mainly protected from the effects of mercury. Effects of mercury exposure should thus be examined after the completion of feather growth in young birds.

Selenium and other metals interact in complex ways. Metals bind selenium and interfere with the action of selenoenzymes involved in normal thyroid function and antioxidant protection of cells. Selenium reduces mercury toxicity, and mercury may, in turn, protect against selenium toxicity when there is a very high molar excess of selenium.

APPENDIX

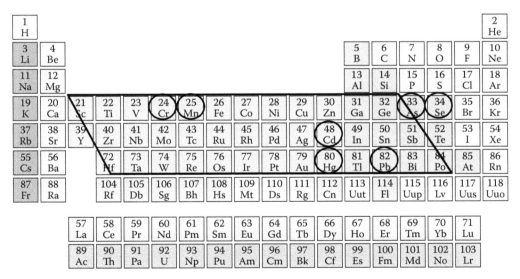

Figure 9A.1 Periodic table of the elements attributed to the Russian chemist Dimitri Mendeelev who began classifying the elements on the basis of their atomic mass. Heavy metals are enclosed in the parallelogram, and the seven elements we studied are circled: Cr = chromium, Mn = manganese, As = arsenic, Se = selenium, Cd = cadmium, Hg = mercury, and Pb = lead.

CHAPTER 10

Heavy Metals in Fish, Lower Trophic Levels, and Passerine Birds

INTRODUCTION

The coastal birds discussed in this book divide their time between land for nesting, water for feeding, and air for migrating. As a group they are predators, feeding on a wide range of invertebrates and fish, and even preying on the eggs and young of other species. Shorebirds, Clapper Rails, and Glossy Ibis feed mainly on invertebrates. Terns are piscivorous, feeding mainly on fish. Gulls, egrets, and Night-Herons are more versatile predators. They switch prey, depending on the environment and the availability of prey. All organisms in an ecosystem are part of a complicated food web, either producing food (plants) or consuming other organisms. Some organisms, however, are more critical than others as prey for coastal birds.

Foraging behaviors differ as well. Terns plunge-dive for fish whereas egrets stand in the water and wait for fish to swim by; Black-crowned Night-Herons and Clapper Rails skulk along the marsh edge looking for Fiddler Crabs (among other prey), whereas shorebirds scurry along the beach or mudflat picking up invertebrates or Horseshoe Crab eggs, or wade in the water, probing the mud for prey. Gulls, on the other hand, forage everywhere by plunge-diving, surface dipping, picking up prey, eating carrion or fish offal, or feeding at garbage dumps. These foraging descriptions are generalizations. Humans participate in these coastal food webs by extracting fish and seafood, or by hunting the birds themselves. Although the commercial slaughter of shorebirds was banned in the United States in the early 1900s, the slaughter continues today on migratory stopover and wintering sites in South America.

In this chapter, we describe and present data on natural history and metal levels in a range of species that contribute to the food web for coastal birds. These species are indicators of metals in themselves, in the plants or prey they consume, and of the potential exposure to predators that consume them. Some species are also sentinels for human health, both through the food chain (people consume shellfish and fish) and through living in the same environments as we do. Sentinel data, although not usually the primary data considered in risk assessments, contribute to a weight-of-evidence approach (van der Schalie et al. 1999). The primary indicators discussed in this chapter are Horseshoe Crabs and Fish.

LOWER TROPHIC LEVELS

Lower trophic levels include plants (primary producers) and animals that are themselves herbivores (these primary consumers feed on phytoplankton and microorganisms). The latter might include omnivores that eat both plant and animal material (often zooplankton and invertebrate larvae). Although we did not focus on these levels, it is instructive to compare mercury levels for

Table 10.1   Mercury and Lead Levels (Means ± SE) (ppb, Wet Weight) in Lower Trophic-Level Organisms from Raritan Bay, New Jersey (2008–2010)

Species	Lead	Mercury
Sea Lettuce—*Ulva lactuca* (*n* = 10)	65 ± 20	6.8 ± 0.9
Rock Weed—*Fucus gardneri* (*n* = 10)	NA	4.1 ± 0.3
Ribbed Mussel—*Geukensia demissa* (*n* = 5)	22 ± 8	11 ± 0.8
Grass Shrimp—*Palaemonetes pugio* (*n* = 26)	32.7 ± 12.7	12.4 ± 0.3
Sand Shrimp—*Crangon septemspinosa* (*n* = 11)	21.2 ± 5.6	32.2 ± 2.3
Green Crab—*Carcinus maenas* (*n* = 9)	846 ± 172	18.6 ± 3.5
Blue Crab—*Callinectes sapidus* (*n* = 5)	60 ± 3.1	86.8 ± 20.5
Striped Killifish—*Fundulus majalis* (*n* = 9)	34.5 ± 12.8	20.7 ± 2.7
Atlantic Silversides—*Menidia menidia* (*n* = 10)	36.8 ± 6.1	44.6 ± 3.8
Young Bluefish—*Pomatomus saliatrix* (*n* = 35)	47.3 ± 12.9	296 ± 45.6

Sources: Burger, J. et al., *J. Toxicol. Environ. Health*, 75, 272–287, 2012e; Burger, J., Unpublished data.
Note: Data presented from lowest to highest trophic level. Fish are given for comparison even though they are discussed in more detail below. NA, not available; SE, standard error.

several different organisms. We use data from Raritan Bay (New York–New Jersey Harbor Estuary) to illustrate differences in mercury and lead for some lower tropic levels (Table 10.1). The data were collected to examine potential ecologic and human exposures to lead from a "Superfund" on Raritan Bay. The data presented in the following sections are from the reference sites far from the contaminated site (Burger et al. 2012e).

The data illustrate that both lead and mercury levels are rather low in these midtrophic-level organisms, except for young Bluefish (mercury) and Green Crabs (lead). Although the superfund site had very high levels of lead, the reference sites had background levels, probably a legacy of the use of lead in gasoline a generation earlier. Mercury mainly enters the system by atmospheric deposition from coal-fired power plants. The large Bluefish, which represent a higher trophic level, were caught by fishermen who were disappointed because the Bluefish were at the lowest end of the "allowable catch size." Bluefish are migratory and whatever levels they possessed would not have been local in origin. Green Crab had higher levels of lead than the other species (including Bluefish) by an order of magnitude. We cannot account for this difference, although it may reflect different food items, different bioavailability, or that there are differences in intracellular partitioning within the prey that ultimately affect trophic availability to predators (Goto and Wallace 2009). Goto and Wallace (2009) found similar differences for lead (as well as cadmium), and suggested that there may be both prey-dependent processes, and predator-dependent processes that may offset or shift the usual tropic transfers among levels. Differences could also be attributable to dietary habits, foraging areas, differences in prey or foods, or size of prey items (Burger and Gochfeld 2001b; Shahbaz et al. 2013).

HORSESHOE CRABS

Horseshoe Crabs as Bioindicators

Horseshoe Crabs can be bioindicators because they serve a pivotal role in coastal bays and estuaries where they spawn. Every stage of their life cycle is a node on the overall food and economic web of bays and local communities (Figure 10.1). Their eggs are eaten by a wide range of shorebirds, gulls, and even Night-Herons. The larvae and juveniles, still bearing paper-thin shells, are eaten by birds, fish, and Diamondback Terrapins. Although the adults are generally immune to predators owing to their hard shell, they are eaten by sea turtles. Once ashore and overturned by a wave, the

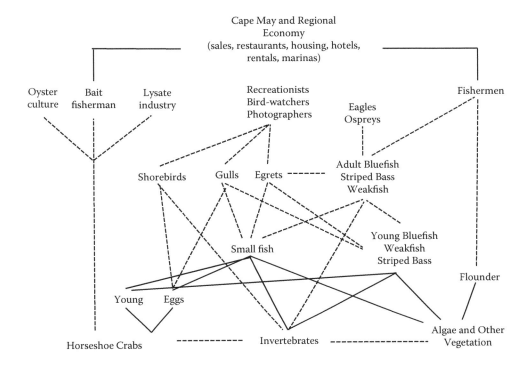

Figure 10.1   Interconnections of economic and ecologic webs linked to Horseshoe Crabs. The crabs serve as an iconic and keystone species in the bay.

adults are helpless until the next high tide allows them to escape. In this stage they are exposed to predators such as gulls (Shuster 1982; Niles et al. 2012) that peck furiously, dismembering the crabs. However, humans are the main predators on adult Horseshoe Crabs, collecting them for fertilizer a century ago, and now for fishing bait and medical purposes. Huge numbers of female Horseshoe Crabs are captured and taken to a laboratory where half their blood is removed to make a substance called *Limulus* lysate, an important biomedical product. Horseshoe Crabs that survive are returned to the water, although there is concern about the effect of such massive bleeding on survival (Hurton and Berkson 2006; Leschen and Correia 2010). Horseshoe Crabs are also collected commercially for sale as conch and eel bait. This practice is currently banned in New Jersey.

The Horseshoe Crab and the food web it supports are not the only victims of overexploitation. Striped Bass feed largely on Menhaden. A recently established large industrial Menhaden fish processing facility in Virginia has considerably depleted the Menhaden, resulting in a decline of Striped Bass from the Hudson River to Chesapeake Bay. This decline from its 2004 recovery peak is attributed to starvation and increased susceptibility to disease (Mason 2004). The same condition, a decline in body lipids, has been implicated in the lower body burdens of polychlorinated biphenyls (PCBs) in Striped Bass from Long Island Sound; PCBs are stored in fat (Skinner et al. 2009).

Natural History

Each spring, thousands of Horseshoe Crabs migrate from the continental shelf of the Atlantic Ocean to bays and estuaries to spawn in the tide-swept sands. Delaware Bay is the center of Horseshoe crab breeding in North America (Shuster 1982; Shuster and Botton 1985; Swan 2005). During high tide, amplexing (mating) groups of Horseshoe Crabs with one to several males gripping the back of a female, deposit and fertilize their eggs in a crude cup that the female digs in the sand. When the crabs are abundant, late arriving females on the same or subsequent nights inadvertently dig up the nests of previous females when they deposit their own eggs (Figure 10.2). The egg masses of previous females are thus brought to the surface, usually broken apart, and the eggs concentrated at the water line are available for shorebirds to eat (Dunne et al. 1982). Shorebirds concentrate along Delaware Bay beaches in direct proportion to the abundance of Horseshoe Crab eggs (Botton et al. 1994). Thousands of shorebirds that migrate through Delaware Bay each spring feast on the Horseshoe Crab eggs (Burger 1986; Castro and Myers 1993; Botton et al. 1994; Burger et al. 1997a; Tsipoura and Burger 1999; Baker et al. 2004; Niles et al. 2008). It is an essential resource to build up fat reserves for migration. Without it, they may not complete migration or may be unable to breed successfully. Shorebird populations take time to respond to food availability and habitat loss. Several species of shorebirds have declined in Delaware Bay and elsewhere in the past 25 years (Morrison et al. 2004, 2007; Mizrahi et al. 2012; Andres et al. 2013). Other species, such as Laughing and Herring Gulls, also rely on crab eggs during the breeding season (Burger 2015a). The stresses on populations are cumulative. During migration, shorebirds are faced with habitat loss, decreases in the prey base, predation, and human disturbance (Goss-Custard et al. 2006; Burger and Niles 2013a,b, 2014).

During the Lenapé Indian period, whole Horseshoe Crabs were buried with corn seeds as fertilizers, and this practice was taken up by the early settlers. In the 1800s, collecting crabs for "commercial" fertilizer was an important industry in South Jersey. Horseshoe Crab populations declined during this period, but they increased in the 1900s when the industry stopped. Beachgoers no longer wanted to smell the piles of rotting Horseshoe Crabs, and alternative fertilizers came into use. Horseshoe Crab populations continued to increase into the 1980s, when new and efficient exploitation began. Populations of Horseshoe Crabs clearly declined, but the magnitude was unclear (Faurby et al. 2010).

In the late 1980s and 1990s, fishermen began using Horseshoe Crabs for bait for eel and conch. The take for bait greatly increased, and trailer trucks came from neighboring states to pick up the crabs while they were spawning (Burger 1996a; ASMFC 1998, 1999). Truckers parked by beaches and literally filled a tractor-trailer in two nights. When the extent of the "take" was recognized, New Jersey banned handpicking of crabs from the beaches every other day, making it economically unfeasible. So the crabbers figured out where to go to fill their trucks in a single night. Then, New Jersey followed with regulations forbidding the taking of Horseshoe Crabs from beaches. Crabbers responded by collecting the crabs by trawling off the spawning beaches. The number of crabs counted at high tide on May and June nights plummeted, and the concentrations of crab eggs in the sand declined as well. Shorebird studies documented fewer birds reaching a desirable premigratory weight (Niles et al. 2008). New Jersey instituted a statewide ban on Horseshoe Crab harvesting or landing, but the crabbers landed their crabs in other states that did not have a ban. There is no evidence that the Horseshoe Crab population has recovered (Dey et al. 2015).

Vocal popular concern about the decline in Horseshoe Crabs along the Atlantic coast of the United States, particularly along Delaware Bay (Burger 1996a; ASMFC 1999; Botton 2000; Botton and Loveland 2001), resulted in the development of a Horseshoe Crab management plan, with limits on take by the Atlantic States Marine Fisheries Council (ASMFC). The decline in spawning Horseshoe Crabs in Delaware Bay was attributed to overharvesting by fishermen that use them as bait (ASMFC 1999; Niles et al. 2008; Kreamer and Michels 2009). The ASMFC plan initially called for a voluntary reporting of take, which later resulted in a quota for each state from Maine to Florida.

This was not sufficient to halt the decrease in the number of breeding Horseshoe Crabs on coastal beaches, and the decrease in excess eggs on the sand. Each year during the 1990s saw fewer shorebirds (particularly Red Knot, Ruddy Turnstone, and Semipalmated Sandpipers) on the Delaware Bay shore. Those birds that did visit had to hustle to store sufficient fat for their northward migration (Burger 1984e, 1986; Botton et al. 1994, 2003). It was a unique move for the Commission

(a)

(b)

(c)

Figure 10.2 Spawning Horseshoe Crabs on Delaware Bay beach (c) and shorebirds foraging at the high tide line (b and a).

(ASMFC) to incorporate birds into a fishery management plan, and signaled a willingness of the ASMFC to consider an ecosystem approach to fisheries management.

Delaware Bay, although the center of Horseshoe Crab breeding, is not the only place where shorebirds, gulls, and other species feed on the excess eggs. At other stopover sites, such as Virginia, Barnegat Bay, and New York–New Jersey Harbor, shorebirds also feed on crab eggs, although the number of Horseshoe Crabs and the number of shorebirds are lower. In many places along Delaware Bay, hundreds of breeding Laughing Gulls line the beaches and sand shoals where there are crabs and eggs. Although Horseshoe Crabs breed from Maine to Florida, the eggs are less of a resource for migrating shorebirds both to the north and to the south because of the timing of peak egg-laying in relation to the time of shorebird migration. Farther north (Massachusetts), the crabs spawn too late to be useful for northbound shorebirds, and to the south (Florida), where many shorebirds spend the wintering months, they have migrated north before the crabs spawn. Gulls, however, are everywhere, all the time, and can always make use of an abundance of Horseshoe Crab eggs.

Once the eggs hatch, the juvenile Horseshoe Crabs are washed out of their nests by receding high tides, where they face predation by a wide range of fish. They gradually make their way to the ocean shelf, where they remain in the bottom muds, growing with each molt of their hard exoskeleton until they return to breed at 10 or more years of age (Shuster 1982; Shuster and Botton 1985).

Atlantic Coast Patterns in Metal Levels

Horseshoe Crabs are obviously an important part of food webs for many coastal bays and estuaries—eggs for birds and other predators, juveniles for fish, and adults for sea turtles, other marine predators, and people. We examined metal levels in eggs and leg muscle of Horseshoe Crabs from Maine to Florida (Burger et al. 2002a). Although these data are from a decade ago, they give an indication of variation in metal levels, particularly as the samples were gathered in the same year, and the analyses were completed in the same laboratory, at the same time, using the same procedures. Unfortunately, this does not occur very often because of time and money constraints. Table 10.2 summarizes the highest and lowest locations for each met al.

Horseshoe Crabs from Maine and Florida were slightly smaller than those from Massachusetts to Maryland. The levels of metals in eggs varied significantly among sites (Figure 10.3), but a pattern is elusive (Burger 2002a). New York and Maryland seem to be spared highest values. Massachusetts had the highest values for mercury, manganese, and arsenic.

One way to evaluate the relative levels of metals at any given place is to compare the levels with the mean levels for the whole sample. Those that are higher represent greater exposure. This is particularly important for eggs, which are such an important food source for shorebirds and other species (Table 10.3).

Table 10.2   High and Low Levels of Metals and Metalloids in Horseshoe Crabs from Different States

	Maine	Massachusetts	New York	New Jersey	Delaware	Maryland	Florida
Number	11	12	24	28	5	8	12
Lead	Highest						Lowest
Mercury		Highest				Lowest	
Cadmium	Lowest			Highest			Highest
Selenium	Lowest				Highest and lowest		
Chromium				Lowest	Highest		Highest
Manganese		Highest		Lowest			Highest
Arsenic		Highest		Highest			Lowest

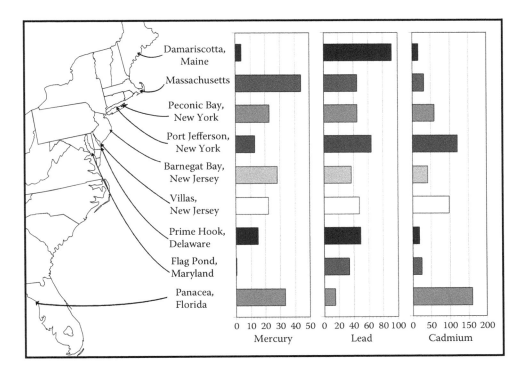

Figure 10.3 Spatial patterns of mercury, lead, and cadmium levels (ppb, wet weight) (mean ± SE) in eggs of Horseshoe Crabs from Maine to Florida. (After Burger, J. et al., *Environ. Res.*, 89, 85–97, 2002b.)

Table 10.3 Mean Levels (± Standard Error) of Metals in Eggs and Leg Muscles of Horseshoe Crabs from Maine to Florida

Metals	Grand Mean (± SE) Maine to Florida[a] 2000	NJ Delaware Bay Mean (± SE) in 2000[b]	NJ Delaware Bay Mean (± SE) in 2012[c]
Eggs			
Lead	45 ± 4	37 ± 4	37.6 ± 6.7
Mercury	22 ± 3	21 ± 2	6.2 ± 1.7
Cadmium	70 ± 14	44 ± 12	1.6 ± 0.7
Selenium	953 ± 46	1007 ± 70	990 ± 98.9
Chromium	46 ± 4	55 ± 7	82.7 ± 14.3
Manganese	2210 ± 158	2071 ± 144	6076 ± 530
Arsenic	10,900 ± 1401	5924 ± 345	3729 ± 263
Leg Muscle			Not analyzed
Lead	41 ± 4	44 ± 4	
Mercury	57 ± 5	100 ± 6	
Cadmium	37 ± 6	29 ± 2	
Selenium	876 ± 39	851 ± 48	
Chromium	73 ± 6	82 ± 10	
Manganese	1710 ± 368	1782 ± 236	
Arsenic	17,400 ± 1430	14,482 ± 685	

Note: These data can be used to evaluate the levels in other bays and estuaries. Results are in ppb (ng/g; wet weight). For eggs, *n* = an egg mass, sample size was 100 for eggs and 100 for leg muscle.
[a] Burger, J. et al., *Environ. Res.*, 90, 227–236, 2002a.
[b] Burger, J. et al., *Arch. Environ. Contam. Toxicol.*, 44, 36–42, 2003b.
[c] Burger, J., Tsipoura, N., *Environ. Monit. Assess.*, 186, 6947–6958, 2014; Burger, J., Unpublished data.

The data from 2000 indicate that the eggs and leg muscle from New Jersey's Delaware Bay were similar to or less than the mean—indicating that contaminants are not unduly high in eggs. In 2012–2015, the metal levels in eggs remained low. Mercury, cadmium, and arsenic declined, whereas chromium and manganese increased (Table 10.3).

Temporal Patterns in Delaware Bay

In the preceding section, spatial patterns were clearly demonstrated among Horseshoe Crab eggs and leg muscles from different locations along the Atlantic Coast. Data from Delaware Bay can be used to assess temporal patterns, and whether the changes have been significant over time (Burger 1997b; Burger et al. 2003b; Burger and Tsipoura 2014). That is, if the levels have remained relatively constant in Delaware Bay, then the data for Maine to Florida in the preceding section are probably representative of what the crabs obtain from water, sediment, and their food at each site.

Overall, metals in eggs of Horseshoe Crabs were relatively high in the 1990s, declined in the late 1990s and in 2000, and remained low from 2012 to 2015 (Figure 10.4). All of the metals have been declining in Delaware Bay, but the main decline happened in the 1990s. Unfortunately,

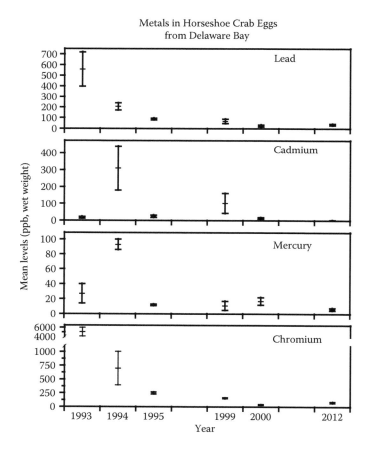

Figure 10.4 Temporal pattern in metal levels (ppb, wet weight) (mean ± SE) in eggs of Horseshoe Crabs from Delaware Bay, 1993 to 2015. (After Burger, J., *Environ. Monit. Assess.*, 46, 279–287, 1997b; Burger, J. et al., *Ecotoxicol. Environ. Saf.*, 56, 20–31, 2003a; Burger, J. et al., *Environ. Res.*, 133, 362–370, 2014; Unpublished data.) These data include eggs from both sides of the bay.

we did not measure arsenic in the early 1990s; however, it declined from the late 1990s to the present (2012–2015). Declining metal levels in sediments have been reported for several of the estuaries, a testimonial to environmental regulation of industrial, agricultural, and urban effluent (Mason 2004). Declining metal levels have also been documented in California (lead and mercury in San Francisco Bay; Hornberg et al. 1999), Europe (e.g., Severn Estuary; Duquesne et al. 2006), and Indonesia, and these are likewise attributed to environmental regulation (Hosono et al. 2011). Vallius (2013) reported that "almost all heavy metal species have declined" in Baltic Sea sediments, except mercury, cadmium, and arsenic.

Spatial Patterns in Delaware Bay

In the early 2000s, Horseshoe Crab eggs were collected from a number of the key spawning beaches in New Jersey and Delaware (Figure 10.5). There were significant differences in metal levels for some metals and tissues (Table 10.4). It is clear, however, that the differences are not very substantial, and may not be very significant in terms of exposure for the developing embryos. However, for shorebirds that feed primarily on crab eggs for a 3-week period, the contaminant levels in their main food may be important. Arsenic, which was higher in Delaware compared to New Jersey for both eggs and leg muscle, was the only metal with a consistent difference between states.

Metal levels in eggs did not vary significantly among sites on the Delaware side, although they were not as far apart as the sampling sites in New Jersey. Metal levels did vary significantly on

(a)   (b)

Figure 10.5 (a) Horseshoe Crabs spawn in a continuous ribbon on beaches with appropriate sand. Female crabs are usually pursued by several males, which are considerably smaller. (b) Ron Porter holding a male Horseshoe Crab.

Table 10.4   Levels of Metals in Horseshoe Crab Tissues for Delaware Bay, for Metals Where There Were Significant Differences

Tissue and Metal	New Jersey Side	Delaware Side	χ^2 (P)
Eggs			
Arsenic	5920 ± 345	6770 ± 478	3.72 (0.05)
Cadmium	44 ± 12	22 ± 3	3.1 (0.08)
Chromium	55 ± 7	65 ± 10	2.92 (0.09)
Mercury	21 ± 2	19 ± 3	2.87 (0.09)
Leg Muscle			
Arsenic	14,500 ± 685	18,100 ± 1490	4.42 (0.04)
Lead	44 ± 4	29 ± 5	9.81 (0.002)
Manganese	1780 ± 236	4120 ± 1970	4.34 (0.04)

Source: Burger, J. et al., *Arch. Environ. Contam. Toxicol.*, 44, 36–42, 2003b.
Note: Given are means ± standard error (in ppb, wet weigh); 2000 data. Levels that differed by $P < 0.1$ are given to avoid type II errors. Sample sizes for eggs were egg masses: NJ sample, $n = 74$, Delaware sample, $n = 40$.

the New Jersey side (Burger et al. 2003b). Again, arsenic showed significant differences among sites, with levels being the highest at the two ends of the sampling region. However, there were no order-of-magnitude differences, and the observed differences might not be biologically significant (Burger et al. 2003b).

For leg muscle, there were few differences in metal levels among the sampling sites in Delaware (except for arsenic, which was twice as high at Prime Hook than those at most of the other sites). However, in New Jersey, there were significant differences in metal levels for all metals, although the means were within an order of magnitude, except for chromium. Chromium varied from a mean of 44 ± 5 ppb (Moore's Beach) to 100 ± 55 ppb (Cape Shore) and 120 ± 23 ppb (Fortescue). Again, the levels were highest at the north and south ends of the sampling sites (see Figure 10.6). Both

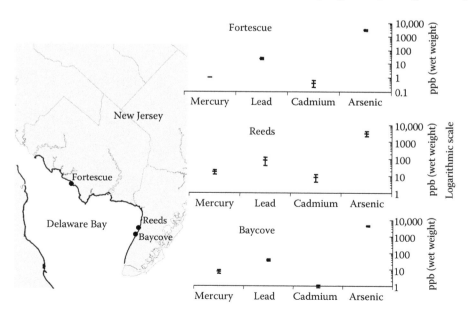

Figure 10.6   Spatial pattern of metal levels (ppb, wet weight) for Horseshoe Crab eggs from three New Jersey locations on Delaware Bay. (After Burger, J. et al., *Ecotoxicol. Environ. Saf.*, 56, 20–31, 2003a.) (Note logarithmic scale.)

Fortescue and Cape Shore are more exposed to the open water of the bay than the other sites. Overall, however, the differences in metal levels within Delaware Bay were not substantial, even when they were statistically significant. The small differences may reflect different subpopulations with somewhat different wintering or staging areas, with exposure occurring during the weeks that the Horseshoe Crabs are inshore waiting to spawn.

Importance and Implications of Metals in Crabs

The eggs of Horseshoe Crabs are an extremely important food source for migrant shorebirds, as well as a range of other species of birds and fish. The eggs must be sufficiently abundant so that Red Knots and other shorebirds can nearly double their body weight, storing fat for the long journey north to breeding grounds, arriving ready to lay eggs (Baker et al. 2004; Karpanty et al. 2006; Morrison et al. 2007; Figure 10.7). The Horseshoe Crab eggs are therefore a critical food source during spring stopover (Atkinson et al. 2007; Morrison et al. 2007).

We considered whether the feathers from adult shorebirds collected from Delaware Bay are within the normal range for feathers of birds, and whether the metals in Horseshoe Crab eggs are within the same order of magnitude as levels in birds (Table 10.5). As is clear, the feather levels for Red Knots and Semipalmated Sandpipers are generally an order-of-magnitude lower than the median value from the literature, although arsenic is similar. Thus, there is no immediate cause for concern.

It was unclear whether the levels of heavy metals in Horseshoe Crab eggs represent very local exposure (e.g., from the water or sediment adjacent to the spawning beaches), more general bay exposure, or exposure on the continental shelf. Because the eggs have a high water content, it is likely that they reflect local exposure as the females spend weeks off the spawning beaches before

Figure 10.7 Comparison of body condition in Red Knots when they first arrive, energy-depleted, after a long migration (Red Knot in Larry Niles' right hand), and after 2–3 weeks of gaining weight and fattening just before they leave to migrate north (Red Knot in his left hand). The Red Knots gain weight from consuming Horseshoe Crab eggs.

Table 10.5 Comparison of Metal Levels in Feathers (ppb, Dry Weight) as Summarized in the Literature, with Levels in Shorebird Feathers (ppb, Dry Weight) and Blood (ppb, Wet Weight), and in Horseshoe Crab Eggs (ppb, Wet Weight) from Delaware Bay

	Arsenic	Cadmium	Chromium	Lead	Manganese	Mercury	Selenium
Feathers in the Literature							
Studies[a] (n =)	6	63	17	69	19	180	42
Low	0.06	ND	ND	ND	ND	ND	300
High	3170	24,000	17,900	32,700	119,900	172,000	54,000
Median	960	100	8100	1600	3400	2100	2200
Shorebirds in the Literature[b]							
Low		75	14,530	1665	1915	21	1301
High		700	26,294	2700	7222	2813	6221
Mean	209	95	12,200	2010	3560	79	2700
Red Knot Feathers (2011–2012, n = 30)[c]							
Low	30	0.01	89	160	NA	1	
High	1000	49	2100	1500		2249	
Mean	448	18	596	496		581	7550
Red Knot Blood (2011–2012, n = 30)[c]							
Low	270	0.01	110	0.08	NA	0.3	NA
High	6150	22	1700	310		64	
Mean	1040	3	484	90		16	
Semipalmated Sandpiper Feathers (2011–2012, n = 30)[c,d]							
Low	0.01	0.01	200	160		21	
High	2500	65	1300	1500		1231	
Mean	846	15	534	414	3390	422	3600
Semipalmated Sandpiper Blood (2011–2012, n = 30)[c]							
Low	75	0.01	38	1	NA	1.1	NA
High	1125	10	1200	320		100	
Mean	381	2	268	60		13	
Horseshoe Crab Eggs (1993–1995, n = 46)[a]							
Low			250	87	4982	12	1965
High			5059	558	18,371	67	3472
Mean			1300	212	8921	37	2772
Horseshoe Crab Eggs (2012, n = 17)[e]							
Low	2200	0.01	23	6.5	3300	0.1	330
High	5900	12	210	130	10,000	22.8	1800
Mean	3729	1.6	82.6	37.6	6076	6.15	990

Note: NA = not analyzed, ND = not detected, below method detection limit.

[a] Burger, J., *Rev. Environ. Toxicol.*, 5, 197–306, 1993.
[b] Burger, J., *Environmental Monitoring and Assessment*, 46, 279–287. 1997b.
[c] Tsipoura et al. (2015).
[d] Burger et al. (2014a).
[e] Burger, J. et al. *Environ. Res.*, 90, 227–236, 2003b; Burger, J., Tsipoura, N., *Environ. Monit. Assess.*, 186, 6947–6958, 2014.

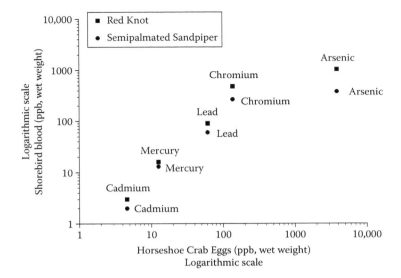

Figure 10.8 Relationship between metal levels (ppb, wet weight) in blood of Red Knots and Semipalmated Sandpipers, and metal levels in Horseshoe Crab eggs. (After Burger, J., Tsipoura, N., *Environ. Monit. Assess.*, 186, 6947–6958, 2014; Tsipoura, N. and J. Burger, Unpublished data.) Sampling sites were from Reeds Beach south to Villas, New Jersey.

they spawn. We hypothesized that the levels in the blood of the shorebirds should reflect the levels of metals in the eggs of Horseshoe Crabs because they are primarily eating them during the stopover (Tsipoura and Burger 1999). In Figure 10.8, we plot the levels of metals in Horseshoe Crab eggs and shorebird blood on a log scale. For most metals, the relationships are very clear. The one metal that does not correspond as clearly is arsenic, which is 10 times higher in the eggs of crabs than in the blood of Semipalmated Sandpiper, and three times higher than in the blood of Red Knots. The difference may reflect the more varied diet of Semipalmated Sandpipers; Knots are eating almost exclusively Horseshoe Crab eggs during this period, and thus might be expected to accumulate higher arsenic levels than the sandpipers. Furthermore, levels of all metals were slightly higher in the blood of Red Knots than of Semipalmated Sandpipers (Tsipoura et al. 2015). Another possibility is that arsenic is differently partitioned intracellularly, which in turn affects the trophic availability to predators, as demonstrated, for example, with Mummichogs (Goto and Wallace 2009).

If the tides are not high enough, or if the waves are too intense, females will not come in to spawn, but instead wait offshore for better conditions. Spawning during windy nights may result in many Horseshoe Crabs, both male and female, being washed up above the reach of the waves. If positioned right-side-up, the crabs can make their way through the wet sand to the bay, but if overturned by wind and wave, they are helpless (Figure 10.9). Many will survive the hours to the next tide, but meanwhile they are exposed to predation by gulls when the tide waters recede. Gulls and other species that feed on the flesh of crabs can also be exposed to metals, particularly if some birds specialize on this abundant food source for the month or more when female crabs are spawning. In this case, the gulls are active predators, eating still living crabs. Because there can be literally hundreds of overturned crabs after particularly strong high tides, the cost can be high for the female crabs (Botton 1984).

It would be useful to know whether the metal levels in flesh are similar (or related) to those in Horseshoe Crab eggs, because the eggs are easy to collect and analyze, and data for many years are available (Figure 10.4). We collected female Horseshoe Crabs as they came ashore to collect an egg sample and a leg sample from the same individuals. There is a modest correlation between the mercury and cadmium in the eggs and flesh of Horseshoe Crabs (Figure 10.10).

(a)

(b)

Figure 10.9 Overturned Horseshoe Crab female being eaten by a young Great Black-backed Gull (a); a Delaware Bay beach scene showing numerous crabs overturned by waves; some of these will survive to escape on the next high tide (b).

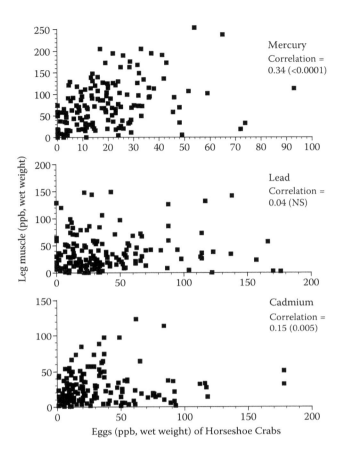

Figure 10.10 Relationship between metal levels in eggs and leg muscle of female Horseshoe Crabs, showing a modest but significant positive correlation (Kendall Tau) between levels of mercury and cadmium in the eggs of a female and her muscle. Each square represents one female sampled.

FISH AS BIOINDICATORS

Fish are the main prey of many coastal birds, including herons, egrets, terns, Black Skimmers, and to a lesser extent, gulls. But fish are also important bioindicators because metal levels in fish reflect the prey they eat, possible adverse effects to themselves, and possible adverse effects in the predators that eat them, including people. Like birds, fish have been useful bioindicators because they span the trophic scale and are located at different points in the food web. Some fish are herbivores (e.g., juveniles), some are carnivores (bass, Bluefish, tuna, shark, Swordfish), and some are omnivorous. Some are small, some are large; some are short-lived and others are long-lived. For our purposes, fish are one of the primary prey of many species of waterbirds, making it important to understand metal levels and possible bioaccumulation and bioamplification.

NATURAL HISTORY BACKGROUND

The life cycles and natural history of fish differ among species, but coastal and estuarine habitats are extremely important as nurseries for both shellfish and finfish. The estuaries in the Middle Atlantic Bight from Cape Cod to Cape Hatteras provide nursery habitat for young fish that inhabit much wider areas as adults (Able and Fahay 1998, 2014; Jivoff and Able 2001). The estuaries provide the juvenile fish to populate much of the offshore coastal waters. The creeks are important nursery habitats for young-of-the year fish (Rountree and Able 1992) (Figure 10.11), and the eelgrass beds are of primary importance during the spawning and early life stages.

It is in the nurseries that the fish undergo the greatest changes in morphology, diet, and habitat. Studies in Barnegat Bay, for example, show that in the winter and spring months, the greatest number of prey fish can be found in the creeks, whereas in the summer and fall (June through October), the greatest number of fish (and of fish species) are in the eelgrass beds (Jivoff and Able 2001), and numbers vary among creeks as well (Neuman et al. 2004a). The diets of juvenile fish differ in different creeks (Nemerson and Able 2003, 2004), providing an opportunity for differential exposure.

To live in these variable environmental conditions, species need to be able to withstand changes in temperature, salinity, and oxygen that occur as a function of living in a zone dominated by saltwater from the ocean and freshwater from rivers and runoff. The number of fish species that live permanently in estuaries and bays is low, but the biomass is substantial because so many fish have their early life stages in bays and estuaries. For most fish, juveniles are more tolerant of environmental variation than adults (Able and Fahay 1998). Thus, the number of species in bays can be quite large, although many are transient (Table 10.6). Although this number may have increased since these data were compiled, it gives a rough estimate of the species numbers that may be available as prey for birds foraging in bays and estuaries. These data also indicate that species diversity is highest in bays and estuaries that are the largest and most complex.

Many estuarine fish remain rather stationary as juveniles, although they may shift sites slightly with very high tides (Able and Fahay 1998, 2014; Hunter et al. 2007). Many adult fish move out of the estuaries into the oceans, where some species migrate up and down the Atlantic coast. For example, most North American Bluefish are migratory, spending their summers in the north, and their winters around Florida and the Gulf Stream (Pottern et al. 1989; Adams et al. 2003). The larger fish move into the bays and estuaries in late spring to spawn, and then move offshore, whereas schools of intermediate-sized fish and smaller fish remain inshore in bays, estuaries, and marsh creeks (Able and Fahay 1998; Neuman et al. 2004). Similarly, the main breeding areas for Striped Bass are the New York–New Jersey Harbor Estuary, Delaware Bay, and the Chesapeake Bay, and they are highly mobile (Bjorgo et al. 2000; Able et al. 2007).

Figure 10.11   Professor Ken Able, director of the Rutgers Marine Laboratory at Tuckerton, at the edge of a salt marsh creek to set a seine for fish in Barnegat Bay; most of the fish species that are seined are also prey fish for birds nesting in the Bay.

Table 10.6 Number of Species Reported for Several Estuarine Systems

Estuary	Number of Species
Nauset Marsh, Massachusetts	35
Narragansett Bay, Rhode Island	99
Connecticut River, Connecticut	44
Great South Bay, New York	29
New York–New Jersey Harbor	113
Barnegat Bay, New Jersey	197
Delaware Bay, New Jersey/Delaware	98
Chincoteague Bay, Virginia	99
Chesapeake Bay, Maryland/Virginia	285

Source: Able, K.W., Fahay, M.P., *The First Year in the Life of Estuarine Fishes in the Middle Atlantic Bight*, Rutgers University Press, New Brunswick, 1998.

PREY FISH

From 2001 to 2010, we collected prey fish left at the nests of Common Terns, Black Skimmers, and Black-crowned Night-Heron, as well as fish from traps and by seines from the shores and shoals of Barnegat Bay. We intended to test the null hypothesis that there were no differences in metal levels of the prey fish caught by birds to feed their mates or chicks, and those caught in nets and seines. We also wanted to determine if the metal levels in prey fish differed as a function of location in the bay, year, and fork length. To initially examine these questions, we computed general linear models (GLM) models (SAS 2005). These questions were best examined with Mummichogs, because we collected and analyzed more than 150 during our studies. Nest/Trap is the variable for whether the fish were collected in a nest or by traps.

For Mummichogs, the best models explaining variation in the different medals were accounted for as follows: arsenic ($r^2 = 47$, nest/trap, year), cadmium ($r^2 = 18$, year, fork length), chromium and lead (no significant model), mercury ($r^2 = 25$, nest/trap, year), selenium ($r^2 = 13$, nest/trap, fork length), and manganese ($r^2 = 9\%$, nest/trap. Thus, overall, metal levels were partly dependent on whether the fish were collected from nests or traps, year, and size (fork length).

Species Comparison for Fish Brought to Nests to Feed Chicks

Terns and skimmers forage on small prey fish, whereas herons and egrets can also catch larger, older fish. Thus, there is some stratification in prey selection, which reduces competition. From 2001 to 2014, we collected fish from the nests of Common Terns and Black Skimmers as an indicator of metal exposure (Figure 10.12). Fish that are too fat or too long to be swallowed will end up adjacent to the Tern or Skimmer nests. Also, chicks of all three species sometimes regurgitated fresh fish (which we collected) when we handled them for weighing or banding. Levels of metals varied significantly by species, with Bay Anchovy (*Anchoa mitchilli*) having the highest mercury and selenium levels, Menhaden (*Brevoortia tyrannus*) having the highest levels of lead, Winter Flounder having the highest levels of arsenic, and Butterfish (*Peprilus triacanthus*) having the highest levels of cadmium (Figure 10.13). Menhaden is a favored prey fish for Striped Bass and may be the major source of lead and mercury in the larger species (Mason 2004).

Mercury is one of the metals of highest concern for marine species, especially in fish that lead to higher trophic levels on the food chain (Krabbenhoft and Sunderland 2013). We examined mercury levels further. The mean mercury levels in Mummichogs and Silversides, two fish commonly caught by Common Terns, Skimmers, and Black-crowned Night-Heron nests varied significantly (Table 10.7). Skimmers brought back the smallest fish with the lowest mercury levels; terns brought

(a)   (b)

Figure 10.12   Common Tern chicks are fed a range of fish, including fish that are too large to swallow immediately. (a) A chick has just been fed a large Sand Eel that it cannot swallow immediately.
(b) Delayed swallowing has allowed the sibling time to grab the fish resulting in a tug of war.

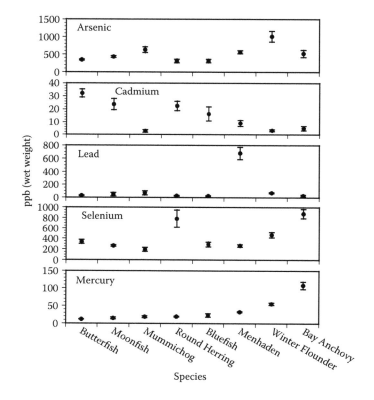

Figure 10.13   Levels of metals (ppb, wet weight) (mean ± SE) in prey fish brought back to the nests to feed to
Common Terns and Black Skimmers (fish collected from nests or chicks from 2001 to 2014 in
Barnegat Bay).

back the intermediate-sized fish, and Black-crowned Night-Herons brought back the biggest fish
with the highest mercury levels. Both Terns and Night-Herons brought back bigger fish than were
caught generally in traps from the same period, and Skimmers brought back smaller fish. Mercury
levels varied significantly among the fish collected from the three bird species and from the traps
(Table 10.7).

Table 10.7   Metal Levels (ppb, Wet Weight) in Fish Brought to Nests by Common Terns, Black Skimmers, and Black-Crowned Night-Herons, and in Fish Trapped or Seined in the Bay, during the Week Chicks Were Being Fed

	Mummichog *Fundulus heteroclitus*	Silversides *Menidia* spp.	Pipefish *Syngnathus* spp.	χ^2
Fish from Tern Nests				
Sample size	12			
Length (cm)	7.1 ± 0.2	None	None	
Lead (ppb)	69.8 ± 31.6			
Cadmium (ppb)	2.7 ± 1.1			
Mercury (ppb)	18.4 ± 2.6			
Fish from Skimmer Nests				
Sample size	26	17	13	
Length (cm)	5.5 ± 0.3	5.0 ± 0.2	13.4 ± 1.3	25.6 (<0.0001)
Lead (ppb)	24.0 ± 2.7	31.4 ± 4.8	92.0 ± 27.7	15.5 (0.0004)
Cadmium (ppb)	4.0 ± 1.4	5.9 ± 1.1	8.4 ± 1.3	14.5 (0.0007)
Mercury (ppb)	16.8 ± 2.6	16.0 ± 2.7	73.8 ± 7.9	15.5 (0.0004)
Fish from Traps (Nesting Period)				
Sample size	84	26	23	
Length (cm)	6.5 ± 0.1	5.2 ± 0.2	10.8 ± 0.6	76.3 (<0.0001)
Lead (ppb)	49.0 ± 8.5	548 ± 131	972 ± 199	87.0 (<0.0001)
Cadmium (ppb)	2.6 ± 0.3	4.7 ± 0.9	24.2 ± 9.6	46.1 (<0.0001)
Mercury (ppb)	41.3 ± =3.7	35.4 ± 5.6	112 ± 8.0	39.5 (<0.0001)
χ^2 Comparing Trapped Fish with Bird-Caught Fish				
Sample size				
Length (cm)	15.6 (0.0004)	NS	2.9 (0.09)	
Lead (ppb)	NS	30.1 (<0.0001)	22.8 (<0.0001)	
Cadmium (ppb)	NS	NS	4.5 (0.03)	
Mercury (ppb)	14.4 (0.0007)	7.5 (0.006)	5.6 (0.02)	
Fish from Black-Crowned Night-Heron Nests				
Sample size	13	16	0	
Length (cm)	7.3 ± 0.3	6.0 ± 0.3	None	4.9 (0.03)
Lead (ppb)	7.8 ± 1.0	6.7 ± 1.9		NS
Cadmium (ppb)	0.8 ± 0.5	1.8 ± 0.3		NS
Mercury (ppb)	61.7 ± 1.9	61.4 ± 4.2		NS
χ^2 Including BCNH				
Length (cm)	18.7 (0.0003)	10.1 (0.006)	2.9 (0.09)	
Lead (ppb)	6.6 (0.09)	49.1 (<0.0001)	22.8 (<0.0001)	
Cadmium (ppb)	NS	10.9 (0.004)	4.5 (0.03)	
Mercury (ppb)	18.1 (0.0004)	27.6 (<0.0001)	5.6 (0.02)	
Fish from Traps from Fall				
Sample size	30	30	30	
Length (cm)	7.3 ± 0.2	8.1 ± 0.1	15.3 ± 0.5	62.8 (<0.0001)
Lead (ppb)	42.4 ± 14.8	127 ± 11.8	3522 ± 430	72.1 (<0.0001)
Cadmium (ppb)	1.9 ± 0.5	4.5 ± 0.8	5.6 ± 1.4	6.0 (0.05)
Mercury (ppb)	77.9 ± 8.6	73.5 ± 3.1	108 ± 8.2	11.5 (0.003)

(*Continued*)

Table 10.7 (Continued) Metal Levels (ppb, Wet Weight) in Fish Brought to Nests by Common Terns, Black Skimmers, and Black-Crowned Night-Herons, and in Fish Trapped or Seined in the Bay, during the Week Chicks Were Being Fed

	Mummichog Fundulus heteroclitus	Silversides Menidia spp.	Pipefish Syngnathus spp.	χ^2
	χ^2 Including BCNH and Fall Traps			
Sample size				
Length (cm)	29.9 (<0.0001)	60.0 (<0.0001)	21.6 (<0.0001)	
Lead (ppb)	8.3 (0.08)	75.6 (<0.0001)	45.3 (<0.0001)	
Cadmium (ppb)	NS	9.7 (0.02)	17.5 (0.0002)	
Mercury (ppb)	40.1 (<0.0001)	51.4 (<0.0001)	6.1 (0.05)	

There were also differences in Silversides (although there was not a large enough sample to examine Silversides for Terns; Table 10.7). Skimmers brought back the same size fish as were caught in the traps, but they captured fish with lower mercury levels. Night-Herons, on the other hand, brought back bigger fish than caught in the traps, and they had higher mercury levels (Table 10.7).

Taken together, it appeared that Common Terns and Black Skimmers were catching similar sized Mummichogs and Silversides as were caught in the traps, but the fish they brought back had lower levels of mercury than those caught in the traps. Because for this comparison we used only fish caught in traps at the same period that the birds were foraging, the differences are not attributable to temporal differences. Black-crowned Night-Herons, however, were mainly catching slightly larger fish than found in the traps, and they had significantly higher mercury levels (Table 10.7). These findings are consistent with the hypothesis that the birds were foraging in different locations or habitats than where the traps were placed. Terns and skimmers were foraging where the levels are lower, and Night-Herons were foraging where they are higher. Terns and skimmers usually forage in the middle of creeks, whereas Night-Herons forage along the banks of creeks and shorelines. The data suggest that exposure of the small prey fish may be greater in the latter habitats (where traps and nets were also placed) than those living in the middle of creeks, bays, and estuaries.

Pipefish are not a preferred fish, and are usually brought back only when other prey fish are not available, and often the chicks do not eat them. Some adults seem to specialize on Pipefish, and one may find a dozen or more uneaten at a nest. Most of the Pipefish collected were from Skimmer nests (Table 10.7). Skimmers brought back larger Pipefish with lower mercury levels than were found in the traps. One explanation is that Pipefish were collected from Skimmer nests early in our sampling, when chicks were smaller and mercury levels happened to be higher. Older chicks may have swallowed the Pipefish or Pipefish were less available when the chicks were older.

There is a great deal of variation in mercury levels by fish species. The Bay Anchovies brought back to the nest were of a rather narrow size range, but they had very variable metal levels. On the contrary, there was more variation in the size of Pipefish and Needlefish (*Strongylura marina*) brought back, but less variation in mercury levels. Round Herring (*Etrumeus teres*) had little variation in both size and mercury level (Figure 10.14).

It is also instructive to look at the variability in mercury levels in fish from nests and fish from the traps. The Mummichogs brought back to Common Tern and Black Skimmer nests generally had lower levels of mercury than those of similar-sized fish caught in traps (Figure 10.15). Most of the fish brought back by the birds to feed their chicks had less than 50 ppb, whereas similar-sized fish from traps or seines had levels as high as 160 ppb. We compared mercury levels in the fish from bird nests to the same species collected in traps in the fall (Table 10.7). For all three species, prey fish in the fall were larger (they obviously grew), and mercury levels were significantly higher (they had longer time to accumulate mercury). Although it is not surprising that mercury levels increased,

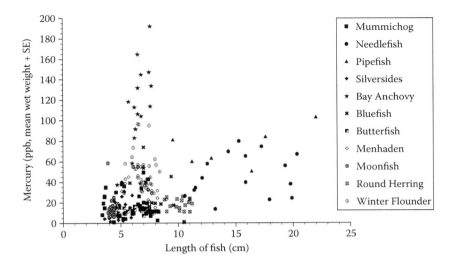

Figure 10.14 Mean mercury levels (ppb, wet weight) (mean ± SE) as a function of fork length for the fish brought back to the nests to feed Common Tern and Black Skimmer chicks. Shown are the values for different species of fish.

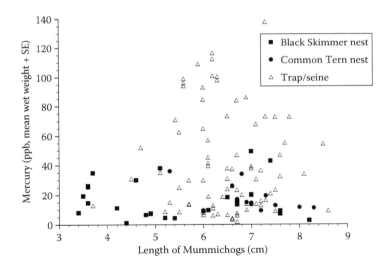

Figure 10.15 Mercury levels (ppb, wet weight) (mean ± SE) in Mummichogs as a function of length of fish and collection method (trap/seine versus nests).

it does indicate that young terns, Skimmers, and Night-Herons, as well as other fish-eating birds, receive higher levels of mercury as the season progresses, until they depart on migration.

Other metals, particularly lead and cadmium, are also important. There were no significant differences in the lead levels of fish brought back to terns and Skimmer nests and those caught in traps or seines for Mummichogs, but lead levels were significantly higher in Silversides from traps compared to those found in nests (Table 10.7). There were significant differences for fish from Night-Heron nests. The levels of lead and cadmium were lower in the Silversides brought back to the nests compared with those taken from traps or seines (Table 10.7). As might be expected, mercury levels increased in the fish caught in traps from June (nesting season) to fall, because of accumulation from their foods. Cadmium and lead, however, did not increase during the season.

Spatial Comparison of Metal Levels within Barnegat Bay

We trapped Mummichogs in the north, middle, and south areas of the Bay. Lead and cadmium levels were significantly higher in the north, and mercury was significantly higher in the south (Table 10.8). In the south bay, Mummichogs had higher mercury and lower lead and cadmium levels than Silversides. One possible explanation is that the source of mercury is atmospheric deposition from remote sources, and that the northern part of the bay has no inlet, and therefore has more freshwater flushing to lower mercury levels. The source of the higher lead and cadmium levels remains elusive, but the differences may reflect a type I error, statistical significance in the absence of environmental or biological significance.

Comparison of Prey Fish from Raritan Bay
(New York–New Jersey Harbor) and Barnegat Bay

Comparing metal levels in prey fish seined from Raritan Bay (New York–New Jersey Harbor) and Barnegat Bay provides some measure of the comparability of the food source for birds. We compared fish of comparable size. For this comparison, we use those caught in traps in the fall to be consistent (Table 10.9). The data indicate that overall the levels are within the same order of magnitude, except for higher lead levels in Silversides at Barnegat Bay, which we cannot account for.

There are comparable data from the New York–New Jersey Harbor for these species (Table 10.10), and several things are clear: (1) levels of mercury were higher for the Koepp et al. (1988) data; (2) cadmium and lead were often at or below detection level; (3) lead and cadmium accumulated in some of the Bluefish and Mummichogs at some locations, particularly near the military base. This suggests to us that collection of prey fish from specific locations around bird colonies might be more useful than comparisons with data from the literature, often with no relation to the bird colonies.

Table 10.8   Comparison of Metal Levels in Prey Fish Seined from the North, Middle, and South of Barnegat Bay

	Mummichog	Silversides
North	$n = 26$	
Mercury	12.8 ± 1.4	None from traps
Lead	52.0 ± 10.3	
Cadmium	3.2 ± 0.4	
Middle	$n = 32$	
Mercury	37.9 ± 4.5	None from traps
Lead	19.7 ± 3.0	
Cadmium	1.0 ± 0.2	
South	$n = 56$	$n = 56$
Mercury	76.5 ± 5.5	54.8 ± 4.0
Lead	47.8 ± 8.7	168.9 ± 13.4
Cadmium	2.8 ± 0.4	4.1 ± 0.5
χ^2 (P)	Kruskal–Wallis test	
Mercury	55.9 (<0.0001)	
Lead	9.6 (0.008)	
Cadmium	12.8 (0.002)	

Note: Given are means ± SE (ppb, wet weight).

Table 10.9 Metal Levels (Mean ± SE) (ppb, Wet Weight) in Prey Fish from Raritan and Barnegat Bays from 2005 to 2012

	Raritan Bay	Barnegat Bay
Mercury		
Bluefish	92 ± 7 (*n* = 20)	26 ± 5 (*n* = 15)
Mummichog	18 ± 3 (*n* = 20)	78 ± 9 (*n* = 155)
Silversides	42 ± 3 (*n* = 20)	74 ± 3 (*n* = 89)
Lead		
Bluefish	47 ± 13	24 ± 8
Mummichog	34 ±12	47 ± 7
Silversides	37 ± 6	322 ± 67

Sources: Raritan Bay data from Burger, J. et al., *J. Toxicol. Environ. Health*, 75, 272–287, 2012e; Unpublished data.

Table 10.10 Comparative Data on Metal Levels in Prey Fish from the New York–New Jersey Harbor Estuary

	Lead	Mercury	Cadmium	Arsenic	Reference
Hudson River (Mile 12) G.W. Bridge					
Bluefish	<MDL	230 ± 10	<MDL	240	Koepp et al. 1988, Table 56
Mummichog	<MDL	140 ± 90	<MDL	930	
Silversides	<MDL	520 ± 30	<MDL	410	
Upper NY Bay (Caven Cove, New Jersey)					
Bluefish	500 ± 100	380 ± 300	380 ± 600	100 ± 100	Koepp et al. 1988, Table 57
Mummichog	<MDL	130 ± 100	<MDL	600 ± 700	
Silversides	<MDL	280 ± 200	290 ± 700	100 ± 80	
Upper NY Bay (Military Operations Terminal)					
Bluefish	<MDL	330 ± 200	<MDL	220 ± 300	Koepp et al. 1988, Table 58
Mummichog	140 ± 300	360 ± 300	1900 ± 4200	160 ± 100	
Silversides	<MDL	470 ± 400	<MDL	300 ± 200	

Source: Koepp, S.J. et al., In: *Fisheries Research in the Hudson River*, ed. C. L. Smith, State University of New York Press, Albany, NY, 1988.

Note: Means ± standard deviations, wet weight converted to ppb wet weight. The method detection limit (MDL) was 10 ppb.

The data show clear instances of cadmium and lead contamination at specific sites, and the findings represent local availability of the metals (Koepp et al. 1988). The Hudson River Mile 12 site was at the George Washington Bridge, and Caven Cove and the Military Operations Terminal are in New York Bay. Of the 13 fish species analyzed, the Mummichog and young Bluefish had the highest cadmium levels, and one Menhaden had a mercury concentration of 990 ppb (wet weight), whereas young-of-the-year Striped Bass and Bluefish averaged almost 400 ppb (wet weight).

Importance of Metal Levels in Prey Fish to Fish Themselves and Avian Predators

Heavy metals can have adverse effects on the fish themselves, and this is usually determined by examining levels in fish tissue (as was done in this study) and comparing them to known adverse levels. For example, Mummichogs raised in an environment polluted with mercury failed to capture prey as quickly, and had higher brain mercury levels than fish from uncontaminated sites (Smith and Weis 1997). The Mummichogs from polluted creeks made fewer attempts to capture prey. Weis and

Weis (1998) showed that exposure to lead decreased prey capture ability, swimming performance, and predator avoidance in larval Mummichogs. Similarly, in laboratory experiments, Weis et al. (2011) showed that Killifish and young Bluefish showed impaired behavioral responses owing to mercury contamination, including poor predatory behavior, poor predator avoidance, reduced rates of feeding, and reduced growth. In a complex system with these and other species, there are cascading effects on both the prey and the predators of these small prey (or forage) fish (Weis et al. 2001, 2011). Given the clear effect of both lead and mercury on prey fish, one might expect that those with higher metal levels might be easier to capture by birds.

The data reported in this book for mercury may well support the hypothesis that the higher the mercury load, the less effective the predator avoidance. Presumably, the fish caught in traps or by seines are fish caught randomly because all fish are equally vulnerable. The prey fish caught by traps had higher levels as the season progressed, which is predictable given the increased exposure time. Fish caught by birds, however, should reflect the capture method. Black-crowned Night-Herons are stand-and-wait predators (at least when catching fish) (Figure 10.16). They stand for hours at the edge of banks, peering into the water, and striking at fish that swim by. They do not, however, catch all the fish they attempt to catch. Impairment due to mercury is consistent with the hypothesis that the higher-mercury Mummichogs have lower predator avoidance (see Table 10.7).

However, the levels of mercury in fish caught by Terns and Skimmers were lower than in the fish caught in traps or seines at the same time. At first glance, this does not support the hypothesis of metal impairment in predator avoidance, as shown by Smith and Weis (1997) in the laboratory. However, during the chick-rearing phase (when these fish were captured), Common Terns are primarily foraging over schools of Bluefish that "force" schools of prey fish to the surface (Safina and Burger 1985, 1988a–c, 1989b) (Figure 10.17). In turn, the arrival of Bluefish in estuaries and along ocean beaches is coincidental with the presence of Atlantic Silversides (*Menidia menidia*) and Bay Anchovy (Taylor et al. 2007). This concentration of prey fish at the surface provides an opportunity for Common Terns to plunge-dive for them, and a feeding frenzy of hundreds of terns can form over such Bluefish feeding schools (Safina 1990). In light of this complex interaction, we suggest that prey fish with lower contaminant levels will be most successful at evading the predatory Bluefish, at the cost of exposing themselves to the hungry terns diving from above (Figure 10.17).

Skimmers that catch Mummichogs while skimming through creeks may well be simply catching the fish available at the surface (similar to a trap or seine in the same area). Figure 10.16 shows

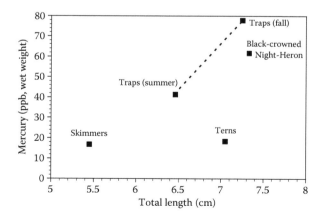

Figure 10.16 Relationships of size and mean mercury levels (ppb, wet weight) for Mummichogs, Silversides, and Pipefish combined, for the fish collected from nests and from traps during the chick-rearing period.

Figure 10.17 Common Terns feed in dense flocks over schools of prey fish chased to the surface by Bluefish predators. This characteristic feeding frenzy attracts terns and human fishermen from a great distance.

that the relationship between size (fork length) and mercury for the Skimmer-caught Mummichogs and those caught by traps (during the breeding season, and in the fall) differed. Skimmers were catching more fish that were smaller and lower in mercury. Black-crowned Night-Herons, on the other hand, were catching the larger Mummichogs available during the nesting season (longer mean lengths than found in the traps or seines), and thus were obtaining fish with higher mercury levels. The hypothesis of mercury-induced impaired predator avoidance has been shown in the laboratory (Smith and Weis 1997), but not in the field, and requires further investigation with larger sample sizes. Our data contrasting Skimmers and Night-Herons indicate the complexity of studying this pattern.

Many bird species are eating small prey fish and are feeding them to their chicks. Terns and Skimmers feed their chicks whole fish. Herons, egrets, and gulls swallow and partially digest fish, and regurgitate to their young. There was little correlation between metals in different prey fish we collected from Tern nests in Barnegat Bay: (1) Bay Anchovy had the highest mean mercury and selenium levels; (2) Menhaden had the highest mean lead levels; (3) Winter Flounder had the highest arsenic levels; and (4) Butterfish had the highest levels of cadmium (Table 10.11).

To evaluate the importance of metal levels, particularly in prey fish, with levels known to cause adverse effects is one method of understanding whether birds are at risk. In Table 10.11, it is clear that some Winter Flounder had arsenic levels above the known effects levels, some Bay Anchovies had selenium levels above the adverse effects levels, and some Menhaden exceeded the effects level for lead. The highest mean mercury level of about 100 ppb is a level usually considered acceptable for humans who eat fish frequently. The other metals were below known effects levels for the fish themselves and for birds that eat them (Table 10.11).

Table 10.11 Mean and Maximum Metal Levels in Prey Fish Brought Back to Nest by Common Terns (n = 224 Fish), and Known Effects Levels (Levels and Effects Levels Given in ppb)

Metal	Species with the Highest Mean	Mean for Highest Prey Fish (ppb)	Maximum for That Species	Adverse Effects Level for Fish that Eat Them (ppb)	Adverse Effects Levels If Eaten by Birds	References
Arsenic[a]	Winter Flounder	1012 ± 152	3664	2000	5000[b]	Eisler 1994; Bears et al. 2006
Cadmium	Butterfish	32 ± 3.2	60	100	1000 ppb	Eisler 1985a
Chromium	Mummichog	98 ± 22	147		10,000 ppb	Eisler 1986c
Lead	Menhaden	680 ± 91			100–500 ppb	Eisler 1988a; Burger 1995a
Selenium	Bay Anchovy	880 ± 87	1374	1000	1000 ppb; 5 ppb	Lemly 1993a; Spallholz and Hoffman 2002
Mercury	Bay Anchovy	108 ±11	192		1100–15,000 ppb (but 20 ppb in sensitive birds) RfD = 21 µg/kg/day	Eisler 1987a; WHO 1990, 1991; Spry and Wiener 1991; Wiener and Spry 1996; Yeardley et al. 1998; EPA 1997d

Note: Given are the levels in prey fish that result in adverse effects in other fish that eat them, or in birds that may eat them.

[a] Most arsenic in fish tissue is organic (they convert inorganic arsenic in water to organic forms), which generally have lower toxicity than inorganic forms (Zhang et al. 2012).

[b] Concentrations of 5000 ppb are associated with increased risk of cancer, damage to organs and cardiovascular disease in mammals (Hughs 2002; Bears et al. 2006).

FINFISH

In keeping with the general decline in metal contamination from industry, agriculture, and combustion, several metals have declined over the four decades of our study. The removal of lead from gasoline has had a major impact on lead levels in all biota, including humans. The increase in mercury emissions from coal-fired power plants has increased mercury accumulation in the Northeast Coast (Chapter 9), but local sources of mercury have been considerably reduced. Recently, Cross et al. (2015) reported that mercury concentrations in Bluefish have declined 43% since the early 1970s.

There are many larger finfish that are eaten whole by herons and egrets, and that are eaten as carrion when they are found dead on the beach. Gulls, shorebirds, crows, and a range of other coastal birds make use of dead fish washed up on the beach or that are discarded by fisherfolk. Humans who eat large finfish are at the top of the aquatic food chain, from commercial, recreational, or subsistence fishing (Figure 10.18) in the New York–New Jersey Harbor to Delaware Bay with potential significant exposure (Burger and Gochfeld 2005a, Burger 2009b,c, 2013b; Burger et al. 2009b, 2011b).

Species Comparisons for Metal Levels in Fish from New Jersey

Many ocean fish are estuarine dependent, moving into estuaries and streams to breed. Some fish species, such as Striped Bass, move up and down the Atlantic Coast, and others remain in estuaries or in nearby offshore waters. We examined metal levels in several species of finfish that are popular commercial and recreational fish (Burger and Gochfeld 2005a). These same fish, however, are captured as juveniles by birds, or eaten as adults by a range of avian and mammalian predators when they wash ashore, and they are eaten as waste or offal thrown from ships, docks, or fish factories.

There is a wide range in metal levels in fish caught along the Atlantic coast (Figure 10.19). Mercury, for example, is low in Ling (*Molva molva*), Menhaden (Bunker), and Flounder, whereas it is relatively high in their predators (Bluefish, Striped Bass, Bluefin Tuna, and Mako Shark (*Isurus oxyrinchus*). Bluefish, however, are eaten by birds at various life stages. As noted in section "Prey Fish," Common Terns and Black Skimmers brought juvenile Bluefish back to their nests to feed chicks. The other issue to note is that there is a range of metal levels for each species. For example, although the mean mercury level for Bluefish was 260 ppb, the levels ranged as high as 1400 ppb; the mean for Striped Bass was 380 ppb but the levels ranged as high as 1300 ppb. Thus, if a Herring Gull or Laughing Gull fed extensively on a washed-up Bluefish with 1400 ppm of mercury, its exposure would be far greater than the average.

As is well known, mercury levels vary by fish size, being higher in older and larger fish (Lange et al. 1994; Bidone et al. 1997; Green and Knutzen 2003; Burger 2009c; Burger and Gochfeld 2011b; Gochfeld et al. 2012). This is true both across species and within a species (Burger 2009c; Gochfeld et al. 2012). As the fish grows it has more time to accumulate mercury, and it can also consume larger prey, thereby moving from a low trophic level as larvae, to higher levels as they age from fingerlings, juveniles, to adults.

Although mercury bioaccumulates with size, metals that are regulated in the body do not (e.g., selenium) (Figure 10.20). Selenium as an essential trace element is more tightly regulated, and both deficiencies (El-Bayoumy 2001; Yang et al. 2008) and excesses (Ohlendorf 2000) can cause problems. Selenium and mercury each can ameliorate the toxic effects of the other, although there is still controversy about what molar ratio could be considered protective (see Chapter 9; Raymond and Ralston 2004; Ralston, 2009; Burger and Gochfeld 2013a,b).

Two fish species that are common along the Atlantic coast for recreational fisheries, commercial fisheries, and as offal eaten by birds, are Striped Bass and Bluefish. For both Striped Bass (Gochfeld et al. 2012) and Bluefish (Burger 2009c), there is a positive relationship between size and mercury levels, but not for selenium (Figure 10.20). If anything, selenium decreases slightly with age, which would reduce any positive effect of selenium on mercury toxicity. Our figures include mercury levels for small fish (captured by New Jersey Department of Environmental Protection trawls). Fish

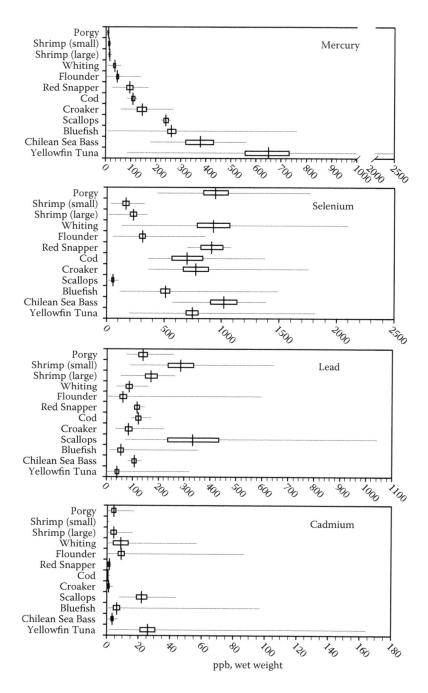

Figure 10.18 Metal levels in recreationally caught and commercially caught seafood, that could also be consumed by gulls as carrion if found dead on the beach. Shown are means (vertical line), standard error (box), and range (horizontal line).

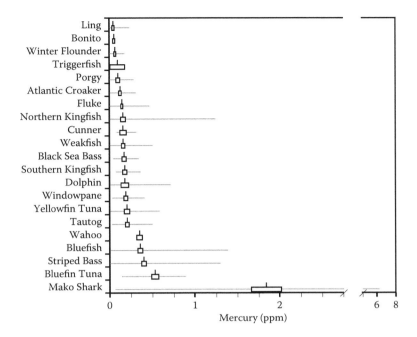

Figure 10.19   Mean (± SE) mercury levels in recreationally caught finfish from the New Jersey coast. (After Burger, J., Gochfeld, M., *Environ. Res.*, 99, 403–412, 2005a.) These fish were all collected from fishermen during interviews and were intended to be consumed. Note: Units in this graph are ppm wet weight.

that are in the 15- to 20-cm range are regularly brought back to nests by Egrets to feed older chicks. In Chesapeake Bay, there is a nearly linear relationship between weight and mercury content of Striped Bass, with an r^2 value of 0.40 for total mercury and 0.23 for methylmercury; total mercury ranged from about 50 to 750 ppb (wet weight) (Mason 2004).

Seasonal Patterns for Indicator Finfish from Barnegat Bay

There are also seasonal patterns in metal levels in finfish, which has implications for wildlife consuming them (often as carrion) and for people consuming them (Figure 10.21). Both Bluefish and Striped Bass are migratory, moving up and down the Atlantic coast. Fishermen and avian predators wait for them wherever and whenever they arrive. In general, mercury levels in Bluefish were lowest in midsummer to early fall, and highest in June when they first come into the region.

The terns benefit from the arrival of Bluefish, which drive prey fish to the surface (Safina and Burger 1985, 1988a–c, 1989b), and in the absence of Bluefish they may take much longer to find food or resort to suboptimal prey such as Pipefish. To some extent, the timing of hatching and chick development is related to the arrival of Bluefish.

Spatial Comparisons for Indicator Finfish from New York–New Jersey Harbor to Delaware Bay

There are also some spatial differences in metal levels in finfish, especially for mercury (Table 10.12). Several aspects of taxonomic or geographic variation include: (1) mercury levels were lowest in Fluke and Windowpane, and higher in Bluefish and Striped Bass; (2) mercury levels were lowest in Central New Jersey for Fluke and Striped Bass; (3) mercury levels were highest in Bluefish from the Central region; (4) selenium levels did not vary by region for Fluke and Bluefish, but did so for

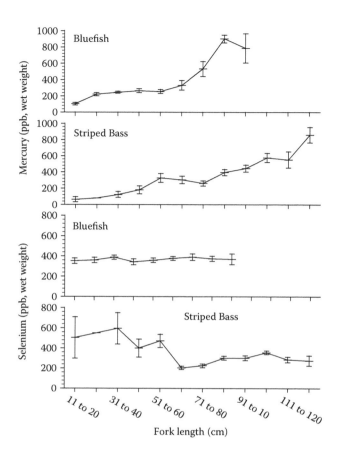

Figure 10.20   Relationship between size (e.g., fork length) and levels of mercury and selenium in Striped Bass and Bluefish from New Jersey. (After Burger, J., *Environ. Res.*, 109, 803–811, 2009c; Gochfeld, M. et al., *Environ. Res.*, 112, 8–19, 2012.) In both species, larger fish have higher mercury levels. Selenium levels generally remain the same with size.

the other two species; and (5) selenium was higher in the south for Windowpane, and was lowest for Striped Bass from the Central region. Bluefish and Striped Bass are migratory, and we expected the levels to be the same throughout the Jersey Shore; this, however, was not the case, although the mean values were within an order of magnitude. Note that Table 10.12 is given in parts per million, because these are the levels used in risk communication for the public.

There are a number of studies on metal levels in fish consumed by people from the bays and estuaries in the Northeast, and some of these are presented in Table 10.13. It is generally estimated that about 90% of the total mercury in fish is in the form of methylmercury. However, Mason (2004) indicates that for Chesapeake Bay at least, this overestimates methylmercury, which averaged 28% of total mercury in White Perch (*Morone americana*), 55% in Blue Crab, 73% in Striped Bass, and 95% in Largemouth Bass (*Micropterus salmoides*). The Food and Drug Administration Action Level for seizing fish in interstate commerce is 1.0 ppm (= 1000 ppb) as methylmercury. European countries have a comparable action level at 0.5 ppm (= 500 ppb). People who eat fish once a month or less are not likely to experience mercury poisoning symptoms. People who eat fish twice a week or more, need to avoid eating any fish with more than 0.1 ppm (= 100 ppb). All four species from all three areas exceed 0.1 ppm as an average, and should not be consumed by people more than once a week.

The Long Island Sound study (Skinner et al. 2009) reported the size–mercury relationship for Striped Bass and Bluefish (Table 10.13), and identified the apparent geographic variability in

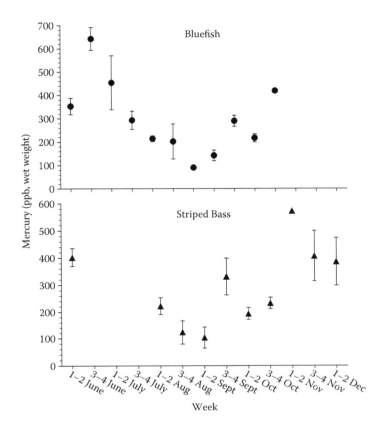

Figure 10.21   Levels of mercury (mean ± SE) in Bluefish and Striped Bass within legal size limits, as a function of season, indicating that potential exposure is not the same throughout the season for birds or people eating them.

mercury concentrations as attributable to regional differences in the size of the fish sampled. This was also reported for the New Jersey coast.

The northern or upper Chesapeake Bay has higher mercury levels in water and sediment than the lower Bay, and this is reflected in concentrations in fish (Mason 2004). In general, fish from lakes, reservoirs, and creeks have higher mercury levels than fish from rivers and bays (Mason 2004). This is reflected in the occurrence of fish advisories based mainly on mercury for the smaller waterbodies, but on PCBs for the larger ones.

Importance of Metal Levels to the Fish Themselves and to Higher Tropic Levels

Fish accumulate methylmercury in aquatic systems through trophic transfer (Monteiro et al. 1996; Downs et al. 1998; Burger et al. 2002b; Burger and Gochfeld 2011b; Crump and Trudeau 2009). This chapter included information on levels of mercury in one of the lowest trophic levels (eggs of Horseshoe Crabs) and in prey fish consumed by birds. Contaminant levels in fish reflect maternal contribution absorbed by embryos from the egg yolk, as well as environmental exposure through diet (Nye et al. 2007). Similarly, metal levels in larger finfish are important for avian predators as well as mammalian predators, including people. Fish are a good example of bioaccumulation because there are so many different foraging patterns, prey types, and prey sizes that fish consume. From larval fish feeding on plankton, to fry, to fingerlings, to intermediate-sized fish that are

Table 10.12 Locational Differences in Mercury and Selenium Levels (ppm, Wet Weight) for Four Finfish in New Jersey

Barnegat Bay Subregion →	North	Central	South	$\chi^2 (P)$
Location	New York–New Jersey Harbor Estuary	Barnegat Bay	Cape May and Delaware Bay	
Mercury				
Fluke (260)	0.14 ± 0.01	0.11 ± 0.02	0.15 ± 0.01	14.9 (0.0006)
Windowpane (48)	0.22 ± 0.02		0.17 ± 0.02	3.9 (0.05)
Bluefish (206)	0.32 ± 0.03	0.49 ± 0.04	0.19 ± 0.02	40.6 (0.0001)
Striped Bass (178)	0.53 ± 0.04	0.24 ± 0.07	0.74 ± 0.02	15.2 (0.004)
Selenium				
Fluke (260)	0.34 ± 0.01	0.34 ± 0.01	0.36 ± 0.01	NS
Windowpane (48)	0.27 ± 0.02	None collected	0.40 ± 0.02	12.9 (0.0003)
Bluefish (206)	0.35 ± 0.03	0.38 ± 0.01	0.36 ± 0.02	NS
Striped Bass (178)	0.33 ± 0.02	0.22 ± 0.02	0.32 ± 0.3	12.8 (0.002)

Sources: Burger, J., *Environ. Res.*, 109, 803–811, 2009c; Burger, J. et al., *J. Toxicol. Environ. Health*, 72, 853–860, 2009b; Gochfeld, M. et al., *Environ. Res.*, 112, 8–19, 2012.

Note: All fish were collected in 2005 to 2008. In this table ppm is used because that is the level used for human risk communication. Sample size is given in parentheses after species name. NS = not significant.

Table 10.13 Comparison of Mercury Levels (ppb, Wet Weight, Mean ± SE) in Fish from Other Northeast Bays and Estuaries (All Fish Are Fillets from Fish for Human Consumption)

Location	Species	Mercury Level (ppb, Wet Weight)	References
Long Island Sound (4 locations)	Striped Bass[a]	270–530[b]	Skinner et al. 2009
	Bluefish < 51 cm	271 ± 94	
	Bluefish > 51 cm[a]	299–400[b]	
Hudson River, Mile 19	Striped Bass	280 ± 210	Koepp et al. 1988
	Bluefish	500 ± 10	
	Menhaden	110 ± 100	
Hudson River, Mile 12	Striped Bass	450 ± 400	Koepp et al. 1988
	Bluefish	210 ± 10	
	Menhaden	140 ± 10	
Hudson River, Mile –2 (Caven Cove)	Striped Bass	430 ± 10	Koepp et al. 1988
	Bluefish	410 ± 60	
	Menhaden	260 ± 80	
Hudson River, Mile –3 (Military Operations Terminal)	Striped Bass	690 ± 180	Koepp et al. 1988
	Bluefish	None captured	
	Menhaden	990 ± 1100	
NY Bight Apex (1993 Collections)	Bluefish	105 ± 20	Deshpande et al. 2002
	Summer Flounder	30 ± 10	
Barnegat Bay	Bluefish	490 ± 40	Burger et al. 2009b; Gochfeld et al. 2012
	Striped Bass	240 ± 70	
Delaware Bay	Bluefish	190 ± 20	Burger et al. 2009b; Gochfeld et al. 2012
	Striped Bass	740 ± 20	
Chesapeake Bay	Striped Bass	260 Max = 750	Mason 2004

[a] Apparent geographic variation attributable to larger-sized fish in certain areas.
[b] Range of means for four sections of Long Island Sound.

eaten by larger tuna, sharks, to humans—these comprise seven trophic levels, with bioaccumulation occurring at each node leading to bioamplification.

In a recent review, Crump and Trudeau (2009) found that both laboratory and field studies supported the hypothesis that mercury causes reproductive impairment in fish. Mercury accumulation in the brain results in reduced neurosecretory material, hypothalamic neuron degeneration, and alterations in neurotransmission, and subsequently reduced gonadotropin-secreting cells, reduced gonadal size, circulating reproductive steroids, reduced gamete production, and spawning success (Crump and Trudeau 2009). Thus, mercury at least, has clear reproductive effects, which presumably increase with bioaccumulation up the food chain.

Trout that died after chronic (100 weeks) exposure to mercuric chloride (0.93 µg/L) accumulated 9.5 ppm (wet weight), with some dying at 5 ppm, equivalent to about 25,000 ppb dry weight (McKim et al. 1976; Armstrong 1979; Wiener and Spry 1996), but sublethal and subclinical effects occur at much lower levels. The levels of mercury in fish are also important because birds are directly feeding on some of these fish, as well as on some as carrion or as offal from fishing vessels. Sensitive birds can experience adverse effects at dietary concentrations of 50–500 ppb. In Table 10.13, mercury levels in some fish averaged between 100 and nearly 1000 ppb, suggesting that sensitive birds could be at risk from eating these species, particularly large individuals that they might eat as carrion.

PASSERINES

Although the focus of this book is on colonial-nesting birds, in this section we briefly report metal levels in some passerines, which were not traditionally considered when examining pollutants, including metals (Evers et al. 2012). However, recently investigators have found that passerines feeding on insects, particularly in aquatic environments, can accumulate high levels of mercury (Jackson et al. 2011). Custer et al. (2007b) reported that Tree Swallows (*Tachycineta bicolor*) and House Wrens (*Troglodytes aedon*) from Carson River (Nevada) had the same levels of mercury as piscivorous birds from the Everglades (a known hotspot for mercury). They found that only about 75% of the eggs hatched, consistent with an adverse effect from mercury exposure.

Here, we present metal levels in eggs and feathers of three passerines from the Meadowlands (New York–New Jersey Harbor Estuary) collected in 2006: Red-winged Blackbirds (*Agelaius phoeniceus*), Marsh Wren (*Cistothorus palustris*), and Tree Swallow (Tsipoura et al. 2008; Table 10.14).

Table 10.14 Levels of Metals (ppb, Mean ± SE, Dry Weight) in Eggs and Feathers of Three Passerines from the Meadowlands (New York–New Jersey Harbor Estuary)

	Red-Winged Blackbird	Marsh Wren	Tree Swallow	Kruskal–Wallis χ^2 (P)
Eggs (sample size)	35	31	Not available	
Arsenic	6.0 ± 1.6	10.1 ± 3.60		5.4 (0.02)
Cadmium	0.26 ± 0.08	0.37 ± 7.06		Not significant
Chromium	120 ± 27.6	59.1 ± 16.9		6.8 (0.009)
Lead	38.5 ± 5.3	34.7 ± 5.28		Not significant
Mercury	48.2 ± 6.4	197 ± 19.2		32.5 (0.0001)
Feathers (sample size)	29	15	5	
Arsenic	142 ± 30.1	7.01 ± 4.27	70.2 ± 39.3	5.34 (0.02)
Cadmium	18.6 ± 6.0	30.9 ± 7.06	11.4 ± 4.28	8.62 (0.01)
Chromium	607 ± 53.2	1040 ± 109	659 ± 219	14.1 (0.0009)
Lead	1080 ± 142	432 ± 73.6	1360 ± 776	7.8 (0.02)
Mercury	826 ± 115	3230 ± 237	2040 ± 355	31.0 (0.001)

Source: Tsipoura, N. et al., *Environ. Res.*, 107, 218–228, 2008.

The levels of mercury in wren eggs from the Meadowlands (mean of 197 ppb, dry weight) were much lower than Custer et al. (2007b) reported for wrens from Nevada (mean of 7340 ppb, dry weight). The levels of mercury in eggs of passerines from the Meadowlands were lower than those we found for colonial waterbirds in Barnegat Bay (see Chapters 11–13). However, the levels of mercury in wren eggs that did not hatch (mean of 300 ppb) were higher than in eggs analyzed randomly (mean of 180 ppb), suggesting the possibility of mercury having an adverse effect on hatchability.

SUMMARY AND CONCLUSIONS

Although this book is largely about birds, birds are only one component in complex food webs. Thus, examining metal levels in other species situated lower on the food chain that are consumed by birds is an integral aspect of understanding contaminant levels in birds. We provide data on plants, mussels, crabs, Horseshoe Crabs, and fish to illustrate contaminant levels in the food web in Barnegat Bay, applicable to other Northeast bays and estuaries. Horseshoe Crab eggs are a good indicator for the estuarine systems in which they occur because their eggs and juveniles are eaten by a wide range of organisms, including fish, birds, and mammals. Metals in Horseshoe Crabs vary by location both among and within bays, although the variation is not substantial. Because the eggs are the main source of food for migrant shorebirds in the spring (including a threatened species—the Red Knot), we examined levels in Horseshoe Crab eggs and shorebirds and found a positive relationship for most metals.

Many birds nesting in these bays and estuaries forage on small fish, and feed them to their young. Metal levels in forage fish vary. Terns bring fish back to the nests that have lower mercury levels than similarly sized fish caught in nets or traps. We suggest this is attributable to the mechanism of Common Terns foraging over Bluefish schools; the Bluefish force the fish to the surface when they are feeding, making them available to the birds. We suggest that the fish with low contaminant levels are best able to escape the Bluefish attacks at the cost of exposing themselves to the birds. Metal levels in larger fish also varied significantly by species, location, and season, suggesting that birds feeding on fish washed up by storms may be exposed when they eat some fish that are larger and higher on the food chain. Similarly, people consuming these fish are exposed to different levels of metals, particularly mercury, depending on the species, season, and location where the fish were caught. Risk to ecoreceptors and humans may exceed regulatory thresholds (see Chapters 10–13).

Most of the literature on contaminants in fish, birds, and other organisms compare known contaminated sites to reference sites, as we did for a Raritan Bay "Superfund Site." It is therefore expected that organisms would be exposed to high levels of several metals at once at the contaminated site, leading to significant correlations between metals. What we found instead in both Horseshoe Crabs and fish were discordant levels. We concluded that the different metals have different distributions and bioavailability in different compartments of the ecosystem, and that fish with different habitats, foraging methods, and metabolism would concentrate one metal, but not another.

CHAPTER 11

Heavy Metal Levels in Terns and Black Skimmers

INTRODUCTION

This is the first of three chapters on metals in Barnegat Bay waterbirds, and this introduction serves for all three chapters. Scientists, environmentalists, managers, public policy makers, and the public are increasingly concerned about contaminants, such as lead and mercury, in the environment. They want reliable, relevant, understandable, cost-effective methods of understanding trends in pollution within their communities. And they want these indicators to be of interest to the public to gain support. Birds are iconic, conspicuous, and interesting to the general public. Some bird species have been used in many different places, over long periods, to track organic and metal contaminants. Heavy metals are released to the environment by mining, milling, manufacturing, combustion, and recycling. Metals occur in air emissions and water effluents of farms, factories, and cities. They are part of fuel cycles and manufacturing cycles. Once released to the environment, they can be transported in the atmosphere or in rivers. Each of the metals has a complex chemistry depending on its ionization potential, its electromotive potential, and its likelihood of forming soluble versus insoluble salts as well as its affinity for sulfur, selenium, and organic complexes. In general, metals are ubiquitous and are present in the food webs of estuarine and coastal ecosystems (see Chapter 8).

Although metals occur naturally from oceanic and geologic processes, most metal pollution comes from urban, industrial, mining, and agricultural emissions and effluents. These sources augment the natural sources leading to levels in the environment that can have adverse consequences for organisms and their food webs (Mailman 1980; Fitzgerald 1989; Fitzgerald and O'Connor 2001; Hoffman et al. 2011). Through atmospheric transport, many contaminants are spread throughout the world. Coastal ecosystems are particularly vulnerable to pollutants, including metals, organic compounds, fertilizers, plastics, pharmaceuticals, and wastes that flow from associated watersheds (Greenberg et al. 2006; Hammerschmidt and Fitzgerald 2006; Pacyna et al. 2006).

Historically, the augmentation of natural concentrations of metals began in the 1500s, increased in the 1700s, and became the dominant source in the 1900s. San Francisco Bay sediments show a clear mercury signal from local mining in the mid-1800s (Hornberger et al. 1999). Metal contamination peaked in the 1980s and 1990s, and there is good news in the slow, but reassuring decline of most metals in several estuaries (Hornberger et al. 1999; Mason 2004; our studies). This can be attributed to reduced use of metals such as mercury, lead, and cadmium, enforced regulations on agricultural and industrial releases, and improved recycling (Vallius 2013). Lead has been removed from gasoline and paints. Reducing reliance on high cadmium fertilizer can be beneficial (Syers and Gochfeld 2001), as grains readily bioconcentrate cadmium from fertilized soil (Grant et al. 2010). Cadmium use in batteries has declined, and many uses of mercury have been phased out.

To summarize from Chapter 8, metals enter the aquatic food chain in different ways, and are distributed among tissues in biota, or are excreted (reviewed in Chapter 8). There is potential for bioamplification at each step of the food chain (Lewis and Furness 1991; Peakall 1992; Burger and Gochfeld 2001b; Becker et al. 2002; Bond and Lavers 2011). This is prominent for mercury, and less prominent for other metals. Long-lived fish-eating or predatory birds at the top of the food chain have the greatest potential to accumulate high levels of persistent pollutants (Furness 1993; Monteiro and Furness 1995; Burger 2002c; Seewagen 2010). Accordingly, most studies of metal levels in birds have examined fish-eating birds, such as Common Loons (*Gavia immer*, Burger et al. 1994a; Burgess et al. 2005), egrets (Frederick et al. 1999, 2004; Spalding et al. 2000a,b), and seabirds (Furness et al. 1995; Burger and Gochfeld 2001b; Cifuentes et al. 2003), although songbirds can have high mercury levels as well (Evers et al. 2012; Custer et al. 2007b).

Fish-eating birds are vulnerable to contaminant effects (e.g., methylmercury) because fish accumulate mercury, and birds may eat larger fish that have high mercury levels (Burger and Gochfeld 1990b; Frederick et al. 1999; see Chapter 9) (Figure 11.1). Females can sequester metals in their eggs at the expense of their embryos (Burger and Gochfeld 1991a, 1996a; Stewart et al. 1994; Nisbet et al. 2002), allowing females to rid their body of contaminants, and providing us the opportunity to use bird eggs as bioindicators. Metals can also be eliminated through the salt gland (Burger and Gochfeld 1985c). Both sexes sequester metals in feathers and feather sheaths during feather development when there is an active blood supply (Burger and Gochfeld 1991a, 1992a). This removes mercury from the body (Sepúlveda et al. 1999a,b), while at the same time allowing feathers to be used for monitoring exposure. Interpreting metals in feathers requires knowing molt patterns and timing of molting as well as the age of the birds (young versus adult) (Bortolotti 2010).

In this and the following chapters, we use colonial waterbirds as bioindicators of metal exposure beginning with metal levels data from Barnegat Bay, followed by data from other bays to provide a picture of trends in metal contamination in the Northeast, supplemented by some data from more distant lands. Where possible, we provide both temporal and spatial patterns of metal levels for Barnegat Bay and the other bays. Together, they will form a pattern of changes in contaminant levels over time and space. We also present data on shorebirds as they are also indicative of global transport via birds.

Each chapter then discusses the important aspects learned from the species group, how the results relate to data from places other than the Northeast, and what they suggest for future research needs. The general discussion among species is presented in Chapter 14. It should be remembered,

(a)   (b)

Figure 11.1 A Common Tern with a Bay Anchovy that will be seized and swallowed in seconds (a). Terns may bring back fish that are too big for chicks to swallow immediately (b), leading to the possibility that an adult or even a sibling will steal this Pipefish out of the chick's beak.

however, that metals are only one class of contaminants that affect coastal birds (Nisbet and Reynolds 1984; Burger 1994b; Burger and Gochfeld 2001b; Burger et al. 2002c; Rattner and McGowan 2007).

In most cases, we present the data in graphic form because it is easier to see the patterns, although in some cases the full data sets are provided in Appendix Tables 11A.1 through 11A.4. This is useful for future researchers, government agencies, regulators, public policy makers, and the public who might want to compare these data with data from other places or other times. Where data are previously published, we refer to the paper for the full data set, but where data are new or augmented by more recent data, the full data sets are provided (see Appendix Tables 11A.1 through 11A.4). To our knowledge, the data presented in Chapters 11 through 13 are the most complete temporal data of metals for colonial waterbirds from the same site in the world. The methods used, as well as species accounts, were described in Chapter 3, although a few salient features are presented here to aid in understanding variations in metal levels.

In this chapter, we examine levels in feathers and eggs of Common, Forster's, and Roseate Terns and Black Skimmers, both temporally and spatially in and beyond Barnegat Bay. In every year, we have obtained the necessary state and federal collecting permits as well as university protocol approvals. Wherever possible, we present spatial and temporal trends from both feathers and eggs. Metal levels in eggs represent female exposure, which partly represents local exposure because parents are near their colony sites for several weeks before egg-laying. Metals in eggs may affect embryonic and chick development. Long-term data sets on metals in eggs indicate trends from female exposure (Burger and Gochfeld 1988d, 2004a; Weseloh et al. 1989).

At hatching, the metals in the chick are derived from the female via the egg. Terns and gulls are covered with down at hatching, while Ardeid chicks are mostly naked (refer to Figures 3.9b and 6.24 for downy chicks, and Figures 3.13 and 5.9 for fully feathered chicks). Down is a sink for mercury, and is likely to have higher concentrations of metals than internal organs (Becker et al. 1994; Burger et al. 2008c). As chicks grow, they receive metals from their diet, and within a few days the amount delivered in food by their parents dwarfs the amount that could have come from the egg. At the same time, their body is growing rapidly so that their initial body burden is rapidly diluted. Feather growth occurs quickly; feather tips are evident on day 5 in Common Terns. The rapidly growing feather becomes a sink for metals, especially for mercury, and much of the dietary burden is transferred to the feathers (Braune and Gaskin 1987; Spalding et al. 2000a). Feather metal levels reflect circulating levels in the week or so before and during the period when a given feather is growing. Correlations between mercury levels in feathers and internal organs will thus be high during feather formation. However, if the feathers are collected months later, they reflect levels when the feathers were growing, and the correlation between feathers and the blood and internal organs is likely to be low because the organs reflect more recent exposures (Eagles-Smith et al. 2008).

Feathers from preflying young or recent fledglings are most useful because they reflect contaminants acquired locally from food gathered by their parents (Burger 1993; Burger and Gochfeld 1997a). We refer to this age group collectively as "young" or "fledglings." They were captured mainly by hand. Collecting feathers from prefledging chicks, however, is not always possible if colonies fail due to mass mortality of young chicks in the first few weeks of life, before they have grown sufficient feathers to analyze. These failures account for some of the gaps in the data sets. We also nest-trapped adults with treadle traps to obtain feathers.

Feathers have been used to examine longer temporal trends. For example, studies of museum specimens of Black-footed Albatross (*Phoebastria nigripes*) spanning 120 years found that methylmercury levels in feathers increased (Elliot-Smith et al. 2000; Olafsdottir et al. 2005; Head et al. 2011; Vo et al. 2011; Weseloh et al. 2011). Therefore, we have archived feathers collected over a 40-year period. This has allowed us to go back to our earlier samples and reanalyze them with current samples and methods to ascertain that unusual peaks and troughs were real, and not attributable to analytic variation. Feather archives are easy to maintain, if free from pests, and can be used by future researchers.

COMMON TERNS

Common Terns are particularly useful as bioindicators, and they have been used in several locations in the Northeast, as well as from the U.S. Great Lakes (Weseloh et al. 1989) and Europe (Guitart et al. 2003), particularly the Wadden Sea (Becker et al. 1985, 1994; Becker and Dittmann 2009; Dittman et al. 2012). Common Terns are indeed common, nesting in most Northeast bays and estuaries, and migrating to South America (Nisbet et al. 2011). They feed almost exclusively by plunge-diving for small prey fish in the open ocean, bays, and salt marsh creeks (Burger and Gochfeld 1991b; Safina and Burger 1985, 1988a; Nisbet 2002; Goyert et al. 2014). On the breeding grounds, members of a pair feed on similar prey (Nisbet et al. 2002). On the wintering grounds they also forage on fish, as well as insects, and will sometimes travel up to 50 km to reach oceanic or estuarine feeding areas (Mauco et al. 2001; Bugoni et al. 2005; Mauco and Favero 2005).

Contaminant data from Common Terns have shown that (1) females transfer metals or other contaminants to their eggs and feathers (Burger 1993; Burger and Gochfeld 1993a, 1996a), (2) feathers grown on the breeding grounds can be used to show local exposure in adults (Burger et al. 1992a), (3) concentrations of some metals in feathers increase with age of the terns (Burger et al. 1994d), (4) concentrations in eggs decline with laying sequence (Becker 1992), (5) developmental defects are associated with different levels of some contaminants (Hays and Risebrough 1972; Gochfeld 1975a,b, 1980b), and (6) a wide range of behavioral deficits that can affect growth and survival are associated with some metals (Burger and Gochfeld 1988c, 2000a; Gochfeld and Burger 1988; Burger et al. 1994e; Frederick and Spalding 1994; Frederick and Jayasena 2010; see Chapter 9).

In the early years of our study, we relied on outside laboratories, including the New Jersey Department of Health. In the mid-1980s, we developed in-house capability first with atomic absorption spectrophotometry and later with a dedicated mercury analyzer. We were able to separate our dissection/sample preparation rooms from our analytic facility to minimize cross-contamination.

Two aspects of metal sequestration in feathers bear mention because they reflect general effects (regardless of the bay or estuary): adult male and female differences (Table 11.1), and levels in regrown feathers (Table 11.2). Females have the opportunity to eliminate metals in their eggs, and because we trapped adults while incubating, this would already have occurred. Levels of metals in feathers of males and females differed for lead, but not for cadmium (Table 11.1). There was a positive correlation between levels in female feathers and their eggs (Burger and Gochfeld 1991a).

There is some question about where adult Common Terns obtain their heavy metals—on the wintering ground before migration or on the breeding ground. Metals derived from their foods first enter the bloodstream, and are then sequestered in different organs. For mercury, more than 90% of the body burden is in the feathers (Furness et al. 1986; Braune and Gaskin 1987). Terns normally molt their body feathers before migrating north in the spring (Nisbet 2002), and thus breast feathers (as generally used in our studies) reflect body burdens on the wintering grounds, months earlier. In collaboration with Ian Nisbet, we addressed this question by trapping Common and Roseate Terns at Cedar Beach (Long Island) and Common Terns on Bird Island (Buzzards Bay). We nest-trapped terns early in incubation, plucked a sample of breast feathers ("arrival" feathers), and retrapped the

Table 11.1   Levels of Cadmium and Lead (ppb, Dry Weight) (Mean ± SE) in Feathers of Paired Adult Common Terns from Barnegat Bay, and in Eggs from These Pairs

	Number	Cadmium	Lead
Male feathers	24	47.1 ± 5.5	1365 ± 101
Female feathers	24	51.1 ± 6.5	1616 ± 115
Eggs	24	12.0 ± 0.6	270.0 ± 3
χ^2 (P) for all three		65.5 (0.0001)	64.2 (0.0001)
χ^2 (P) for male versus female		8.84 (0.01)	3.81 (0.05)

Source: Burger, J., Gochfeld, M., *Environ. Monit. Assess.*, 16, 253–258, 1991a.

Table 11.2   Metals Levels (ppb, Dry Weight) (Mean ± SE) in Feathers of Adult Common and Roseate Terns from Cedar Beach (Long Island, New York) and Bird Island (Buzzards Bay) That Were Grown on Wintering Grounds (Initial "Arrival" Feathers) and Those Regrown on Breeding Grounds

Species and Location	Stage	Lead	Cadmium	Mercury	Selenium
Common Terns					
Cedar Beach	Initial versus	3600 ± 760	150 ± 37	5000 ± 1100	2900 ± 300
(n = 15)	regrown	3600 ± 1100	120 ± 30	8300 ± 1100	2100 ± 190
	χ^2 (P)	Not significant	Not significant	5.1 (0.02)	Not significant
Bird Island	Initial versus	1500 ± 560	120 ± 47	1800 ± 1200	1100 ± 120
(n = 7)	regrown	100 ± 340	420 ± 310	11,800 ± 1468	2000 ± 420
	χ^2 (P)	3.2 (0.07)	Not significant	2.4 (0.007)	Not significant
Roseate Terns					
Cedar Beach	Initial versus	1030 ± 340	85 ± 14	2700 ± 970	1100 ± 130
(n = 8)	regrown	1500 ± 460	160 ± 34	7500 ± 1200	3900 ± 860
	χ^2 (P)	Not significant	3.1 (0.07)	4.3 (0.03)	9.2 (0.002)

Source: Burger, J. et al., *Toxicol. Environ. Health*, 36, 327–338, 1992a; Unpublished data.

birds at hatching about 3 weeks later. By this time, the bare, plucked patches had new regrown feathers that were slightly cream pink in Common Terns and strikingly pink in Roseates. We analyzed both the initial ("arrival") feathers and the regrown feathers. Although there were no consistent differences for most metals, mercury levels were significantly higher in regrown feathers for both species, for both locations (Table 11.2; Burger et al. 1992a). Roseate Terns accumulated higher selenium levels at Cedar Beach than on the wintering ground (or than Common Terns), and Common Terns at Bird Island had lower levels in regrown feathers. These findings require further study.

Nisbet et al. (2002) then used stable isotopes to investigate individual differences in diet that might suggest where the terns were getting the mercury. They found that the high mercury levels were a result of consumption of inshore prey at low trophic levels in small areas of Buzzards Bay. Together, these two types of data (mercury levels in feathers, stable isotope analysis of diets) demonstrated that mercury exposure could be attributed to specific locations (both locally in Buzzards Bay and between the wintering and breeding grounds). Cotin et al. (2011) used stable isotopes to show differences in the foraging behavior of Common Terns between freshwater and salt pans. Subsequently, Lavoie et al. (2014, 2015) used stable isotopes and mercury levels in tissues of Double-crested Cormorants and Caspian Terns to illustrate that exposure to mercury on the wintering grounds also affects mercury toxicokinetics in birds on the breeding grounds (suggesting a rapid uptake followed by a slow depuration rate). Stable isotopes have also been used to examine age-related differences in breeding efficiency of Common Terns (Galbraith et al. 1999).

In the graphs that follow, the correlation between metal level and year is shown with the non-parametric Kendall tau based on all of the individuals analyzed. The curved lines represent a polynomial best fit performed with Deltagraph. Actual values for the Barnegat Bay birds are given in Appendix Tables 11A.1 and 11A.2.

Barnegat Bay Egg Patterns

Metals in Common Tern eggs from Barnegat Bay were examined from 1971, 1982, and then annually from 1987 until 2015 (Burger and Gochfeld 2004a, unpublished data; Appendix Table 11A.1). This was our largest and longest sample. We aimed to sample at least one colony from the north, middle, and south parts of Barnegat Bay, and aimed for a sample of 12 eggs and 12 fledglings for each year. This goal was not always achievable.

Lead and cadmium declined dramatically, both showing the greatest decline from 1971 to the early 1990s (Figure 11.2; Appendix Table 11A.1). Mean mercury levels remained relatively constant

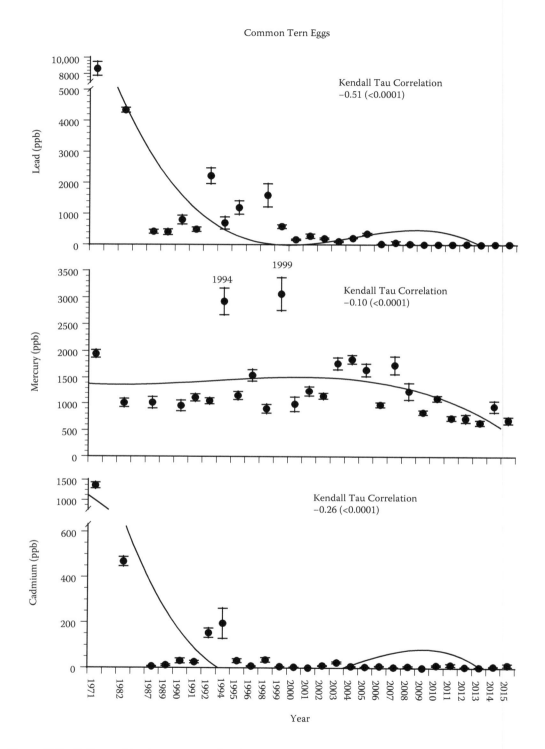

Figure 11.2 Patterns of lead, mercury, and cadmium levels (ppb, dry weight) (mean ± SE) in Common Tern eggs from Barnegat Bay (1971–2015). The declines are all statistically significant (because of the large sample size), but the mercury decline is much weaker than for lead or cadmium. Based on 1414 eggs representing at least three colonies in most years.

between 1990 and 2000 ppb, except for some peak years in the 1990s, followed by a slight decline (Figure 11.2). If our mercury series had begun with the 1994 data, it would show a steep, but illusory, decline. Arsenic and chromium (only analyzed from 1987 to 2015) show very slight, but statistically significant, declines (Figure 11.3). Selenium and manganese, both essential trace elements subject to homeostatic regulation, showed no change (Figure 11.3). The levels of metals have remained relatively stable over the past 3 years. Lead levels have averaged less than 17 ppb; mean mercury have ranged from 639 to 944 ppb, and mean cadmium levels have ranged from 1.7 to 6.9 ppb. These are all very low compared to the 1970s and 1980s.

There were some differences in metal levels as a function of location in Barnegat Bay, where the North includes eggs from the Lavallette Islands; the Middle eggs were from islands around Barnegat Inlet, and the South eggs included those collected from colonies south of the Manahawkin Bridge (Burger and Gochfeld 2003). We began this part of the study in 1990. Patterns were variable. In the early years (after 1990), lead levels were higher in the northern end of the bay, but since 2000 there have been no regional differences. Mercury levels were lower in the north throughout the 1990s, but since 2000, the fluctuating levels have been similar throughout the three regions, consistent with the presumed remote source—atmospheric transport and deposition from coal-fired power plants in the Midwest. Cadmium levels were consistently low in the south, but were higher in the north and middle in the mid-1990s (Figure 11.4). Selenium, chromium, manganese, and arsenic levels have fluctuated, but for the most part the three regions are concordant (Figure 11.5).

For the entire 40+-year period (1971 to 2015), there were no locational differences in lead, cadmium, and selenium. However, there were small but significant differences for the other metals (Table 11.3). Lead, cadmium, and chromium levels were strongly right-skewed as indicated by the large difference between arithmetic mean and geometric mean. Common Terns foraging in the ocean in the north of the bay are feeding closer to the New York metropolitan area, although differences in lead and cadmium were only apparent in the early years. The Environmental Protection Agency (EPA) instituted a partial ban on leaded gasoline effective in 1975, but through the 1980s leaded gasoline was still in use, and lead contamination was widespread, related to traffic density. The widespread restoration of bridges, involving deleading and repainting, occurred in the early 1990s as well. These would have impacted birds in the New York–New Jersey Estuary more than in Barnegat Bay. This is probably sufficient to account for the higher lead levels in the North, and the subsequent decline in lead concentrations. The main source of chromium contamination in the mid-twentieth century were the nearly 200 chromium ore processing residual sites in Hudson County, northern New Jersey. Chromite ore processing slag had been distributed free and was widely used as fill for building lots in Jersey City and neighboring towns, resulting in residential exposure (Stern et al. 2013) and water pollution. The chromium peaks in the early 1990s correspond to a period of extensive remediation of many chromium-contaminated sites in Hudson County, with transient releases of dust and runoff into the New York–New Jersey Estuary. The terns provide the only evidence (circular at best) that this contamination was not limited to the immediate vicinity of the source.

Barnegat Bay Feather Patterns

Feathers of prefledging Common Terns in Barnegat Bay were collected from 1978 to 2014, in those years where young fledged (Appendix Table 11A.2). Because of the excessively high flood tides, or heavy rains that could last for 2–3 days, Common Terns did not fledge any young in some years (see Chapters 5 and 7). As a familiar theme, mean lead levels declined significantly over the years primarily as a result of removal of lead from gasoline. The peaks in the early 2000s are unexplained, but several metals show a 2003 peak. Mercury and cadmium show very slight declines, with unexplained peaks. The 1995 cadmium analyses were rerun to ascertain that this was not due to

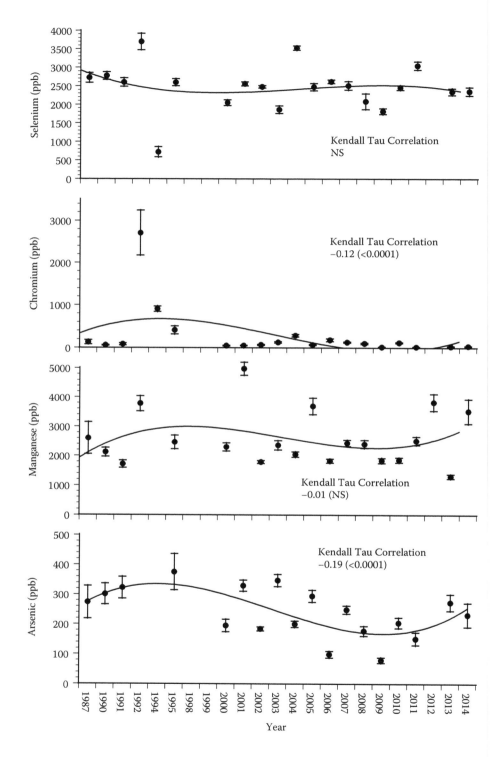

Figure 11.3 Patterns of selenium, chromium, manganese, and arsenic levels (ppb, dry weight) (mean ± SE) for Common Tern eggs from Barnegat Bay. There have been slight declines for chromium and arsenic and no change for selenium and manganese. Based on 1414 eggs.

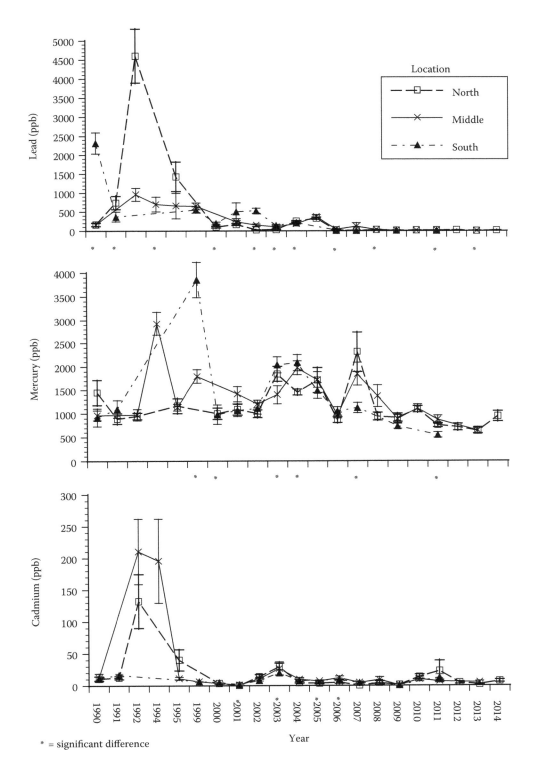

Figure 11.4 Locational differences in lead, mercury, and cadmium (ppb, dry weight) (mean ± SE) in Common Tern eggs from Barnegat Bay. Based on at least 12 eggs per location each year.

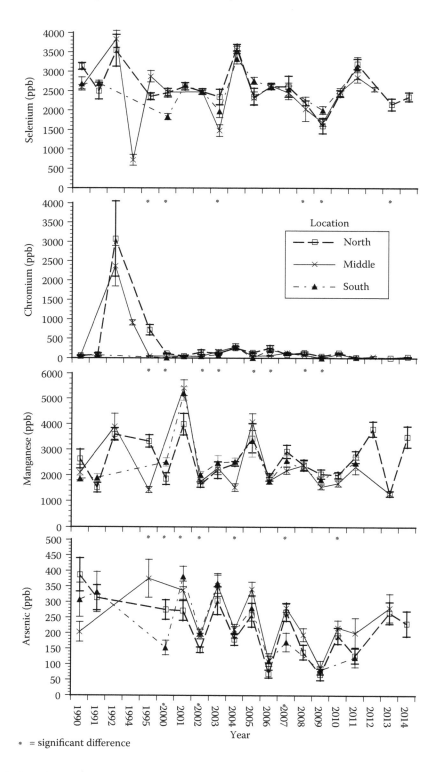

* = significant difference

Figure 11.5 Locational differences in arsenic, chromium, manganese, and selenium (ppb, dry weight) (mean ± SE) in Common Tern eggs from Barnegat Bay. Based on at least 12 eggs per location each year.

Table 11.3 Concentrations of Metals (ppb or ng/g, Dry Weight) in Eggs of Common Terns from New Jersey (All Years Combined)

	n	North Barnegat 272	Middle Barnegat 358	South Barnegat 237	χ^2 (P)
Lead	813	349 ± 66	196 ± 21	250 ± 32	2.2 (NS)
		25	23	25	
Mercury	812	1130 ± 35	1420 ± 48	1480 ± 72	10.2 (0.006)
		1017	1180	1123	
Cadmium	815	22.3 ± 8.7	21.4 ± 3.6	8.5 ± 0.9	4.6 (NS)
		0.7	1.4	1.3	
Selenium	610	2640 ± 48	2533 ± 53	2480 ± 51	2.4 (NS)
		2515	2299	2340	
Chromium	646	292 ± 57	200 ± 30	102 ± 15	34.8 (<0.0001)
		69	42	20	
Manganese	527	2560 ± 72	2280 ± 81	2440 ± 84	20.7 (<0.0001)
		2307	1893	2151	
Arsenic	495	195 ± 9	235 ± 8	208 ± 10	12.5 (0.002)
		110	143	113	

Note: Data are presented as means ± SE and geometric means are given below. Comparisons are made with Kruskal–Wallis one-way ANOVA, yielding a χ^2 statistic. NS, not significant.

analytic error (Figure 11.6). Levels of manganese also declined significantly (Figure 11.7). Selenium, chromium, and arsenic showed no significant pattern, with highs in the late 1990s and early 2000s (Figure 11.7).

We also analyzed levels of metals in feathers of Common Tern fledglings from the different regions of the Bay, and there were some differences in certain years (Figures 11.8 and 11.9). However, the patterns were not clear when examined over a long period. Thus, if we had only studied terns for 3 years, we would have been (and were at first) confused by what appeared to be clear patterns. Several conclusions can be drawn: (1) for most metals and years, the levels for any metal varied concordantly across years although the south was generally lowest; (2) there was a general peak in several metals in 2003; (3) the levels were often highest in fledglings from the middle of the Bay; and (4) for lead, cadmium, selenium, chromium, and manganese, the levels were generally low from the mid-2000s to the present in mid-bay. We reanalyzed the 2003 samples to verify that these peaks were not due to analytic errors.

Other Northeast Bays and Other Locations

There are some metal results for Common Terns from other Northeast bays and estuaries. Some of these were published on a wet weight basis. To convert wet weight to a comparable dry weight basis, we multiplied by 4, based on the mean moisture content of tern eggs we dried in our laboratory. Thus, a published report of 0.25 ppm (wet weight) would be converted to 1000 ppb (dry weight). Levels of mercury in eggs of Common Terns from Cedar Beach, New York, averaged 1100 ± 200 ppb in 1995 (Table 11.4; Burger and Gochfeld 1997a), comparable to the means that ranged from 1000 to 1500 ppb for Barnegat Bay in the same period. Since Common Terns became listed as a threatened species in New York, we did not continue to collect eggs from terns in New York.

One of the difficulties of comparing egg or feather levels among geographical regions is that they were not necessarily collected in the same period, and as is clear from the data from Barnegat Bay, levels vary significantly among years. For example, mean mercury levels in feathers from the early 1990s averaged 1800 ppb (Bird Island) to 5000 ppb (Cedar Beach, Table 11.4), which was much higher

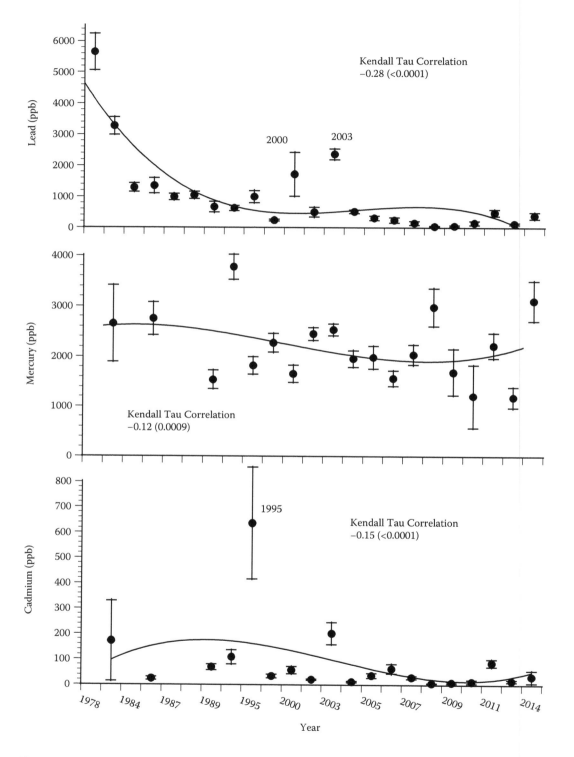

Figure 11.6 Patterns of lead, mercury, and cadmium levels (ppb, dry weight) (mean ± SE) Common Tern fledgling feathers from Barnegat Bay. Based on 513 feather samples.

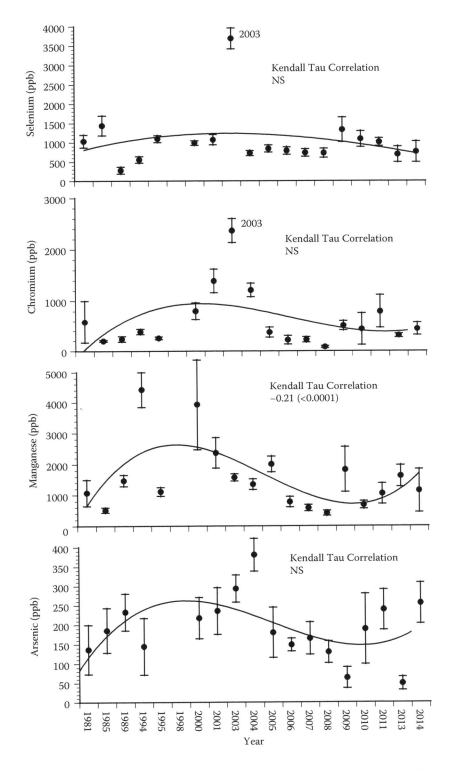

Figure 11.7 Patterns of selenium, chromium, manganese, and arsenic levels (ppb, dry weight) (mean ± SE) in Common Tern fledgling feathers from Barnegat Bay. Based on 513 feather samples.

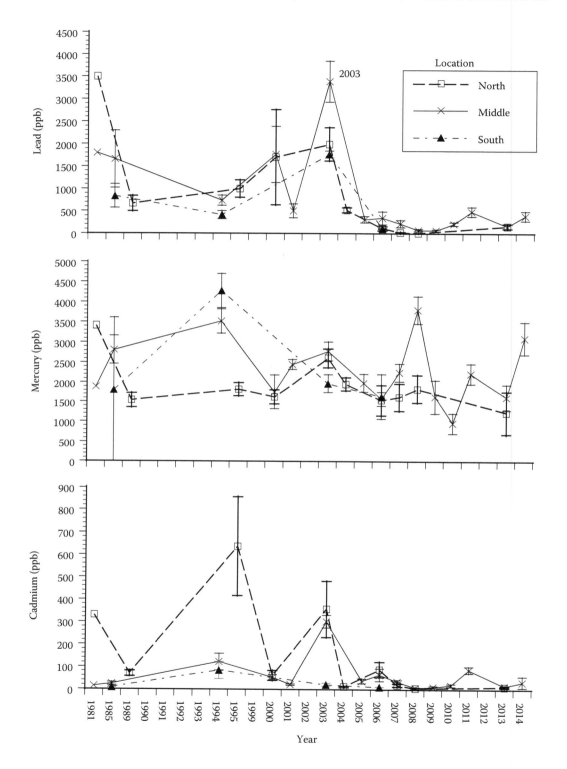

Figure 11.8 Locational differences in lead, mercury and cadmium (ppb, dry weight) (mean ± SE) in Common Tern fledgling feathers in Barnegat Bay. Based on at least 12 feather samples per year.

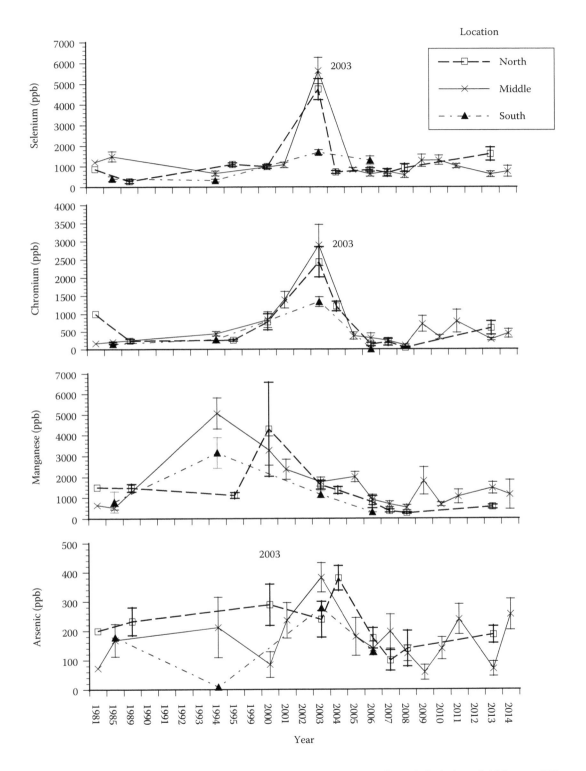

Figure 11.9 Locational differences in selenium, chromium, manganese, and arsenic (ppb, dry weight) (mean ± SE) Common Tern fledgling feathers in Barnegat Bay. Based on at least 12 feather samples per year.

Table 11.4 Levels of Metals (ppb, Dry Weight) in Eggs of Common Terns from the Northeast Bays and Elsewhere in the World (All Converted to Dry Weight)

Location	Year	Lead	Mercury	Arsenic	Cadmium	Selenium	Source
Northeast Bays							
Islands in Maine	2005		440	280 ± 80		2400	Mierzykowsk et al. 2005, 2008
Seal Island, Massachusetts	2005		270	240		2680	Mierzykowski 2008
Monomoy NWR, Massachusetts[a]	2005	Below detection	440	320	Below detection	2440	Mierzykowski 2008
Great Gull Island, New York	1971		360 (range of 80–1080)				Connors et al. 1975
Cedar Beach, New York	1989	389 ± 110			12 ± 5		Burger and Gochfeld 1993a
Cedar Beach	1995		1100 ± 200				Burger and Gochfeld 1997a
Other Locations							
Winnipeg	1971		2320 ± 360				Vermeer 1973
Athabasca, Saskatchewan			1500 and 930 (SD)	80 ± 30 and 90 ± 30 (SD)			Hebert et al. 2011
Wadden Sea, Germany	1988		2360 ± 800				Becker 1992
Wadden/North Sea	1991		1800 ± 640				Becker et al. 1993
Wadden Sea	1981–1998		Means of 2100 to 9600				Becker and Dittmann 2009
Western Wadden Sea	1998–2008		Means of 600–2440				Becker and Dittmann 2009
Eastern Wadden Sea	1998–2008		Means of 800–4800				Becker and Dittmann 2009
Wadden Sea	2008–2010		1250 ± 215				Dittman et al. 2011
Greece	1997	Mean of 200			Mean of 28		Goutner et al. 2001
Ebro Delta, Spain			Albumen = 2200 Yolk = 548			Albumen = 1864 Yolk = 2632	Guitart et al. 2003
Ebro Delta, Spain	2006–2008		Mean = 1136				Cotin et al. 2011
Poland	2010–2012		1239 ± 572				Grajewska et al. 2015

Note: Conversion of wet to dry weight was wet weight concentration × 4 = dry weight concentration.
a Samples were composites. Manganese levels for Monomoy were 1600 ppb, and 2000 ppb for five islands in Maine (Mierzykowski 2008; Mierzykowski et al. 2008).

Table 11.5 Levels of Some Metals (ppb, Dry Weight) in Breast Feathers of Adult and Prefledging Chicks of Common Terns

Location	Year	Age	Lead	Mercury	Cadmium	Selenium	Reference
				Northeast Atlantic Coast			
Machias, New Brunswick, Canada	2006	Adult		1380 ± 991 (SD)			Bond and Diamond 2009
Bird Island, Massachusetts	1990	Adult	2030 ± 851	1010 ± 142	390 ± 320	2540 ± 424	Burger and Gochfeld, unpublished data
		Chick	4340 ± 761	4390 ± 352	384 ± 117	2150 ± 643	
	1991	Adult	1500 ± 560	1800 ± 1200	120 ± 47	1100 ± 120	Burger et al. 1992a
	1992	Adult		2800 ± 2400 (SD)	108 ± 12	2040 ± 173	Nisbet et al. 2002; Burger and Gochfeld, unpublished data
Falkner Island, Connecticut	1990	Chick	4310 ± 851		400 ± 154		Burger and Gochfeld, unpublished data.
Cedar Beach, Long Island, New York	1971–72	Chick		1270 ± 510			Gochfeld 1980b
Cedar Beach, Long Island, New York	1978	Adult	5500 ± 500				Burger et al. 1994e
Cedar Beach, Long Island, New York	1980	Adult	3300 ± 300				Burger et al. 1994e
Cedar Beach, Long Island, New York	1984	Adult	1200 ± 100		480 ± 60		Burger et al. 1994e; Burger and Gochfeld 1991a
Cedar Beach, Long Island, New York	1985	Adult	1100 ± 75				Burger et al. 1994e
Cedar Beach, Long Island, New York	1986	Adult	1000 ± 50				Burger et al. 1994e
Cedar Beach, Long Island, New York	1987	Adult	1000 ± 50				Burger et al. 1994e
Cedar Beach, Long Island, New York	1989	Adult	3300 ± 700				Burger et al. 1994e
Cedar Beach, Long Island, New York	1989	Chick	950 ± 80				Burger et al. 1994e
Cedar Beach, Long Island, New York	1990	Adult	1600 ± 100				Burger et al. 1994e

(Continued)

Table 11.5 (Continued) Levels of Some Metals (ppb, Dry Weight) in Breast Feathers of Adult and Prefledging Chicks of Common Terns

Location	Year	Age	Lead	Mercury	Cadmium	Selenium	Reference
Cedar Beach, Long Island, New York	1991	Adult	3700 ± 700	5000 ± 1100	150 ± 37	2900 ± 300	Burger et al. 1994e, Burger et al. 1992a
Cedar Beach, Long Island, New York	1992	Adult	2080 ± 213	5400 ± 665	75.5 ± 11	2590 ± 209	Burger and Gochfeld, unpublished data
		Chick	1240 ± 121	2350 ± 215	68 ± 19	1320 ± 132	
Cedar Beach, Long Island, New York	1994	Chick		1650 ± 207	59 ± 31	1270 ± 106	Burger and Gochfeld, unpublished data
Other Areas							
Michigan	1890–1919 1920–1949 1950–1979	All ages		5500 ± 2300 (SD) 2300 ± 2200 1300 ± 120			Head et al. 2011[a]
German seacoast	<1940 >1940	Young and adult		6060 (y) 2940 (a) 18,430 (y) 10,410 (a)			Thompson et al. 1993
German North Sea	1991	Chicks		3260 ± 700			Becker et al. 1994
Iran	2007	Adult		11,530 ± 6100 (SD)			Zamani-Ahmadmahmoodi et al. 2014

Note: New York–New Jersey Harbor was mainly Cedar Beach and West End; these colonies were eventually deserted because of predation. Bird Island is in Buzzards Bay. In some cases, levels are provided from other places for comparison. Data are presented as means ± standard error (ppb or ng/g, dry weight).

[a] There was a significant temporal trend (decreasing) for the Common Terns feathers from Michigan, although Head et al. (2011) stated that the trend may be the result of inorganic mercury contamination of feathers owing to museum preservation techniques.

than the levels reported for eggs from the 2000s (see Figure 11.4). However, levels from the German North Sea from 1991 averaged 1800 ppb (Becker et al. 1993), which was similar to the levels found in Bird Island at the same period.

There are data for metal levels in feathers of young Common Terns from some other Northeast bays (Table 11.5). These are indicative of local exposure as parents feed the growing chicks fish caught in the vicinity of the breeding colony. For mercury, most results were in the 1000 ppb range. Furthermore, there has been a general decline worldwide, although in any given year and location, levels can be high (see Table 11.5).

FORSTER'S TERN

Much less information is available about metal levels in Forster's Terns, although understanding changes in metal levels is informative because, at least in the Northeast, they often nest in the same marshes, in the same colonies as Common Terns. Unlike Common Terns, Forster's Terns forage mainly over pools in salt marshes, creeks, and bays, and over freshwater marshes and impoundments (McNicholl et al. 2001, personal observations). They forage on fish that they capture by plunge-diving and on polychaetes that they snatch from the surface (personal observation; Figure 11.10). Their nesting colonies are often overlooked on remote salt marsh islands or among the more numerous and noisy Common Terns (Nisbet et al. 2013). Forster's Tern is the only tern restricted to North America throughout the year (McNicholl et al. 2001). It is vulnerable to the same threats as Common Terns: habitat loss, flooding, predators, and contaminants.

Barnegat Bay Egg Patterns

Eggs were collected beginning in 1990 when Forster's Terns began to increase generally in Barnegat Bay. It was often difficult to collect their eggs because the colonies they selected were located in the southern sections of Barnegat Bay where the bay waters are very shallow, making boat access difficult. Overall, lead decreased dramatically, and cadmium, chromium, and manganese decreased slightly in eggs, but there was no consistent pattern in mercury, selenium, or arsenic levels (Figures 11.11 and 11.12; Appendix Table 11A.3). Chromium and manganese showed slight decreases.

Figure 11.10 Photo of a flying Forster's Tern looking for small fish over a salt marsh pond.

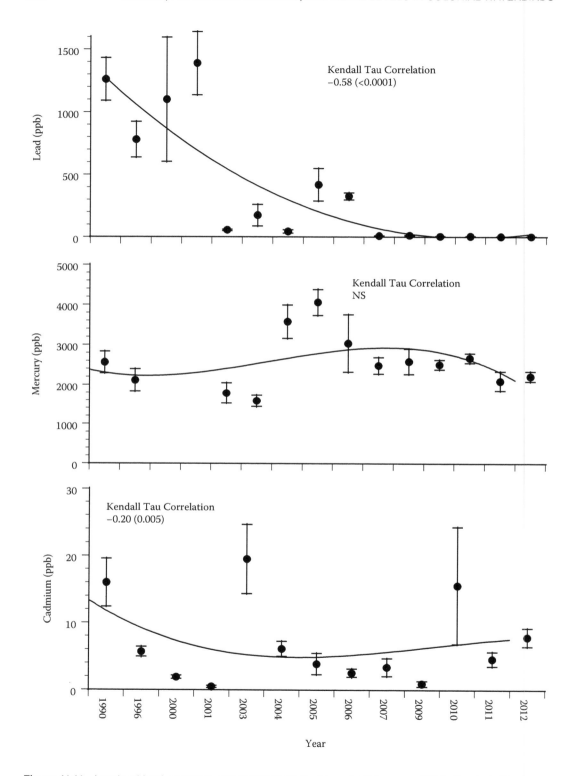

Figure 11.11 Levels of lead, mercury, and cadmium (ppb, dry weight) (mean ± SE) in Forster's Tern eggs from Barnegat Bay. Based on 237 egg samples.

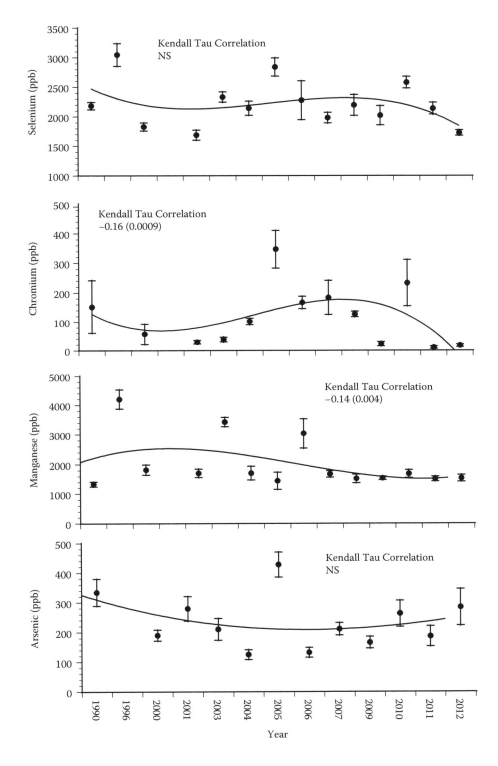

Figure 11.12 Levels of selenium, chromium, manganese, and arsenic (ppb, dry weight) (mean ± SE) in Forster's Tern eggs in Barnegat Bay. Based on 237 egg samples.

Barnegat Bay Feather Patterns

Forster's Terns usually have lower reproductive success than Common Terns because they often nest in lower parts of the marsh, building up nests in *Spartina alterniflora*; unlike Common Terns they usually do not nest on wrack, which is usually higher on the marsh than the sites selected by Forster's Terns at the edge of salt marsh pools. Although we have continued to search for Forster's Tern fledglings, they have not been successful at raising their young in recent years. For the period when levels in feathers were examined (1989–2005), there were significant changes. The apparent increase in lead is spurious because of the 2003 sample when lead levels temporarily increased (Figure 11.13). There was also a 2003 peak in Common Tern feathers. Mercury decreased from 1995 to 2005, back to the 1990 level. Cadmium decreased significantly. Figure 11.14 shows increases and decreases in chromium, and no evident patterns for selenium, manganese, and arsenic. In part, these differences among yearly metal levels relate to the relatively short temporal span (15 years), relatively small sample sizes, and the infrequent analyses (only 6 of 15 years). This suggests the importance of extended temporal sequences.

Other Northeast Bays and Other Regions

Only a few Forster's Terns have nested north of New Jersey. We did not find data on metal levels for Forster's Tern eggs or chick from other Northeast bays or estuaries. Mercury levels in eggs from two locations in LaVaca Bay, Texas, averaged 1600 ± 640 and 880 ± 680 ppb, and selenium averaged 2840 ± 400 and 2720 ± 480 ppb (King et al. 1991). Mercury levels in Forster's Tern eggs from San Francisco Bay, California, averaged 3600 ± 280 ppb (Ohlendorf et al. 1988). Mercury levels in the blood of chicks averaged 330 ng/g (wet weight) and in their breast feathers averaged 6440 μg/g (Ackerman et al. 2008), but there was no relation between these levels and postfledging survival. Thus, the levels we found in Forster's Terns from Barnegat Bay were higher than those reported in Texas, but lower than in San Francisco Bay.

ROSEATE TERNS

Roseate Terns (Figure 11.15) are federally endangered in the Northeastern United States, where the species range has shrunk and population numbers declined. Roseate Terns no longer nest in New Jersey. There is a Northeastern population breeding from the Canadian Maritimes to Long Island, and a Caribbean population that reaches Florida (Gochfeld and Burger 1996). Although the Northeastern population is listed as endangered, the Roseate Tern is a pantropical species, and its worldwide population is not threatened. Roseate Terns feed in a similar manner to Common Terns, and often feed in the same flocks (Safina 1990; Nisbet et al. 2014). We studied Roseate Terns on Long Island. Because of its endangered status, our data set on metal levels for this species is much more limited than for the other terns, but data are presented to allow comparisons among terns.

Roseate Tern Egg Patterns

Metal levels were analyzed from Cedar Beach, from 1989 to 1994 (the last date Roseate Terns nested at the colony; Gochfeld and Burger 1998). Even in this short period, lead declined sharply; and mercury decreased slightly, whereas the other metals showed increases in 1991–1993, followed by declines (Figure 11.16). We are not confident about comparing Cedar Beach to Barnegat Bay, but the increased cadmium level in the early 1990s and a lead peak in 1992 in Roseate Tern eggs can

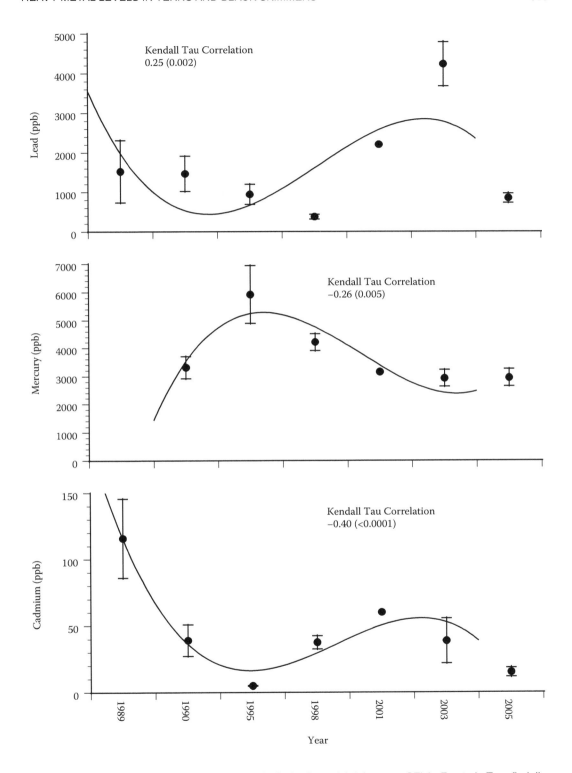

Figure 11.13 Lead, mercury, and cadmium levels (ppb, dry weight) (mean ± SE) in Forster's Tern fledgling feathers from Barnegat Bay. Based on 98 feather samples.

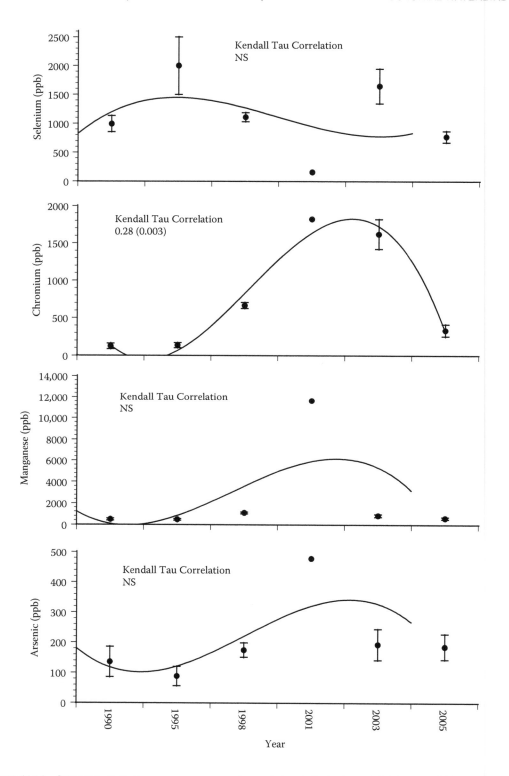

Figure 11.14   Selenium, chromium, manganese, and arsenic levels (ppb, dry weight) (mean ± SE) in Forster's Tern fledgling feathers from Barnegat Bay. Based on 98 feather samples.

Figure 11.15 Photo of an adult Roseate Tern above a nest site.

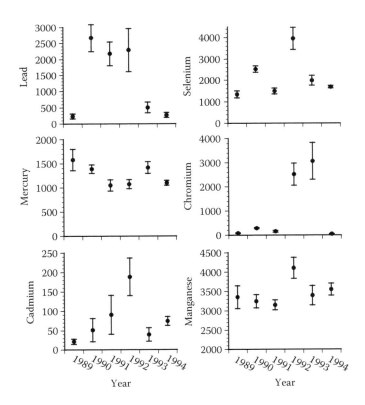

Figure 11.16 Metals levels (ppb, dry weight) (mean ± SE) in Roseate Tern eggs from Cedar Beach (1989–1994). Shows a peak for several metals in 1992. Based on 70 samples.

be seen in the Common Tern egg data as well (Figure 11.2). Common and Roseate Terns overlap on the wintering grounds off South America, which could account for prearrival exposure reflected in egg-laying. The Roseate Tern data set is too short to see other patterns clearly.

The data comparing levels of Roseate and Common Terns from Cedar Beach are particularly interesting because they illustrate two things: (1) for most metals, there were interyear differences, but the differences were not directional (i.e., no evident trend); (2) lead levels were low initially, then rose and declined sharply to the 1989 levels (Figure 11.16). This pattern was initially difficult to understand, until we talked to a New York State Department of Transportation employee, who noted that the period of high lead levels corresponded to the removal of lead paint from New York City bridges. Although lead paint removal was somewhat contained to prevent environmental contamination, the procedures are not perfect. The levels of lead in eggs of Roseate Terns in 1989 and in 1994 were similar to those of Common Terns from about the same period. Mierzykowski (2008) reported levels of contaminants in one composite of eggs from Minimoy (Massachusetts).

Roseate Tern Feather Patterns

We also analyzed feathers from adult and young Roseate Terns from Bird Island (Buzzards Bay), Falkner Island (Connecticut), and Cedar Beach (Long Island) from 1989 to 1994 (Figures 11.17 and 11.18). These figures show levels for both adults and young from the three colonies. There were

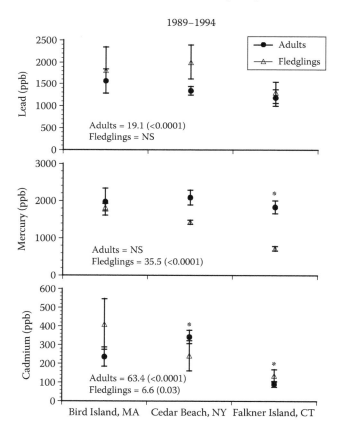

Figure 11.17 Levels of lead, mercury, and cadmium (ppb, dry weight) (mean ± SE) in feathers of adult and fledgling Roseate Terns from three Northeast nesting colonies. Based on feather samples from 239 adults and 103 fledglings.

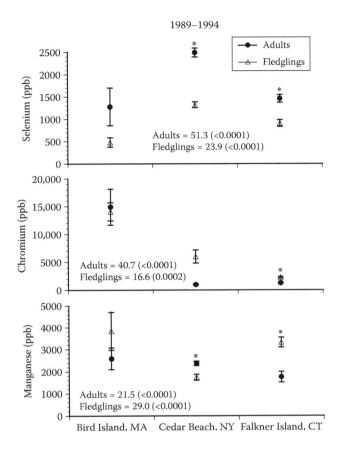

Figure 11.18 Levels of selenium (ppb, dry weight) (mean ± SE), chromium, and manganese in feathers of adult and fledglings of Roseate Terns from three Northeast nesting colonies. Based on feather samples from 239 adults and 103 fledglings.

significant intercolony differences in feathers of adult Roseate Terns for all metals except mercury, and significant intercolony differences in feathers of chicks for all metals except lead (Kruskal–Wallis χ^2 tests). Falkner Island Roseates, both adults and fledglings, tended to have the lowest levels for most metals. For lead, fledglings at Cedar Beach had the highest levels. For mercury, fledglings had lower levels at Falkner. For cadmium, fledglings had higher levels at Bird Island and fledglings had lower levels at Cedar Beach. There was a surprising but consistent difference in selenium levels with adults having higher levels than fledglings at all three sites (Figure 11.18).

After the breeding season, Roseate Terns from Long Island and New England gather in a large, premigratory assemblage off the Massachusetts coast. They migrate to an ill-defined, wintering area off Brazil. We did not expect that adult Roseate Terns in different colonies would have different metal levels in feathers grown on the wintering grounds they are believed to share. On the other hand, we did expect that chicks might accumulate different levels of metals related to local conditions at the time they are being fed from food near the colony. However, these are long-lived birds, and their exposure in 1 year increases their body burden, and metals may be mobilized from tissue stores during molt in a subsequent year. Thus, the metal differences could reflect exposure differences in the previous year. We predicted that the adults and young would show relatively concordant patterns. This is apparent for cadmium (Figure 11.17), and for selenium and chromium (Figure 11.18).

We also studied Roseate Terns in Puerto Rico. Metal levels in feathers of adult Roseate Terns from Culebra, Puerto Rico, averaged 2440 ± 563 ppb for lead, 95 ± 32 ppb for cadmium, and 2240 ± 310 ppb for mercury (Burger and Gochfeld 1991e). These cadmium levels are lower than the levels found in adult Roseate Terns from the Northeastern United States (refer to Table 11.4). However, the lead and mercury levels are within the range of the Northeastern birds. Culebra served for many years as a U.S. Navy artillery and bombing target and waste disposal site, and the land and waters are contaminated with lead, mercury, and "military ordnance."

BLACK SKIMMERS

Black Skimmers are endangered in New Jersey, Delaware, and Maryland. They forage on prey fish by skimming along the water with their bottom mandible slicing the water surface (Burger and Gochfeld 1990b; Favero et al. 2001). When they encounter a fish the bill snaps shut, and the bird veers off to swallow the fish or carry it back to the colony (Figure 11.19). This fishing method lends itself to calm waters, and Skimmers usually forage in salt marsh creeks and the edges of bays, and less frequently in the open bay. They can catch larger prey fish than either Common or Forster's Tern, and vary the size of the fish they bring back to feed their chicks according to chick size (Burger and Gochfeld 1990b).

There is little information on contaminant levels in Black Skimmers because they are relatively wary, do not tolerate much disruption by investigators entering colonies, and have declined in several places (Burger and Gochfeld 1990b). However, there are some data from Barnegat Bay and from western Long Island. Information on Black Skimmers is of interest because they have a unique foraging method (skimming), which could result in their capture of different prey items (or sizes) from different places (see Chapter 10).

Barnegat Bay Egg Patterns

We collected Skimmer eggs intermittently from 1989 to 2000. The species is listed as "Endangered" in New Jersey, and by 2000, the decline in population in Barnegat Bay dissuaded us from collecting further samples. The patterns were generally not clear because of the low number of years eggs were collected (Figures 11.20 and 11.21). The decline in mercury is not real, but is attributable to the small sample sizes. The higher levels of chromium in the early 1990s corresponded to the aggressive remediation of contaminated sites in Hudson County (see section "Common Terns"). The trends in selenium and manganese are probably artifacts of small size. These data are presented, however, because they will be useful for future work.

Barnegat Bay—Feather Patterns

Data on heavy metals in Skimmers comes mainly from feathers of prefledging chicks collected in Barnegat Bay from 1989 to 2004 (Figure 11.22). Levels of lead declined significantly with the highest value in year 1 and the lowest in year 7, although the data were noisy enough that the overall correlation (Kendall tau) was not significant. Mercury, cadmium, and manganese fluctuated with a peak in 1992 and then much lower values in the past 2 years. As with terns, mercury levels were variable in feathers, although they peaked in 1992. Selenium may have increased, and there was no significant pattern for chromium (Figure 11.23).

Feathers were also collected from adult Skimmers nesting in Barnegat Bay in 1989 (Gochfeld et al. 1991; Table 11.6). In the following discussion, we provide comparisons for the same years for adults and fledglings. It is important to compare them for the same year of collection because metal levels may vary over time (lead and cadmium have generally decreased in the environment, Chapter 8).

(a)

(b)

Figure 11.19 Black Skimmers forage in marsh creeks and at the edges of marshes by skimming along the surface (a). Their eggs and chicks are cryptic on a sandy substrate (b).

In Table 11.6, it is clear that Barnegat Bay adults had higher lead levels than young Skimmers, but lower cadmium levels—suggesting that the lead burden of adults may have been obtained elsewhere, but the cadmium levels were obtained by fledglings from the bay.

The keratin structural protein of feathers, rich in disulfide bonds, is a sink for metals. Black feathers have melanin pigment, which protects them from physical abrasion. We wondered whether

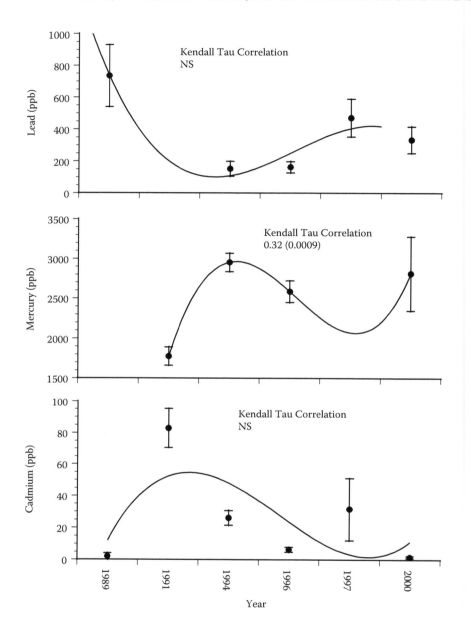

Figure 11.20 Levels of lead, mercury, and cadmium (ppb, dry weight) (mean ± SE) in Black Skimmer eggs in Barnegat Bay. Based on 121 eggs.

there might be higher metal levels in black feathers than in white ones, but we found no differences in metal levels of black and white body feathers collected from adult Skimmers (Gochfeld et al. 1991).

Black Skimmers display an extreme degree of sexual dimorphism (Burger and Gochfeld 1992c), leading us to suspect that males and females would have significantly different exposure. Feathers from 1989 to 1991 were analyzed to examine gender differences. These data were from adults. Females had significantly higher concentrations of lead (mean of 1600 versus 1160 ppb) and cadmium (98 versus 31 ppb) than males, but there were no gender differences for mercury (mean of 12,400 ppb), selenium (1220 ppb), chromium (10,040 ppb), manganese (2520 ppb), and copper (27,430 ppb). Overall, the

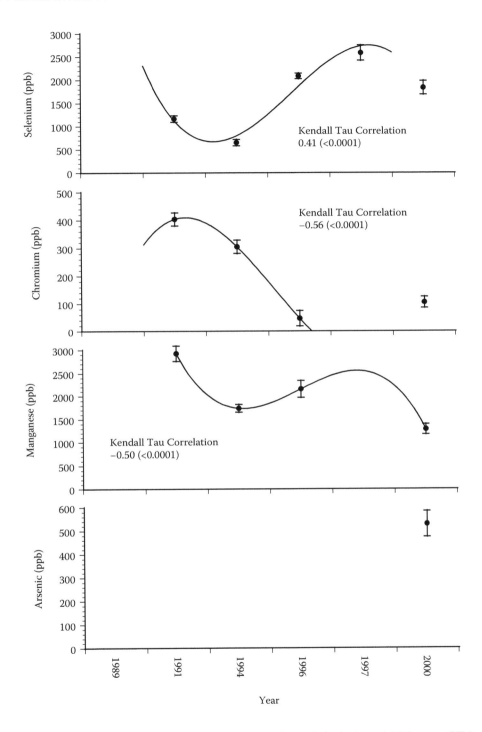

Figure 11.21 Levels of selenium, chromium, manganese, and arsenic (ppb, dry weight) (mean ± SE) in Black Skimmer eggs from Barnegat Bay. Based on 121 eggs.

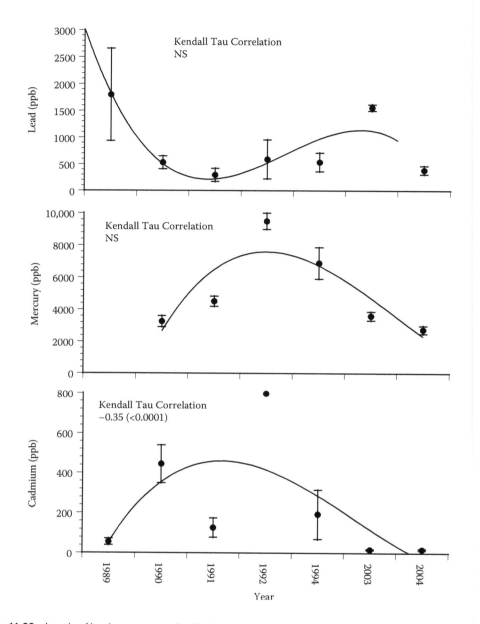

Figure 11.22 Levels of lead, mercury, and cadmium (ppb, dry weight) (mean ± SE) in Black Skimmer fledgling feathers from Barnegat Bay. Based on 111 feather samples.

levels of cadmium and lead in feathers of adults were lower in adults from Cedar Beach (New York) than those from Barnegat Bay (see Table 11.6). The mercury level was particularly high.

Henny et al. (2008) analyzed the levels of organochlorine pesticides, polychlorinated biphenyls (PCBs), and selenium in Skimmer eggs from the Salton Sea in California (1993), and found that mean levels of selenium ranged from 4650 to 6010 ppb. The lowest selenium concentration causing effects is 6000 ppb (dry weight; Henny et al. 2008), which indicates that some Salton Sea birds exceed this level and might be vulnerable to adverse effects.

Although it is important to continue to examine metal levels in terns, it is very important for Black Skimmers, given the declines in their populations in a number of regions. Although it is

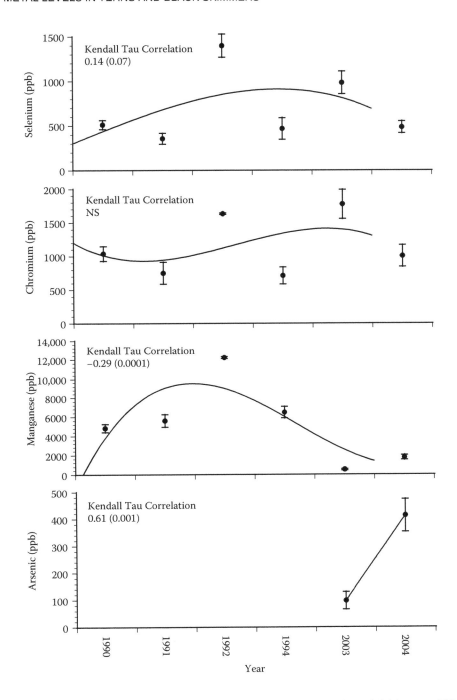

Figure 11.23 Levels of selenium, chromium, manganese, and arsenic (ppb, dry weight) (mean ± SE) in Black Skimmer fledgling feathers from Barnegat Bay. Based on 111 feather samples.

Table 11.6   Levels of Lead and Cadmium (ppb, Dry Weight) in Feathers of Adult and Young Black Skimmers from Barnegat Bay and Long Island (New York) from 1989 and 1990

	Sample Size	Lead	Cadmium
Barnegat Bay, New Jersey Prefledging chicks	25	997 ± 76	465 ± 36
Barnegat Bay, New Jersey Adults	24	2730 ± 982	112 ± 29
Cedar Beach, New York Adults	31	1370 ± 134	67 ± 22

Sources: Gochfeld, M. et al., *Arch. Environ. Contam. Toxicol.*, 20, 523–526, 1991; Burger, J., Gochfeld, M., *Arch. Environ. Contam. Toxicol.*, 23, 431–434, 1992c; Burger, J., Gochfeld, M., this book.

Note: Data are presented as means ± SE.

difficult to collect eggs from dense colonies without causing disturbances (Figure 11.24), given their wariness, it is important to determine if metals could be causing sublethal effects in this species.

DISCUSSION

In this section, we discuss issues not discussed separately for each species. There are more data for Common Terns from other sites than there are for the other species—thus, much of the discussion relates to Common Terns. Other species are mentioned where relevant. As our analyses accumulated over 40 years, we found inconsistencies in the data that were difficult to explain. In order to make our data meet the EPA's rigorous criteria, we had instituted a quality assurance program, which included consistent analytic quality control of every batch of samples. We considered ourselves fortunate in that over more than 30 years of running our own laboratory, we had only four primary analytic technicians, and only two pieces of analytic equipment, although we upgraded our earlier AA when Zeeman correction became affordable. We also were fortunate in being able to design our laboratories in the planning stages of the Environmental and Occupational Health Sciences building, so we could separate dissection and digestion rooms from the spectrometer laboratory to avoid background cross-contamination. Despite a rigorous quality assurance program throughout (see Chapter 3), we had a lingering suspicion that some unexplained peaks or discrepancies might have been analytic. Therefore, we took many of the samples that had been archived (feathers and dried egg contents) and reran them in multiyear batches, which allowed us to confirm that the earlier analyses were valid.

Temporal Patterns in Terns

Our most complete data set of metal levels in feathers and eggs from Barnegat Bay comes from Common and Forster's Terns. More data were available for eggs because they are easier to collect than feathers. Furthermore, eggs are laid early in the season and we timed our visit to early incubation. In contrast, chicks have to survive incubation, hatching, and the chick-rearing phase to reach the age of 2–3 weeks when they have body feathers that would have accumulated metals and were developed enough to pluck (e.g., no blood supply remaining). Thus, there was a 6-week interval during which storms and flooding could wipe out a colony.

For analyzing factors influencing trends, we developed multiple regression models to examine the importance of species, location, and year in accounting for variations in metal levels (Table 11.7). Results differed by metals: (1) for lead, cadmium, chromium, manganese, and arsenic, year was the most significant factor explaining variations; and (2) for mercury and selenium, species was the

Figure 11.24 Black Skimmers nesting in the Cedar Beach colony, Long Island. Once occupied by 6000 pairs of Common Tern and 200+ pairs of Skimmers, the colony was abandoned because of fox predation.

Table 11.7 Multiple Regression Models on Differences in Metal Levels in Common Tern ($n = 979$) and Forster's Tern ($n = 226$) Eggs Collected from Barnegat Bay, New Jersey

	Lead	Mercury	Cadmium	Selenium	Chromium	Manganese	Arsenic
Common Tern, mean ± SE	357 ± 30	1319 ± 27	24 ± 3.9	2557 ± 29	272 ± 28	2498 ± 45	214 ± 5.2
Forster's Tern, mean ± SE	193 ± 29	2670 ± 89	9.0 ± 1.2	2233 ± 45	308 ± 65	1876 ± 73	200 ± 9.6
Model							
F	94.4	84	16.6	11.7	32.2	17.2	16.7
df	4	4	4	4	4	4	4
P	<0.0001	<0.0001	<0.0001	<0.0001	<0.0001	<0.0001	<0.0001
r^2	0.27	0.24	0.06	0.04	0.11	0.06	0.06
Factors entering (F, P)							
Species	0.3 (NS)	248 (<0.0001)	0.3 (NS)	16.4 (<0.0001)	0.4 (NS)	29.7 (<0.0001)	0.4 (NS)
Location	13.0 (<0.0001)	11.7 (<0.0001)	4.0 (0.02)	5.5 (0.004)	13.7 (<0.0001)	3.6 (0.03)	6.1 (0.002)
Year	362 (<0.0001)	16.5 (<0.0001)	59.4 (<0.0001)	6.4 (0.01)	108 (<0.0001)	18.5 (<0.0001)	55.9 (<0.0001)

Note: df, degrees of freedom; NS, not significant; P, probability.

most significant variable. As mentioned earlier, the species difference may well relate to the locations where they feed: Common Terns feed in the ocean a significant percent of the time whereas Forster's Terns normally feed in inland ponds or in pools on the marshes. Temporal trends for the terns are given in Table 11.8.

It is easier to see the patterns in individual graphs for temporal trends, but in Table 11.8 we summarize these patterns. The curves fitted to the data show overall patterns, which have changed over the years. Even so, the overall trend of declining metal concentrations is clear in the largely negative correlation coefficients, which are reported in these tables. The r^2 value is the percentage of variance in the metal concentrations accounted for by the entire model of species, location, and year.

Since cadmium, lead, and mercury are the significant contaminants in marine ecosystems, it is not surprising they were found in tern eggs and feathers (Mailman 1980; Thompson et al. 1998b). Declines in levels of lead in eggs of Common and Forster's Terns, and in feathers in Common Terns, correlate with the removal of lead from gasoline (Annest et al. 1983). Although cadmium declined in eggs and feathers of both species, and lead declined in eggs of both species, Forster's Tern feather did not show a consistent decline in lead, because of the unexplained high levels in 2003.

Lead and cadmium levels have declined in the environment generally from regulations concerning cadmium in batteries, and the removal of lead from paint and gasoline (Mailman 1980; ATSDR 2007a). The declines in lead levels in Common Tern eggs and feathers were also mirrored in declines in blood lead levels in children (Annest et al. 1983). Lead in estuaries, and in the associated food chains, are affected by runoff from urban and suburban development (Kennish 2001a, 2001b), suggesting that lead levels in runoff into Barnegat Bay have declined substantially from 1971 to the present. Declines in cadmium tracked declines in lead, although the declines were not

Table 11.8   Summary of Trends in Metal Levels in Common and Forster's Tern Eggs and Feathers in Barnegat Bay

	Common Terns	Forster's Terns
Lead	Eggs: −0.50 (P < 0.0001); steady decline to early 2000, stable thereafter Feathers: −0.28 (P < 0.0001); similar levels since 2003	Eggs: −0.58 (P < 0.001); steady decline Feathers: NOTE +0.25 (P < 0.002), but a decline in 2005, and no fledged young available in later years
Mercury	Eggs: −0.14 (P < 0.0001); Variable with some high years, significant declines since 2000 Feathers: −0.12 (P < 0.0009), but very variable for most of the period	Eggs: Not significant Feathers: −0.26 (P < 0.005), decline from 1995 to 2005
Cadmium	Eggs: −0.28 (P < 0.0001); rather steady decline, with stable levels after 1995 Feathers: −0.15 (P < 0.0001), but based on high levels in early 1990s	Eggs: −0.20 (P < 0.005); very variable, with some high years Feathers: −0.40 (P < 0.0001), overall decline by variable
Selenium	Eggs: Not significant, variable Feathers: No significant change	Eggs: Not significant, variable Feathers: Not significant, variable
Chromium	Eggs: −0.12 (P < 0.0002), some highs early, but low levels since 2000s Feathers: No significant change, but highs in the early 2000s	Eggs: −0.16 (P < 0.0009), decline in later years Feathers: 0.28 (P < 0.003); increase followed by a decrease in 2005
Manganese	Eggs: No significant trend Feathers: −0.21 (P < 0.0001), but levels higher 2009–2014 than before	Eggs: Not significant Feathers: Not significant; generally low, with one high year (2001)
Arsenic	Eggs: −0.19 (P < 0.0001); very variable, but steady decline Feather: Not significant	Eggs: Not significant; variable Feathers: Not significant, but high year in 2001

Note: Given is the overall trend using the Kendall Tau Correlation option in SAS PROC CORR. Mercury, lead, and cadmium were analyzed for a longer period (1971 or 1978 to present). Given is overall correlation coefficient (Kendall Tau) and P value.

as substantial (Bouton et al. 1991). Cadmium is not sequestered in eggs in large quantities, although there is clearly some transference from the female to eggs (Burger and Gochfeld 1991a).

Long-term trends in mercury levels in birds, particularly gulls and terns, have been used to examine patterns of local, regional, and global exposure. Weseloh et al. (2011) found steady declines in mercury in eggs of Herring Gulls from the mid-1970s to 2011 from the Great Lakes (Chapter 12). Mercury levels in eggs of Common Terns from the Wadden Sea (German North Sea) generally decreased from 1985 to 1995, and then remained stable thereafter, although some colonies initially had higher levels (Becker and Dittmann 2009). The declines were attributed to decreases in local mercury sources. These studies that focused on mercury and organics did not provide data on other heavy metals.

In the present study, mercury levels in Common Tern eggs varied over the 40+-year period, appearing to decline for some years (e.g., 1971 to 1982), but then increased in others (e.g., 2000 to 2005). Levels of mercury in chick feathers declined in both terns, although the declines were small for Common Terns. The main source of mercury in these coastal, aquatic ecosystems is anthropogenic deposition (Fitzgerald 1989; Fitzgerald and O'Connor 2001) mainly from coal-fired power plants to the west of the state (NJ Mercury Task Force 2001). The EPA enforced new regulations, the *Mercury and Air Toxics Standards*, to reduce mercury emissions from power plants, but these have been challenged by the industry and delayed by a Supreme Court decision. The EPA has greatly reduced its air deposition tracking system over the past 15 years, yet tracking mercury levels is important to provide early warning of increases that might affect humans, birds, and other components of ecosystems.

Fewer data are available from the literature for other metals. However, from Barnegat Bay:

1. Selenium showed no significant pattern for both eggs and feathers for both species, which is consistent with its regulation by the body as an essential trace element (Eisler 2000a,b).
2. Chromium declined in eggs of both species, but showed no pattern for Common Tern feathers, and increased for Forster's Tern feathers. We cannot account for the increase in chromium, which—being internally regulated—suggests a local source.
3. Manganese, also an essential element, showed no significant pattern for eggs and an increase in Common Tern feathers.
4. Arsenic levels generally showed no significant trend, except for a suggestive increase Common Tern eggs, but we had a much shorter time series for arsenic.

Locational Differences in Eggs of Common Terns

There were significant, albeit small, locational differences in metal levels in eggs and feathers of Common Tern from Barnegat Bay. Barnegat Bay is 67 km long, and has two main ocean inlets (in the middle and the lower end of the Bay) as well as two major rivers flowing into it (Toms River slightly above Barnegat Inlet, and the Mullica River that flows into Great Bay), and then into the Atlantic Ocean in the south. Although geographical differences in mercury levels have been recognized for some time (Evers et al. 2012), as well as locational differences at varying distances from a point source, local differences within the 67-km-long bay with no clear point source for mercury were unexpected. However, there are several possible explanations for such differences: (1) the northern area of the bay (Lavallette islands) has no inlet, and Toms River flows into the northern bay, through the city of Toms River (which brings urban contaminants); (2) the middle part of Barnegat Bay has Barnegat Inlet, which opens directly to the Atlantic Ocean (allowing rapid dilution); and (3) the Mullica River flows into Great Bay at the southern end of Barnegat Bay, and both the southern end of Barnegat Bay and Great Bay open to the Atlantic Ocean through Beach Haven (Little Egg) Inlet. Thus, the northern part of the bay has a major runoff site (Toms River) and no outlet to the Atlantic Ocean, the middle part has no major runoff site but has a major outlet to the Atlantic Ocean, and the southern part of the bay has only a small river but has an outlet to the Atlantic Ocean.

The variability in metal levels by location indicates that many of the metals were highest in fledgling feathers in the mid-2000s, and many metals peaked at this time. Levels were generally highest in mid-bay, and lowest in the southern part of Barnegat Bay, a pattern that was similar for eggs as well. In birds, selenium, chromium, and manganese are essential trace elements (Drown et al. 1986; Eisler 2000a,b), and thus should show less variation in levels than metals that are not essential and do not occur naturally in the body (mercury, lead, cadmium, arsenic). Even essential elements can be present at toxic levels, such as selenium (Ohlendorf et al. 1986a, 1989, 1990; Ohlendorf 2000).

Interspecies Comparisons

Metals in birds reflect trophic levels (Becker et al. 2002), which can be confirmed by stable isotope analysis (Hobson et al. 1994, 2000; Thompson and Furness 1995; Thompson et al. 1998a). Common Terns eat fish, exposing them to contaminants. Courtship feeding contributes significantly to female nutrition during egg-laying, influencing the size of eggs and ultimately fledging success (Nisbet 1978). Levels in eggs appear to reflect female exposure over a comparatively short period (Sanpera et al. 2000). Furthermore, male Common Terns courtship feed their mates before and during egg-laying, and they often feed their mates large fish (Burger and Gochfeld 1991b), thus potentially exposing them to higher contaminant levels than females would normally accumulate on their own. They adjust their feeding rates to the size of the fish (Morris 1986). Any contaminants in these large "courtship" fish can then be sequestered in their eggs, potentially resulting in adverse effects to the developing embryos, as well as predators that eat eggs. Thus, female Common Terns are at the top of their trophic chain (Gochfeld and Burger 1982b).

We had both egg and feather samples for Common and Forster's Terns and Black Skimmers in the early 2000 period, although not all from the same year. The sample size of Skimmer eggs was small. We analyzed eggs for 2000–2003 and feathers for 2003–2005. There were significant differences among species in metal levels in eggs, with Common Terns having the highest levels of all metals except mercury. Black Skimmer had an average mercury level of 2810 ppb, putting many of the eggs above an effects level (Table 11.9). After returning to the bay from their wintering grounds, all three species spend several weeks before they start to lay eggs. The metal levels in eggs largely reflect this recent exposure (Sanpera et al. 2000), particularly the nutrition obtained in the few days before egg-laying (Nisbet 1978). However, there is also a contribution from exposures and body burden that would have occurred before arrival.

Table 11.9 Levels of Metals (ppb, Dry Weight) (Mean ± SE) in Eggs from Barnegat Bay Terns and Skimmers for 2000–2003 (Years When We Could Compare Levels)

	Common Tern (n = 252)	Forster's Tern (n = 47)	Black Skimmer (n = 12)	χ^2 (P)
Lead	185 ± 19	90 ± 29	334 ± 83	15.5 (0.0004)
Mercury	1330 ± 51	2370 ± 220	2810 ± 466	38.8 (<0.0001)
Cadmium	11.5 ± 1.5	7.8 ± 2.2	1.7 ± 0.6	14.5 (0.0007)
Selenium	2250 ± 42	2060 ± 69	1830 ± 146	15.1 (0.0005)
Chromium	77 ± 7	58 ± 7	105 ± 20	5.5 (0.06)
Manganese	2730 ± 107	2250 ± 158	1280 ± 112	16.8 (0.0002)
Arsenic	266 ± 10	225 ± 20	529 ± 55	20.4 (<0.0001)

Note: Kruskal–Wallis nonparametric ANOVA was used to compare levels among species for each metal.

Table 11.10   Comparison of Metal Levels (ppb, Dry Weight) (Mean ± SE) in Fledgling Feathers for Species Nesting in Barnegat Bay from 2003 to 2005 (a Period When Feathers Were Available for All Three Species)

	Common Tern (n = 155)	Forster's Tern (n = 29)	Black Skimmer (n = 24)	χ^2 (P)
Lead	2050 ± 173	3200 ± 495	1270 ± 116	8.8 (0.01)
Mercury	2380 ± 96	2930 ± 211	3390 ± 224	19.0 (<0.0001)
Cadmium	164 ± 32	30 ± 11	18 ± 2.3	17.9 (0.0001)
Selenium	3110 ± 247	1310 ± 204	859 ± 105	35.0 (<0.0001)
Chromium	1970 ± 180	1130 ± 172	1580 ± 180	6.7 (0.03)
Manganese	1642 ± 101	744 ± 93	815 ± 138	39.3 (<0.0001)
Arsenic	302 ± 27	190 ± 35	176 ± 40	7.3 (0.03)

Note: Kruskal–Wallis nonparametric ANOVA used to compare levels among species for each metal.

We also compared metals in feathers (Table 11.10). The years are different so the results are not directly comparable, but the pattern was somewhat different than for eggs. Common Terns had the highest levels of cadmium, selenium, and manganese, but not selenium. Black Skimmers had the highest levels of lead, mercury, and arsenic. These relationships did not hold in all years, reaffirming the importance of comparing results within the same time frame.

Comparisons of Levels from Other Northeast Bays and Regions

Examining metal levels in eggs of Common Terns is useful because levels have been published from other locations, although not necessarily from the same period. At this point, it is essential to mention the importance of the timing of collection, as well as the feathers collected, and the age of birds when feathers were collected. In the preceding sections, we have demonstrated that there is a difference in metal levels by age (fledgling feathers, adult feathers). Thus, most of our data are from fledgling breast feathers for consistency across periods, species, and locations. Even within the 3–4 weeks of the prefledging period, metal levels in tissues may vary (Ackerman et al. 2011). Metal concentrations may also vary in different parts of a single feather (our unpublished data), which is one reason we prefer to use pooled breast feathers.

However, the effect of year of collection is dramatic for many metals, particularly lead and cadmium (which declined over time), and even for mercury, which varied among years. In most models, year was the factor that contributed the most to variations in metal levels. Thus, although we make a few comparisons of metal levels from other geographical regions, we note that the most appropriate comparisons are for the same years (and age, breast feathers). This may be true especially for mercury, because atmospheric deposition of mercury varies among years, but might be similar among Northeast bay regions in the same year.

Table 11.11   Comparison of Metal Levels (ppb, Dry Weight) (Mean ± SE) in Feathers of Young Common Terns from Several Northeast Estuaries (1990–1992)

	Lead	Mercury	Cadmium	Selenium
Bird Island, Massachusetts (n = 25)	4340 ± 761	4390 ± 352	384 ± 117	
Falkner Island, Connecticut (n = 25)	4310 ± 851	Not analyzed	400 ± 154	
Cedar Beach, New York (n = 25)	1240 ± 121	2520 ± 235	68 ± 19	1320 ± 132
Barnegat Bay, New Jersey (n = 45)	931 ± 121	2100 ± 210	112 ± 20	437 ± 110

Source: Burger, J., Gochfeld, M., Unpublished data.

Levels of mercury in Common Tern eggs (Table 11.4) and feathers (Table 11.5) are shown from other Northeast bays and other colonies elsewhere. Chick breast feathers are the best for comparison among Northeast bays because they represent local exposure (Table 11.11).

These data suggest that lead, mercury, and chromium levels were lower in the New York–New Jersey Harbor Estuary and in Barnegat Bay than farther north. The feathers of chicks from Barnegat Bay had lower levels of selenium than those from Cedar Beach. For the same year of collection (1991), levels from adult Common Terns at Cedar Beach were higher for lead, mercury, and selenium than they were for adults from Bird Island (Table 11.2). We suggest that efforts be made to sample regularly from several colonies, either annually or periodically in the same year, so the possibility of year × location interactions can be addressed.

Effects Levels

Effects levels are difficult to determine because there are differences and individual variability in exposure, toxicokinetics, and toxicity among species and life stages. There are various sources of uncertainty in this data set. Effects levels are method-dependent, and effects of one substance may be enhanced or decreased by coexposures. Toxicity was discussed in Chapters 8 and 9, as were effects generally, and in this section we discuss a few issues specific to Common Terns and Skimmers. There are many publications relating dose to outcome, but many of these made no measurements of tissue concentrations or measured levels in organs (liver, kidney). Egg and feather effects data are wanting. It should be remembered, however, that different metals or even forms of metals are stored in different organs and exert toxicity on different organs. Although metals are not formed or destroyed in the body, the metal compounds may be altered in ways that enhance or reduce toxicity. Demethylation of methylmercury and methylation of arsenic are two such reactions. Oxidative–reductive reactions influence the kinetics and toxicity of chromium. These make it impossible to pick a single toxic effects threshold. Moreover, interactions among contaminants influence toxicity. There is extensive literature on mercury × selenium interactions (Chapter 9). Eagles-Smith et al. (2009a,b) proposed that after mercury is demethylated, selenium may precipitate the inorganic mercury, thus reducing the potential for secondary toxicity by methylation.

Lead is a neurotoxin that causes cognitive and behavioral deficits in vertebrates (Weber and Dingel 1997), and can cause decreases in survival, growth, learning, and metabolism (Eisler 1988a; Burger and Gochfeld 1997b, 2000a). Dietary levels as low as 100–500 ppb can cause learning deficits in some vertebrates (Eisler 1988a). However, most effects levels are higher than this. Current lead levels are generally below the toxic effects levels, in contrast to levels in the 1970s and early 1980s.

The cadmium levels in eggs that are toxic to the developing embryos are uncertain, but young birds are more susceptible to cadmium poisoning than adults (Wren et al. 1995). Bird predators are adversely affected by cadmium levels of 1000 ppb or higher (Eisler 1985a). Current cadmium levels in eggs are well below this level, mostly below 100 ppb.

For mercury, adverse effects for developing embryos, including mortality, lowered hatching rates, higher chick defects, and other neurobehavioral deficits, can occur when egg levels are as low as 500 ppb (wet weight), and more severe effects usually occur at 1000–2000 ppb (wet weight; Eisler 1987a, 2000a,b; Burger and Gochfeld 1997a). Thompson (1996) suggested that mercury levels of greater than 200 ppb (600 ppb dry weight) in eggs are likely to have deleterious effects, although there are clearly interspecific differences (Heinz et al. 2009). Egg concentrations of mercury seem to be the best predictor of mercury risk to avian reproduction (Wolfe et al. 1998). The LC_{50} for Common Tern eggs injected with mercury was 870 ppb (wet weight = 3500 ppb dry weight; Heinz et al. 2009). The level that kills 50% of embryos could certainly be significant for a population, and even an LC_{10} causing 10% hatching failure is important. However, hatching rates vary from year to year so that levels below LC_{10} might get lost in the "noise." There is no fixed relationship between

LC_{50} and LC_{10}, but typically the latter is at least half or more below the LC_{50}. Common Tern eggs from Barnegat Bay exceeded 3000 ppb dry weight in 1994 and 1999, which would have been above an LC_{10}, and were in the 1500–2000 ppb (dry weight) range in several years. Even at the current 1000-ppb (dry weight) concentrations, sublethal embryotoxicity may occur.

In the 1960s and 1970s, there were a number of Common Tern colonies (Great Gull Island, Cedar Beach, Bird Island) with low hatching rates and other abnormalities, and investigators attributed these to PCBs and/or mercury. In the Great Gull Island, mercury levels in eggs in 1971 averaged 360 ppb (dry weight, range of 80–1080 ppb; Connors et al. 1975). Mercury levels in the liver of embryos that died in a colony with abnormal young terns from Great Gull Island (New York) in 1969–1970 were 3096 ppb (dry weight, with some 1-day-old chick having levels as high as 4640; Hays and Risebrough 1972). At Cedar Beach (Gochfeld 1980b), mercury levels in liver averaged 6660 ppb (dry weight) for abnormal chicks versus 3180 ppb in normal chicks ($P < 0.05$). Abnormalities in chicks included skull and eye defects, crossed bills, undeveloped lower mandible, sparse down, and abnormal feather loss, although the abnormalities may have been attributable to either mercury or organics (Hays and Risebrough 1972; Gochfeld 1980b). Unfortunately, metal levels were not measured in the feathers of all these chicks. The causes were likely to be PCBs, mercury, or a combination of both (see Chapter 9; Gilbertson et al. 1991). The Great Lakes embryo mortality, edema, and deformities syndrome (Gilbertson et al. 1991) was also recognized in the early 1970s, resulting in mortality, reproductive failure, and developmental defects. Ludwig et al. (1995) provided evidence of a causal role of planar chlorinated hydrocarbons, particularly some PCB and dioxin compounds. They also emphasized that contaminants might have potent toxic impacts on individuals, without necessarily being disastrous for populations (Ludwig et al. 1995).

Mercury levels in eggs can also have effects on the predators that eat them, and levels of 200–400 ppb (dry weight in diet) are known to affect sensitive birds (Eisler 1987a). At the other end of the spectrum are some marine seabirds that tolerate relatively high concentrations of mercury (Thompson and Furness 1989a,b; Thompson et al. 1991). Mammals are sensitive to mercury at levels as low as 3300 ppb dry weight (Eisler 1987a; WHO 1990, 1991). Average levels of mercury in eggs of Common Terns in this study ranged as high as 4000 ppb in the early years of the study, suggesting they might have posed a risk to sensitive predators. Recent levels, however, are much lower, in the range of 1000–2000 ppb.

Recently, considerable attention has been devoted to the potential for selenium to moderate the effects of mercury toxicity (see Chapter 9), either by sodium selenite precipitating mercury selenide (Sell 1977; Sell and Magat 1979) or by the binding of methylmercury to selenium compounds such as selenoprotein P (Ralston and Raymond 2010), reducing toxicity while perhaps causing selenium deficiency. However, recent studies with breeding Mallards (*Anas platyrynchus*) exposed to methylmercury and selenium in combination resulted in more deformities than either compound by itself (Heinz and Hoffman 1998; Heinz et al. 2011). The ameliorating effect of selenium and mercury on toxicity in wild birds requires considerably more research.

Selenium toxicity in birds was found in Kesterson Reservoir in California, where high levels of selenium resulting from agricultural runoff were associated with hepatic lesions, liver changes, congenital malformations, lowered reproductive success, and adult mortality (Ohlendorf et al. 1986a,b, 1990). Selenium concentrations of 6000 ppb (dry weight) in eggs was associated with reproductive impairment in Black-necked Stilts (*Himantopus mexicanus*) (Ohlendorf and Heinz 2011), including decreased hatchability and deformed embryos. Levels above 12,000 ppb (dry weight) in eggs cause deformities in several species (Spallholz and Hoffman 2002). Selenium toxicity in bird eggs is 6000 ppb (Chapter 9), so terns are generally not affected. If the eggs themselves are not affected, selenium concentrations of 1000 ppb (Lemly 1993b) or more than 3000 ppb (Eisler 2000; Ohlendorf and Heinz 2011) are toxic to other wildlife that consume them.

The effects of chromium and manganese are less clear because there are few laboratory or field experiments with these metals. We found that neurobehavioral deficits in gulls injected with chromium or manganese were similar to lead (Burger and Gochfeld 1995a–c, 2000a). Manganese is an essential micronutrient that serves as an important cofactor in various metabolic reactions (Drown et al. 1986), although high exposures can lead to adverse neurological and respiratory health effects in humans and decreases in motor activity, learning disabilities, and decreased fertility and mortality in other mammals (Laskey et al. 1982; Ingersoll et al. 1995; Senturk and Oner 1996).

The question of relying on mean values assessing exposure and risk bears examination. In most reports, and in this chapter, means are generally presented. For risk assessment from contaminants, variation among individuals is critical. That is, it is important to know what percent of a population falls above a toxic threshold, whether acute or chronic, or lethal or sublethal. Human risk assessors deal with this aspect more frequently because for humans, every individual is important. This topic is discussed more fully in Chapter 14 for all species at once.

SUMMARY AND CONCLUSIONS

Common Terns and Black Skimmers have been used as bioindicators to assess exposure to metals because they eat fish, and represent relatively high trophic levels of coastal ecosystems. They forage almost exclusively on small fish mostly captured within a few kilometers of their colonies. While levels in eggs represent both local exposure and exposure on the wintering grounds, levels in fledgling feathers represents local exposure. We monitored the levels of metals in Common Tern eggs from 1971 to the present, in Common Tern feathers from 1978 to the present, and in the other species for varying periods.

In general, levels of lead and cadmium declined steadily with an upward blip corresponding to bridge deleading activities. Mercury declined slightly with closure of mercury-polluting industries offset by increased emissions from coal-fired power plants. Chromium showed slight declines. Manganese and selenium generally showed no pattern. The decline in lead and cadmium reflects declines in the use of leaded gasoline and cadmium in batteries. Metal levels were sometimes higher at Bird Island and Falkner Island than on the bays farther south (New York–New Jersey Harbor, Barnegat Bay), and the use of regrown feathers demonstrated that mercury, at least, was obtained in Buzzards Bay, and not just on the wintering grounds. In the 1970s, tern eggs had sufficiently high levels of some metals to pose a risk to predators, but levels of most metals are currently below toxic thresholds for the birds themselves and for predators that eat eggs. This is examined more thoroughly in Chapter 14 for the range of indicator species.

APPENDIX

Table 11A.1   Metal Levels (ppb, Dry Weight) in Common Tern Eggs from Barnegat Bay

Year	n	Lead Mean ± SE GM	Mercury Mean ± SE GM	Cadmium Mean ± SE GM	Selenium Mean ± SE GM	Chromium Mean ± SE GM	Manganese Mean ± SE GM	Arsenic Mean ± SE GM
1971		8640 ± 853	1944 ± 72.0	1365 ± 74.0				
1982		4338 ± 81.0	1011 ± 82.0	468 ± 20.0				
1987		42.6 ± 13.4	1016 ± 111	6.5 ± 1.3	2725 ± 135	129 ± 51.7	2600 ± 546	274 ± 54.9
1989	8	411 ± 93.1		11.7 ± 3.6				
		152		6.2				
1990	27	803 ± 144	957 ± 101	30.5 ± 11.0	2965 ± 254	57 ± 7.0	2117 ± 150	301 ± 35.3
		363	972	3.9	2754	52	2061	276
1991	38	492 ± 69.6	1115 ± 72.1	26.4 ± 5.9	1798 ± 144	81 ± 27	2760 ± 223	323 ± 36.6
		220	1045	11.5	1582	165	2500	300
1992	46	2224 ± 254	1050 ± 57.6	154 ± 20.4	3991 ± 128	2705 ± 534	4122 ± 146	
		1750	903	135	3532	1772	3595	
1994	10	700 ± 198	2926 ± 250	196 ± 66.2	722 ± 139	906 ± 64.8		
		270	2827	141	623	898		
1995	40	1198 ± 219	1155 ± 78.7	31.2 ± 8.3	2512 ± 71	410 ± 97.4	2765 ± 187	375 ± 60.8
		647	1023	14.2	2475	259	2498	256
1996	26	237 ± 33.0	1544 ± 106	8.1 ± 1.7	2693 ± 87	41 ± 32	3135 ± 218	
		188	1452	5.7	2656	0.6	2966	
1998	13	1597 ± 375	899 ± 83.7	35.4 ± 9.7				
		1245	848	26.1				
1999	26	588 ± 61.0	3063 ± 306	6.2 ± 0.6				
		526	2737	5.6				
2000	35	164 ± 25.4	990 ± 137	4.2 ± 0.6	2046 ± 80	45.3 ± 9.7	2290 ± 139	195 ± 21.0
		119	759	3.2	1984	7.9	2149	112
2001	56	279 ± 67.0	1242 ± 91.1	0.7 ± 0.1	2555 ± 54.9	48.4 ± 5.8	4970 ± 219	329 ± 18.9
		127	1068	0.1	2517	18.7	4697	292
2002	85	200 ± 29	1145 ± 58.8	10.7 ± 1.3	2476 ± 37	65.1 ± 12.6	1782 ± 51.1	184 ± 6.7
		47	1037	5.3	2446	43.4	1722	172

(Continued)

Table 11A.1 (Continued) Metal Levels (ppb, Dry Weight) in Common Tern Eggs from Barnegat Bay

Year	n	Lead Mean ± SE GM	Mercury Mean ± SE GM	Cadmium Mean ± SE GM	Selenium Mean ± SE GM	Chromium Mean ± SE GM	Manganese Mean ± SE GM	Arsenic Mean ± SE GM
2003	75	104 ± 17	1765 ± 111	24.0 ± 4.3	1860 ± 103	126 ± 17.9	2354 ± 165	347 ± 20.5
		43	1460	9.5	1584	94.8	2004	290
2004	119	210 ± 8.3	1842 ± 82.7	7.4 ± 1.5	3525 ± 58	283 ± 25.8	2038 ± 92	200 ± 10
		190	1658	1.5	3449	204.6	1682	163
2005	37	350 ± 30.1	1644 ± 117	5.1 ± 0.8	2472 ± 99	67 ± 12.6	3681 ± 281	294 ± 20
		330	1507	3.0	2345	24.3	3150	258
2006	82	21.0 ± 5.4	976 ± 46.0	8.5 ± 1.3	2616 ± 47	178 ± 31.8	1822 ± 70	99 ± 11
		12.8	814	0.6	2581	79.8	1632	23
2007	44	62.7 ± 49.8	1735 ± 166	3.6 ± 1.1	2507 ± 116	126 ± 10.5	2428 ± 111	248 ± 14
		9.2	1493	0.1	2381	93.9	2317	228
2008	41	24.6 ± 5.2	1236 ± 163	7.0 ± 3.3	2080 ± 213	97.9 ± 16.3	2392 ± 137	177 ± 16
		17.5	938	0.1	1863	8.1	2238	140
2009	50	6.7 ± 1.2	832 ± 41.9	0.6 ± 0.2	1814 ± 79	14.8 ± 7.3	1834 ± 87	79 ± 9
		8.2	788	0.0	1717	0.5	1743	17
2010	48	9.2 ± 2.4	1101 ± 54.0	11.6 ± 2.6	2451 ± 64	117 ± 12.1	1852 ± 85	204 ± 17
		7.1	1039	0.4	2414	55.8	1764	97
2011	36	8.1 ± 2.0	726 ± 43.8	13.8 ± 5.7	3043 ± 112	16.8 ± 2.1	2506 ± 147	151 ± 21
		8.3	676	0.4	2902	9.5	2371	63
2012	15	17.0 ± 2.2	718 ± 74.7	4.3 ± 1.6			3813 ± 292	
		15.2	670	0.2			3669	
2013	23	1.0 ± 0.5	641 ± 46.9	3.2 ± 1.1	2341 ± 91.6	25.4 ± 9.2	1297 ± 74	273 ± 28
		5.7	607	0.2	2293	7.8	1247	236
2014	12	9.4 ± 2.3	944 ± 107	6.9 ± 2.3	2350 ± 113	34.6 ± 6.9	3508 ± 413	231 ± 40
		8.0	875	0.9	2319	26.5	3254	109
2015	8	9.4 ± 3.0	685 ± 61	13.1 ± 6.4				
		4.5	671	2.6				
χ^2 comparison among years		620 (<0.0001)	319 (<0.0001)	404 (<0.0001)	429 (<0.0001)	489 (<0.0001)	333 (<0.0001)	256 (<0.0001)

Note: Data are presented as means ± SE and geometric means are given below. Comparisons are made with Kruskal–Wallis one-way ANOVA, yielding a χ^2 statistic. All values are expressed in ng/g (ppb dry weight). NS, not significant.

Table 11A.2   Metal Levels (ppb, Dry Weight) in Common Tern Fledgling Feathers from Barnegat Bay

Year	n Pb	n Others	Lead Mean ± SE	Mercury Mean ± SE	Cadmium Mean ± SE	Selenium Mean ± SE	Chromium Mean ± SE	Manganese Mean ± SE	Arsenic Mean ± SE
1978	20	0	5653 ± 587						
			5037						
1981	18	2	3276 ± 283	2654 ± 761	172.0 ± 158.0	1035 ± 165	585 ± 415	7825 ± 7175	137 ± 64
			3075	2542	68.0	1022	412	3122	121
1984	25	0	1293 ± 139						
			1143						
1985	28	12	1303 ± 233	2753 ± 325	21.4 ± 5.8	1303 ± 234	198 ± 14	572 ± 95	170 ± 49
			1006	1137	5.2	1088	193	501	26
1987	11	0	993 ± 103						
			948						
1988	15	0	1045 ± 117						
			972						
1989	30	12	669 ± 175	1541 ± 186	69.4 ± 11.6	280 ± 92	237 ± 57	1474 ± 185	233 ± 47
			269	1405	15.0	49	193	1330	109
1994	30	30	629 ± 87	3775 ± 250	108.9 ± 27.8	556 ± 86	387 ± 56	4426 ± 574	145 ± 73
			393	3504	33.2	360	241	2960	16
1995	17	17	1000 ± 195	1821 ± 172	636.5 ± 220.7	1095 ± 90	255 ± 28	1112 ± 140	
			795	1689	256.4	1018	232	950	
1998	24	24	239 ± 29	2275 ± 185	34.1 ± 7.1				
			107	2081	9.7				
2000	17	15	1722 ± 705	1660 ± 170	57.8 ± 14.2	988 ± 62	802 ± 163	3936 ± 1466	217 ± 53
			892	1472	41.6	956	651	2436	33
2001	25	25	502 ± 156	2452 ± 124	21.0 ± 2.8	1072 ± 133	1391 ± 228	2366 ± 486	236 ± 60
			7	2376	13.1	859	1123	1792	7

(Continued)

Table 11A.2 (Continued) Metal Levels (ppb, Dry Weight) in Common Tern Fledgling Feathers from Barnegat Bay

Year	n Pb	n Others	Lead Mean ± SE	Mercury Mean ± SE	Cadmium Mean ± SE	Selenium Mean ± SE	Chromium Mean ± SE	Manganese Mean ± SE	Arsenic Mean ± SE
2003	117	112	2369 ± 175	2523 ± 118	202.0 ± 43.0	3880 ± 294	2311 ± 228	1670 ± 125	295 ± 34
			1976	2187	54.4	2961	1565	1332	66
2004	26	26	520 ± 54	1959 ± 167	11.9 ± 2.7	727 ± 67	1215 ± 134	1353 ± 177	381 ± 42
			428	1616	2.4	649	1059	809	302
2005	12	12	312 ± 73	1986 ± 226	36.5 ± 9.9	840 ± 93	374 ± 103	2008 ± 250	180 ± 65
			235	1828	14.8	784	241	1792	10
2006	14	14	245 ± 88	1571 ± 154	63.3 ± 18.9	789 ± 96	219 ± 87	782 ± 165	149 ± 17
			90	1466	39.5	712	51	622	134
2007	12	12	150 ± 60	2040 ± 200	29.4 ± 5.5	737 ± 98	227 ± 55	587 ± 102	165 ± 42
			22	1923	23.3	651	139	496	114
2008	12	12	45 ± 18	2981 ± 377	6.5 ± 1.4	728 ± 118	81 ± 20	418 ± 86	130 ± 28
			3	2679	2.3	612	38	358	66
2009	12	12	56 ± 25	1691 ± 463	8.9 ± 3.6	1279 ± 295	719 ± 225	1813 ± 664	58 ± 25
			3	1242	1.2	368	517	1097	2
2010	12	12	156 ± 64	961 ± 244	12.6 ± 7.4	1299 ± 228	338 ± 62	679 ± 91	140 ± 38
			189	692	7.4	680	287	619	22
2011	12	12	487 ± 108	2219 ± 253	86.5 ± 14.8	1013 ± 104	793 ± 324	1048 ± 332	239 ± 52
			355	2048	73.0	959	468	790	110
2013	17	17	130 ± 27	1180 ± 211	15.6 ± 4.9	696 ± 205	309 ± 40	1628 ± 350	91 ± 24
			48	1195	2.6	316	287	1011	13
2014	7	7	390 ± 105	3105 ± 397	31.9 ± 24.8	764 ± 269	1859 ± 1525	1149 ± 698	256 ± 53
			305	2945	4.2	522	500	561	195
χ^2 comparison among years			351 (<0.0001)	98.4 (<0.0001)	137 (<0.0001)	208 (<0.0001)	234 (<0.0001)	120 (<0.0001)	47.7 (<0.0001)

Note: Data are presented as means ± SE and geometric means are given below. Comparisons are made with Kruskal–Wallis one-way ANOVA, yielding a χ^2 statistic. All values are expressed in ng/g (ppb dry weight). NS, not significant.

Table 11A.3   Metal Levels (ppb, Dry Weight) in Forster's Tern Eggs from Barnegat Bay

Year	n	Lead Mean ± SE GM	Mercury Mean ± SE GM	Cadmium Mean ± SE GM	Selenium Mean ± SE GM	Chromium Mean ± SE GM	Manganese Mean ± SE GM	Arsenic Mean ± SE GM
1990	11	1261 ± 171	2564 ± 273	16.0 ± 3.6	2182 ± 61.5	151 ± 90.8	1330 ± 83.0	334 ± 45.5
		1099	2446	11.1	2173	63.7	1305	279
1992	13	781 ± 141	2114 ± 280	57.3 ± 8.0	3045 ± 192	3206 ± 738	4201 ± 327	
		623	1898	49.9	2968	2288	4065	
1996	6	1100 ± 496		5.7 ± 0.7	1827 ± 68.8	56.7 ± 35.5	1814 ± 172	
		611		5.4	1820	11.4	1772	
1998	4	1389 ± 252		88.0 ± 25.4		2.1 ± 0.9	17.2 ± 5.0	
		1319		73.5		1.5	14.8	
2000	15	56.3 ± 7.2	1787 ± 258	1.9 ± 0.3	1688 ± 85.7	28.4 ± 5.8	1702 ± 143	190 ± 18.3
		49.8	1504	1.6	1653	9.5	1632	176
2001	15	174 ± 86.3	1587 ± 149	0.5 ± 0.2	2331 ± 86.8	37.6 ± 8.3	3428 ± 156	279 ± 41.5
		70.4	1486	0.1	2310	18.4	3381	153
2003	17	44.7 ± 13.7	3578 ± 416	19.5 ± 5.1	2142 ± 119	102 ± 11.5	1701 ± 230	210 ± 36.6
		21.7	3282	3.9	2091	89.4	1484	162
2004	24	419 ± 130	4061 ± 324	6.1 ± 1.1	2840 ± 154	347 ± 63.6	1440 ± 293	125 ± 16.8
		285	3739	1.9	2730	256	988	81.8

(Continued)

Table 11A.3 (Continued)   Metal Levels (ppb, Dry Weight) in Forster's Tern Eggs from Barnegat Bay

Year	n	Lead Mean ± SE GM	Mercury Mean ± SE GM	Cadmium Mean ± SE GM	Selenium Mean ± SE GM	Chromium Mean ± SE GM	Manganese Mean ± SE GM	Arsenic Mean ± SE GM
2005	8	326 ± 28.6 316	3037 ± 718 501	3.9 ± 1.6 1.4	2278 ± 328 1653	167 ± 21.3 157	3030 ± 501 2892	428 ± 42.1 413
2006	34	12.0 ± 1.5 9.9	2482 ± 207 1382	2.5 ± 0.6 0.2	1982 ± 91.2 1906	183 ± 57.6 67.1	1673 ± 113 1554	132 ± 16.9 47.2
2007	24	15.9 ± 2.7 11.4	2582 ± 316 2067	3.4 ± 1.3 0.1	2197 ± 177 1968	127 ± 9.2 121	1514 ± 147 1007	211 ± 21.2 181
2009	24	8.1 ± 0.9 6.5	2505 ± 124 2431	0.9 ± 0.4 0.0	2023 ± 164 1851	21.9 ± 7.3 3.0	1530 ± 60.1 1505	165 ± 19.8 107
2010	10	8.5 ± 1.2 7.8	2670 ± 119 2645	15.6 ± 8.7 0.5	2580 ± 98.7 2562	233 ± 78.2 155	1680 ± 132 1634	263 ± 43.9 230
2011	12	6.8 ± 1.6 3.9	2094 ± 238 1952	4.6 ± 1.1 1.0	2142 ± 100 2117	9.8 ± 4.2 2.9	1508 ± 86 1486	186 ± 33.8 92.0
2012	11	9.0 ± 1.0 8.6	2209 ± 123 2175	7.8 ± 1.3 4.3	1727 ± 48.8 1720	16.5 ± 4.7 11.8	1527 ± 118 1489	284 ± 61.7 207
χ^2 comparison among years		179 (<0.0001)	58.9 (<0.0001)	108 (<0.0001)	77.1 (<0.0001)	150 (<0.0001)	91.3 (<0.0001)	50.6 (<0.0001)

Note: Data are presented as means ± SE and geometric means are given below. Comparisons are made with Kruskal–Wallis one-way ANOVA, yielding a χ^2 statistic. All values are expressed in ng/g (ppb dry weight). NS, not significant.

Table 11A.4   Metal Levels (ppb, Dry Weight) in Forster's Tern Fledgling Feathers from Barnegat Bay

Year	n	Lead Mean ± SE GM	Mercury Mean ± SE GM	Cadmium Mean ± SE GM	Selenium Mean ± SE GM	Chromium Mean ± SE GM	Manganese Mean ± SE GM	Arsenic Mean ± SE GM
1989	17	1528 ± 791.32 439		116 ± 29.757 74.1661				
1990	7	1446 ± 378.34 1117.68	3331.4 ± 328.71 3229.46	38 ± 10.001 12.2303	992.86 ± 116.78 932.33	122.4 ± 31.731 102.94	475.714 ± 82.833 427.48	133.1 ± 42.504 87.641
1995	2	945 ± 255 910	5923 ± 1023 5834	4.9 ± 0.3 4.9	2000 ± 500 1936	135 ± 35 130	470 ± 80 463	88 ± 32 82.0
1998	30	376.3 ± 62.435 285.33	4215 ± 308.84 3929.21	37.4 ± 5.0027 32.0737	1109.4 ± 77.838 1018.01	665.6 ± 38.993 637.56	1092.34 ± 86.282 999.95	174.4 ± 23.674 51.812
2001	1	2209 2209	3150 3150	60 60	158 158	1821 1821	11,642 11,642	478 478
2003	19	4241 ± 557.11 3648.69	2925.5 ± 292.22 2673.42	38.5 ± 16.832 10.6461	1644.8 ± 300.18 1408.69	1620 ± 199.15 1489.01	832.444 ± 134.67 705.1	193 ± 51.304 28.217
2005	11	851.3 ± 120.72 783.57	2945.5 ± 302.66 2797.7	15 ± 3.4763 6.0217	770.91 ± 98.336 711.42	334.1 ± 80.263 259.61	598.182 ± 97.326 529.19	185.4 ± 42.21 135.788
χ^2 comparison among years		54.7 (<0.0001)	29.6 (<0.0001)	47.1 (<0.0001)	24.2 (0.0002)	66.5 (<0.0001)	29.4 (<0.0001)	5.3 (0.3851)

Note: Data are presented as means ± SE and geometric means are given below. Comparisons are made with Kruskal–Wallis one-way ANOVA, yielding a χ^2 statistic. All values are expressed in ng/g (ppb dry weight). In 2001, we had feather from only one bird. NS, not significant.

CHAPTER 12

Heavy Metal Levels in Gulls

INTRODUCTION

Gulls have often been used as bioindicators of environmental conditions and contaminant exposure because they are colonial, abundant, and have diverse foods and feeding methods. Some species have a wide distribution, allowing broad geographic comparisons. Many also exhibit colony site stability; samples can be collected over many years, providing temporal trends in reproductive success, colony numbers, and contaminant levels. As they are abundant and usually lay three eggs (but often raise fewer than three young), collecting an egg from a clutch or a hatchling from a brood does not greatly affect reproductive success.

In this chapter, we examine levels of metals in eggs and feathers of gulls. We first examine metal levels by species nesting in Barnegat Bay, present metal levels from the same species from other Northeast bays, and then compare metal levels across regions and species. Because the foods of gulls are much more variable than those of terns or herons (Burger and Gochfeld 1996c; Gochfeld and Burger 1996), it is more difficult to interpret metal levels. For example, in some parts of their range Herring Gulls eat a high proportion of garbage or offal; in other areas, they specialize on fish and crabs, and in other places a mixture of available prey (Pierotti and Good 1994). Moreover, gulls are very opportunistic, switching quickly from one type of prey to another that becomes superabundant and available.

Like terns, gull chicks are covered with down at hatching, but rely on their parents for food for several months, even after becoming capable of flight. Feathers from prefledging chicks not yet able to fly or from fledglings able to fly short distances are particularly useful because they reflect contaminants acquired locally from food gathered by their parents (Burger 1993). Herring Gull parents and fledglings continue to return to their nest sites for feeding sessions well after fledging (Burger 1980c, 1981d).

Collecting feathers from chicks, however, is not always possible if colonies fail (as sometimes happens in salt marsh colonies; Burger and Gochfeld 1991b). Herring Gulls are partially migratory, many moving to southern states in the winter, but few reach tropical areas. The gulls return to their colonies in Barnegat Bay by March to establish and defend their territories. The females are on site for weeks before egg-laying begins. Therefore, metal levels in eggs largely represent recent and local female exposure. Egg levels also indicate natal exposure that the hatchlings obtain from their mother. In the graphs that follow, the correlation between metal level and year was performed with the nonparametric Kendall tau based on all of the individuals analyzed. The curved lines represent a polynomial best fit performed with Deltagraph.

GREAT BLACK-BACKED GULL

Great Black-backed Gulls was a northern species largely limited to breeding in Canada and northern Eurasia. The species extended its range southward in the mid-1900s, peaking at about 35,000 pairs in North America by 1995 (Anderson et al. 2013). In many places, they began to

(a)

(b)

Figure 12.1 Great Black-backed Gulls nest in colonies with low bushes in the New York–New Jersey Harbor Estuary, Barnegat Bay, and Chesapeake Bay (a). They usually nest with Herring Gulls, and family groups are often close to the water's edge so chicks have an easy escape (b).

decline because of the closing of landfills and loss of fish due to overfishing (Cotter et al. 2013). The first Great Black-backed Gulls bred in Barnegat Bay in the early 1970s (Burger 1978a). In many Northeast bays and estuaries, they nest on dry rocky islands or sandy beaches. In New Jersey, they nest on the highest places in salt marshes, and can outcompete Herring Gulls for these secure sites (Burger and Gochfeld 1983a; Figure 12.1). They are voracious predators, and will kill apparently healthy ducks or young gulls on the water (Burger and Gochfeld, unpublished data in New Jersey; Burger and Gochfeld 1984a). They also eat a variety of fish, invertebrates, garbage, and offal, and some individuals dine on Horseshoe Crab eggs (Pierotti and Good 1994; Cotter et al. 2013).

Barnegat Bay Egg Patterns

The eggs of Great Black-backed Gulls were not routinely collected for analysis in our Barnegat Bay studies. In part, this reflects our assumption that the levels would track those of Herring Gulls as they nest and forage in the same habitats. However, 24 eggs collected in 1999 and 2000 had the following mean levels (± SE, dry weight): lead, 323 ± 78 ppb; mercury, 439 ± 110 ppb; cadmium, 8 ± 3 ppb; selenium, 1560 ± 143 ppb; chromium, 26 ± 2 ppb; manganese, 1690 ± 112 ppb; arsenic, 125 ± 72 ppb (after Burger 2002c). Eggs collected in 2015 had the following mean levels: lead, 18.5 ± 10.2 ppb; mercury, 918 ± 248 ppb; cadmium, 10.1 ± 5.9 ppb. Thus, lead and cadmium apparently declined, whereas mercury increased.

Barnegat Bay Feather Patterns

We analyzed feathers from Great Black-backed Gulls chicks from 1995 to 2014. In general, levels of lead, cadmium, and manganese declined strongly with year, whereas mercury and chromium showed a weaker decline (Appendix Table 12A.1). Several metals showed unusually high values in 2003. This was seen in terns as well. The analyses were rerun with other more recent batches to verify the original results (which were correct). There was no apparent temporal pattern for selenium and arsenic, and the levels varied markedly among years (Figures 12.2 and 12.3).

Other Northeast Bays

There are only limited data on Great Black-backed Gulls from the other bays and estuaries—mainly from the New York–New Jersey Harbor estuary (Table 12.1). We analyzed eggs from the New York–New Jersey Harbor in 2012 and 2013 to assess the impact of Hurricane *Irene* (2011) and Superstorm *Sandy* (2012) (Burger and Elbin 2015b; Table 12.1). Superstorm *Sandy* hit the Northeast coast in late October 2012, after we had collected gull eggs. We collected eggs again in 2013. Levels of mercury and selenium were significantly lower in 2013 than in 2012, and lead was significantly higher (Burger and Elbin 2015b). The potential effects of *Sandy* are discussed in the following sections.

HERRING GULL

The Herring Gull has been used as a bioindicator of contamination, particularly in Europe (Thompson et al. 1993; Garthe 2013; Grajewska et al. 2015), the Great Lakes and Canada (Weseloh et al. 2011; Burgess et al. 2013) and in the Northeast (Burger 1997c; Grasman et al. 2013; Burger and Elbin 2015a,b). In 1940–1980, Herring Gull populations increased and expanded southward, both in North America and Europe. With open-faced garbage dumps, Herring Gulls increased throughout their range because young of the year had a steady source of food, reducing postfledging losses due to learning foraging sites and techniques (Burger 1981c; Pierotti and Good 1994). In the 1980s, the closing or covering of garbage dumps removed this abundant food source, contributing to

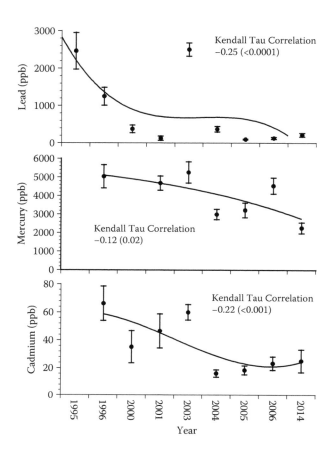

Figure 12.2　Levels of lead, mercury, and cadmium (ppb, dry weight) (mean ± SE) in feathers from fledgling Great Black-backed Gulls from Barnegat Bay. Based on a total of 245 samples.

subsequent population declines throughout much of their range in both North America and Europe (Anderson et al. 2013). However, Herring Gulls eat a wide variety of organisms gleaned from rocky or sandy shores, as well as offal from fishing boats and processing plants. Some specialize on the eggs and chicks of other species, but do not generally kill fully grown birds. They forage along shores and in impoundments, follow fishing boats, scavenge along beaches, and forage at land-fills; garbage made up a significant portion of their diet (particularly for young of the year; Burger 1981c,d; Pierotti and Good 1994).

The North American Herring Gull has been split taxonomically from the European Herring Gull and is now named *Larus smithsonianus*, although we will consider them together for geographical comparisons. Data on contaminants in eggs and feathers of Herring Gulls provides long-term data sets that allow temporal and spatial comparisons among sites. The small gulls, such as Laughing Gull along the Atlantic coast and Franklin's Gull in interior North America, have more limited distributions and do not occur in Europe or Asia, and there are limited long-term data.

Herring Gulls were relatively newcomers as breeders in Barnegat Bay (Burger 1977a), although they quickly expanded at the expense of Laughing Gulls (Burger and Shisler 1978c; Burger 1979a, 1983; see Chapter 5). They claimed the highest places in *Spartina* marshes where there were bushes (Figure 12.4), until Great Black-backed Gulls moved in to displace them.

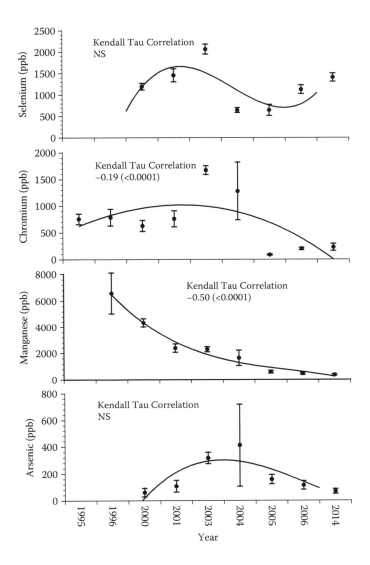

Figure 12.3 Levels of selenium, chromium, manganese, and arsenic (ppb, dry weight) (mean ± SE) in feathers of fledgling Great Black-backed Gulls from Barnegat Bay. Based on 245 samples.

Table 12.1 Metals Levels (ppb, Dry Weight) (Mean ± SE) in Eggs of Great Black-Backed Gulls from the New York–New Jersey Harbor Estuary

Great Black-Backed Gull	2012	2013	χ^2 (P)
Number	11	14	
Lead	62.7 ± 22.5	260 ± 55.6	10.3 (0.001)
Mercury	1050 ± 172	504 ± 66.2	10.1 (0.002)
Cadmium	4.51 ± 1.7	3.16 ± 1.1	0.9 (NS)
Chromium	55.3 ± 16.0	96.3 ± 27.7	1.0 (NS)
Selenium	2160 ± 112	1860 ± 66.8	3.7 (0.05)

Source: Burger, J., Elbin, S., *Ecotoxicology*, 24, 445–452, 2015b (see Figure 12.9).

(a)   (b)

Figure 12.4   Herring Gulls normally eat fish, offal, and a variety of other items, including the eggs of neighboring gulls (a). A gull in hunting mode at the edge of a tern nesting colony (b). Each year, only a few gulls specialize on tern eggs or chicks (b).

Barnegat Bay Egg Patterns

Levels of heavy metals in eggs of Herring Gulls were examined in Barnegat Bay from 1989 to 2015 (Figures 12.5 and 12.6). During this period, the results were similar to those observed in Great Black-backed Gulls. Lead, cadmium, manganese, and chromium declined sharply, and mercury declined slightly. There was even a slight decline in selenium. The data for arsenic were insufficient to detect a trend (Figure 12.6). The declines in mercury and selenium are not very substantial, and both show a great deal of variation. In contrast, lead shows a clear and expected decline through 2006 (Appendix Table 12A.2).

Barnegat Bay Feather Patterns

Levels of metals in feathers of young Herring Gulls from Barnegat Bay for 1989 to 2006 are presented in Figures 12.7 and 12.8. There were significant declines in lead, cadmium, and manganese (except for a 2003 peak), and a weak decline in chromium. Those results are concordant between eggs and feathers. However, there was no decline in mercury, selenium, or arsenic in feathers (Appendix Table 12A.3).

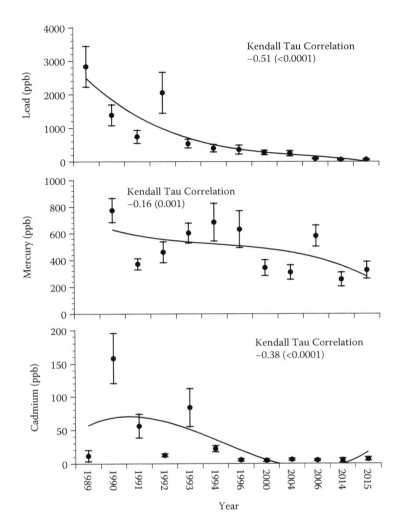

Figure 12.5 Levels of lead, mercury, and cadmium (ppb, dry weight) (mean ± SE) in eggs of Herring Gulls from Barnegat Bay, New Jersey. Based on a total of 238 eggs, about 20 per sample year.

Other Northeast Bays and Elsewhere

Over the years, we also analyzed metal levels in Herring Gull feathers from other Northeast bays and estuaries, particularly the New York–New Jersey Harbor Estuary. From 1980 to 2013, metal levels in eggs of Herring Gulls from the estuary showed significant declines for lead and cadmium, weak declines for selenium and chromium, and no trend for mercury or manganese (Figure 12.9).

Gull and tern chicks are covered with down when they hatch. Any metals in the down derive from the yolk and ultimately from the female. In 1993, we compared metal levels in down (representing exposure from the female during egg development), levels in prefledging chick feathers (representing local exposure from diet), and adults (representing exposure before the

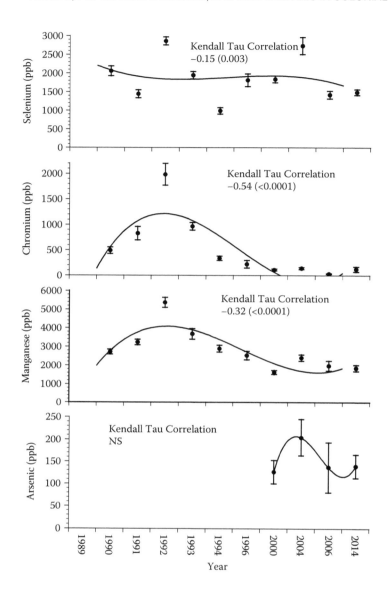

Figure 12.6 Levels of in selenium, chromium, manganese, and arsenic (ppb, dry weight) (mean ± SE) in eggs of Herring Gulls from Barnegat Bay. Based on a total of 209 eggs, about 20 per sample year.

breeding season). Birds were collected at Captree, Long Island (Table 12.2). We found significant differences for all metals. Except for selenium and chromium, adults had the highest levels. Interestingly, down had significantly higher levels of selenium than chick or adult feathers. Thompson et al. (1993) also reported that mercury concentrations were higher in adult feathers than in juvenile feathers.

Temporal trends for metals in feathers of both adult and young Herring Gulls from Captree were examined for a few years in the 1980s and early 1990s (Burger 1995b). We consider this a relatively short temporal data set, and thus not necessarily indicative of an overall trend. However, it allows us to compare levels in adults and young. Against the overall trend, lead appears to have increased

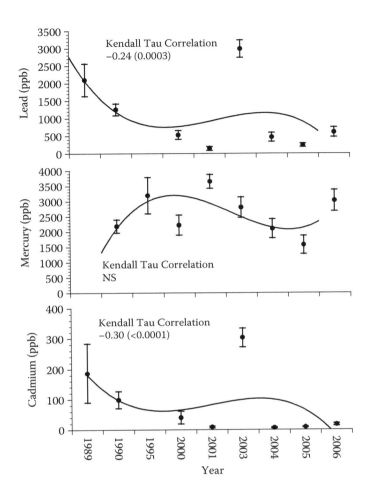

Figure 12.7 Levels of lead, mercury, and cadmium (ppb, dry weight) (mean ± SE) in feathers of fledgling Herring Gulls from Barnegat Bay. Based on a total of 132 samples.

during this period for both adults and young, corresponding to bridge deleading activities in New York State. Adults had significantly higher lead levels than chicks (Figure 12.10). There was also a significant difference between the levels in feathers of adults and young for mercury, selenium, and manganese in all years (Figures 12.10 and 12.11). Cadmium, however, showed no age-related difference except for 1993, when chicks had much higher levels than adults. In the same year, chicks also had higher selenium and chromium, but not manganese levels. Chick feathers represent local sources, and this suggests that in 1993 parents were bringing back food with much higher cadmium content than usual.

We also compared levels of metals in feathers of Herring Gull chicks from Eastern Long Island to Virginia collected in 1990 (Burger 1997c). This kind of comparison is useful as feathers were collected in the same year (reducing the potential for some temporal differences, especially in exposure from atmospheric deposition). There were significant differences in metal levels among locations (Kruskal–Wallis one-way analysis of variance, χ^2 tests), but the differences were not concordant (Figures 12.12 and 12.13). The concentration of lead was highest at Prall's Island in the highly polluted Arthur Kill. We would have expected all metals to be highest there.

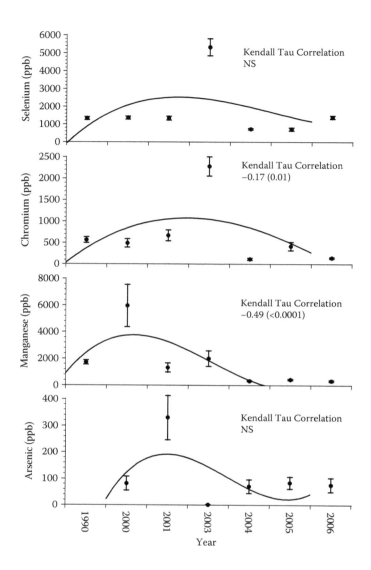

Figure 12.8   Levels of selenium, chromium, manganese, and arsenic (ppb, dry weight) (mean ± SE) in feathers of fledgling Herring Gulls from Barnegat Bay. Based on a total of 132 samples.

However, mercury was low at Prall's Island and high at Shinnecock, Huckleberry, and Harvey
Sedge. Cadmium was also low at Prall's and highest at Huckleberry Island. Chromium was highest at Huckleberry. Manganese was highest at Captree. Huckleberry Island is probably the southern outpost of islands with rocky intertidal (as opposed to sandy intertidal zones). It lies 1 km
east of David's Island, a former highly contaminated military base, and is shown on some maps
as Whortleberry Island. In the early 1900s, Huckleberry Island was among the targets of artillery
practice from batteries on David's Island.

Levels of metals, particularly mercury, have been examined in Herring Gull eggs and feathers
from places other than the Northeastern United States (Table 12.3). Although these data are interesting, comparisons are difficult because data were not collected in the same year. However, the data

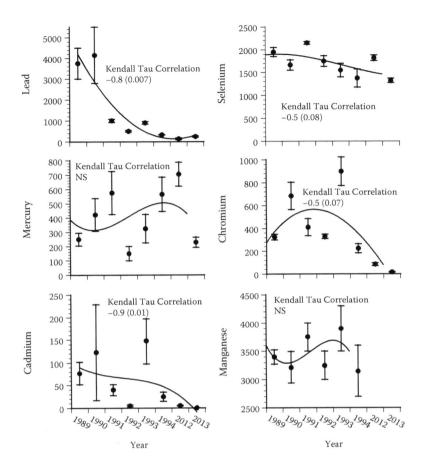

Figure 12.9 Temporal variations in levels of lead, mercury, cadmium, selenium, chromium, and manganese (ppb, dry weight) (mean ± SE) for Herring Gulls eggs from the New York–New Jersey Harbor Estuary. Based on 267 samples. (After Burger, J., Gochfeld, M., *Arch. Environ. Contam. Toxicol.*, 29, 192–197, 1995d; Gochfeld, M., *Arch. Environ. Contam. Toxicol.*, 33, 63–70, 1997; Burger, J., Elbin, S., *J. Toxicol. Environ. Health*, 78, 78–91, 2015a; Burger, J., Elbin, S., *Ecotoxicology*, 24, 445–452, 2015b.)

Table 12.2 Metal Levels in Down (ppb, Dry Weight), Chick Breast Feathers, and Adult Breast Feathers from Herring Gulls Nesting at Captree, Long Island, in 1993

	Down (*n* = 12)	Chicks (*n* = 20)	Adult (*n* = 45)	χ^2 (*P*)
Mercury	1105 ± 200	1800 ± 105	3810 ± 266	24.1 (0.0001)
Lead	1580 ± 209	1950 ± 193	4100 ± 256	13.3 (0.001)
Cadmium	1200 ± 328	1150 ± 275	1370 ± 89	29.7 (0.001)
Chromium	786 ± 101	2820 ± 606	1280 ± 209	23.5 (0.0001)
Manganese	2410 ± 217	3690 ± 318	6780 ± 568	22.7 (0.0001)
Selenium	1720 ± 131	1430 ± 53	906 ± 43	6.5 (0.04)

Source: Burger, J., *Environ. Monit. Assess.*, 38, 37–50, 1995b.

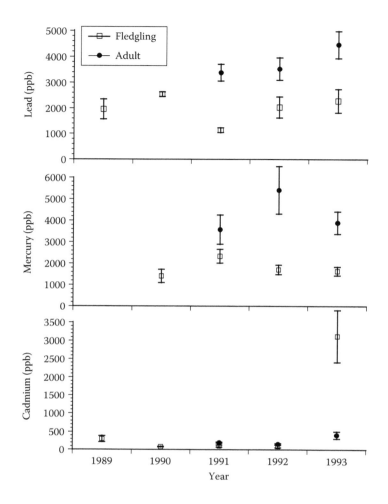

Figure 12.10 Temporal variations in levels of lead, mercury, and cadmium (ppb, dry weight) (mean ± SE) in feathers of adult and fledgling Herring Gulls from Captree, Long Island, in the New York–New Jersey Harbor Estuary. Based on 45 adults and 12 fledglings. (After Burger, J., *Environ. Monit. Assess.*, 38, 37–50, 1995b.)

illustrate a number of points: (1) eggs were used as indicators more often than feathers; (2) declines in one region do not occur everywhere; (3) declines may occur generally in an area, but not in all colonies; (4) mercury levels appear to be higher during the Second World War, than at other times; and (5) significant trends (e.g., declines) may be attributable to differences in diet over the years, rather than exposure from the same food items.

LAUGHING GULL

Laughing Gulls have a disjunct distribution of nesting colonies, and are a common breeder along the New Jersey coast from Stone Harbor (New Jersey) north to Barnegat Bay, Jamaica Bay (Long Island), and then sporadically farther north to Maine (Burger 2015a). They nest on salt marsh islands in the southern part of their Northeast Atlantic coastal range (to Jamaica Bay), and on rocky islands

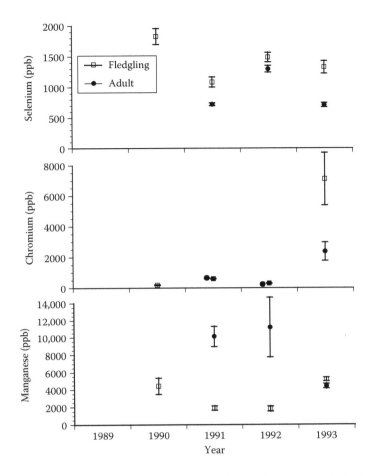

Figure 12.11 Temporal variations in levels of selenium, chromium, and manganese (ppb, dry weight) (mean ± SE) in feathers of adult and fledgling Herring Gulls from Captree, Long Island, in the New York–New Jersey Harbor Estuary. Based on 45 adults and 12 fledglings. (After Burger, J., *Environ. Monit. Assess.*, 38, 37–50, 1995b.)

farther north. They are omnivorous foragers, feeding on fish, insects, invertebrates, Horseshoe Crab eggs (Figure 12.4), and all manner of garbage and offal. They sometimes engage in aerial acrobatics catching swarming insects, or visit suburban parking lots to panhandle for food. They engage in synchronous activities at breeding colonies (Burger 1976), and breeding success is dependent on nest site choices because low elevation nests are vulnerable to high tides (Burger and Lesser 1979; Burger and Shisler 1980a). In Barnegat Bay, the invasion of Herring and Great Black-backed Gulls has forced Laughing Gulls to nest on lower places on salt marsh islands, leading to lowered reproductive success and eventual desertion of some nesting islands (Burger 1979a; see Chapters 4, 5, and 7).

Barnegat Bay Egg Patterns

Levels of lead, cadmium, chromium, and arsenic decreased in eggs of Laughing Gulls nesting in Barnegat Bay from 2003 to 2011. There is a slight downward trend for selenium from 2003 to 2011, and the arsenic data also suggest a decrease from 2003 (Appendix Table 12A.4). As with many of the other species nesting in the bay, mercury did not decline in eggs, although the trend from 1996 onward is down (Figures 12.15 and 12.16).

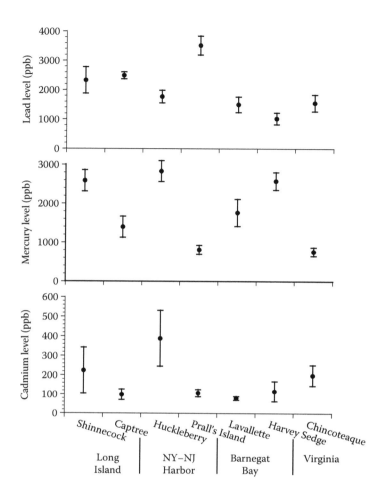

Figure 12.12   Levels of lead, mercury, and cadmium (ppb, dry weight) (mean ± SE) in feathers of Herring Gulls from several northeast nesting colonies. Based on 106 feather samples. (After Burger, J., *Environ. Monit. Assess.*, 48, 285–296, 1997c.)

Barnegat Bay Feather Patterns

Levels of most metals, including mercury, declined in the feathers of young Laughing Gulls, but there was no significant temporal trend for selenium and arsenic (Figures 12.17 and 12.18). The decline in mercury is unusual, given the patterns for other species from the bay (see the following discussion). However, this may relate to too few years of data on feather levels (8 years, and an unusual peak in 2003).

Other Northeast Bays

Generally, data on metal levels in Laughing Gulls are not available for fledglings from nesting colonies in other Northeastern bays and estuaries. However, beginning in 1991 there was a massive culling (i.e., shooting) program at John F. Kennedy (JFK) International Airport to reduce the threat of bird strikes. This was in response to the rapid increase in Laughing Gulls breeding at the end of one of the airport runways (Dolbeer et al. 1993). We took advantage of the mass shootings,

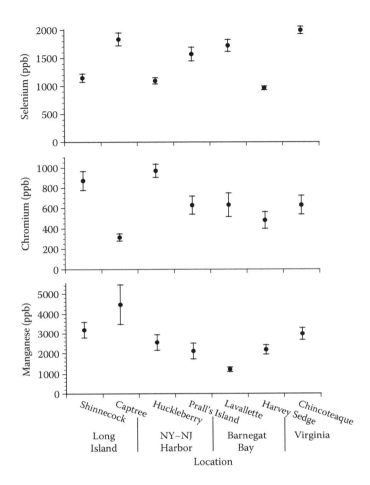

Figure 12.13 Levels of selenium, chromium, and manganese (ppb, dry weight) in feathers of Herring Gulls from several Northeast nesting colonies. Based on 106 feather samples. (After Burger, J., *Environ. Monit. Assess.*, 48, 285–296, 1997c.)

and analyzed metal levels in Laughing Gulls as a function of gender and age. A high percentage of the Laughing Gulls that were shot at JFK Airport had been banded in Barnegat Bay as chicks, and therefore were of known age and location. For the sample analyzed, shooting occurred over a period of months (May–August 1992). The large mortality of New Jersey birds was followed by a drastic decline in New Jersey's breeding population of Laughing Gulls (see Chapter 5). We analyzed feather levels only in birds hatched in New Jersey. Males had significantly higher lead in feathers than did females. Otherwise, there were no significant sex differences in metal levels (Gochfeld et al. 1996).

After the breeding season, many Laughing Gulls move north from New Jersey to Long Island. The 1992 cohort of Barnegat Bay fledglings did not reach Long Island until after the 1992 shooting had ceased. Therefore, our sample had birds from the 1991 cohort (1-year-olds) and earlier. For economy, we chose to analyze every other year age classes: 1-year-olds (now called "subadults"), and adults were 3-, 5-, and 7-year-olds. The patterns as a function of age (1, 3, 5, 7 years old) are complex (Figure 12.19). For most metals, there was not a clear increase with age. For cadmium, chromium, selenium, and manganese, 7-year-old birds had lower levels than most other age groups.

Table 12.3　Levels of Mercury (ppb, Dry Weight) in Eggs and Feathers of Herring Gulls from Places Other Than the Northeastern United States

Location	Year	Mercury	Comments	Source
		Eggs		
Atlantic Canada	1972–2008	Range from low of about 200 ppb to 800 ppb (derived from graphs)	Unadjusted values show a significant decline, but after adjusting for dietary shifts, environmental mercury has remained constant at most Canada sites over the past 36 years.	Burgess et al. 2013
Great Lakes	1972–1992	Overall means ranged from 480 ppb (1985) to 3520 ppb (1988)	Peak mercury levels were in 1982; declines for most colonies.	Koster et al. 1996
Lake Superior	1974–2009	1324 (1974) to 424 (2009)	Significant temporal declines in 10 of 15 colonies, but no significant declines in the past 15 years.	Weseloh et al. 2011[a]
Lake Michigan	1976–2009	1868 (1976) to 388 (2009)	Same as above.	Weseloh et al. 2011[a]
Lake Heron	1976–2009	696 (1974) to 256 (2009)	Same as above.	Weseloh et al. 2011[a]
Great Slave Lake, NWT, Canada	1995	Ranged from 920 to 8168 ppb	Selenium[b] and mercury were correlated in eggs from every lake.	Wayland et al. 2003
Northern Norway	1983–2003	360 ± 160 400 ± 240	No significant trend in mercury levels over time in either study.	Barrett et al. 1985, Helgason et al. 2008
Wadden Sea	1988	1720 ± 72	Mercury decreased with egg laying order.	Becker 1992
Gulf of Gdansk, Poland	2011	560 ± 301	Mercury levels were higher in albumen (mean of 1503 ppb) than yolk (mean of 86 ppb).	Grajewska et al. 2015
		Feathers		
Michigan (State)	1890–1919 1920–1949 1950–1979 1980–2009	8500 ± 2700 14,000 ± 13,200 10,800 ± 4100 11,600 ± 5800	Museum specimens were used. No linear trend was noted for gulls, although levels in 1920–1949 period were higher.	Head et al. 2011
Bay of Fundy	2000s	366–8483	Concentrations did not exceed effects levels.	Otorowski 2006
German North Sea Coast	1880–1990	Peak of 12,000 ppb[c]	High values were during World War II.	Thompson et al. 1993
Baltic Sea	2009–2010	3020 (range, 1250–4340	Contour feathers were higher than primaries.	Szumilo et al. 2013

Note: All values converted to dry weight (for gulls, Wet weight × 4 = Dry weight for eggs; see Table 3.4).

[a] Values were presented for several colonies on each Lake, for much of the temporal sequence. Data are summarized to give an overview.

[b] Selenium ranged from 1360 to 4350 ppb, dry weight.

[c] These high levels were recorded during the 1940s and thought to be due to World War II activities.

(a)　　　　　　　　　　　　　　　　　　　(b)

Figure 12.14　In the spring, Laughing Gulls forage on Horseshoe Crab eggs in Delaware Bay (a), where it is easy to catch them in cannon nets to obtain adult feathers for analysis. Clive Minton holding gulls (b).

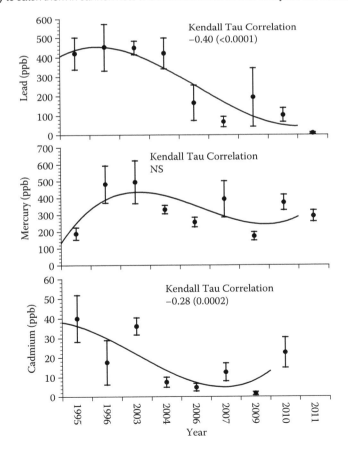

Figure 12.15　Levels of lead, mercury, and cadmium (ppb, dry weight) (mean ± SE) in eggs of Laughing Gulls breeding in Barnegat Bay, New Jersey. Based on a total of 129 eggs, average of 14 per year.

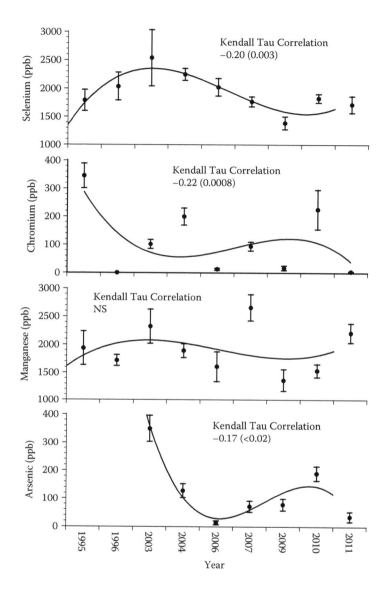

Figure 12.16   Levels of selenium, chromium, manganese, and arsenic (ppb, dry weight) (mean ± SE) in eggs of Laughing Gulls breeding in Barnegat Bay, New Jersey (same basis as Figure 12.15).

Particularly surprising, selenium levels decreased significantly with age. These findings show the value of known-age birds and data sets where age-related differences can be evaluated.

DISCUSSION

Temporal Patterns

One advantage of a long-term study of multiple species is that long-term trends can be compared among closely related species to obtain general patterns. Table 12.4 provides information on metal levels in the three gull species nesting in Barnegat Bay.

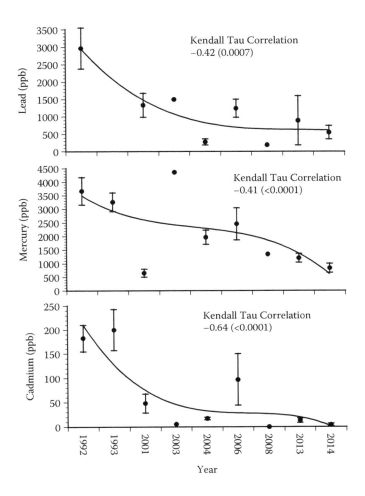

Figure 12.17 Levels of lead, mercury, and cadmium (ppb, dry weight) (mean ± SE) in feathers of fledgling Laughing Gulls from Barnegat Bay. Based on a total of 44 feather samples.

There were some common trends across species and some trends within species (Table 12.4). Lead showed significant declines (or no trend) for all three gulls (both eggs and feathers). For mercury, feathers showed a significant decline, but eggs did not (except for Herring Gull). The latter difference may be attributable to the time Herring Gulls spend on the colony sites before egg-laying, making it possible for eggs to reflect local patterns. Cadmium showed declines (except for Great Black-backed Gulls where we had fewer eggs than the other species). Selenium generally showed no significant patterns, whereas chromium generally showed a significant pattern (except for Great Black-backed eggs). The patterns for manganese and arsenic were very mixed among tissues and species. Overall, however, there were significant declines where there were long temporal data sets.

In any given year and tissue for trends data from Barnegat Bay, the distributions for most metals was relatively tight, yet some years lie outside otherwise smooth trends. The year 2003 was characterized by concordant peaks of several metals in gulls, as well as terns (Chapter 11). We reanalyzed samples from 2003 along with more recent samples, and showed that the levels were correct. Periodic storms and wave action can disturb the surface sediments, exposing lower sediments and resuspending contaminants. This would allow for concordant increases (or decreases) of several metals at the same time. Such events may happen multiple times in a year, but what would

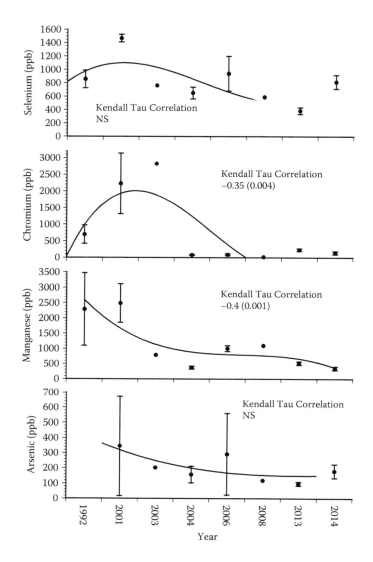

Figure 12.18   Levels of selenium, chromium, manganese, and arsenic (ppb, dry weight) (mean ± SE) in feathers of fledgling Laughing Gulls from Barnegat Bay. Based on a total of 44 feather samples.

have happened between the 2002 season and 2003 season to upset the trends in Barnegat Bay is unclear. According to the Wikipedia website ("New Jersey Hurricanes"), New Jersey experienced 11 hurricanes in the 1980s including the famously devastating Hurricanes *Gloria* (1985) and *Hugo* (1989). There were 13 hurricanes in the 1990s including *Floyd* (1999). This included a "Perfect Storm" (October 31, 1991), which could have accounted for the high lead levels in 1992 in areas where bridge deleading was not occurring. There were 20 hurricanes in the 2000s, but none of note in 2002–2003. In 2011, Hurricane *Irene* produced heavy rains inland washing contaminants, particularly lead, off the land into rivers and down to the bay. By contrast, *Sandy* was characterized by storm surge and sand migration, covering sediment, rather than scouring and resuspension (Burger and Elbin 2015b; Figure 12.9; Table 12.1).

There are also trends in metal levels in gulls from other regions. In the New York–New Jersey Harbor Estuary (islands in the New Jersey harbor and associated rivers), eggs of Herring Gulls showed

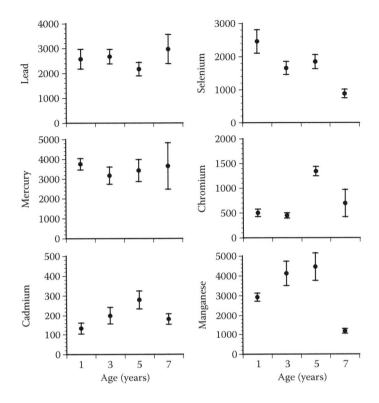

Figure 12.19 Levels of metals (ppb, dry weight) (mean ± SE) in feathers of subadult and adult Laughing Gulls banded as chicks in Barnegat Bay and later killed at JFK International Airport (May–August 1992). We analyzed feathers and tissues of five males and five females for each of the four age classes. (After Gochfeld, M. et al., *Environ. Toxicol. Chem.*, 15, 2275–2283, 1996.)

clear and significant declines in lead, cadmium, selenium, and chromium (Burger and Gochfeld 1995d; Burger and Elbin 2015a,b). However, feathers from Herring Gulls breeding at Jamaica Bay (New York–New Jersey Harbor) showed an increase in lead in adults and chicks, a decrease in selenium, and no clear patterns for the other metals (Burger 1995b). The increase in lead in the feathers may reflect lead paint chip exposure because many of the New York bridges were being repainted (lead paint stripped off) during this period (see section "Common Terns," Chapter 11).

In the Great Lakes, Koster et al. (1996) and Weseloh et al. (2011) reported steady declines in mercury in eggs since the early 1970s, but there were no significant declines in the past 15 years. However, when Burgess et al. (2013) considered dietary shifts that have occurred in Herring Gulls over this period, the declines in mercury in birds did not signify a decline in environmental levels. This implies that Herring Gulls are fishing down the trophic scale—that is, with overfishing and depletion of high trophic levels, birds are foraging at lower and lower trophic levels where we would expect the mercury levels to be lower. Pauley et al. (1998) demonstrated this phenomenon for marine food webs, which has implications for human consumption patterns as well. In Michigan Herring Gull colonies, Head et al. (2011) did not find a linear trend in mercury in feathers, suggesting that long-term data sets are needed to determine mercury exposure, along with dietary information.

In Europe, Thompson et al. (1993) examined changes in mercury contamination in feathers of Herring Gulls and Common Terns from the German North Sea coast, and found that mercury levels increased during the 1940s, probably because of increased military ordnance explosions during the Second World War. Fulminate of mercury (mercury cyanate) was widely used in detonators and

Table 12.4 Temporal Trends in Metal Levels in Feathers and Eggs of Gulls from Barnegat Bay

	Great Black-Backed Gull	Herring Gull	Laughing Gull
Lead	Eggs: Declined Feathers: −0.25, $P < 0.0001$, significant decline	Eggs: −0.51, $P < 0.0001$, significant decline Feathers: −0.24, $P < 0.003$, significant decline	Eggs: −0.40, $P < 0.0001$, significant decline Feathers: −0.42, $P < 0.0007$, significant decline
Mercury	Eggs: Increased Feathers: −0.12, $P < 0.02$, slight significant decline	Eggs: −0.16, $P < 0.001$, significant decline, Feathers: No significant pattern	Eggs: No significant pattern Feathers: −0.41, $P < 0.0001$, significant decline
Cadmium	Eggs: Declined Feathers: −0.22, $P < 0.001$, significant decline	Eggs: −0.38, $P < 0.0001$, significant decline Feathers: −0.30, $P < 0.0001$, significant decline	Eggs: −0.28, $P < 0.0002$, significant decline Feathers: −0.64, $P < 0.0001$, significant decline
Selenium	Eggs: No temporal data Feathers: No significant pattern	Eggs: −0.15, $P < 0.003$, initial increase but significant decline Feathers: No significant pattern	Eggs: −0.20, $P < 0.003$, significant decline Feathers: No significant pattern
Chromium	Eggs: No temporal data Feathers: −0.19, $P < 0.0001$, stable, then significant decline	Eggs: −0.54, $P < 0.0001$, significant decline Feathers: −0.17, $P < 0.01$, slight decline, but mainly stable	Eggs: −0.22, $P < 0.0008$, significant decline Feathers: −0.35, $P < 0.004$, initial increase, but significant decline
Manganese	Eggs: No temporal data Feathers: −0.50, $P < 0.0001$, straightforward significant	Eggs: −0.32, $P < 0.0001$, significant decline Feathers: −0.49, $P < 0.0001$, significant decline	Eggs: No significant pattern Feathers: −0.40, $P < 0.001$, significant decline
Arsenic	Eggs: No temporal data Feathers: Not significant	Eggs: No pattern Feathers: No significant pattern	Eggs: −0.17, $P < 0.02$, significant decline Feathers: No significant pattern

Note: Shown are Kendall Correlation Tau values and associated P values (for probability of no trend).

blasting caps. Subsequently, mercury levels declined in the 1950s and then increased to a peak in the 1970s before beginning to decrease once more (Thompson et al. 1993). The observed patterns fit with known information about discharges from the rivers Elbe and Rhine, suggesting that mercury exposure is from local inputs rather than from atmospheric deposition (Thompson et al. 1993).

Locational Differences within Barnegat Bay

Locational differences can be examined only when metals were analyzed from about the same time in different places; otherwise, spatial differences could be swamped by temporal or dietary patterns. We did not routinely take eggs of Herring Gulls from different parts of the bay, because initially Herring Gulls nested only in the northern section of Barnegat Bay, and we continued to collect them from this part of the bay. Furthermore, the shifts in colony establishment from the northern end of the bay, where they first established colonies (e.g., Lavallette Islands), to the southern end of the bay involved variations in age structure. That is, newly established colonies had a high proportion of 3-, 4-, and 5-year-old Herring Gulls (and later younger classes of Great Black-backed Gulls) than expected. Over the years, fewer and fewer very young Herring Gulls could establish themselves in these colonies, and younger birds moved farther south in the bay to establish new colonies.

Trophic Level Relationships

We expected to find that metals that bioaccumulate (especially mercury) would be higher in large gulls, given the large size variation in gulls. Larger gulls have a wider food niche. Both large species (Herring Gulls, Great Black-backed Gulls) eat large dead fish washed up on shore, and Great

Table 12.5   Metal Levels (ppb, Dry Weight) (Mean ± SE) for Feathers of Gull Fledglings

	Laughing Gull	Herring Gull	Great Black-Backed Gull	χ^2 (P)
Sample size >	14	35	46	
Lead	414 ± 124	441 ± 67.9	252 ± 43.9	8.8 (0.01)
Mercury	2030 ± 237	2240 ± 208	3450 ± 220	17.7 (0.0001)
Cadmium	28.5 ± 9.8	10.9 ± 1.7	18.5 ± 2	13.0 (0.002)
Selenium	693 ± 83.8	954 ± 69.5	773 ± 57.1	6.4 (0.04)
Chromium	82 ± 11.6	229 ± 41.3	737 ± 293	18.4 (0.0001)
Manganese	464 ± 74.2	366 ± 26.8	1110 ± 306	11.1 (0.004)
Arsenic	178 ± 55.3	77.2 ± 14	279 ± 160	1.6 (NS)

Note: Kruskal–Wallis one-way ANOVA. Only values from 2004 to 2006 are given because levels were available for all three species.

Black-backed Gulls are clearly active predators. In contrast, insects make up a significant part of the diet of Laughing Gulls (Dosch 2003; Burger 2015a).

Because feather levels in chicks represent local exposure, comparing feather levels provides the best indication of metal levels. We compare levels only for 2004 to 2006, when we had a sample of chicks from all three species. In general, the larger gulls had higher levels of most metals (Table 12.5). Laughing Gulls, however, had higher levels of cadmium than the large gulls, which may relate to their insect diet (Dorsch 2003). There were no significant species differences in arsenic levels in young gulls.

These data provide some indication of trophic level. It indicates that Great Black-backed Gulls had significantly higher levels of mercury, chromium, manganese, and arsenic, but not of the other metals. Their diets differ (as noted earlier), as does the size of organisms they can eat. Laughing Gulls catch smaller fish than the other species, for example, and they also eat insects. All three species eat garbage, although they do so mainly outside of the breeding season. Stable isotope analysis could clarify the trophic levels, if they were used for a range of species within the same ecosystem.

Grajewska et al. (2015) used stable isotopes to determine trophic level relationships in Herring Gulls in relation to mercury levels in eggs. Mercury received from the albumen and yolk was most effectively removed from developing embryos into down. Thus, even embryos can sequester their mercury load into down. As chicks grow in size, there is dilution of the original exposure, and to some degree, of local exposure because they are growing rapidly and their developing feathers become a sink for mercury.

Comparisons with Other Regions

The eggs of Herring Gulls have been used as bioindicators of contaminant exposure, particularly for mercury and organics in many places in the Northern Hemisphere (Table 12.3). The most useful comparisons are for samples obtained in the same year. We provide regional comparison for eggs collected in Barnegat Bay (Figures 12.5 and 12.6) and in the New York–New Jersey Harbor Estuary (Figure 12.9). Comparable data are available for 1989 to 2013–2014. To facilitate comparisons, we graph lead, mercury, and cadmium from the two bays in Figure 12.20. There are several clear similarities: (1) lead uniformly declined during the period, (2) mercury was more variable in both estuaries, (3) cadmium was concordant; both estuaries had variable and high levels in 1990 and again in 1993 with declines in later years. This suggests a regionwide event—perhaps a storm that resuspended sediments containing cadmium or some other cause. The 1992 lead peak in Barnegat Bay was not apparent in the New York–New Jersey Harbor, suggesting that bridge deleading was not the only contributor. We anticipated that the 2013 season might reveal a signal from Superstorm *Sandy*. However, the data and other observations indicate that the storm surge from *Sandy* in some

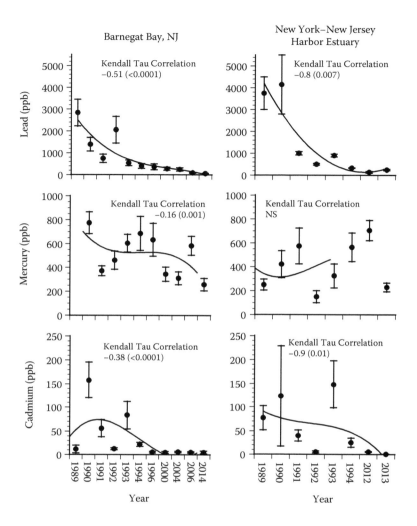

Figure 12.20 Levels of lead, mercury, and cadmium in eggs (ppb, dry weight) (mean ± SE) of Herring Gulls from Barnegat Bay (see Figure 12.5) and from the New York–New Jersey Harbor Estuary (see Figure 12.9) comparing early 1990s to more recent samples. Based on 480 eggs.

bays was more likely to cover contaminated sediment with fresh ocean sand than to scour the sediment, resuspending contaminated strata.

Comparisons with other, more distant regions are also interesting, but only for the same periods. Levels of mercury in eggs during 1972–2008 from the Atlantic Canadian region were similar to those from Barnegat Bay and the New York–New Jersey Harbor Estuary (ranging from 200 to 800 ppb; Burgess et al. 2013), as did those from the Great Lakes (Koster et al. 1996; Weseloh et al. 2011). Mercury levels in eggs in 1988 from the Wadden Sea (Becker 1992; Table 12.4) were 1720 ± 72 ppb, higher than those in either of the two Northeast bays.

Potential Effects

The most important metal contaminants in marine and coastal waters are lead and mercury, followed by cadmium (Eisler 1987a; Thompson 1996). Because they are common, nonessential, and highly toxic, there have been numerous studies conducted on these three metals (Wolfe et al. 1998; see Chapters 8 and 9). Most studies on metal levels in the eggs and feathers of gulls are for Herring Gulls, largely because of their abundance and circumpolar distribution. Although there are many studies of levels, effects studies are rare, except for those conducted with mercury and lead (see Chapter 9).

Mercury is the metal of greatest concern, partly because it is methylated in the environment and bioamplified up the food chain to humans. It is present in both eggs and feathers of Herring Gulls. The usual mercury effects level in the literature for eggs is 5000 ppb (Eisler 1987a; Burger and Gochfeld 2004a); however, this is highly variable by species, effect, and life stage (Table 9.6). Effects at these levels include slower growth, cognitive and behavioral deficits in feeding behavior or locomotion. However, Heinz et al. (2009) gave a level of 280 ppb (wet weight = ~1120 ppb dry weight) as reducing egg hatchability, and others have noted effects at 750 ppb (Thompson 1996). The effects level for mercury in eggs can be as low as ~750 ppb (dry weight) as the LC_{50} (lethal concentration that kills 50% of test subjects) for Snowy Egrets. Severe effects were reported at 1000–2000 ppb wet weight (Eisler 1987a; Thompson 1996). It is likely that Herring Gulls are not overly impacted, however, because Heinz et al. (2009) found that they were in the intermediate category with respect to egg sensitivity to mercury (LC_{50} ~1300 ppb dry weight). They reported that Laughing Gulls were even less sensitive (LC_{50} ~5000 ppb dry weight).

Mercury in feathers clearly reflects local exposure because chicks are fed entirely by parents while still in the nest. The level of mercury in feathers usually associated with adverse effects is 4000–5000 ppb (Eisler 1987a), although Jackson et al. (2011) demonstrated effects at levels below this. Some marine birds have the capability of demethylating mercury, which may reduce their response to high levels of total mercury (Eisler 1987a; Henny et al. 2002; Eagles-Smith and Ackerman 2010). In this study, mercury levels in Herring Gull feathers in Barnegat Bay averaged from 1500 to 3500 ppb; however, feather levels for some individual birds were higher, suggesting some cause for concern. In the early years, Great Black-backed Gull fledgling feathers averaged about 5000 ppb, with many individuals exceeding this value—clearly a cause for concern for these individuals. The population, however, was expanding; garbage was abundant, and we saw many Great Black-backed Gull pairs accompanied by three full-grown young, resulting in a steadily increasing population (see Chapter 5). Levels of mercury in Barnegat Bay Laughing Gull chicks have been declining, and are currently well below any effects levels.

Current levels of mercury in feathers from other bays have not been examined, although levels in eggs of Herring and Great Black-backed Gulls from the New York–New Jersey Harbor averaged below 800 ppb in 2013 and 2014 (Burger and Elbin 2015b). This suggests that they are not impaired currently.

The proposed toxic effects level for lead in avian feathers is 4000 ppb (Eisler 1988a; Custer and Hoffman 1994; Custer et al. 2007a), although some seabirds can tolerate higher levels (Burger and Gochfeld 2000a). In Barnegat Bay, the levels of lead in feathers of fledgling Great Black-backed Gulls from Barnegat Bay averaged from about 2500 ppb (1995, 2003) to less than 100 ppb in recent years (Figure 12.2). For Herring Gulls, lead in feathers peaked at nearly 3000 ppb in 2003 and declined to less than 500 ppb currently (Figure 12.7), suggesting that they are currently not at risk from lead. Levels in feathers of fledgling Laughing Gulls from Barnegat Bay remain higher than those of the larger gulls (Figure 12.17), which is curious given their lower trophic status.

Cadmium is the third metal of concern for estuarine and coastal ecosystems. The effects level for cadmium in feathers is in the 210–2000 ppb range, depending on species (see Chapter 9). Cadmium levels in feathers from young Great Black-backed Gulls from Barnegat Bay averaged below 70 ppb. However, Herring Gull feathers averaged as high as 300 ppb, although levels have been below any effects threshold since 2004. Similarly, average cadmium levels in feathers of Laughing Gull chicks were below 50 ppb for most years since 2001 (except for 2006). This suggests that young gulls from Barnegat Bay are not at risk from cadmium.

The selenium level in feathers known to be associated with adverse effects is 5000 ppb (Heinz 1996; Eisler 2000b) (Chapter 9). In the present study, average selenium levels were well below the threshold for all three gulls, suggesting little cause for concern from the toxicity of this essential element.

The question of examining means for exposure, however, bears examination. For risk assessment from contaminants, variation among individuals is critical. That is, to some extent the percentage of a population that has levels above an effects level is one of the critical factors. This is discussed more fully in Chapter 14 for all species at once.

SUMMARY AND CONCLUSIONS

Gulls are generally omnivores. They feed on a wide variety of invertebrates (particularly crabs), vertebrates (particularly fish) and carrion, offal, and garbage. They may also hawk for insects or feed on fruits and berries. Thus, their metal levels reflect a wide trophic space and a wide foraging habitat space (from oceanic to terrestrial). Because their populations were increasing rapidly in the 1970s, Herring Gulls were selected as indicators when biomonitoring studies were initiated, and have continued to serve this role. Levels of lead and cadmium have generally declined. Mercury has varied (and declined only slightly when it has declined). Manganese declined in the larger gulls. The patterns for other metals were less consistent.

Adults generally had higher levels of metals in breast feathers than young birds, although selenium levels were significantly higher in down and chick feathers than for adults. The larger gulls (Herring and Great Black-backed Gulls) generally had higher levels of most metals than Laughing Gulls, although Laughing Gulls had the highest levels of cadmium, which generally concentrates and bioaccumulates in insects. The feathers of some species of gulls were above known adverse effects levels for lead (Herring and Great Black-backed Gulls), mercury (Great Black-backed Gull), and cadmium (Herring Gull). Although this is useful, feathers represent local exposure to young birds, whereas levels in eggs reflect female exposure over a period of weeks. Mercury in eggs of Herring Gulls showed similar patterns as those from several different places around the world, when the same periods were examined. Eggs, rather than feathers, have generally been used as the indicator, particularly when hatching failures or developmental defects were discovered. One difficulty with examining temporal trends is that they reflect exposure, but exposure may not reflect similar environmental sources if birds shift their diets with time. Despite limitations, gulls have proven useful indicators of environmental change.

APPENDIX

Table 12A.1 Metal Levels (ppb, Dry Weight) in Great Black-Backed Gull Feathers from Barnegat Bay

Year	n	Lead Mean ± SE GM	Mercury Mean ± SE GM	Cadmium Mean ± SE GM	Selenium Mean ± SE GM	Chromium Mean ± SE GM	Manganese Mean ± SE GM	Arsenic Mean ± SE GM
1995	8	2462 ± 489				753 ± 98.7		
		2188				707		
1996	17	1252 ± 241	5042 ± 624	66.2 ± 12.2	241 ± 150	783 ± 157	6576 ± 1534	
		951	4393	52.2	146	665	4440	
2000	14	381 ± 106		35.2 ± 11.5	1194 ± 76.6	626 ± 103	4342 ± 324	60.1 ± 32.3
		154		12.6	1163	524	4137	0.5
2001	10	134 ± 58.1	4684 ± 385	46.6 ± 12.2	1457 ± 150	756 ± 150	2377 ± 313	107 ± 43.9
		38.7	4525	35.4	1314	651	2214	11.7
2003	138	2483 ± 176	5182 ± 565	61.2 ± 5.6	2065 ± 117	1655 ± 82.7	2263 ± 190	315 ± 41.0
		1921	2970	37.0	1779	1369	1590	37.3
2004	24	375 ± 75.0	2993 ± 277	16.2 ± 2.6	647 ± 57.4	1278 ± 542	1636 ± 567	411 ± 306
		238	2632	7.6	600	485	882	8.8
2005	10	100 ± 8.8	3222 ± 393	18.4 ± 3.2	647 ± 130	85.3 ± 9.8	602 ± 108	158 ± 35.3
		96.7	3020	15.5	309	80.7	527	68.3
2006	12	133 ± 27.0	4538 ± 428	23.3 ± 4.8	1128 ± 102	198 ± 24.6	472 ± 89.0	114 ± 32.1
		86.9	4317	18.3	1077	182	379	20.1
2014	13	219 ± 49.9	2265 ± 293	25.1 ± 8.1	1411 ± 96.8	227 ± 67.1	371 ± 72.7	68 ± 18.5
		143	2055	6.0	1372	168	289	15.5
χ^2 comparison among years		142 (<0.0001)	18.1 (0.0061)	43.9 (<0.0001)	107 (<0.0001)	125 (<0.0001)	97.3 (<0.0001)	21.8 (0.0013)

Note: Given are means ± SE and geometric means below. Comparisons are made with Kruskal–Wallis one-way ANOVA, yielding a χ^2 statistic. All values are in ng/g (ppb dry weight).

Table 12A.2   Metal Levels (ppb, Dry Weight) in Herring Gull Eggs from Barnegat Bay

Year	n	Lead Mean ± SE GM	Mercury Mean ± SE GM	Cadmium Mean ± SE GM	Selenium Mean ± SE GM	Chromium Mean ± SE GM	Manganese Mean ± SE GM	Arsenic Mean ± SE GM
1989	12	2837 ± 611		11.5 ± 8.2				
		2267		18.0				
1990	28	1391 ± 309	774 ± 90.6	158.0 ± 37.7	2059 ± 133	492 ± 64.6	2723 ± 130	
		683	652	89.4	1944	442	2643	
1991	15	746 ± 194	371 ± 41.2	56.1 ± 17.9	1444 ± 109	829 ± 132	3233 ± 147	
		352	345	28.5	1386	748	3187	
1992	27	2055 ± 607	461 ± 77.7	12.3 ± 2.0	2859 ± 106	3083 ± 451	8888 ± 798	
		920	299	8.1	2804	2344	7953	
1993	15	539 ± 129	605 ± 74.4	83.7 ± 28.6	1945 ± 96	966 ± 75.2	3680 ± 287	
		379	545	53.4	1911	930	3566	
1994	24	400 ± 112	686 ± 141	21.8 ± 4.5	989 ± 87	339 ± 42.2	2893 ± 181	
		200	449	13.7	884	300	2755	
1996	13	360 ± 132	633 ± 138	5.4 ± 1.7	1813 ± 169	221 ± 73.6	2530 ± 226	
		164	486	3.5	1711	35.5	2408	
2000	12	273 ± 69.4	344 ± 59.2	4.8 ± 2.1	1836 ± 91	110 ± 17.4	1622 ± 108	126 ± 26.4
		194	286	2.5	1806	94.4	1587	54.6
2004	18	247 ± 78.1	310 ± 54.2	5.9 ± 1.6	2731 ± 228	140 ± 19.8	2404 ± 173	204 ± 40.8
		141	265	3.7	2577	120	2296	139
2006	27	89.8 ± 16.4	583 ± 79.1	5.1 ± 1.3	1420 ± 103	32.0 ± 7.4	2318 ± 426	137 ± 56.4
		49.1	453	0.7	1221	18.2	1466	7.0
2014	15	59.2 ± 12.8	258 ± 51.7	4.8 ± 3.0	1487 ± 75.5	123 ± 49.6	3593 ± 1750	139 ± 26.3
		39.9	223	0.1	1461	59.6	2123	77.9
2015	25	62.3 ± 20.5	326 ± 63.0	6.6 ± 2.6				
		15.8	238	1.7				
χ^2 comparison among years	129 (<0.0001)	45.9 (<0.0001)	132 (<0.0001)	109 (<0.0001)	160 (<0.0001)	103 (<0.0001)	7.2 (0.0661)	

Note: Given are means ± SE and geometric means below. Comparisons are made with Kruskal–Wallis one-way ANOVA, yielding an χ^2 statistic. All values are in ng/g (ppb dry weight).

Table 12A.3 Metal Levels (ppb, Dry Weight) in Herring Gull Feathers from Barnegat Bay

Year	n	Lead Mean ± SE / GM	Mercury Mean ± SE / GM	Cadmium Mean ± SE / GM	Selenium Mean ± SE / GM	Chromium Mean ± SE / GM	Manganese Mean ± SE / GM	Arsenic Mean ± SE / GM
1989	10	2101 ± 465		187 ± 96.3				
		1648		95.6				
1990	29	1256 ± 166	2187 ± 216	98.7 ± 27.8	1331 ± 89.0	565 ± 68.5	1716 ± 159	
		981	1860	63.6	1257	474	1520	
1995	11		3193 ± 594					
			2865					
2000	15	533 ± 126	2222 ± 329	40.2 ± 21.2	1364 ± 86.9	491 ± 97.9	5963 ± 1586	81.7 ± 26.4
		285	1954	5.7	1322	342	4500	4.3
2001	17	150 ± 54.9	3654 ± 227	8.9 ± 3.1	1341 ± 107	671 ± 127	1320 ± 343	330 ± 83.2
		39.3	3522	1.2	1275	533	1003	14.9
2003	15	2993 ± 253	2804 ± 335	303 ± 31.3	5333 ± 474	2280 ± 223	2004 ± 582	0.9 ± 0.3
		2854	2526	287	5033	2136	1355	0.4
2004	11	471 ± 128	2105 ± 314	6.1 ± 1.2	729 ± 45.0	114 ± 25.7	330 ± 30.1	70.8 ± 26.0
		339	1803	4.2	715	98.3	315	11.4
2005	12	242 ± 48.2	1572 ± 299	8.8 ± 1.4	724 ± 86.9	419 ± 97.5	431 ± 42.6	84.2 ± 22.6
		195	1158	7.6	424	329.1	413	15.9
2006	12	612 ± 139	3027 ± 347	17.3 ± 3.9	1392 ± 89.0	144 ± 17.0	334 ± 57.6	76.2 ± 26.1
		404	2813	8.1	1358	134	292	7.4
χ^2 comparison among years	76.3 (<0.0001)	30.8 (<0.0001)	87.7 (<0.0001)	66.7 (<0.0001)	69.3 (<0.0001)	78.7 (<0.0001)	8.2 (0.1438)	

Note: Given are means ± SE and geometric means below. Comparisons are made with Kruskal–Wallis one-way ANOVA, yielding a χ^2 statistic. All values are in ng/g (ppb dry weight).

Table 12A.4   Metal Levels (ppb, Dry Weight) in Laughing Gull Eggs from Barnegat Bay

Year	n	Lead Mean ± SE / GM	Mercury Mean ± SE / GM	Cadmium Mean ± SE / GM	Selenium Mean ± SE / GM	Chromium Mean ± SE / GM	Manganese Mean ± SE / GM	Arsenic Mean ± SE / GM
1995	11	420 ± 81.5 / 205	188 ± 36.8 / 151	60.0 ± 22.8 / 36.4	1789 ± 188 / 1666	345 ± 44.0 / 329	1935 ± 300 / 1705	
1996	8	453 ± 278 / 66.3	483 ± 112 / 372	17.5 ± 11.3 / 5.3	2037 ± 247 / 1905	0.2 ± 0.0 / 0.2	1719 ± 98.1 / 1698	
2003	7	449 ± 34.4 / 440	495 ± 128 / 415	36.0 ± 4.3 / 34.3	2540 ± 495 / 2171	102 ± 15.9 / 95.1	2324 ± 304 / 2206	350 ± 46.0 / 331
2004	40	422 ± 78.5 / 178	330 ± 24.9 / 292	7.4 ± 2.5 / 2.4	2250 ± 104 / 2145	232 ± 43.9 / 122	1893 ± 122 / 1739	128 ± 25.0 / 30.5
2006	12	167 ± 90.3 / 44.1	256 ± 28.1 / 237	4.8 ± 1.9 / 0.2	2025 ± 156 / 1961	12.9 ± 3.4 / 4.4	1608 ± 267 / 1309	13.7 ± 6.8 / 0.7
2007	12	70.4 ± 26.9 / 33.5	393 ± 108 / 243	12.6 ± 4.6 / 1.8	1767 ± 87 / 1742	93.9 ± 15.8 / 83.9	2658 ± 233 / 2545	70.9 ± 19.8 / 13.7
2009	12	195 ± 151 / 30.8	172 ± 25.0 / 157	1.6 ± 0.9 / 0.1	1387 ± 113 / 1329	17.1 ± 8.2 / 0.9	1364 ± 196 / 1268	77.8 ± 20.9 / 24.9
2010	12	103 ± 35.0 / 43.3	372 ± 45.8 / 346	22.6 ± 7.8 / 2.7	1825 ± 77 / 1808	225 ± 70.5 / 164	1535 ± 116 / 1485	189 ± 25.7 / 173
2011	12	13.9 ± 3.9 / 10.8	292 ± 33.5 / 274		1716 ± 149 / 1622	4.5 ± 0.9 / 3.2	2208 ± 171 / 2133	34.5 ± 17.3 / 2.9
χ^2 comparison among years		42.6 (<0.0001)	21.9 (0.0051)	40.3 (<0.0001)	24.6 (0.0017)	82.7 (<0.0001)	26.1 (0.001)	46.3 (<0.0001)

Note: Given are means ± SE and geometric means below. Comparisons are made with Kruskal–Wallis one-way ANOVA, yielding a χ^2 statistic. All values are in ng/g (ppb dry weight).

CHAPTER 13

Heavy Metal Levels in Herons, Egrets, Night-Herons, and Ibises

INTRODUCTION

Great Egrets, Snowy Egrets, and Black-crowned Night-Herons are members of the Heron family (Ardeidae) often referred to as "long-legged waders." They are used as bioindicators of contamination and overall aquatic ecosystem health because they forage in freshwater and estuarine ecosystems around the world (Erwin and Custer 2000). Many Ardeid species are conspicuous parts of aquatic ecosystems in suburban, farming, and pristine areas, and even in urban wetlands. Ardeids are of interest because several species nest together in the same mixed-species colonies (heronries) (Figure 13.1), but they eat different foods and are at different nodes of the food web (Golden and Rattner 2003). Larger species such as Great Blue Heron and Great Egret consume some of the same large fish eaten by people— hence, they can serve as sentinels for biomonitoring ecosystem quality and human health (Fox 2001).

Ardeids are also useful because while they are raising chicks they forage rather close to their nesting colony (Brzorad et al. 2015), and even in coastal environments they do not venture into the ocean (unlike terns that can reflect contaminants in both bays and the ocean). Most species nest in both inland and estuary colonies, making them ideal for comparing coastal and freshwater habitats (Maccarone et al. 2012). Different Ardeid species use slightly different habitats and behaviors for feeding (Maccarone and Brzorad 2005), partly related to differences in leg length (Master et al. 2005). Habitat selection for feeding by species and individuals are influenced by tidal regimes and dry-downs of wetlands (Maccarone and Brzorad 2005; Pierce and Gawlick 2010) and water management practices. Different species of herons and egrets feed on different organisms (Kushlan 1976, 1978), allowing for examination of different trophic levels. Finally, levels found in the wild are sometimes well above the threshold level that can cause adverse effects in birds (Spahn and Sherry 1999; De Luca-Abbott et al. 2001; Malik and Zeb 2009), and adverse effects have been noted (Spahn and Sherry 1999; Frederick and Jayasena 2010) although effects have not always been demonstrated in the field (Sepúlveda et al. 1999a).

Ardeids are exposed to a wide range of chemicals and occupy high trophic levels, making them susceptible to bioaccumulation of pollutants (Walsh 1990; Lewis and Furness 1991; Peakall 1992; Nygard et al. 2001; Henny et al. 2002; Hoffman et al. 2009). Many herons and egrets have high colony site fidelity, and thus can be followed for many years (Custer 2000; Kushlan and Hafner 2000a; Scheifler et al. 2005). Studies in some areas have shown declines in metals, including mercury (Heath and Frederick 2005). There are studies of metals in herons and egrets from many parts of the world. For example, metal levels in feathers of Great Egret have been examined from the Everglades in Florida (Sepúlveda et al. 1999b; Rumbold et al. 2001, 2006; Herring et al. 2009), Georgia and South Carolina (Bryan et al. 2012), California (Henny et al. 2008), and Korea (Honda et al. 1986, 1990; Kim et al. 2009). Both field and experimental work with metals have been conducted on foraging, reproduction (Frederick et al. 1999; Sepúlveda et al. 1999b; Rumbold et al. 2001; Rumbold 2005; Herring et al. 2009), and survival, with mixed results (Chapter 9). In the Old World, Little Egrets, similar in size and diet to Snowy

(a)

(b)

Figure 13.1   Colony of egrets and Night-Herons (not visible) in bushes in Barnegat Bay. When people approach, the white egrets fly up from their nests and perch near the top of the vegetation (a). In Delaware Bay, herons, egrets, and cormorants nest in trees on an island in the Heislerville Impoundment (b). Note that cormorant defecations have killed the trees.

Egrets, have been used extensively as bioindicators (Burger and Gochfeld 1993d; Goutner and Furness 1997; Boncompagni et al. 2003; Zhang et al. 2006; Kim et al. 2009; Shahbaz et al. 2013).

In this chapter, we examine levels in eggs and feathers both temporally and spatially. We first examine metal levels by species from Barnegat Bay, then present metal levels from the same species from other Northeast bays, and then compare metal levels across species. Wherever possible, we present spatial and temporal trends from both eggs and feathers. Feathers from fledglings not yet able to fly more than a few meters are the most useful because they reflect contaminants acquired locally from food gathered by their parents (Burger 1993). Some studies analyzed metals in wing feathers (e.g., Hoffman and Curnow 1979), but primary molt pattern is sometimes unknown, and removal of a primary from fledglings is not a good idea. Metal levels in eggs represent female exposure, which usually represents local exposure as parents are near their colony sites for many weeks before egg-laying. Egg levels also indicate natal exposure from mother to chick.

GREAT EGRET

Great Egrets are the largest Ardeid that breeds in Barnegat Bay, although Great Blue Herons breed in other coastal colonies. They eat mainly fish, but are opportunistic in preying on invertebrates, small mammals, frogs, and bird eggs (McCrimmon et al. 2011). Great Egrets normally feed within 15 km of their nesting colonies during the chick phase (Bancroft et al. 1994; McCrimmon et al. 2011; Brzorad et al. 2015). They are a conspicuous sight in the estuary as they feed in marshes and on mudflats ahead of the receding tide (Figure 13.2a). They also capture larger fish than the

(a)

(b)

Figure 13.2 Great Egrets and Snowy Egrets hunt for small fish on the edges of shoals (a); back at the nest, the half-grown chicks await food (b). The oldest chick is in front and the smaller chick is behind.

other Ardeids, Terns, and Skimmers. In New Jersey, prey fish were mainly in the 6 to 10 cm range (Recher and Recher 1980). Fish may make up 95% or more of the diet. Adults regurgitate boluses of partially digested small fish onto the nest floor for the young to pick up (Figure 13.2b; McCrimmon et al. 2011).

In Barnegat Bay, as in other places in the United States, Great Egret populations are increasing at a rapid rate, whereas the other heron species are all declining (see Chapter 6; McCrimmon et al. 2011). Great Egrets nest on salt marsh islands in bushes and *Phragmites*. As they are the largest species, and arrive at the colonies a few days before the other species, they obtain the highest and most secure nest sites, forcing the smaller species to nest lower (Burger 1978c). Because there is asynchrony among species nesting in mixed-species heronries, and entering colonies when any species has small young can cause disruptions (Parsons and Burger 1982), we collected eggs and feathers intermittently. Significant studies on the effects of mercury on Great Egrets have been conducted in the Florida Everglades by Marilyn Spalding, Peter Frederick, and colleagues, and these were described in Chapter 9 (Frederick et al. 1999).

For Figures 13.3 through 13.16, the correlation between metal level and year was performed with the nonparametric Kendall tau based on all of the individuals analyzed. The curved lines represent a polynomial best fit performed with Deltagraph.

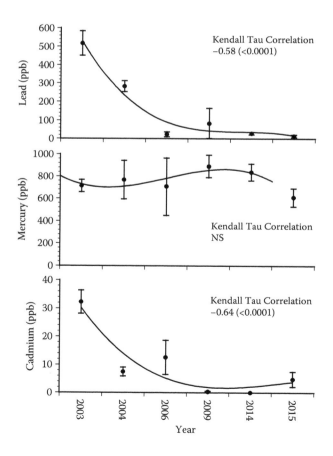

Figure 13.3 Levels of lead, mercury, and cadmium (ppb, dry weight) (mean ± SE) in eggs of Great Egrets from Barnegat Bay. Based on 44 samples.

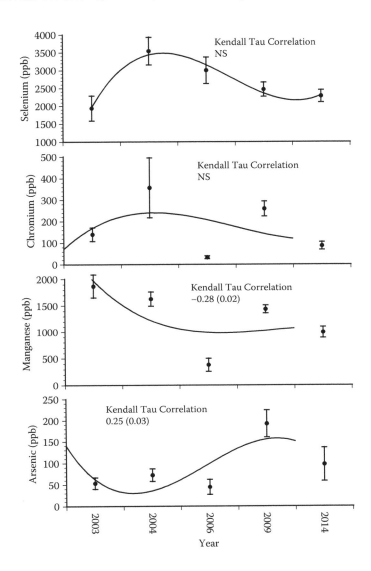

Figure 13.4   Levels of selenium, chromium, manganese, and arsenic (ppb, dry weight) (mean ± SE) in eggs of Great Egrets from Barnegat Bay. Based on 44 samples.

Egg and Feather Patterns from Barnegat Bay

Levels of metals in eggs were analyzed from 2003 to 2015 (Figures 13.3 and 13.14). Levels of lead, cadmium, and manganese deceased significantly during this time, although there were no temporal differences in mercury levels. Levels of arsenic increased slightly, with the 2014 sample significantly lower than the 2009 sample (Figure 13.4).

Levels of lead, cadmium, mercury, and manganese decreased significantly in feathers of fledgling Great Egrets from 1989 to 2014, although the former two decreased in a smoother pattern (Figures 13.5 and 13.6). The other metals showed no significant temporal pattern. Both selenium and chromium were at their highest in 2003 (Appendix Table 13A.1), a phenomenon noted also for some of the tern samples.

Over the 25 years that we examined, metal levels in prefledging egrets from Barnegat Bay, there were significant locational differences (Table 13.1). Except for arsenic, metal levels were higher in

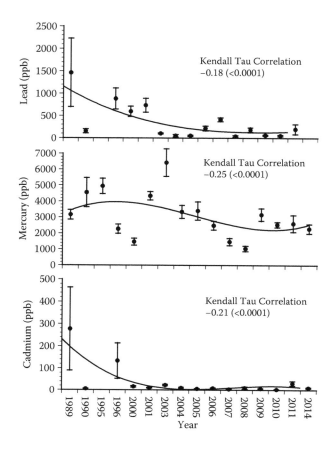

Figure 13.5   Levels of lead, mercury, and cadmium (ppb, dry weight) (mean ± SE) in feathers of fledgling Great
Egrets from Barnegat Bay. Based on 315 samples.

colonies located in the north and central parts of Barnegat Bay. This may reflect their closer location
to the more contaminated New York–New Jersey Harbor Estuary. These patterns were not as clear as
for Common Terns (see Chapter 11), however, perhaps reflecting the different prey and foraging locations (Great Egrets are strictly bay foragers, rather than feeding in the open ocean as terns often do).

Other Northeast Bays and Other Locations

There are remarkably few data points for metals in Great Egrets from the other Northeast Bays
and estuaries, although there are data from more distant places (Table 13.2), including the Everglades
of Florida and Asia (Korea, Hong Kong). Mercury levels in feathers of Great Egrets have declined in
the Everglades, especially because of the efforts of the South Florida Water Management District to
lower mercury levels in the Everglades (SFWMD reports, 2014 and earlier). Levels of mercury were
higher in the good feeding/good breeding years, than when food was scarce, which may reflect the
concentration of fish in small isolated pools or ponds (allowing the birds to catch more and bigger
fish) (Herring et al. 2009).

Hoffman and Curnow (1979) analyzed mercury in the flight feathers of Great Egrets from the
Lake Erie region. They found that mercury levels ranged from 2860 to 28,280 ppb in adults, 2640 to
4960 ppb in juveniles, and 1280 to 43,670 ppb in nestlings (age undefined). Overall, the differences
between flight feather levels of mercury in Great Egret and Black-crowned Night-Heron was small.

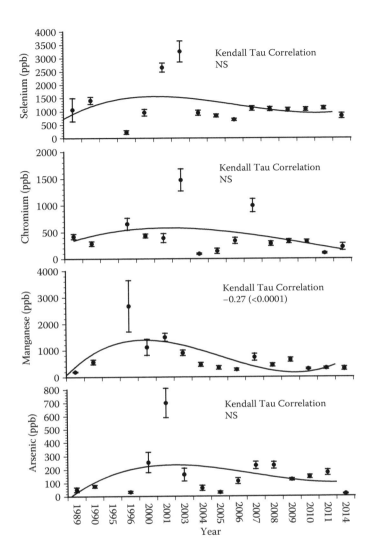

Figure 13.6 Levels of selenium, chromium, manganese, and arsenic (ppb, dry weight) (mean ± SE) in feathers of fledgling Great Egrets from Barnegat Bay. Based on 284 samples.

Table 13.1 Metal Levels (ppb, Dry Weight) in Feathers of Prefledging Great Egrets as a Function of Location in Barnegat Bay

	North Barnegat	Mid-Barnegat	South Barnegat	χ^2 (P)
Sample size	57	149	73	
Lead	732 ± 159	300 ± 3.6	52.5 ± 8.0	52.6 (0.0001)
Mercury	4400 ± 402	3090 ± 205	2240 ± 199	18.5 (0.0001)
Cadmium	104 ± 47	8.2 ± 0.90	11.8 ± 2.94	31.8 (0.0001)
Selenium	813 ± 139	1530 ± 99.2	966 ± 39.5	15.8 (0.0004)
Chromium	495 ± 68	569 ± 53.9	203 ± 21.1	19.2 (0.0001)
Manganese	1750 ± 530	708 ± 58	362 ± 28.0	37.4 (0.0001)
Arsenic	73 ± 18	225 ± 27	143 ± 13.5	7.2 (0.03)

Source: Burger, J., *Environ. Res.*, 122, 11–17, 2013a.

Table 13.2 Mercury Levels (ppb, Dry Weight) in Great Egret Eggs and Feathers from Other Northeast Bays and Other Places

Location	Year	Mercury Levels	Comments	Reference
		Eggs		
Salton Sea, California	1985–2004	Not analyzed	Selenium: 1985 = 3530 (up to 4510) (S end) 1992 = 4950 (up to 6170) (S end) 1993 = 6690 (up to 7900) (N end) 2004 = 3020 (up to 4900) (N end)	Henny et al. 2008
Florida Everglades	1993 1999 2000	1950 ± 950 (up to 4300)	No colony differences in mercury levels	Rumbold et al. 2001
Lower Laguna Madre, Texas	1993	350 (up to 8500)	Levels were not of concern for health effects. Selenium = 1500, chromium = 550	Mora 1996
		Feathers		
Lake Erie	1973	29,150 in adults; 13,200 in nestlings	These were primary feathers, but are some of the earliest data available	Hoffman and Curnow 1979
Florida Everglades	1910–1980	2770 ± 3370 (SD)	Samples from species during the 1990s showed a 4–5× increase over pre-1980 levels	Frederick et al. 2004
Florida Everglades	1999 2000	1400–8600 (range)	Nestling feather levels were positively correlated to egg levels from the same clutch	Rumbold et al. 2001
Florida Everglades	2006, 2007	"Good" year = 6250 ± 810 (SD) "Bad" year = 1600 ± 110	Mercury levels were higher in the good condition year, compared to the bad (= poor breeding success)	Herring et al. 2009
Georgia	2003	Inland = 6380 ± 550; coastal = 200 ± 750	Preflight collection. Those from freshwater had higher levels and higher trophic status by stable isotope analysis	Bryan et al. 2012
Korea	1981	3120	Adult egrets	Honda et al. 1985
Korea	Mid-1980s	Down: 390 ± 90; Fl.: 183 ± 21; Ad: 795 ± 285	Mercury levels increased with age	Honda et al. 1986
Hong Kong	1992	270 ± 33 (y) 1500 ± 420 (a)	Fledglings and adult levels differed significantly for all metals Lead = 1500 ± 400 (c), 4800 ± 670 (a), cadmium = 72 ± 14 (c), 120 ± 12 (a), selenium = 1300 ± 130 (c), 1800 ± 210 (a), chromium = 8500 ± 1300 (c), manganese = 4200 ± 440 (c)	Burger and Gochfeld 1993d

Note: Wet weight converted to dry weight for eggs (wet weight × 5 = dry weight for eggs and Night-Herons; see Table 3.4).

[a] 85% was methylmercury.

[b] Chicks ranged in age from 11 to 31 days (our samples are from fledglings >30 days old).

SNOWY EGRET

Snowy Egrets nest in mixed-species colonies with Great Egrets, Black-crowned Night-Herons, Glossy Ibis, Little Blue Herons, and Tricolored Herons (Parsons and Master 2000). Being smaller than Great Egrets, they are usually forced to nest lower, and sometimes even nest on the ground if sturdy branch supports are limited (Burger 1978c). They forage in similar habitats as Great Egrets, but may be more active foragers. Their shorter legs and smaller bills mean that they have a more restricted foraging habitat range, and catch smaller fish than Great Egrets (Kushlan 1978; Powell 1987; Hancock and Kushlan 2010). Snowy Egret populations are decreasing in Barnegat Bay (see Chapter 5; Figure 13.7).

Figure 13.7 Snowy Egrets and Great Egrets often forage along creek banks when the tide is rapidly going out, exposing some small fish.

Egg and Feather Patterns from Barnegat Bay

We analyzed eggs of Snowy Egret in several years from 1994 to 2009 (Figures 13.8 and 13.9) but with only six data points, it is difficult to document trends. Egg levels decreased for lead, cadmium, manganese, and arsenic significantly, and mercury declined as well. We collected relatively few eggs of Snowy Egrets because in most years they lay later than Great Egrets, and we did not enter the colonies when there was a high likelihood of disturbing young Great Egrets. However, the nestling phase is shorter for Snowy Egrets, so that fully feathered young were available about the same time as fledgling Great Egrets (Figures 13.10 and 13.11). Levels of lead and cadmium in feathers declined precipitously after 1989, then remained low through the early 1990s, and then stabilized. Mercury shows a real increase through the 1990s, declining only after 2000. Selenium levels were relatively consistent with an outlier year (2001). Chromium and manganese levels were lowest before 1992, with several unexplained years of high variability thereafter. Arsenic levels were generally low, except for 2003 (Figure 13.11; Appendix Table 13A.2).

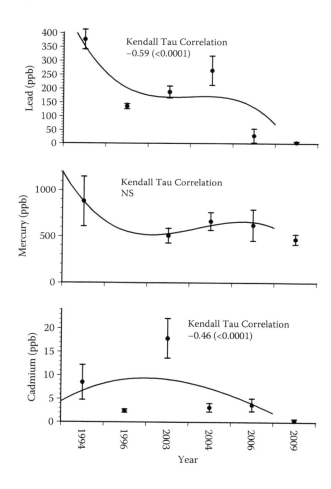

Figure 13.8 Levels of lead, mercury, and cadmium (ppb, dry weight) (mean ± SE) in eggs of Snowy Egrets from Barnegat Bay. Based on 47 samples.

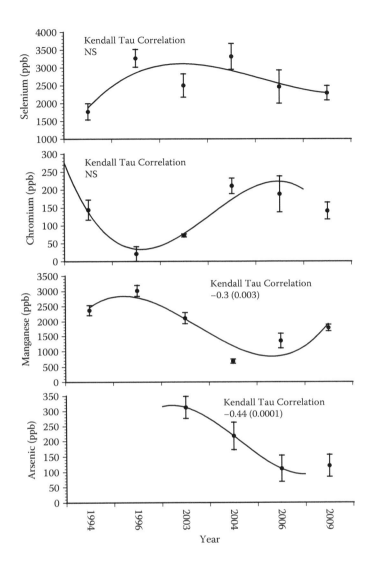

Figure 13.9   Levels of selenium, chromium, manganese, and arsenic (ppb, dry weight) (mean ± SE) in eggs of Snowy Egrets from Barnegat Bay. Based on 47 samples.

Other Northeast Bays and Other Locations

There are few data on levels of metals in Snowy Egrets from other Northeast bays and estuaries, although we did collect feathers from fledgling Snowy Egrets in the New York–New Jersey Harbor Estuary in the late 1980s and early 1990s (Table 13.3). There were two sites in Jamaica Bay (Carnarsie Pol and Goose Island) and one in the Kill Van Kull (Isle of Meadows). There was variation among colonies, with mercury being higher in Canarsie Pol (in Jamaica Bay) than in other places. There was also a significant difference in lead levels, with lead being higher in Snowy Egret feathers from Isle of Meadows in the Kill van Kull, compared to levels from Jamaica Bay.

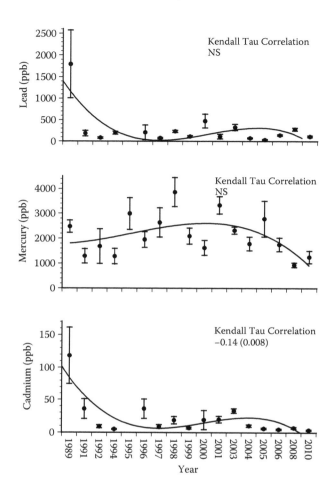

Figure 13.10   Levels of lead, mercury, and cadmium (ppb, dry weight) (mean ± SE) in feathers of Snowy Egret fledglings from Barnegat Bay. Based on 182 samples.

Snowy Egret chicks fed a diet high in mercury showed a metabolic shift, catabolizing protein as evidenced by enrichment of the N-15 stable isotope of nitrogen (Shaw-Allen et al. 2005). This can serve as a biomarker of stress, and proved more sensitive than measuring increases in glutathione, a marker of oxidative stress (Shaw-Allen et al. 2005).

There are few data on Snowy Egrets from other places. In Georgia, Bryan et al. (2012) reported mercury levels in feathers of Snowy Egrets as 810 ± 200 ppb, with a maximum of 1710 ppb. Henny et al. (2002) reported on mercury levels in birds at a Nevada site that had been highly contaminated with mercury from the 1860s to 1890. In 1998, mercury levels in feathers of fledgling Snowy Egrets averaged 30,640 ppb (compared with only 7770 ppb in a reference site). This was much higher than those found at many other sites. Hoffman et al. (2009) found levels in the feathers that ranged from 20,500 to 36,400 ppb at the same contaminated site in 2002–2003, but only 6600 ppb at the reference site. These are quite high levels, and Henny et al. (2002) hypothesized that the tolerance was attributable to a threshold-dependent demethylation coupled with sequestration of resultant mercury. The potential for demethylation as a protective mechanism has been previously noted (Eisler 1987a; Henny et al. 2002; Eagles-Smith and Ackerman 2010), but this requires more species-specific studies.

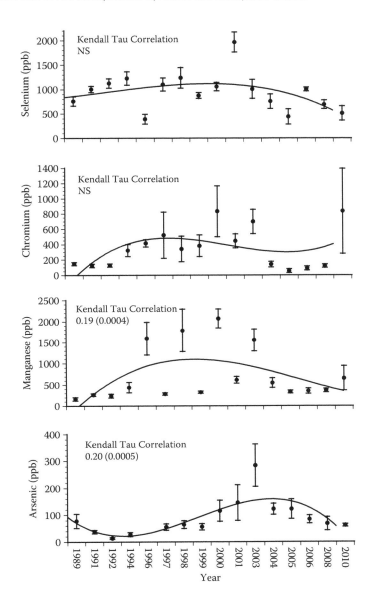

Figure 13.11 Levels of selenium, chromium, manganese, and arsenic (ppb, dry weight) (mean ± SE) in feathers of Snowy Egret fledglings from Barnegat Bay. Based on 169 samples.

Table 13.3 Levels of Metals (ppb, Dry Weight) in Snowy Egret Feathers from the New York–New Jersey Harbor Estuary (Burger, Unpublished Data from 1989 to 1990) and Barnegat Bay

Location	N	Lead	Mercury	Cadmium	Selenium	Arsenic
Isle of Meadows (1989)	18	2180 ± 914		444 ± 253		
Canarsie Pol	15	1490 ± 442	1000 ± 167	86.3 ± 16	1300 ± 196	
Goose Island	18	<MDL of 100	338 ± 33	10 ± 10		
(*P*)		< 0.01	<0.01	< 0.05		

Note: The Isle of Meadows is in the Kill van Kull, and the other two are in Jamaica Bay.

At the Nevada site, egg levels were higher at the contaminated site (2700 ppb) than at the reference site (mean of 850 ppb; Henny et al. 2002). Because of the high levels in eggs and feathers, Hill et al. (2008) examined the relationship between mercury and drought along the Carson River from 1997 to 2006. Over the 10-year period, levels varied in eggs from 1150 ppb in 1997 to 9660 ppb in 2006 (Hill et al. 2008). They found that there may be a variable threshold of tolerance to mercury associated with habitat quality (food type and abundance), which clearly requires follow-up. Mercury decreased and then increased, depending on the drought conditions (Hill et al. 2008).

Mercury levels are also available for Snowy Egret eggs from California (mean of 1050 ppb; Ohlendorf et al. 1988), Texas (mean of 350 ppb; Mora 1996), and Colombia (mean ± SD, 135 ± 30; Olivero-Verbal et al. 2013). Selenium was also measured in eggs from the Salton Sea in California, wherein levels ranged from means of 3930 to 4970 ppb (Henny et al. 2008), and from the Laguna Madre of Texas (mean of 1500 ppb; Mora 1996). In the same Laguna Madre colonies, Mora (1996) found that mercury levels averaged 720 ppb in Tricolored Heron.

BLACK-CROWNED NIGHT-HERON

Black-crowned Night-Herons are largely crepuscular and nocturnal, and can be found on piers, creek beds, mudflats, and on banks of any waterway, or stalking through salt marsh and beach vegetation, watching and waiting for prey to move by (Figure 13.12). Night-Herons can be serious predators at tern colonies (Collins 1970). Of all the herons, egrets, and ibises, Black-crowned Night-Heron has the most varied diet, including invertebrates, as well as frogs, fish, and anything else they can catch (Kim and Koo 2007; Hothem et al. 2010).

Their arrival at nesting colonies varies each year, and they may lay eggs before, at the same time, or later than the Great Egrets. They nest in mixed-species colonies of herons, egrets, and ibises, but they also nest in small isolated colonies, which are more difficult to locate (also because they are dark colored). In Barnegat Bay, they have nested in small groups (n = 3–8) in clumps of

(a)   (b)

Figure 13.12 Adult Night-Heron foraging at an outfall (a). Black-crowned Night-Heron chicks on their nest (day-old chick asleep) (b).

Iva or *Baccharis* bushes on salt marsh islands. Without walking through these clumps, the Night-Herons can be missed because they do not flush until approached closely. The nests are in the low branches so that incubating adults are not usually visible from outside the colony.

Egg and Feather Patterns from Barnegat Bay

We were able to obtain Night-Heron eggs in 5 years when pairs laid eggs as early as Great Egrets. Since 1991, levels of lead, chromium, and manganese declined significantly (Figures 13.13 and 13.14). Mercury and selenium show no trend. Cadmium levels were low except for 1994 when the mean was 260 ppb. The cadmium peak in 1994 was verified and probably represents a dietary shift to a temporarily abundant high cadmium-abundant food source. Only three inconsistent points are available for arsenic. The few data points show no real evidence of trend for mercury, but the increased level in the last sample (2009) bears watching.

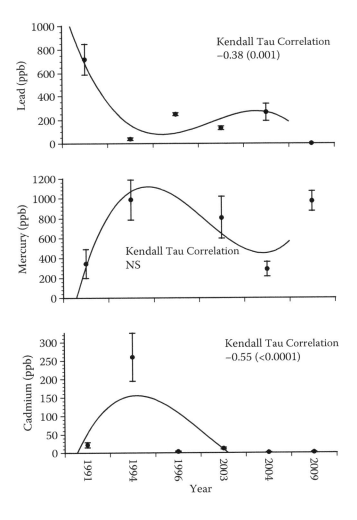

Figure 13.13 Levels of lead, mercury, and cadmium (ppb, dry weight) (mean ± SE) in eggs of Black-crowned Night-Heron from Barnegat Bay. Based on 45 samples.

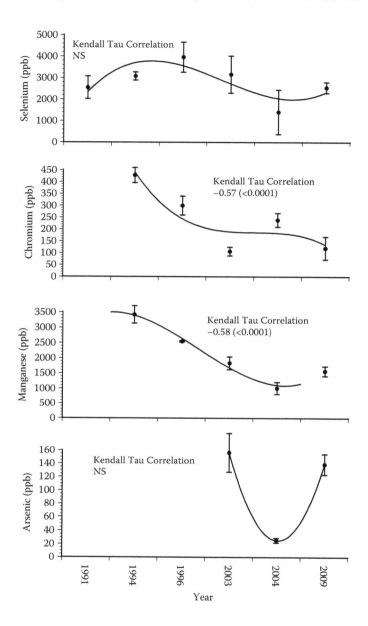

Figure 13.14 Levels of selenium, chromium, manganese, and arsenic (ppb, dry weight) (mean ± SE) in eggs of Black-crowned Night-Herons from Barnegat Bay. Based on 45 samples.

Feathers were collected in 9 years between 1991 and 2014 (Figures 13.15 and 13.16). Lead and cadmium declined, but both showed spikes in 2001. Mercury, selenium, and chromium showed no trend. Manganese declines, and arsenic levels were unstable but showed a slight increase in arsenic (tau = 0.34, P = 0.003); however, this is apparent only from 2001 to 2011. Arsenic levels dropped in 2014 to earlier levels. Feathers and eggs tell a similar story, except that chromium in eggs showed a significant decline (tau = −0.57; P < 0.001), which is not evident in the feather data set.

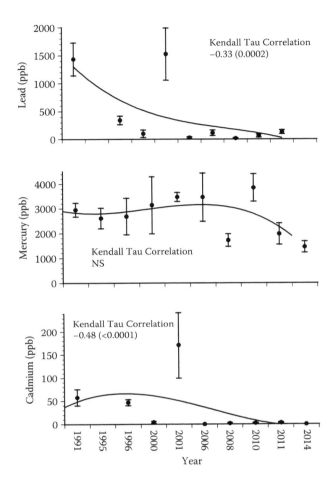

Figure 13.15 Levels of lead, mercury, and cadmium (ppb, dry weight) (mean ± SE) in feathers of Black-crowned Night-Heron fledglings from Barnegat Bay. Based on 67 samples.

Other Northeast Bays and Other Locations

In the late 1970s, Hoffman and Curnow (1979) analyzed mercury in the flight feathers of some herons and egrets from Lake Erie. They found that mercury levels ranged from 2150 to 18,650 ppb in adult Black-crowned Night-Herons, from 3330 to 5420 ppb in juveniles, and from 2250 to 4250 ppb in nestlings (age undefined) (Table 13.4).

Although we do not specifically address other contaminants, it should be noted that most of the bays examined have high levels of other contaminants, such as polychlorinated biphenyls (PCBs), dioxins, polyaromatic hydrocarbons (PAHs), as well as fire retardants, pharmaceuticals and their metabolites, and personal care products. Significant work with Black-crowned Night-Herons and Herring Gulls in the New York–New Jersey Harbor Estuary indicated that skin responses to phyto-hemagglutinin were 70–80% suppressed, and related to dioxins and PCBs—providing strong evidence that these chemicals contribute significantly to immunosuppression in waterbirds in the harbor (Grasman et al. 2013).

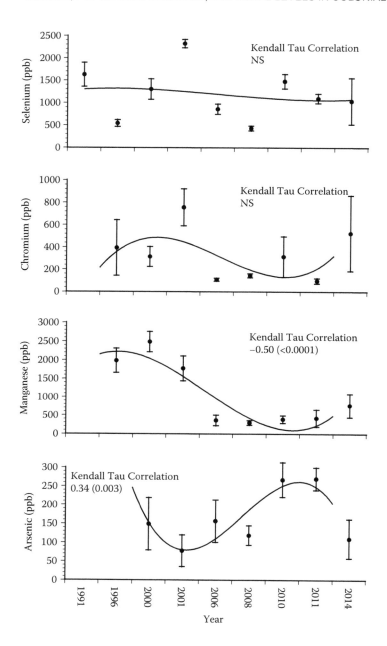

Figure 13.16 Levels of selenium, chromium, manganese, and arsenic (ppb, dry weight) (mean ± SE) in feathers of Black-crowned Night-Heron fledglings from Barnegat Bay. Based on 74 samples.

Table 13.4 Levels of Mercury (ppb, Dry Weight) in Eggs and Feathers of Black-Crowned Night-Herons from Northeast Bays and Other Places

Location	Year	Mercury Levels	Comments	Reference
			Eggs	
St. Lawrence R.	1991–1994	Means ranged from 610 to 1370	Egg levels varied among colonies	Champoux et al. 2002
Tennessee Valley	1980–1981	Means ranged from 800 to 1900 (max = 4400)	There were significant variations among colonies. Chromium ranged from 500 to 1550	Fleming et al. 1984
Agassiz NWR, Minnesota	1994	986 ± 201	Cadmium = 297 ± 195, selenium = 3070 ± 190, chromium = 487 ± 66, manganese = 3419 ± 283	Burger and Gochfeld 1996b
Minnesota	2001	Geometric mean of 290 (up to 390)	Concentrations were lower than reported in 1994	Custer et al. 2007a
Nevada	1997–2006	Mean of 1100 in 1997 to 5050 in 2006.	Levels studied at a contaminated site. Night-Herons were less sensitive to mercury than Snowy Egrets	Hill et al. 2008
California	1982	1400 ± 200 to 2250 ± 300	There were no differences in mercury levels in random or failed eggs	Ohlendorf et al. 1988
Salton Sea, California	1985–2004	Not analyzed	Selenium: 1985 = 5880 (up to 7490) 1991 = 5270 (up to 6500) 1992 = 6180 (up to 7850) 2004 = 1770 (up to 5030)	Henny et al. 2008
Edwards Air Force Base, California	1996–1999	292 (up to 890) in 1996, 202 (up to 1100) in 1999	There were no differences in levels in eggs that failed to hatch compared to random eggs (given here); Selenium levels were 2780 (up to 3800)	Hothem et al. 2006
Nevada	1997	1997 = 2550; 1998 = 2350 (geometric mean)	Values given were for contaminated site; reference site = 850 ppb	Henny et al. 2002
Hong Kong	2000–2003	520 ± 187	Metal levels in eggs were related to those in coastal sediments. Lead = 7 ± 7, selenium = 3274 ± 428, chromium = 318 ± 67, manganese = 1720 ± 932, cadmium was ND	Lam et al. 2005
			Feathers	
New York–New Jersey Harbor Estuary	2004–2005	1233 ± 105	Lead = 2015 ± 183, cadmium = 145 ± 19, chromium = 1411 ± 271, arsenic = 208 ± 30	Burger and Gochfeld, Unpublished data
New York–New Jersey Harbor Estuary	2004–2005	Means ranged from 289 ± 85 to 2027 ± 340	There were differences in metal concentrations among islands. Lead = 1030 ± 168 to 3210 ± 518, cadmium = 62.3 ± 12 to 234 ± 70, chromium = 456 ± 94 to 1270 ± 141, arsenic = 22 ± 14 to 457 ± 91	Padula et al. 2010

(Continued)

Table 13.4 (Continued) Levels of Mercury (ppb, Dry Weight) in Eggs and Feathers of Black-Crowned Night-Herons from Northeast Bays and Other Places

Location	Year	Mercury Levels	Comments	Reference
Baltimore Harbor and Holland Island	1998	805 (up to 1380) and 810 (up to 1120)	14–16-day-old chicks: lead levels were 110 and 320 ppb; cadmium levels were ND to 38; selenium levels were 2110 (up to 2530) and 2180 (up to 3480)	Golden et al. 2003b
Pea Patch Island (Delaware Bay)	1998	1250 (up to 5180)	14–16-day-old chicks: lead levels were 420 (up to 3770); cadmium levels were ND to 1200; selenium levels were 2160 (up to 3570); Levels do not threaten the species	Golden et al. 2003b
Agassiz NWR, Minnesota	1994	3630 ± 244	Lead = 400 ± 110, cadmium = 1300 ± 206, selenium = 1070 ± 166, chromium = 540 ± 142, manganese = 3590 ± 578	Burger and Gochfeld 1996b
Agassiz, Minnesota	1999	1690 (up to 1,95)	10-day-old chicks. For most elements, levels in feathers did not relate to levels in water or sediment; Lead levels were <900; cadmium levels were 150 (up to 190); selenium was 2100 (up to 2370)	Custer et al. 2008
Lake Erie	1973	Median of 34,000 (A) 13,700 (C)	They analyzed primary feathers	Hoffman and Curnow 1979
Agassiz, Minnesota	1998, 1999, 2001	1600 (up to 1800)	10–15-day-old chicks. Selenium = 2200 (up to 2500); manganese = 2600 (up to 3400)	Custer et al. 2007a
Nevada	1998	35,250 ppb in contaminated site, 5510 in reference site	High levels because of historic exposure from 1859 to 1890. Lack of effects may be attributable to threshold-dependent demethylation and sequestration	Henny et al. 2002
Northern Italy	1994	1982 (up to 2640)	Cadmium levels were.553 (up to 862) and lead levels were 3360 (up to 6092) (20-day-olds)	Fasola et al. 1998
Greece	1993, 1994	Median of 2110 in 1993 and 3010 in 1994	Mercury content was negatively correlated with the size of the chick	Goutner and Furness 1997
Szechuan, China	1992	2300 ± 470 (fledgling)	Lead = 5600 ± 670, cadmium = 220 ± 40, selenium = 2000 ± 470, chromium = 20,800 ± 2200, manganese = 42,400 ± 7300	Burger and Gochfeld 1993d
Hong Kong	1992	840 ± 38 (fledglings)	Levels differed significantly by location. Lead = 9100 ± 2200, cadmium = 140 ± 51, selenium = 2800 ± 870, chromium = 17,100 ± 5900, manganese = 45,100 ± 2000	Burger and Gochfeld 1993d
Hong Kong	2000	Means of 300 and 1200 (a)	Molted primaries of adults were collected. Lead = 700, and 2600 ppb, cadmium = 40 and 60 ppb, chromium = 900 and 1100 ppb, and manganese = 4000 and 13,900 ppb	Connell et al. 2002
Korea	2001	Mercury was not analyzed	Acute local contamination. Lead = 330 ± 110, cadmium = 116 ± 55, manganese = 27,000 ± 29,800	Kim and Koo 2008
Korea	2008	Mercury was not analyzed	24–26-day-olds; Lead = 2570 ± 1490, cadmium = 900 ± 179	Kim and Oh 2014

Note: Given are means for fledglings, unless otherwise noted as (a) for adult. Published data have been converted to ppb, dry weight using a conversion factor of 5×. ND, not detected.

OTHER SPECIES

Several other species nest in the heronries in Barnegat Bay, but their numbers are far lower, they are less synchronous in the timing of egg-laying (making it more difficult to collect a sample), and their population vulnerabilities have caused us to collect eggs and feathers only sporadically. We present them here (Table 13.5, after Burger et al. 2012f).

From Table 13.5, it is clear that there were interspecific differences for all metals. Mercury, the metal that bioaccumulates and biomagnifies the most, was highest in Great Egrets and Black-crowned Night-Herons. However, other metals were higher in the other species. Notably, Snowy Egrets had much higher levels of lead than the other species, but this may be a function of years of collection (see the following discussion). Cadmium and manganese were much higher in Glossy Ibis, a species that eats more insects and invertebrates than the other species (Davis and Kricher 2000). Chromium and arsenic were higher in Little Blue Herons, and selenium was higher in Tricolored Herons compared to the other species (Table 13.5).

A species that is not very common today in any of the colonies is Cattle Egret; they increased substantially into the 1980s and then began to decline (Figure 13.17). It is useful to compare levels of metals in feathers of fledglings from several parts of the world, collected and analyzed in the same manner and in the same laboratory (Burger et al. 1992b; Figure 13.18). There were significant differences as a function of location, with fledglings from Pea Patch and the Arthur Kill having the highest concentrations of mercury in their feathers. Metals varied significantly among colonies (e.g., geographically). Lead levels were 41 times higher in fledglings from Cairo (Egypt) than the other colonies, which we attributed to the continued use of leaded gasoline and the heavy traffic in that city. The lead levels were in the range known to cause adverse effects in birds (see Chapter 9 and the following discussion). Mercury levels were highest at Aswan, Egypt. The Puerto Rican colony had the highest levels of cadmium. These comparisons are useful because they were completed in generally the same period, with the same methods by the same investigators.

Bryan et al. (2012) found that inland-nesting Cattle Egrets from Georgia had lower levels of mercury in feathers, compared to those nesting on the coast (640 ± 80 versus 1100 ± 140 ppb). Most papers on metals in Cattle Egret analyze only internal tissues (Hulse et al. 1980). However, Henny et al. (2008) reported levels of selenium in eggs from Nevada as averaging 3600 ppb (up to 5400 ppb), and Shahbaz et al. (2013) reported that lead levels in eggs in Pakistan averaged 370 ± 260 and 440 ± 330 ppb at two colonies. Cadmium levels averaged 180 ± 100 and 140 ± 90 ppb in eggs from two different colonies (Shahbaz et al. 2013).

In Pakistan, levels of cadmium, lead, and chromium in feathers were well above the effects threshold (Malik and Zeh 2009). Lead levels averaged 37,500 ± 10,700 (SD) to 76,500 ± 8600 ppb in three colonies; cadmium averaged 2400 ± 10,700 to 3100 ± 500 ppb; manganese averaged 15,300 ± 2000 ppb; chromium averaged 5380 ± 1000 to 7100 ± 2200 ppb (Malik and Zeh 2009).

DISCUSSION

Temporal Patterns of Metals in Ardeids

Other than Barnegat Bay, there are no long-term studies of metal levels in herons, egrets, and ibises from other Northeast bays. Data from episodic studies, both ours and others, are given in Tables 13.2 and 13.4. There are long-term data for Great Egrets from the Everglades from the work of Spalding, Frederick, and colleagues (Frederick et al. 2002). In the Everglades, mercury levels in nestling feather levels declined in all colonies studied from 1994 to 2000, with a minimum colony mean of about 3280 ppb and a maximum of 29,200 ppb. Not only have levels decreased in recent decades, but Frederick et al. (2004) compared feathers from museums (before 1910 through 1980)

Table 13.5 Concentrations of Metals in Feathers of Young Egrets, Ibises, and Herons from New Jersey (All Years)

	Tricolored Heron	Little Blue Heron	Glossy Ibis	Snowy Egret	Great Egret	Black-Crowned Night-Heron	χ^2 (P)
Sample size	6	10	3	70	305	45	
Lead	818 ± 181	1130 ± 418	1060 ± 147	2280 ± 530	278 ± 30.3	746 ± 220	184 (<0.0001)
	730	746	1040	57.5	29.0	45.9	
Mercury	1630 ± 428	1910 ± 351	2350 ± 612	2470 ± 202	3130 ± 143	3150 ± 242	47 (<0.0001)
	1420	1430	2180	2050	2150	2750	
Cadmium	31.1 ± 10.3	33.3 ± 19.3	150 ± 103	64.9 ± 15.4	20.1 ± 5.67	84.7 ± 31.8	248 (<0.0001)
	20.9	8.90	69.5	13.8	1.20	11.5	
Selenium	1840 ± 509	1090 ± 203	1770 ± 185	1460 ± 220	1250 ± 59.6	1570 ± 122	19 (<0.0001)
	1610	888	1750	621	858	1290	
Chromium	513 ± 126	2940 ± 2050	1270 ± 292	1070 ± 209	448 ± 33.3	510 ± 101	58 (<0.0001)
	314	1130	1200	627	160	213	
Manganese	5750 ± 1780	1240 ± 208	9690 ± 3180	2140 ± 275	725 ± 94.9	1640 ± 208	188 (<0.0001)
	4270	1150	8770	1470	471	1160	
Arsenic	460 ± 206	981 ± 159	14.1 ± 13.97	236 ± 50.3	180 ± 15.9	141 ± 31.0	34 (<0.0001)
	253	849	0.75	23.6	36.7	5.70	

Source: After Burger, J. et al., *Int. J. Environ. Sci. Eng. Res.*, 3, 147–160, 2012f.
Note: Given are arithmetic means ± SE over geometric means. Interspecies comparisons are made with Kruskal–Wallis one-way ANOVA, yielding a χ^2 statistic. All values are in ng/g (ppb dry weight.) NS, Not significant.

(a)

(b)

(c)

Figure 13.17 Cattle Egrets normally forage on insects on grassy fields (a) or behind animals (cows in the New World [b]). They have also adapted to foraging in garbage dumps (c).

with current levels and found no change in mercury levels from 1910 to the 1970s, but a large and significant increase in the 1990s. Pre-1980s mean mercury levels for Great Egret feathers were 2770 ppb, compared to 19,840 ppb after 1990 (Frederick et al. 2004). Henny et al. (2008) also mentioned that between 1992 and 2003, selenium levels in Night-Herons and Great Egret eggs from Whitewater River Delta of the Salton Sea (California) decreased by 91% and 55%, respectively.

All of the NJ egg and feather samples we collected from these species were collected from the same mixed-species colonies in Barnegat Bay. Therefore, differences in metal levels should be attributable to species differences in diet, proportion of different prey items, size of prey items, location where prey items were captured, and toxicodynamics within each species. This assumes,

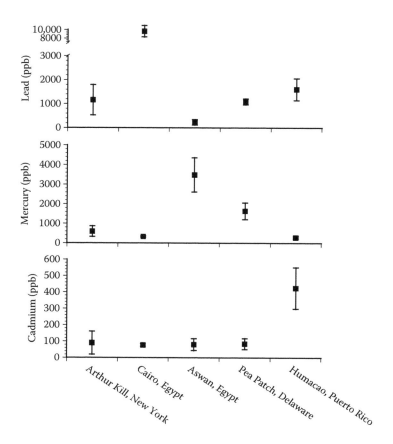

Figure 13.18   Metals in Cattle Egret feathers from several colonies, including the New York–New Jersey Harbor Estuary (Arthur Kill) and Delaware Bay (Pea Patch) based on 57 samples. (After Burger, J. et al., 1992b.)

of course, that diet does not deviate markedly from year to year (Rumbold et al. 2001), which is a big assumption for herons and egrets (and all other coastal birds). This is especially true as many of the larger prey items Great Egrets and Night-Herons bring back to the nests are migratory fish (e.g., Bluefish, Striped Bass) that may come into the bays at different times in different years. Great Egret diets have been shown to differ among years in the Everglades as well (Frederick et al. 1997).

However, long data sets, such as the 25+-year feather data for Great Egrets, can partly account for these differences. If a pattern persists over this period, it is likely real. Differences in diet, however, can account for the years when a mean mercury level, for example, is very different from the general curve (e.g., 2003, Figure 13.3, Great Egret; 2001, Figure 13.9, Snowy Egret); 2003 was also a year of metal peaks in terns and gulls.

The temporal trends described in this chapter are summarized in Table 13.6. As is clear, trends varied somewhat. Although there are sometimes differences between eggs (primarily female exposure coupled with sequestration in eggs) and feathers (local exposure), the following patterns emerge: (1) lead and cadmium levels declined, (2) mercury generally did not decline, (3) there was no pattern for selenium or chromium, (4) manganese decreased in most species (but increased in feathers of Snowy Egret), and (5) arsenic increased in some samples and decreased in others. Thus, the clearest patterns were the decreases for lead and cadmium.

Table 13.6   Temporal Trends in Metal Levels in Eggs and Feathers of Herons and Night-Herons in Barnegat Bay

	Great Egret	Snowy Egret	Black-Crowned Night-Heron
Lead	Eggs: −0.58, $P < 0.0001$ Feathers: −0.18, $P < 0.0001$	Eggs: −0.59, $P < 0.0001$ Feathers: Declined, but not significant	Eggs: −0.38, $P < 0.001$ Feathers: −0.33, $P < 0.0002$
Mercury	Eggs: Not significant Feathers: −0.25, $P < 0.0001$	Eggs: Not significant Feathers: Not significant	Eggs: Not significant Feathers: Not significant
Cadmium	Eggs: −0.64, $P < 0.0001$ Feathers: −0.21, $P < 0.0001$	Eggs: −0.46, $P < 0.0001$ Feathers: −0.14, $P < 0.008$	Eggs: −0.55, $P < 0.0001$ Feathers: −0.48, $P < 0.0001$
Selenium	Eggs: Not significant Feathers: Not Significant	Eggs: Not significant Feathers: Not significant	Eggs: Not significant Feathers: Not significant
Chromium	Eggs: Not significant Feathers: Not significant	Eggs: Not significant Feathers: Not Significant	Eggs: −0.57, $P < 0.0001$ Feathers: Not significant
Manganese	Eggs: −0.28, $P < 02$ Feathers: −0.27, $P < 0.0001$	Eggs: −0.30, $P < 0.003$ Feathers: +0.19, $P < 0.0004$	Eggs: −0.58, $P < 0.0001$ Feathers: −0.50, $P < 0.0001$
Arsenic	Eggs: +0.25, $P < 0.03$ Feathers: Not significant	Eggs: −0.44, $P < 0.0001$ Feathers: +0.20, $P < 0.0005$	Eggs: Not significant Feathers: +0.34, $P < 0.003$

Note: NA = not available, NS = not significant.

Trophic Level Relationships

Great Egrets (primarily piscivorous) and Night-Herons (varied diet) represent different trophic levels because of their diets. Many studies comment that mercury bioaccumulation in organs is reflected in feathers and eggs. Few authors comment on cadmium or selenium accumulation, but these exhibit low accumulation and are not efficiently transferred to eggs (White et al. 1978; Custer 2000). We found substantial levels of selenium in the Ardeid eggs. Actually, it might be problematic for embryo development if mercury was transferred but not selenium, because the latter can moderate the effects of the former (Ralston 2008). How well this protection works for developing embryos is not known.

Egrets and Night-Herons derive their metal exposure primarily from food. This is often difficult to track because all three species have a wide and varied diet, which changes seasonally and yearly (depending on prey availability). Regurgitated food samples can be used to assess diet and to correlate metals levels in prey items with those in young. Frederick et al. (1999) used regurgitated food samples to examine food intake and comparative mercury exposure from different prey items from Great Egret chicks in the Everglades. They found that during the 80-day nestling period, Great Egret chicks consumed a cumulative total of 4.32 mg of mercury, a level that is at the potential effects level. For an approximately 1 kg bird, this translates into a daily dose of 54 μg/kg of body weight. This exceeds the avian Reference Dose of 21 μg/kg/day (EPA 1997d).

Although all three species eat both fish and invertebrates, Snowy Egrets eat more fish, Great Egrets eat larger fish, and Black-crowned Night-Herons eat more crabs and other invertebrates (Kushlan 1976, 1978; Parsons and Master 2000; Hothem et al. 2010; McCrimmon et al. 2011). Furthermore, Great Egrets are much larger than Night-Herons. Thus, we expected that Great Egrets would generally have higher levels of all metals. When we examined overall metal levels in the three species (Table 13.7), we found that Great Egret had higher levels of lead and cadmium; Black-crowned Night-Herons had higher levels of mercury and arsenic, and there were no differences in the other metals. However, as mentioned in other chapters, it would be most valuable in the future to have systematic collecting within the same periods. Collecting the samples has costs, both in terms of boats and personnel, but also in terms of disturbance of the nesting colonies. In studying the impact of contaminants on reproduction and populations, we wanted to be sure that our intrusion did not become part of the problem.

Table 13.7   Species Differences in Metal Levels in Feathers for All Years Examined (ppb, Dry Weight)

	Great Egret	Snowy Egret	Black-Crowned Night-Heron	χ^2 (P)
Lead	139 ± 29.2	107 ± 13.4	28 ± 14.5	10.1 (0.006)
Mercury	2920 ± 245	1905 ± 200	3464 ± 980	5.6 (0.06)
Cadmium	7.6 ± 1.1	6.6 ± 0.9	0.1 ± 0	7.3 (0.03)
Selenium	800 ± 35.6	850 ± 64.7	857 ± 115	1.9 (NS)
Chromium	237 ± 36.4	100 ± 18.8	108 ± 13.3	2.2 (NS)
Manganese	349 ± 34.5	411 ± 52.6	360 ± 143	3.7 (NS)
Arsenic	78.5 ± 12.9	101 ± 11.7	157 ± 157	11.5 (0.003)

Note: Data are presented as arithmetic means ± SE.

Unlike the terns and gulls, we had less complete temporal data for Ardeids, especially in eggs. However, when we compared the metal levels in eggs of Great Egret, Snowy Egret, and Black-crowned Night-Heron for 2003–2006, we found no significant differences in any metal levels, except for arsenic. Arsenic levels were highest in Snowy Egret Eggs (mean of 244 ± 72 ppb), compared to Great Egrets (mean of 62 ± 10 ppb) and Black-crowned Night-Heron (mean of 124 ± 30 ppb).

In Table 13.8, we show the metal levels in fledgling feathers for three different years with comparable data. Interestingly or frustratingly, the patterns varied between years, and there were no clear differences that persisted. In 1996, Great Egrets had the highest arithmetic mean levels of lead, cadmium, chromium, and manganese, and Night-Herons had the highest levels of selenium. In 2001, Great Egrets had the highest levels of mercury, selenium, and arsenic, Night-Herons had the highest levels of lead and chromium, and Snowy Egrets had the highest levels of cadmium. In 2010–2011, Great Egrets had the highest levels of mercury and selenium, Night-Herons had the highest levels of arsenic, and Snowy Egrets had the highest levels of lead.

Table 13.9 summarizes Table 13.8, showing the maximum geometric mean (GM) values. This shows that in various years Great Egrets had the highest values for all metals except cadmium (Snowy Egret in 1996) and manganese (Night-Heron in 1996). The data illustrate the interyear variation in dietary sources of metals within and among species. However, as with risk assessment for people, it may be that the average consumption for a period of days, weeks, or years is critical to potential effects, not consumption on 1 day or breeding season. We conclude that both Great Egrets and Night-Herons are useful as bioindicators. In some egret studies, mercury bioaccumulated, but selenium and cadmium showed low accumulation, and arsenic and chromium did not bioaccumulate (Boncompagni et al. 2003). We found that these species bioaccumulated lead and mercury, the contaminants of greatest concern.

Comparisons of Levels with Other Regions

Exposure of birds to mercury is largely through their diet, which accounts for trophic level differences. There are also locational differences, both within a region and among different regions. Results have been mixed, however. We found locational differences in Barnegat Bay, with Great Egrets from the North Bay having the highest levels of mercury, lead, cadmium, manganese, and arsenic—quite a toxic load.

Locational differences are to be expected between adjacent bays, and certainly for bays and estuaries separated by great distances. We report such interbay differences in this book, and others have reported them also (e.g., Ohlendorf et al. 1988). In the New York–New Jersey Harbor, Padula et al. (2010) found locational differences in metal levels among colony locations, but no one colony had the highest level of all metals. In Florida, Rumbold et al. (2001) did not find differences in mercury levels in Great Egret chicks from different colonies in the Everglades, whereas

Table 13.8 Concentrations of Metals in Feathers of Young Egrets from New Jersey

	n	Lead Mean ± SE	Mercury Mean ± SE	Cadmium Mean ± SE	Selenium Mean ± SE	Chromium Mean ± SE	Manganese Mean ± SE	Arsenic Mean ± SE
1996								
Snowy Egret	16	208 ± 171	1950 ± 316	35.9 ± 14.8	390 ± 100	416 ± 47.7	1590 ± 395	
		4	1580	18	113	385	1140	
Great Egret	15	886 ± 234	2270 ± 285	134 ± 80.6	222 ± 69.2	654 ± 113	2670 ± 973	
		177	2090	59	30	540	1650	
Black-crowned Night-Heron	11	341 ± 74.9	2690 ± 738	46.5 ± 6.43	541 ± 76.2	39 ± 250	1980 ± 329	
		161	2030	43	485	60	1750	
χ^2 comparison among species		13.8 (0.001)	1.3 (NS)	13.6 (0.001)	8.2 (0.02)	9.0 (0.01)	3.8 (NS)	
2001								
Snowy Egret	13	359 ± 119	3210 ± 326	41.5 ± 11.8	2040 ± 181	619 ± 99.7	1640 ± 587	137 ± 49.3
		14	2950	16	1930	513	972	8.7
Great Egret	24	738 ± 155	4340 ± 259	9.76 ± 1.66	2650 ± 170	390 ± 84.7	1490 ± 154	700 ± 110
		478	4090	3	2540	61	1340	410
Black-crowned Night-Heron	19	1530 ± 466	3470 ± 186	171 ± 71.5	2320 ± 93.1	756 ± 166	1770 ± 333	77.9 ± 42.6
		294	3370	50	2290	504	1310	0.96
χ^2 comparison among species		6.2 (0.04)	10.4 (0.005)	19.8 (<0.0001)	5.3 (0.07)	4.7 (NS)	3.3 (NS)	29.1 (<0.0001)
2010–2011								
Snowy Egret	6	114 ± 19.9	1250 ± 248	3.6 ± 0.97	510 ± 148	836 ± 557	655 ± 284	61.2 ± 5.5
		105	1105	2	385	247	448	60
Great Egret	29	59.1 ± 10.4	2570 ± 264	15.4 ± 5.6	1100 ± 46.2	216 ± 26.5	315 ± 25	161 ± 12.8
		12	2266	2	1074	168	289	146
Black-crowned Night-Heron	9	26 ± 11.3	3430 ± 515	2.4 ± 0.8	1391 ± 137	265 ± 143	394 ± 85.6	268 ± 35.9
		2	3130	1	1330	143	326	249
χ^2 comparison among species		9.0 (0.01)	11.1 (0.004)	1.1 (NS)	15.1 (0.0005)	0.7 (NS)	0.7 (NS)	18.8 (<0.0001)

Source: Burger, J. et al., *Int. J. Environ. Sci. Eng. Res.*, 3, 147–160, 2012f.
Note: Given are means ± SE and geometric means below. Comparisons are made with Kruskal–Wallis one-way ANOVA, yielding a χ^2 statistic. All values are in ng/g (ppb dry weight). NS = not significant.

Table 13.9   Summary of Table 13.8 Showing the Year with the Maximum Geometric Mean Value in Parentheses for Each of the Metals in the Three Species over Three Sampling Years

	Lead	Mercury	Cadmium	Selenium	Chromium	Manganese	Arsenic
Snowy Egret			1996 (59)				
Great Egret	2001 (478)	2001 (4090)		2001 (2540)	1996 (540)		2010 (410)
BCN Heron						1996 (1750)	

Frederick et al. (2002) were able to find differences. Boncompagni et al. (2003) also did not find differences in several metals among different local colonies of Little Egret in Pakistan, although Barata et al. (2010) did find intercolony differences for the same species in Spain. This suggests that it is not a safe assumption that egrets at several colonies within a bay are sampling the same foods, even within the same year. Although atmospheric deposition may be similar in a given area, other sources of metals may exist, metals may be retained in sediments differently (Atkeson et al. 2003), and the prey base may vary.

In this book, we examine levels of metals in eggs and feathers of waterbirds because they provide different information. Egg may be the best predictor of mercury risk of embryotoxicity, developmental effects, and overall avian reproduction (Wolfe et al. 1998), as well as being indicative of female exposure for those birds that remain in the vicinity of the colony during egg formation. Feathers, in contrast, provide a record of exposure at the time of feather formation, which for prefledging young means while they are still in the nest or on the territory being fed by their parents. Thus, fledgling feathers provide a record of exposure of the young from local food sources as parents normally forage near the breeding colonies to provision chicks. The difficulty with feathers arises when scientists collect the feathers of young at different developmental stages. Frederick et al. (2002) attended to this issue by adjusting a young bird's mercury level by its bill size to obtain a standardized concentration.

Choice of feathers is also important. Spalding et al. (2000a) dosed birds and found that mercury levels were highest in growing scapular feathers, then powder down, and then mature scapular feathers. This suggests that growing breast feathers would have higher levels than mature breast feathers of fledglings because they also contain blood, which would contribute to the levels. It is for this reason that we always took mature breast feathers (even from herons and egrets).

As is clear from Tables 13.2 and 13.4, biologists collect feathers at varying developmental stages, from when they are covered with down to young almost ready to fledge (and sometimes able to fly short distances; our studies). Feathers from fledglings are no longer growing, and the blood supply is atrophied. Because the mercury levels in feathers formed at different times during the prefledging period can vary (Ackerman et al. 2011), ideally comparisons should be made for the same developmental stage. Thus, caution must be used in generalizing from studies in different areas.

There have been few studies that measured exposure rates (or intake rates) for mercury or other contaminants. A rather direct comparison can be made with Great Egrets from Florida, at least with respect to exposure. Frederick et al. (1999) examined exposure by determining mercury in regurgitated food samples of Great Egret chicks at a breeding colony in Florida. Mercury concentrations in the fish fed to chicks ranged from 40 to 1400 ppb (wet weight), giving a total mercury ingestion of 410 ppm (0.41 mg/kg)—a level that would be associated with reduced nestling mass, increased lethargy, and possibly lowered survival (Frederick et al. 1999). Their samples were taken from 1993 to 1996.

In comparison, mercury in the prey fish consumed by egrets and Black-crowned Night-Herons in Barnegat Bay ranged from means of 16.8 to 77.9 ppb (wet weight) for prey fish (Table 10.7), and 490 ppb (wet weight) for Bluefish (Table 10.12). Great Egrets may capture juvenile Bluefish, or may find large Bluefish as carrion and bring back pieces to feed their chicks. These levels are lower than those in Florida. The mercury levels in fledgling feathers from Great Egret from the same years indicate that those from Barnegat Bay averaged 3100 ppb, and those from the Everglades averaged 1400 to 8600 ppb (Rumbold et al. 2001). Thus, some young birds from the Everglades had higher levels—although it should be noted that feathers were collected from chicks whose feathers were not fully grown (in contrast to our studies, where we collect feathers from fledglings). This makes the comparison difficult because feathers that are not fully developed still have some circulating blood levels and may receive additional metals before the blood supply atrophies. Analyzing feathers that are still growing results in measuring the levels of metals in both blood and feathers, making it difficult to compare these values with those in fully formed feathers.

Effects Levels

Effects levels for egrets have been discussed much more fully in Chapter 9. Ardeids exposed to mercury have been shown to exhibit oxidative stress responses (Hoffman et al. 2005) as well as other sublethal effects in the field (Fimreite 1974; Frederick et al. 2011). Spalding et al. (2000a) reported that the same level of mercury produced more severe effects in young birds in the laboratory compared with those in the field (an effect we found for lead as well; Burger and Gochfeld 1997b). This is an important finding because it suggests that it is critical to examine effects in wild-reared nestlings, even if the exposure is administered as it would be in the laboratory. There may be something about the diversity of prey parents feed chicks, parental persistence in feeding them, or growth trajectories that account for this finding.

The studies of Heinz et al. (2009) clearly established in laboratory experiments that there were susceptibility differences to mercury among the eggs of different species of birds. They discussed the problems without drawing conclusions about thresholds from such studies, but noted that Great Egret was in the intermediate vulnerability group, whereas Snowy Egret was in the high vulnerability group (they did not test Black-crowned Night-Heron). There is no reason to assume that such sensitivity differences would not occur for the nestling phase as well.

The lead level in feathers known to cause sublethal adverse effects is 4000 ppb (Scheuhammer 1987; Eisler 1988a; Burger and Gochfeld 2000a). Lower effect levels have been estimated at 2400 ppb (Chapter 9) and 2000 ppb (Golden et al. 2003a,b; Chapter 9). The mean levels of lead for the three Ardeids studied extensively in Barnegat Bay averaged well below these values. After 2000, all three species averaged below 500 ppb. The highest mean lead was 1530 ppb (GM = 294 ppb) in Night-Herons (2001), suggesting no cause for concern. However, before 1989 historic lead levels would have been much higher; hence, a historic impact of lead is likely to have occurred.

The mercury level in feathers known to cause sublethal adverse effects is 4500–5000 ppb, although physiologic changes may occur at lower levels in vulnerable species (Eisler 1987a; Thompson 1996). Common Tern abnormalities occurred in birds with a median feather mercury level of 1800 ppb, and a lowest level of 800 ppb. Arithmetic means and GMs for the three Ardeids were below 5000 ppb, but in 2001 the GMs were 4090 ppb for Great Egret feathers, 2950 ppb for Snowy Egrets, and 3370 for Night-Herons. Thus, some individuals at the high end of the distribution would have been above the least conservative mercury effects level of 5000 ppb, and some would have been below 1800 ppb as well. Furthermore, Heinz et al. (2009) reported that Snowy Egret eggs were much more vulnerable than those of Great Egrets. Thus, the lower levels of mercury in Snowy Egret eggs (mean of 1905 ppb in 2004–2006) may still be a cause for some concern, particularly for sublethal effects that might ultimately affect survival and recruitment.

The cadmium level in feathers known to cause sublethal adverse effects has a large range (210–2000 ppb) depending on species (Abduljaleel and Shuhaimi-Othman 2013; Chapter 9). In 2003–2006, Great Egrets had an arithmetic average of <20 ppb (GM = 59). In 1996, Night-Herons averaged 171 ppb (GM = 50; no Night-Heron feathers were available in 2003–2006). No birds exceeded the lower estimate for adverse effects. Cadmium levels have declined since 2001, and the 2010–2011 levels suggest no current cause for concern.

For most metals, the arithmetic means are much higher than the GMs, indicating right skewed data as is commonly seen in environmental contamination data sets. However, for selenium, the arithmetic means and GMs are quite close, indicating a more normal distribution, presumably because of the homeostatic regulation of selenium levels. However, this would predict that selenium levels should be more or less constant from year to year, which is not the case. The selenium level in feathers known to cause sublethal adverse effects is 5000 ppb (Chapter 9). In 2001, the GM values for all three species did not exceed 5000 ppb: Snowy Egrets, GM = 1930; Great Egret, GM = 2540; Night-Heron, GM = 2290. The complexity of selenium toxicity in relationship to mercury toxicity makes interpretation of toxicity levels difficult (Chapter 9 and Appendix Figure 9A.1).

There are few controlled laboratory experiments for the other metals. However, we found behavioral deficits for both chromium and manganese at about the same dose (and feather levels) as with lead. Our experiments suggested that all three act on the nervous system to produce cognitive and locomotion deficits (Burger and Gochfeld 1995b). Whether these three metals act independently or are synergistic is unknown (especially for these species). The lack of experiments with these two metals is a serious deficit in our understanding. The average levels of chromium and manganese were below the levels achieved in our studies.

It should be noted, however, that Ardeids and the other species we studied are exposed to other contaminants, such as PCBs, dioxins, and PAHs, which can clearly affect behavior, physiology, reproductive success, and survival. For example, Grasman et al. (2013) examined the immunological and reproductive health of Herring Gulls and Black-crowned Night-Herons in the New York–New Jersey Harbor Estuary and reported strong negative correlations between T cell function and PCBs levels, suggesting that these chemicals contribute to immunosuppression in New York Harbor waterbirds. This once again illustrates the importance of examining as many contaminant levels as possible, both fate (levels in tissues) and effects.

SUMMARY AND CONCLUSIONS

A suite of Ardeids (egrets and Night-Herons) are good bioindicators because they eat different foods and are at different nodes on the food chain. The samples collected from these species in Barnegat Bay were collected from the same mixed-species colonies each year, and thus differences among species do not represent yearly differences or local differences, but rather interspecific differences in prey types or sizes. Our data show significant declines in levels of lead and cadmium for eggs and feathers for all three species. Mercury showed mixed results, but generally showed no significant pattern. Selenium and chromium showed no significant trends (expected for a trace element that is regulated in the body). However, manganese (also an essential trace element) showed erratic changes, and arsenic (nonessential) showed no overall pattern. The clearest patterns in Ardeids as well as terns and gulls, were (1) the declines in lead and cadmium (mirroring the removal of lead from gasoline, removal of cadmium from batteries) and regulation of industrial emissions and effluents; and (2) the lack of a significant pattern for mercury, selenium, and chromium.

Great Egrets and Black-crowned Night-Herons had the highest levels of most metals, except for arsenic (Snowy Egret eggs had higher levels). Metal levels were sometimes higher in colonies located in the north and central sections of Barnegat Bay. This may reflect their closer location to New York City and the high contaminant loads, and perhaps contaminants being carried by currents drifting south from the New York–New Jersey Harbor. Some of the mean metal levels in fledgling feathers were above effects thresholds for mercury, meaning that many individuals had levels for mercury that may result in adverse effects. This may be especially true for Snowy Egrets, given their greater sensitivity to mercury than Great Egrets, at least in the egg phase (see Heinz et al. 2009).

APPENDIX

Table 13A.1 Metal Levels (ppb, Dry Weight) in Feathers of Great Egret Fledglings from Barnegat Bay

Year	n	Lead Mean ± SE	Mercury Mean ± SE	Cadmium Mean ± SE	Selenium Mean ± SE	Chromium Mean ± SE	Manganese Mean ± SE	Arsenic Mean ± SE
1989	4	1470 ± 765	3160 ± 305	277 ± 187	1070 ± 435	420 ± 50.0	200 ± 10.0	50.0 ± 17.0
		634	3150	12.2	972	417	200	47.0
1990	15	155 ± 46.7	4555 ± 914	5.16 ± 1.67	1411 ± 134	285 ± 44.5	565 ± 88.0	75.9 ± 10.0
		45.1	3489	0.48	1322	247	493	58.8
1995	34		4950 ± 480					
			3980					
1996	15	886 ± 234	2271 ± 285	134 ± 81	222 ± 69	654 ± 113	2669 ± 973	32.0 ± 7.3
		177	2092	59.3	29.5	540	1651	10.1
2000	4	600 ± 114	1472 ± 217	15.6 ± 3.83	960 ± 123	432 ± 38.3	1113 ± 303	254 ± 74.9
		565	1428	14.2	939	428	1015	196
2001	24	738 ± 155	4340 ± 259	10 ± 1.66	2654 ± 170	390 ± 85	1487 ± 154	700 ± 110
		478	4091	3.03	2537	61.3	1342	410
2003	16	103 ± 12.4	6430 ± 873	9.8 ± 2.9	3253 ± 394	1474 ± 92	898 ± 104	163 ± 48.0
		94.3	5722	19.8	2722	1328	787	15.5
2004	15	52.4 ± 27.5	3340 ± 411	9.8 ± 3	948 ± 95	92 ± 13.7	468 ± 71	61.7 ± 19.9
		2.4	2912	5.79	883	69.1	403	5.32
2005	31	54 ± 15.0	3404 ± 575	5.34 ± 1.32	839 ± 58	144 ± 49.7	358 ± 66	29.8 ± 8.09
		8.3	2366	1.20	789	14.4	279	2.58

(Continued)

Table 13A.1 (Continued)   Metal Levels (ppb, Dry Weight) in Feathers of Great Egret Fledglings from Barnegat Bay

Year	n	Lead Mean ± SE	Mercury Mean ± SE	Cadmium Mean ± SE	Selenium Mean ± SE	Chromium Mean ± SE	Manganese Mean ± SE	Arsenic Mean ± SE
2006	49	219 ± 52.9	2485 ± 272	8.35 ± 1.85	693 ± 44	339 ± 59.9	283 ± 41.8	114 ± 22.7
		14.3	1861	1.03	650	128	226	13.8
2007	31	416 ± 43.5	1471 ± 224	3.59 ± 1.98	1113 ± 84.3	996 ± 125	735 ± 133	231 ± 27.5
		218	652	0.03	793	762	556	100
2008	23	40.4 ± 18.1	1042 ± 170	9.75 ± 5.46	1087 ± 79.7	282 ± 44.7	446 ± 61.5	231 ± 25.7
		0.67	732	0.06	990	140	361	146
2009	15	186 ± 50.4	3163 ± 394	6.33 ± 0.96	1055 ± 72	326 ± 39.3	637 ± 78.0	125 ± 8.8
		126	2827	5.06	1020	299	567	121
2010	15	63.5 ± 12.5	2537 ± 162	2.30 ± 0.63	1081 ± 68	322 ± 30.0	303 ± 29.6	146 ± 13.5
		23.8	2450	0.41	1049	297	279	137
2011	14	54.3 ± 17.3	2610 ± 529	30.5 ± 10.7	1124 ± 64	102 ± 12.5	329 ± 43.0	177 ± 21.9
		5.4	2085	8.0	1100	91.0	300	156
2014	15	200 ± 110	2285 ± 294	9.9 ± 2.4	829 ± 103	220 ± 66	317 ± 71.6	17.4 ± 3.0
		16.3	1951	1.8	727	164	245	8.8
χ^2 comparison among years		120 (<0.0001)	111 (<0.0001)	112 (<0.0001)	146 (<0.0001)	129 (<0.0001)	135 (<0.0001)	130 (<0.0001)

Note: Given are means ± SE and geometric means (GMs) below. Comparisons are made with Kruskal–Wallis one-way ANOVA, yielding a χ^2 statistic. All values are in ng/g (ppb dry weight).

Table 13A.2 Metal Levels (ppb, Dry Weight) in Feathers of Snowy Egret Fledglings from Barnegat Bay

Year	n	Lead Mean ± SE GM	Cadmium Mean ± SE GM	Selenium Mean ± SE GM	Chromium Mean ± SE GM	Manganese Mean ± SE GM	Arsenic Mean ± SE GM	Mercury Mean ± SE GM
1989	18	1792 ± 784	118 ± 43.0	759 ± 95.0	147 ± 18.8	162 ± 38.5	77.8 ± 26.1	2479 ± 249
		283	16.0	696	139	128	9.5	2378
1991	20	185 ± 66.9	35.7 ± 15.3	1003 ± 64.5	123 ± 22	261 ± 31.7	37.6 ± 6.3	1290 ± 291
		111	10.0	958	91.8	234	11.6	906
1992	10	82.3 ± 16.8	8.9 ± 2.4	1125 ± 95.2	127 ± 19.4	236 ± 45.1	13.6 ± 3.3	1683 ± 695
		59.6	6.0	1086	115	208	4.7	1016
1994	10	198 ± 34.3	4.9 ± 0.7	1226 ± 137	322 ± 77.7	438 ± 124	27.4 ± 7.4	1281 ± 310
		172	4.4	1158	246	352	7.6	1037
1995	10							2991 ± 904
								2171
1996	16	208 ± 171	35.9 ± 14.8	390 ± 100	416 ± 47.7	1592 ± 395		1954 ± 316
		4.0	18.5	113	385	1143		1578
1997	15	76.7 ± 15.0	8.9 ± 2.9	1101 ± 133	521 ± 301	282 ± 29	55.3 ± 11.9	2638 ± 584
		42.5	5.8	952	207	262	31.6	2074
1998	6	233 ± 25.8	18.5 ± 5.5	1237 ± 210	341 ± 166	1785 ± 505	64.2 ± 14.7	3858 ± 587
		224	14.7	1134	198	1384	57.2	3633
1999	10	115 ± 16.8	6.8 ± 1.4	874 ± 60	382 ± 142	322 ± 28	56.0 ± 11.8	2096 ± 325
		102	5.9	853	202	311	20.0	1863
2000	10	474 ± 160	19.0 ± 14.4	1058 ± 86	833 ± 333	2071 ± 220	115 ± 39.3	1621 ± 301
		17.8	0.4	1026	510	1970	8.0	1376
2001	9	111 ± 55.0	19.5 ± 5.0	1967 ± 206	446 ± 92.0	617 ± 79.4	146 ± 66.4	3327 ± 353
		14.1	7.4	1877	381	573	6.2	3119

(Continued)

Table 13A.2 (Continued) Metal Levels (ppb, Dry Weight) in Feathers of Snowy Egret Fledglings from Barnegat Bay

Year	n	Lead Mean ± SE GM	Cadmium Mean ± SE GM	Selenium Mean ± SE GM	Chromium Mean ± SE GM	Manganese Mean ± SE GM	Arsenic Mean ± SE GM	Mercury Mean ± SE GM
2003	8	328 ± 73.1	32.7 ± 3.3	1008 ± 194	698 ± 159	1554 ± 265	284 ± 78.0	2322 ± 142
		255	31.6	792	607	1415	217	2291
2004	9	75.0 ± 11.5	10.0 ± 1.8	757 ± 149	140 ± 36.2	548 ± 117	122 ± 20.6	1785 ± 272
		67.5	8.3	642	101	466	105	1599
2005	4	36.3 ± 13.1	6.1 ± 1.0	440 ± 157	55.8 ± 25.7	333 ± 35.4	122 ± 36.2	2783 ± 715
		24.5	5.9	107	40.2	326	106	2304
2006	16	143 ± 18.6	4.7 ± 1.0	1004 ± 40	88.4 ± 25.9	354 ± 64.5	83.7 ± 15.1	1752 ± 268
		128	2.2	992	46.7	303	51.4	1370
2008	11	281 ± 33.5	7.0 ± 1.2	691 ± 90	118 ± 22.3	367 ± 44.6	67.7 ± 25.0	932 ± 91.8
		245	5.8	625	102	339	23.8	886
2010	6	114 ± 19.9	3.6 ± 1.0	510 ± 148	836 ± 557	655 ± 285	61.2 ± 5.5	1251 ± 248
		105	2.1	385	247	448	60.0	1105
χ^2 comparison among years		57.3 (<0.0001)	50.9 (<0.0001)	65.5 (<0.0001)	78.8 (<0.0001)	105 (<0.0001)	42.1 (0.0001)	57.8 (<0.0001)

Note: Given are means ± SE and geometric means (GMs) below. Comparisons are made with Kruskal–Wallis one-way ANOVA, yielding a χ^2 statistic. All values are in ng/g (ppb dry weight).

PART **IV**

Implications, Conclusions, and the Future

CHAPTER 14

Heavy Metals, Trophic Levels, Food Chains, and Future Risks

INTRODUCTION

In birds, metal levels, especially mercury, vary with age, trophic level, foods, and molt (Fimreit 1979; Braune and Gaskin 1987; Stewart et al. 1997; Burger and Gochfeld 2000a,b; Nygard et al. 2001). Levels may differ between years and locations (Burger and Gochfeld 2001b, this volume; Wayland et al. 2005). On an individual level, metal levels are influenced by physiological condition, and clutch size and sequence (Rumbold et al. 2001). Gilbertson et al. (1987) suggested that contaminant levels in marine and coastal birds have lower coefficients of variation than levels in fish or marine mammals. Although the toxic impacts on birds relate to the metal levels in blood, brain, or other organs, eggs and feathers are used as noninvasive indicators of metal contamination because: (1) birds sequester metals in their eggs and feathers (Braune and Gaskin 1987; Burger and Gochfeld 1991); (2) egg levels relate to embryotoxicity and hatching (Heinz et al. 2009); (3) the proportion of body burden that is in feathers is relatively constant for each metal, and related to internal tissue levels (Burger 1993; Golden et al. 2003); (4) a relatively high proportion of the body burden of certain metals is stored in the feathers (e.g., mercury; Furness et al. 1986); and (5) there is often a positive correlation between levels of contaminants in the diet of seabirds and levels in their feathers (Monteiro and Furness 1995).

In this book, we primarily examine levels of metals in eggs and feathers of colonial waterbirds. Although egg levels are thought to be the best predictor of mercury risk to avian reproduction (Wolfe et al. 1998), we suggest that eggs and feathers of prefledging young each provide different information. Eggs provide information on female exposure during egg formation (from circulating levels at the time of egg formation) and eggs represent exposure of the developing embryo. Feathers represent local exposure during the prefledging period because parents of our estuarine species bring back foods from the local area to feed chicks. Chapter 9 provides an overview of effects from metals. In the chapters for each species group (Chapters 10 through 13), we describe levels and discuss potential effects for each species.

In this chapter, we examine effects levels of several of our primary species, individual variations, and issues related to understanding effects levels. We also examine food chain effects, including those for humans, and discuss lessons learned from our 40+ years of study with habitat, populations, and metal analyses.

EFFECTS LEVELS AND INDIVIDUAL VARIATION

Understanding effects levels in wild birds is difficult because it is a three-part process: (1) identifying the tissue levels associated with specific toxic effects in the laboratory (requires analyzing tissue not just dose), (2) analyzing tissue levels in the field compared to the laboratory tissue levels, and (3) relating levels found in the field with effects observed in the field. These three are seldom performed together, or in the same species for the same metals (see also Chapter 9). Stebbins et al. (2009) developed an excellent nonlethal microsampling technique to test for the effects of mercury on wild birds. They showed that taking a microsample from an egg, resealing it, and returning it to the field for normal incubation and hatching resulted in only a slight decline in hatching rate (92% versus 100% for controls). This suggests that scientists can now take a microsample (to test for the female component of mercury in the egg), inject different doses into eggs, reseal the eggs, and obtain an effects level from field-incubated eggs.

Over the years of our study, there have been major declines in metal levels worldwide because of laws and regulations. For example, lead and cadmium have declined in the environment generally, representing industrial ecology successes in eliminating major sources and reducing emissions. Lead has declined in the blood of adults, and lead poisoning in children has declined (Annest et al. 1983; Leafe et al. 2015). Our studies of lead in birds are consistent with these declines in both eggs and feathers for most species. Trends in mercury have not been as clear. Mercury has declined in fish (Cross et al. 2015), and in feathers and eggs of some birds (Atkeson et al. 2003). On a global scale, mercury has declined in air in the Canadian Arctic (Li et al. 2003), but we did not document consistent mercury declines (graphs in this volume).

These declines in metal levels are good news, while at the same time making effects on individuals and populations difficult to document (Herring et al. 2009), even though many species have declining populations. Absence of evidence is not evidence of absence of effects. We must be cleverer in designing methods to discover effects and determine potential population changes. This requires more focused observations to increase the sensitivity of our studies for detecting toxic effects.

Part of our long-term data set goes back to the "dark days" of the 1970s when the Environmental Protection Agency (EPA) was in its infancy and contaminant levels were high. In part, the difficulty is that it may be challenging to identify the individuals in the field that have high levels or that are behaving abnormally. For example, for many, many years we observed the behavior of Herring Gulls from blinds while studying territoriality (Burger 1984c). During this time, we noticed that some chicks wandered into neighboring territories and were killed by those neighbors. We had just ascribed this to bad luck on the part of the chick, overreaction on the part of a neighbor, or bad parenting (absent parent that would otherwise call back its chick). It was only when we were studying the effects of lead on Herring Gull chicks that we found that chicks with high lead exhibited delayed parental recognition, which in the wild meant they wandered into a neighbor's territory without recognizing that the adult was not their parent (see Chapter 9). The nonparent neighbors often killed chicks that wandered into its territory. But it took us many years to put this picture together (Figure 14.1). Furthermore, anomalies and effects turn up periodically and may require years of testing and experimentation to determine the cause. For example, Arnold et al. (2015) reported abnormal feather loss among Common Terns at a small colony in northern Lake Ontario, Canada. Chicks lost all their wing, tail, head, and body feathers. A similar syndrome was reported in Common Terns on Long Island in the early 1970s (Hays and Risebrough 1972; Gochfeld 1975, 1980b). Then it was attributed to PCBs, whereas Gochfeld suggested that it may have been a combination of PCBs and mercury. This 2015 epidemic, still under investigation, highlights the need to remain vigilant.

To illustrate individual variation and potential implications for effects, we use data from Barnegat Bay because it is the most complete data set, but the observations and conclusions can apply to other areas as well. We argue that although means (or medians) are very useful for comparing among areas, and for examining temporal trends, for the species itself, understanding variation and the percentage of any population that falls above an effects level is important. That is, the mean may be below an effects threshold, but if a significant proportion of the population has levels above an effects level, it could suggest problems for the population. In this chapter, we examine variation, and discuss effects levels where known. We recognize that species differ in their susceptibility to each contaminant, and thus the effects levels vary (Heinz et al. 2009), but for comparative purposes

Figure 14.1 Gulls are very territorial, and sometimes kill chicks that wander into their territories. Our experiments indicated that many of these chicks were behaviorally impaired because of lead exposure. Here, a Herring Gull is chasing away an intruder (a Great Black-backed Gull), while its three chicks have wandered to the territory of a neighboring gull (to the left), who may attack the three stranger chicks.

we use a variety of published effect levels (Chapter 9). We mainly use fledgling feathers to illustrate these points, and consider the levels from 2003 to 2006, the years when sufficient fledgling feathers were available for the full range of species. As with our other figures, we arranged the species for all graphs in descending order of mercury levels. Because metal levels are not necessarily correlated within individuals, the species are not in decreasing order for the other metals.

LEVELS IN FEATHERS ASSOCIATED WITH EFFECTS

Effects levels are probably species-specific, as illustrated by various reviews (Ohlendorf et al. 1986a,b; Eisler 1987a, 1988a; Burger and Gochfeld 1988c, 1993c, 1997a,b, 2001b; Frederick et al. 1999, 2002, 2011; Heinz et al. 2009; and many others). However, there are few studies that have systematically tested the effects from one chemical on many species using the same methodology and approaches (but see Heinz et al. 2009). Thus, the literature is a composite of effects levels for different species, derived from different experimental protocols or field observations. The following discussion therefore has a caveat of individual species variations not yet fully explored.

We use the term "fledgling" for birds that are already flying or are almost capable of flight. These birds are at least 3 weeks old (terns) or 6 weeks old (gulls, Ardeids). In Figures 14.2 through 14.8, we present average metal levels in feathers from the period 2003–2006, when we had sufficient samples from nine of our primary species, and the Little Blue Heron. We did not have any feather data from Least Terns. We show the mean (vertical line), standard error (box around the mean), and range (horizontal line for each species). We also indicate the 75th (e.g., 25% of the birds exceeded this level) and the 90th percentiles. This provides a picture of variation over a 4-year period, as well as how many fledgling exceeded certain levels. Where values are available from the literature (Chapter 9), we show the lowest effect levels (dotted line), and the range of effects levels reported (shaded). Without further comparative studies, it is difficult to determine the appropriate effects level for each species.

The commonly used level of lead in feathers often associated with adverse effects in birds is 4000 ppb. That is, if levels in feathers exceed 4000 ppb, then birds may exhibit some sublethal behavioral effects (Eisler 1988a; Custer and Hoffman 1994; Burger and Gochfeld 2001a). However, Golden et al. (2003a) and Spahn and Sherry (1999) suggested that effects could occur at as low as 2000 and 1000 ppb of lead in feathers. We did not find useful information on an effects level for lead in eggs.

Several conclusions can be drawn from Figure 14.2. Although the mean lead level for all species was below the 4000 ppb effects level in feathers, more than 25% of Forster's Tern fledglings and more than 10% of Common Tern and Great Black-backed Gull fledglings exceeded 4000 ppb. However, if for some of these species and endpoints the effects level is as low as 1000 ppb, then most of the birds of most species could be experiencing some sublethal effects from lead. This certainly raises some cause for concern for these species, but also suggests that Snowy Egret, Great Egret, and Black-crowned Night-Heron in Barnegat Bay are not currently at risk from lead.

Toxic chemicals may cause or contribute to mortality. While the birds are still on or near the colony (and accessible to us), we observed many documentable causes of mortality: starvation, violence, cold wet weather, flooding, and disease outbreaks. Each of these had characteristic signs: no breast muscle or fat, bleeding wounds on head and back, soaked bedraggled corpses, yellow diarrheal staining around the cloaca. Toxicity could have impaired parental feeding, inhibited recognition of danger, retarded escape to high ground or cover, and impaired immune response. We also saw dead and live chicks that did not fit these categories. These included developmental defects of the skull, skeleton, and extremities (not always fatal). Some birds exhibited splayed legs and could not stand. Some hatched without down; others experienced broken feathers. We have no doubt that contaminants played a role in these outcomes. The sudden rise and decline of developmental defects in the early 1970s is consistent with a chemical cause. Like the connection with smoking and lung cancer, no single study proves the cause, but many cumulative studies leave us with little doubt.

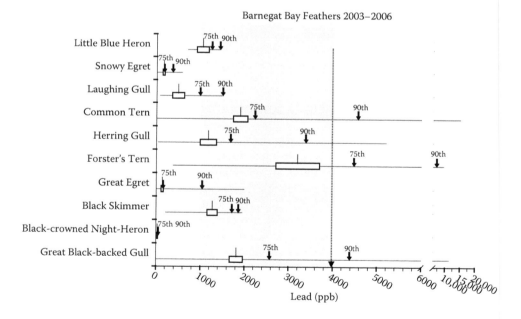

Figure 14.2 Individual variation in lead levels (ppb, dry weight) in feathers of several species from Barnegat Bay. The graph shows mean (vertical bar), standard error (box), and range (horizontal bar), as well as the 75% and 90% of exposure. This indicates what percent of the birds are exposed to higher levels. The dotted line indicates the effects levels normally used, with the shaded area indicating a range of effects levels. The arrow indicates the median level lead level for effects in Common Tern chicks. (From Gochfeld, M., *Mar. Pollut. Bull.*, 11, 362–377, 1980b. With permission.)

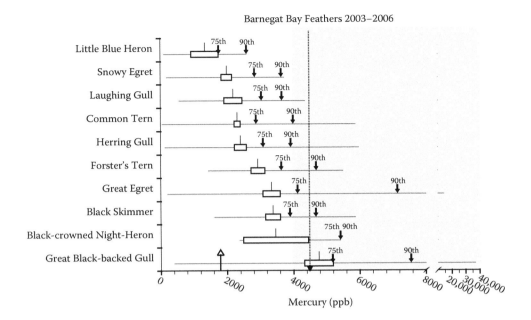

Figure 14.3   Individual variation in mercury levels (ppb, dry weight) in feathers of several species from Barnegat Bay. The graph shows mean (vertical bar), standard error (box), and range (horizontal bar), as well as the 75% and 90% of exposure. This indicates what percent of the birds are exposed to higher levels. The dotted line indicates the effects levels, with the shaded area indicating a range of effects levels.

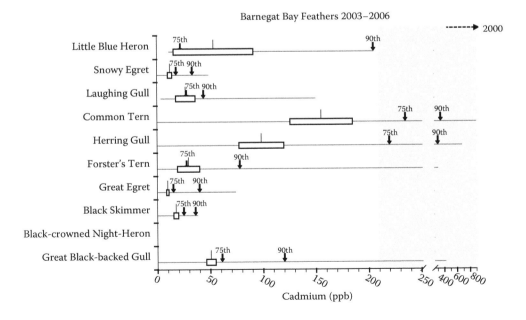

Figure 14.4   Individual variation in cadmium levels (ppb, dry weight) in feathers of several species from Barnegat Bay. The graph shows mean (vertical bar), standard error (box), and range (horizontal bar), as well as the 75% and 90% of exposure. This indicates what percent of the birds are exposed to higher levels. The shaded area indicating a range of effects levels (there is no clear effects level used by researchers).

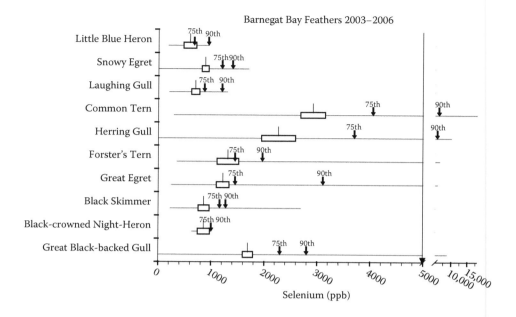

Figure 14.5 Individual variation in selenium levels (ppb, dry weight) in feathers of several species from Barnegat Bay. The graph shows mean (vertical bar), standard error (box), and range (horizontal bar), as well as the 75% and 90% of exposure. This indicates what percent of the birds are exposed to higher levels. The dotted line indicates the effects levels (no range of effects is known).

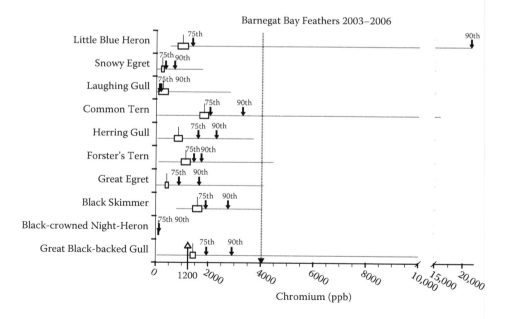

Figure 14.6 Individual variation in chromium levels (ppb, dry weight) in feathers of several species from Barnegat Bay. The graph shows mean (vertical bar), standard error (box), and range (horizontal bar), as well as the 75% and 90% of exposure. This indicates what percent of the birds are exposed to higher levels. The dotted line indicates an effects levels (no range of effects is known). Arrow indicates possible effects at 1200 ppb.

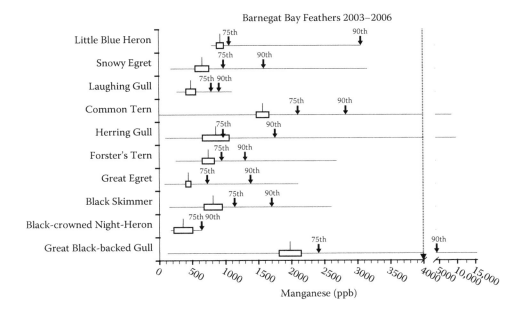

Figure 14.7 Individual variation in manganese levels (ppb, dry weight) in feathers of several species from Barnegat Bay. Graph shows mean (vertical bar), standard error (box), and range (horizontal bar), as well as the 75% and 90% of exposure. This indicates what percent of the birds are exposed to higher levels. The dotted line indicates the effects levels (no range of effects is known).

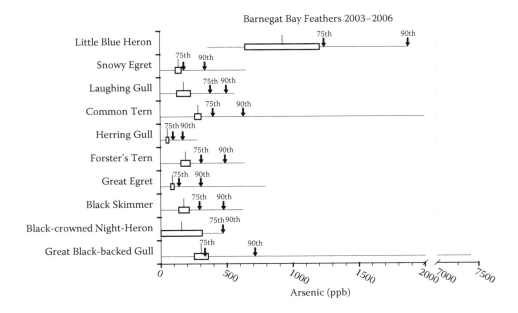

Figure 14.8 Individual variation in arsenic levels (ppb, dry weight) in feathers of several species from Barnegat Bay. Graph shows mean (vertical bar), standard error (box), and range (horizontal bar), as well as the 75% and 90% of exposure. This indicates what percent of the birds are exposed to higher levels. The dotted line indicates the effects levels (no range of effects is known).

There is a high mortality in most birds during the first year of life (particularly in the first few months after fledging; Burger 1980c). The percent of young birds we examined with levels of lead above the effects threshold suggests that some of the postfledging mortality they subsequently experienced may have been due to lead. This aspect needs to be examined. We did find that lead-injected Herring Gull chicks in nature had a higher mortality rate during the prefledging phase compared with control birds (Burger and Gochfeld 1994a, 2000a).

Understanding the effects from mercury is more difficult because of the possibility of demethylation of mercury in the liver (Eagles-Smith et al. 2009a), given that we only measured total mercury. The effects level for mercury in feathers is often given as 4500 ppb (Eisler 1987a; Wolfe et al. 1998; Evers et al. 2008). However, some studies have suggested that the effects level ranges from 4000 to 40,000 ppb, depending on species (Evers et al. 2008). Gochfeld (1980b) reported a median feather mercury level of 1800 ppb in abnormal Common Tern chicks. For mercury, more than 35% of fledgling Great Black-backed Gulls had levels exceeding 4500 ppb (Figure 14.3), and Great Egret, Common Tern, Herring Gull, and Black Skimmer also had some fledglings with levels exceeding the 4500-ppb effects level.

Cadmium clearly causes adverse effects in birds (Eisler 1985a; Scheuhammer 1996; Burger 2008a; see Chapter 9). Currently, the effects level for cadmium in birds ranged from 210 to 2000 ppb (Chapter 9). Overall, the species with the highest means and the greatest variation were Common Tern and Herring Gull's, both having some individuals in the possible effects range (Figure 14.4).

For selenium, the levels in feathers associated with adverse effects is 5000 ppb (Eisler 2000b; Ohlendorf and Heinz 2011, Chapter 9). No species had a mean above this effects level (Figure 14.5). However, some individual Common Terns and Herring Gulls were above the effects threshold. This would suggest cause for concern, but selenium and mercury levels need to be considered together as each may ameliorate the effects of the other. However, mercury levels were not the highest for these two species (refer to Figure 14.3).

Effects levels associated with feathers for chromium and manganese are unknown (Figures 14.6 and 14.7), except that Burger and Gochfeld (1995b) found similar neurobehavioral and locomotory deficiencies in Herring Gulls exposed to manganese and chromium as occurred for lead (e.g., 4000 ppb). If this is generally the case, then most species from Barnegat are not at risk because the levels fell well below this threshold for both chromium and manganese. However, some Little Blue Herons and Common Terns exceeded 4000 ppb for chromium. About 10% of Great Black-backed Gulls exceeded 4000 ppb of manganese.

Effects thresholds are unknown for arsenic (Figure 14.8). Much of the arsenic occurs in the form of arsenobetaine and other organics that have low toxicity. Arsenic levels were rather tight for the species examined in Barnegat Bay. Little Blue Heron, Common Tern, and Great Black-backed Gull had the highest mean levels and the greatest variation (Figure 14.8). The high level of arsenic in Little Blue Heron is puzzling, particularly because mercury, lead, and most other metals were relatively low in this species.

These graphs generally show that some species have large variations in metal levels among individuals, whereas others do not. For example, Forster's Tern showed the greatest variation and range for lead, whereas Great Egret and Great Black-backed Gull showed the greatest variation and range for mercury (Table 14.1). We suggest that these variations need to be considered for risk assessments of possible effects of metals on fledglings and on other age groups as well. If the mean value for a metal in a population is below the effects levels, but a proportion of the fledglings exceed that level, then the population could be adversely affected by the metals, even though one would not guess that from considerations of means alone.

LEVELS IN EGGS ASSOCIATED WITH EFFECTS

Analyzing eggs provides information on female exposure, both from local exposure and longer-term exposure, as well as information on exposure of developing embryos. The relative levels of

Table 14.1　Order from Lowest to Highest Mean Mercury and Lead Levels in Eggs and Fledgling Feathers in Barnegat Bay

Mercury—Eggs	Mercury—Fledgling Feathers	Lead—Eggs	Lead—Fledgling Feathers
Laughing Gull	Little Blue Heron	Common Tern	Black-crowned Night-Heron
Glossy Ibis	Snowy Egret	Glossy Ibis	Great Egret
Herring Gull	Laughing Gull	Herring Gull	Snowy Egret
Snowy Egret	Common Tern	Forster's Tern	Laughing Gull
Great Egret	Herring Gull	Black-crowned Night-Heron	Little Blue Heron
Little Blue Heron	Forster's Tern	Snowy Egret	Herring Gull
Black-crowned Night-Heron	Great Egret	Great Egret	Black Skimmer
Common Tern	Black Skimmer	Little Blue Heron	Common Tern
Forster's Tern	Black-crowned Night-Heron	Laughing Gull	Great Black-backed Gull
	Great Black-backed Gull		Forster's Tern

mercury in eggs did not follow the same pattern as that of fledgling feathers. For example, the lowest level of mercury in feathers was for Little Blue Heron, whereas for eggs it was Laughing Gulls (Table 14.1). The graphs for eggs are shown in Appendix Figures 14A.1 through 14A.5, except for mercury (Figure 14.9) and selenium (Figure 14.10). One of the interesting aspects is that the same species do not have the highest or lowest levels of the same metals in eggs and fledglings (Table 14.1).

The differences in the relative order are attributable to (1) differences in trophic levels, either on the breeding or wintering grounds; (2) shifts in diet between intake of food while egg-laying and intake of food by fledglings; (3) differences in fledgling period (terns spend less time prior to fledging than do herons); and (4) sequestration in eggs differs for different metals. What is clear, however, is that the metal levels vary among species, and generalizations cannot be drawn from one metal to another.

There are not many laboratory and field experiments examining the effects of metal levels on developing embryos. However, there are experiments with mercury and with selenium. The levels of mercury in eggs known to cause adverse effects on the developing embryos range from 750 to 5000 ppb dry weight, depending on species, age, and conditions (Eisler 1987a; Thompson 1996; Heinz et al. 2009; Herring et al. 2010). In Barnegat Bay, average mercury levels in eggs for nine species (eggs from 2003 to 2006 when we had eggs from all species) ranged up to 3600 ppb for Forster's Terns. The nine species showed a rather large range, with Laughing Gulls having the lowest mean (Figure 14.9). Often, the level considered to cause effects in eggs is 500 ppb (wet weight) or 1500 ppb (dry weight). Using this effects threshold would result in risk to several species, including Little Blue Heron, Common Tern, and Forster's Tern. Several species had 10% of the fledgling population with levels above this threshold. The difficulty lies in the interspecific differences in embryo sensitivity (Heinz et al. 2009). Laughing Gulls had the lowest sensitivity. In Figure 14.9, we have indicated the Heinz et al. (2009) effects level for the species they examined as a dotted arrow. Using these values, Herring Gull, Snowy and Great Egrets, and Common Tern would all have some fledglings that could have been impaired by mercury during their embryonic development.

Effect thresholds on developing embryos have been examined for selenium, mainly because of the selenium exposure at Kesterson (see Chapter 9; Ohlendorf 2000; Ohlendorf and Heinz 2011). The proposed effects level range from 6000 to 9000 ppb (Ohlendorf et al. 1986a,b; Ohlendorf and Heinz 2011). However, if the eggs themselves are not affected, selenium concentrations of 1000 ppb (Lemly 1993b) or 3000 ppb (Eisler 2000; Ohlendorf and Heinz 2011) are toxic to other wildlife that eat them. Hence, we show a selenium effect level at 3000 ppb. Mean selenium levels in eggs of terns are still

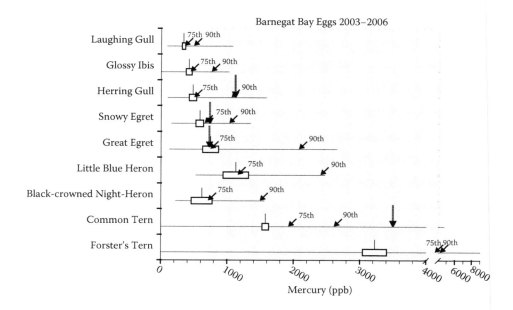

Figure 14.9   Individual variation in mercury levels (ppb, dry weight) in eggs of several species from Barnegat Bay. The graph shows mean (vertical bar), standard error (box), and range (horizontal bar), as well as the 75% and 90% of exposure. This indicates what percent of the birds are exposed to higher levels. The dotted arrows for some species indicate the effects levels as determined by Heinz et al. (2009) from controlled laboratory experiments. The shaded area indicates the range of effects levels given in the literature (refer to Table 9.8).

above this level, suggesting a cause for concern. Using this effects level range, Great Egret, Little Blue Heron, and Black-crowned Night-Herons are most at risk (Figure 14.10). Figures for the other metals in eggs of species from Barnegat Bay are shown in Appendix Figures 14A.1 through 14A.5.

The previous discussion on feather and egg levels included samples from 2003 to 2006, because those were the years with data from the greatest number of species. However, it is also useful to examine data from 2012 to 2015 (Figures 14.11 and 14.12). Comparative egg data from 2012 to 2015 showed similar relative relationships among the species for which we had eggs. Herring Gulls had the lowest average levels of mercury, and Forster's Terns had the highest. Adverse weather and flooding precluded obtaining samples from the full range of species.

Overall, then, the data show that there is a great deal of variation within a 4-year period in the levels of metals and in the variation among individuals within a species and across species. Most species have some fledglings that are above the adverse effects levels for lead and mercury, and even selenium. The question remains about interactions among the metals, and whether having a high level of more than one of these neurotoxins is additive, supra-additive, or protective.

Comparative information on a number of species is not generally available for the other Northeast bays and estuaries. However, Burger and Elbin (2015a) examined metal levels in eggs of a number of different species from the New York–New Jersey Harbor Estuary in 2012. As with the birds in Barnegat, there was little concordance in which species were highest and lowest for any of the metals (Figure 14.10). Given the wide variation in effects levels for mercury, it is difficult to ascribe adverse effects (Table 14.2). However, at the lowest effects level (600–750 ppb; Thompson 1996; Heinz et al. 2009), several species are potentially at risk (e.g., Great Black-backed Gull, Double-crested Cormorant, Black-crowned Night-Heron).

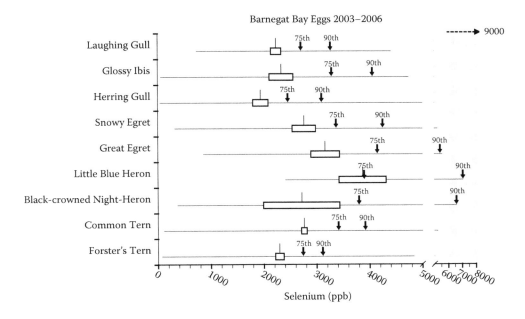

Figure 14.10   Individual variation in selenium levels (ppb, dry weight) in eggs of several species from Barnegat Bay. The graph shows mean (vertical bar), standard error (box), and range (horizontal bar), as well as the 75% and 90% of exposure. This indicates what percent of the birds are exposed to higher levels. The shaded area equals the effects levels suggested by Ohlendorf et al. (1986a,b) and Heinz (1996) (refer to Table 9.10). Accepted effects in birds themselves occur at 6000 ppb, although 3000 ppb is a toxic level for consumers (predators) of eggs.

BUILDING FOOD CHAINS

Metals accumulate in biota, and some bioamplify at each successive level of the food chain. The complexity of food webs is substantial, as food and prey items shift, in species, sizes, and quantities, with year, season, and stage in the reproductive cycle. For example, terns and skimmers bring back large fish to courtship feed mates and very small fish for recently hatched chicks, and they increase fish size as the chicks grow (Burger and Gochfeld 1990b, 1991b). Usually, researchers examine one species of bird, as well as the foods they consume, but there are many different ways of examining food chains. Metal levels may change from year to year and even within a breeding season. Eagles-Smith and Ackerman (2009a,b) showed that mercury in prey fish could rise very rapidly in the spring and decline rapidly during the summer.

Using data presented in Chapters 10 through 13, we developed food chains from Barnegat Bay for lead, mercury, and cadmium—the metals of primary concern—as well as selenium because of its possible ameliorating effects on mercury and cadmium. The base of the generalized food web is invertebrates, as well as phytoplankton, zooplankton, and algae. We analyzed a range of invertebrates, as well as some marine plants (algae) and Horseshoe Crab eggs. These are portrayed at the base of the Barnegat food chain (Figures 14.13 through 14.16).

Several observations can be made from these figures: (1) levels in fledgling feathers are higher than levels in eggs; (2) even though several species are eating prey fish, the levels in eggs and feathers vary; (3) levels in eggs and feathers are similar for most birds; and (4) metals do not accumulate equally (e.g., more mercury ends up in feathers compared to lead or cadmium).

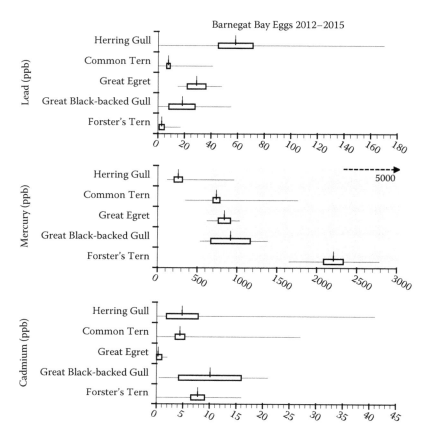

Figure 14.11   Individual variation for levels of lead, mercury, and cadmium (ppb, dry weight) in eggs from 2012 to 2015 for Barnegat Bay. Because of the inclement weather and storm tides, we did not collect eggs of the full range of species.

IMPLICATIONS FOR PEOPLE

Information on populations and metal levels in lower trophic levels, fish, and birds can serve several indicator purposes for humans: (1) they can be sentinels of exposure, as birds and people often live in the same physical space; (2) they can be indicators of exposure as top trophic level fish and birds eat the same foods as humans; and (3) they can be direct indicators of potential exposure to humans through consumption (Peakall and Burger 2003).

The data we provided in Chapters 4 and 7 indicate changes in habitat that can affect people. With sea level rise, salt marsh islands and other low-lying habitats will disappear, reducing available space for housing and for recreational habitat, nursery areas for commercial fish, and shellfish harvesting for commercial interests. Increased frequency and intensity of storms will destroy marshes and sand dunes that protect human communities, to name a few effects.

The population data provided in Chapters 5 and 6 indicate cause for concern because population declines indicate degradation of the environments we also live in. The decline of bird populations may also be an indicator of declining prey fish populations, and prey fish are consumed by the large predatory fish people eat (e.g., Striped Bass, Bluefish, Tuna). Declines in some bird populations deprive hunters (e.g., ducks). Declines in bird populations also decrease the "shore experience" for many people, deprive birdwatchers of their hobby, and reduce the opportunities for photography.

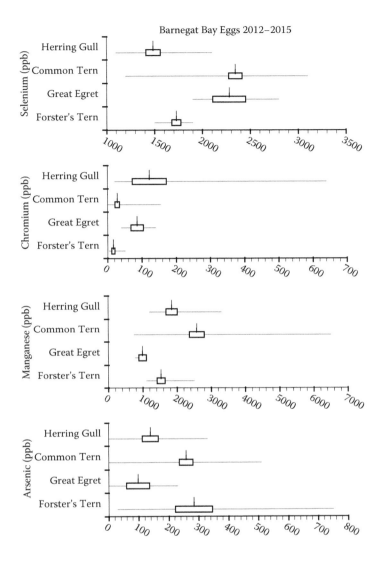

Figure 14.12 Individual variation for levels of selenium, chromium, manganese, and arsenic (ppb, dry weight) in eggs from 2012 to 2015 for Barnegat Bay. Because of the inclement weather and storm tides, we did not collect eggs of the full range of species.

For some, it is the existence value of thriving bird populations that they appreciate, even if they do not "use" them.

The data on metals in Chapters 10 through 13 provide direct information on exposure of birds through the food chain, the same food chain that we are part of. And finally, invertebrates (e.g., crabs, clams), fish, and bird eggs are eaten by people. Metal levels in birds provide information on direct exposure for people (through consumption), as sentinels (because they live in the same environments people live and play in) and as indicators of the health of the food chain leading to people.

Of the species discussed, exposure for most people comes from eating fish (see Chapter 10; Figure 14.17), and for a subset, it comes from eating bird eggs. Both people and birds eat some of the same fish, but birds generally eat the juvenile fish (bait fish or prey fish). However, gulls may eat larger fish as carrion. Thus, metal levels in even the larger individual fish are relevant for both birds

Table 14.2   Comparative Order of Metal Levels in Eggs from New York–New Jersey Harbor Estuary from High to Low

Lead	Mercury	Cadmium	Selenium	Chromium	Arsenic
Great Egret	Canada Goose	Great Egret	Canada Goose	Black-crowned Night-Heron	Great Egret
Double-crested Cormorant	Herring Gull	Great Black-backed Gull	Herring Gull	Great Black-backed Gull	Double-crested Cormorant
Great Black-backed Gull	Great Egret	Herring Gull	Great Black-backed Gull	Canada Goose	Black-crowned Night-Heron
Herring Gull	Great Black-backed Gull	Canada Goose	Double-crested Cormorant	Double-crested Cormorant	Canada Goose
Black-crowned Night-Heron	Double-crested Cormorant	Double-crested Cormorant	Great Egret	Herring Gull	Herring Gull
Canada Goose	Black-crowned Night Heron	Black-crowned Night-Heron	Black-crowned Night Heron	Great Egret	Great Black-backed Gull

Source: Burger, J., Elbin, S., *J. Toxicol. Environ. Health*, 78, 78–91, 2015a.
Note: Given from the lowest (top of list) to the highest (bottom).

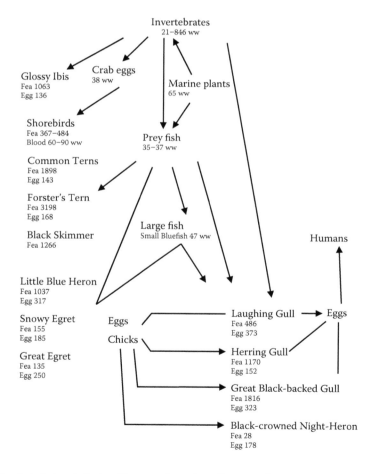

Figure 14.13   Food chain levels for lead for organisms from Barnegat Bay. Levels are ppb, dry weight unless noted (as ww = wet weight).

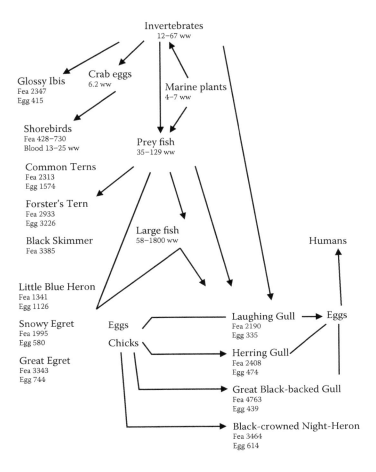

Figure 14.14 Food chain levels for mercury for organisms from Barnegat Bay. Levels are ppb, dry weight unless noted (as ww = wet weight).

and people. Legal size limits for some fish such as Striped Bass (more than 43 inches = 107 cm) may result in people eating individual fish that have the highest mercury levels (Gochfeld et al. 2012).

HUMAN EXPOSURE TO MERCURY FROM FISH

Methylmercury is toxic to all organisms including humans at all ages. However, it is particularly hazardous for the developing fetus (Axelrad et al. 2007). It appears that methylmercury ingested by the mother is preferentially transferred to the fetus, which has higher blood levels of mercury than the mother (Stern and Smith 2003). Fish consumption is the only significant route of exposure for people, and fish-eating individuals and populations have higher mercury levels than people who seldom eat fish. A biomonitoring study in New York City (McKelvey et al. 2007) found that a quarter of adults living in New York City have elevated blood mercury levels, and the proportion exceeds 50% in Asian people who tend to consume more fish, thereby posing a hazard to infants (Rogers et al. 2008).

Determining risk for people requires knowing the consumption rates and the contaminant levels (Burger 2013b). We interviewed fishermen at various points along the Jersey Shore and obtained a variety of fish specimens intended for consumption (Figure 14.17). We analyzed samples for each

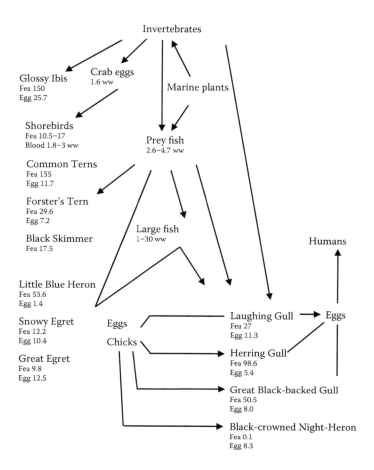

Figure 14.15 Food chain levels for cadmium for organisms from Barnegat Bay. Levels are ppb, dry weight unless noted (as ww = wet weight).

species and computed an arithmetic average total mercury consumption, 90% of which is assumed to be methylmercury. We asked them how much and how often they would eat each particular species, in each month of the year. We accumulated estimates of the mean consumption (grams/day) and the mean concentrations in the fish (micrograms/kilogram wet weight or ppm). In this section, we use ppm wet weight to correspond to risk literature and advisories. We assumed that all our participants had a standard lean weight of 70 kg. We were then able to calculate an average daily intake of mercury (micrograms of mercury per kilogram of body weight per day [µg/kg/day]) for people consuming each type of fish (Figure 14.18). Not surprisingly, the large predatory Mako, Swordfish, and Tuna, with their very high mercury levels, and frequently large meal sizes, resulted in the greatest average consumption and the greatest risk. We compared this to the EPA's Reference Dose (RfD) of 0.1 µg/kg/day (horizontal dotted line), which is the level that people can consume each day without incurring adverse effects.

The EPA defines the Reference Dose as: "An estimate (with uncertainty spanning perhaps an order of magnitude) of a daily oral exposure to the human population (including sensitive subgroups) that is likely to be without an appreciable risk of deleterious effects during a lifetime" (IRIS 2001). The Agency for Toxic Substances and Disease Research, a branch of the U.S. Centers for Disease Control, established Minimal Risk Levels (MRLs), which are analogous to the EPA RfD. "An MRL is an estimate of the daily human exposure to a hazardous substance that is likely to be without

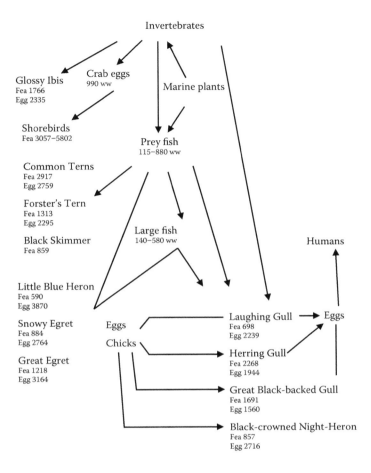

Figure 14.16 Food chain levels for selenium for organisms from Barnegat Bay. Levels are ppb, dry weight unless noted (as ww = wet weight).

appreciable risk of adverse noncancer health effects over a specified duration of exposure" (ATSDR 2015). The chronic MRL for mercury is 0.3 µg/kg/day—3 times higher than the RfD. It is based on a study from the Seychelles Islands (Davidson et al. 2006).

As part of the food chain approach, we examine the relative risk to people from consuming the larger fish, similar in size to what egrets or gulls would eat as carrion (Figure 14.18). Figure 14.18 shows for each fish species the average daily intake based on the annual consumption (lower panel), and the highest monthly intake compared to the RfD of 0.1 µg/kg body weight/day (upper panel). This is based on average consumption by our interviewees and underestimates the exposure to high-end consumers. For several fish species, the people we interviewed were exceeding the EPA RfD. This is not a toxicity or human effects level. The RfD is calculated by dividing a no observed adverse effect level (NOAEL) by uncertainty or protective factors. The mercury RfD is based on human data from a long-term study on Faroe Island (Grandjean et al. 2014). Therefore, some high-end consumers or unusually susceptible people may be experiencing mercury-related symptoms.

The levels are of interest because of the potential effect on people. Methylmercury is the contaminant in fish that usually poses the greatest risk. The U.S. Food and Drug Administration

(a)   (b)

Figure 14.17   People in Barnegat Bay and in other Northeast estuaries are avid fisherfolk and catch some of
the same fish that birds eat as young fry or as carrion washed up on shore. (a) Tony Raniero
holding a Fluke and (b) Joanna Burger holding Bluefish.

action level (a regulatory standard) is 1 ppm (or 1000 ppb wet weight of methyl mercury; USFDA 2001, 2005, 2011), although some countries and states use levels of 0.5 or 0.3 ppm to indicate an unacceptable level (Burger 2013b). High-end fish consumers—people who eat fish more than twice a week—can accumulate toxic levels of mercury, even eating fish with 0.3 ppm (300 ppb wet weight). A person who eats a daily meal (8 oz = 227 g) of fish containing 0.1 ppm of mercury receives a dose of 22.7 μg/day. This is divided by our hypothetical 70 kg body weight, resulting in a dose of 0.32 μg/kg/day—thereby exceeding both the RfD and the MRL. Even if this is the 99th percentile of fish consumption, it would amount to 55,000 adults in New Jersey alone with excessive exposure to mercury from fish.

In addition, some people illegally eat bird eggs, including eggs of gulls and terns. We examined the potential risk from consumption of bird's eggs from both Barnegat Bay and the New York–New Jersey Harbor Estuary. In both places, we have observed people collecting gull and tern eggs to eat. We examined the percentage of eggs that would be above these levels for eggs collected in the past 5 years for some species in the New York–New Jersey Harbor Estuary, and in Barnegat Bay (Table 14.3). Table 14.3 shows that most Great Black-backed Gull and Forster's Tern eggs in Barnegat Bay have exceeded these levels.

We have often been interviewed about the risks and benefits of eating fish. We have concluded that for people who consume fish rarely, eating fish weekly probably confers a health benefit. For people who eat fish frequently (more than twice a week) choosing only low mercury fish would be wise. Information on mercury content of fish is now readily available on the web.

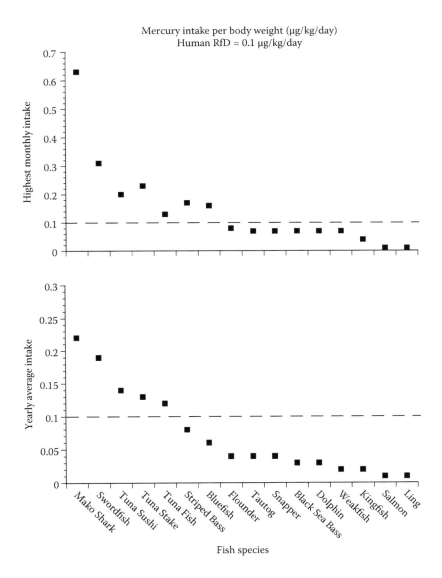

Figure 14.18 Average daily mercury intakes from consuming each species of fish, based on interviews with Jersey Shore fisherfolk, and fish obtained from them for mercury analysis. (After Burger, J., *J. Risk Res.*, 16, 1057–1075, 2013b.) We multiplied the average consumption (grams/day) and average mercury concentration (micrograms/kilogram) to compute the intake (micrograms/kilogram of body weight/day) for each fish species (assuming a 70 kg adult). The dotted line indicates the EPA reference dose (the dose below which a person could be exposed without sustaining adverse effects) (IRIS 2001).

LESSONS LEARNED

The process of examining 40 years of our data on habitats, populations, and metal levels has left us with several clear lessons, were we to start over: These are in no particular order, but bear consideration for future studies by much younger biologists who can devote another 40 years to their projects. They are both comments on what we would like to do with another 40 years, and provide ideas for future students of colonial birds.

Table 14.3   Percentage of Eggs above Regulatory and Advisory Levels of Mercury for Human Consumption

Species	Number of Eggs Analyzed	% Above 300 ppb (0.3 ppm)	% Above 500 ppb (0.5 ppm)	% Above 1000 ppb (1 ppm)
		New York–New Jersey Harbor		
Canada Goose	9	0	0	0
Great Egret	6	0	0	0
Double-crested Cormorant	10	10	0	0
Herring Gull	49	14	2	0
Great Black-backed Gull	25	27	9	0
Black-crowned Night-Heron	10	50	10	0
		Barnegat Bay		
Laughing Gull	25	0	0	0
Herring Gull	17	0	0	0
Great Black-backed Gull	3	100	100	33
Great Egret	6	0	0	0
Common Tern	141	17	1	0
Forster's Tern	35	94	54	0
Black-crowned Night-Heron	12	100	92	50

Sources: Burger, J., Elbin, S., *J. Toxicol. Environ. Health*, 78, 78–91, 2015a; Burger, J., Elbin, S., *Ecotoxicology*, 24, 445–452, 2015b.

Note: Data from the New York–New Jersey Harbor Estuary and from Barnegat Bay (this book).

With our focus on metal contaminants, we did not address other hazards, such as organics or plastic pollution. While we were studying metals in seabirds on Midway Atoll (Burger and Gochfeld 2000b) including cases of frank fatal lead poisoning (Sileo and Fefer 1987; Work and Smith 1996; Burger and Gochfeld 2000c), our attention was riveted on the plastic debris, including discarded cigarette lighters that adult albatrosses fed to their chicks. This was a far greater cause of mortality in albatrosses. There were far too many skeletal outlines on the sand, where chicks had once stood patiently, now surrounding a circle of plastic debris that had once obstructed their stomach and intestine leading to slow starvation. While the lead poisoning on Midway is being ameliorated, plastic pollution of the ocean and seabird mortality are increasing (Lavers et al. 2014; Wilcox et al. 2015).

Our generation of ecotoxicologists has focused attention on heavy metals, chlorinated hydrocarbons (including pesticides, PCBs, and dioxins), and acute and chronic oil exposure and polyaromatic hydrocarbons. To be sure, there remain many questions regarding these contaminants. Recent attention has focused on flame retardants, polyfluorinated compounds (perfluorooctanoic acid, perfluorooctane sulfonate; Chu et al. 2015), pharmaceuticals, and personal products rich in nanoparticles (Osachoff et al. 2014). These compounds are already widespread in water and food chains. Water treatment facilities are not yet designed to capture them. Research on exposure, risk, and effects is in its infancy. There is plenty for ecologists and ecotoxicologists to do.

Here are several suggestions:

1. Do not expect to prove anything, even to prove that a situation is getting better or worse. We do not prove anything in science, but we do erase doubt and increase confidence. Remember the definition of statistics is "never having to say you're certain." *P* is for probability.

2. The best plan for tracking populations is to have a set protocol that is amenable to changing climate conditions and can be followed by well-trained biologists in different places.

3. Deciding the overall sampling plan early on is essential, but it must be flexible. We added some species along the way (egrets, Night-Herons), and deleted others because some populations declined drastically (Least Tern) or they became more vulnerable (Black Skimmer),

4. The best comparisons for metals or other contaminants are those done by the same individuals, the same laboratories, using the same methods year after year. Failing that, procedures and protocols need to be codified and quality assurance standardized and easy to follow for future generations. Along the same lines, all samples need to be archived so that analyses can be repeated and verified in the future, or other contaminants can be examined.

5. There still exists in the literature a dual system of some people using dry weight, some wet, and in some cases, it is not clear what was used. There is a need for a common usage, and common conversions between tissues for wet and dry weight. Our wet/dry conversions are based on very large egg sample sizes. The problem does not exist for feathers, which again makes them very useful as bioindicators of exposure. Conversion from dry weight to weight wet (usually done by a factor of 3 for liver) is not appropriate for all species and tissues. Based on actual determination of moisture content, we found that the conversion for eggs was 4 for gulls and terns but 5 for Ardeids and fish. It may be very different for the eggs of other species.

6. Be inventive in devising methods to examine mixtures, particularly of chemicals known to interact. This is especially true for mercury and selenium, as well as lead with other neurotoxins, and the relative effects of such exposures in the wild.

7. Entertain multiple hypotheses. Exhibiting a temporal decline in metal levels over time clearly reflects a decrease in exposure, but the decrease may not reflect a real change in metal levels, but rather a change in dietary habits. That is, if Herring Gulls were to eat the same prey items, at the same sizes, then a decline in levels in their eggs and feathers would reflect a real decline in overall ecosystem levels. However, if over time, Herring Gulls are no longer able to catch as many top trophic level fish or other organisms, but instead eat more organisms that are lower on the food chain, then they are obviously exposed to less mercury. However, it is not necessarily because there has been a decline in mercury in top trophic level fish, but a change in dietary habitats. Burgess et al. (2013) showed that despite the apparent decline in mercury levels over time in the eggs of Herring Gulls in the Atlantic Canadian region, when dietary shifts are accounted for, the decline disappears. The decline in mercury, for the most part, is not very clear for gulls and terns nesting in Barnegat Bay.

8. Sufficient habitat use and preference data are needed to design restoration and creation of nesting and foraging habitat to replace that lost to development and sea level rise. Such data cannot be acquired over a short period. In our experience, conditions change from year to year, and the variations need to be incorporated into planning and management documents. Furthermore, social factors need to be considered. (Is the habitat large enough to support a critical number of individuals? Can it support multiple species? Is it too close to people? Does it have easy predator access?)

9. Do not give up on banding or on the incorporation of new tracking technologies. They cause disturbance. They are costly. But the return on investment is high. We learned a lot about birds, migration, and exploitation from our banding, color banding, wing tagging, and geolocator studies.

10. Obtain adequate samples. Large samples are power—literally. Variation must always be considered: individual variation, gender and age variation, population variations, and variations in competitors, predators, human disturbance, and weather events, as well as the secondary variations that come from interactions among these variables and with the physical environment. There are also variations in our methods, in censusing, observations, samples collected, species selected for study, and so on.

11. Convince regulators and budget people (or a wealthy benefactor) that ecological services and resources have intrinsic as well as extrinsic value, and that support for long-term studies and monitoring will be amply repaid.

12. Contribute population and analytic data to a common national or international database to facilitate generating regionwide assessments as well as to generate hypotheses.

13. Both sensitivity and specificity of an investigation must be considered. In Chapter 3, we mentioned the importance of sensitivity and specificity of any metric for detecting a phenomenon, such as a population trend or a contamination trend versus mere fluctuations. There are very few sensitive measure of the impact of metal toxicity on bird reproduction. Moreover, any change that we do find lacks specificity, for there are other factors besides toxicity that may impair behavior, reproduction, and survival. We can confidently say that many individual birds have had levels of one or more metals that exceed published effects levels. We can also say that effects were much more likely in days when metal concentrations, particularly lead and cadmium, were an order-of-magnitude higher than

today. We join others (Nisbet, Scheuhammer) in being uncertain whether metal toxicity is contributing in a major way to the declining populations of fish-eating birds today—not just the colonial waterbirds that we study, but species such as the Bald Eagle, now recovering from the impacts of chlorinated pesticides, and the Common Loon, a species with high mercury levels that is still in decline (Scheuhammer 1987; Depew et al. 2013).

14. Do not get trapped in fruitless controversies—whether or not something is an endocrine disruptor, for example. As with the above, DDT (dichlorodiphenyltrichloroethane) was the first widely recognized endocrine disruptor, long before the term became fashionable. Endocrine disruption is very common; anything that is toxic to the brain or gonads can be an endocrine disruptor.

15. Do not give up on contaminants, just because PCBs and metals have declined in the environment and in birds. Society is cleverly designing new substances, for example, nanoparticles in personal care products, the consequences of which are little known. PCBs caused a lot of damage to fish, birds, and people, but it was not the answer to every problem as Nisbet et al. (1996) showed in feminized Common Terns.

16. Do not plan on a 1-year study to reveal great truths, or a 2-year study, or …. We are still learning something new each year or finding puzzles that may not be unraveled until far in the future.

Integrating all the factors we have examined is difficult, and we find we can mainly compare the importance of effects among categories (e.g., habitat, contaminants), rather than between them. Often, we found that things looked bleak after 2 or 3 years of downturn, only to be uplifted by a change of fortune, which itself may be short-lived. There is no absolute beginning or end to a trend, save when we start and stop counting. We read that terns once numbered in the millions. Could that have been true? Long-term data sets are really the key, although getting funding beyond 1 year seems hopeless. Annual surveying, censusing, and monitoring (each slightly different) maximizes the likelihood of detecting and documenting trends, but failing that, monitoring at regular intervals is essential. Monitoring needs to be comprehensive to detect whether birds have moved rather than declined. Finally, anything or everything may be due to, or be influenced by, chance, so leave as little as possible to chance.

Lest we be accused of rearranging the deck chairs on the *Titanic*, we feel it necessary to look to the future (see Chapter 15). As to the *Titanic*, had we been there, we would have immediately removed all the doors, freed everyone, and used the doors as floatation devices for the thousands who needlessly perished because of the inability to recognize imminent danger, and act outside of the box. A few hours in icy waters would have been a small price to pay for a future. The goal is to not only understand the factors affecting populations, but to maintain healthy populations.

To maintain healthy, stable, sustainable populations that are adaptable and resilient is a key goal for a wide range of scientists of different disciplines—including social scientists, engineers, public policy makers, and the public. We will succeed only if we face the major twin problems of overpopulation of coastal areas and sea level rise. We can argue over the causes of each, but it is a fact that more people are moving to (and developing) coastal areas of the world, and that the seas are rising. This, too, results in less available and suitable habitat for the range of coastal species, both solitary and colonial nesters, for coastal ecosystems generally, and for human communities living along coasts. "Climate deniers" may reluctantly accept that climate is warming and sea level rising without accepting responsibility. More and more, we hear policymakers giving up on carbon emission and focusing on adaptation—building higher seawalls or growing gills. The waterbirds have no say in the matter, but we do. And they cannot build higher salt marsh islands.

SUMMARY AND CONCLUSIONS

The ultimate indicator of individual health, reproductive success, and population sustainability is a level or growing population over a period of several bird generations. This requires that each pair of breeding adults raise enough young during their lifetime so that at least two will survive long enough to reach reproductive age and return to the breeding range. This successful recruitment of replacement adults is difficult to measure directly, particularly for the many species where young breeders settle far from their natal territory (outside the researchers' study area). Measuring population size to document trends is subject to uncertainties related to finding breeding birds and counting them. This is much easier for colonial nesting species. Black Skimmers, for example, can be completely enumerated by aerial photography probably using drones in the future. In general, however, measuring recruitment is impractical, and detecting declining populations would lack sensitivity of quickly determining a toxic effect of a contaminant. And, as we have noted repeatedly in this volume, many of the trends are short-lived, and many factors contribute to a population decline, so that a declining population is a nonspecific measure of a contaminant effect.

Examining the overall trends and possible effects on populations requires examining individual variability within and among populations. Mean values are useful in comparing across years and populations, but it is individual variation that contributes to risk. If a significant proportion of the population is above risk thresholds, even though the mean is well below the threshold levels for effects, a significant proportion of the population may be at risk and vulnerable to behavioral impairments that lead to death or lowered reproductive success, which can affect future population levels. We found that for many species and metals, a significant proportion of the population was above the threshold effect levels (even currently), although clearly levels of several metals have declined significantly over the 40 years of our study.

Food chain effects can occur for many metals, and birds are part of a complex food web, where they both eat and are eaten by higher trophic levels. The waterbirds we studied are indicators of their own health and well-being, that of prey that they eat, and that of predators that eat them. They are potential sentinels for people because they live in the same environments, and they are indicative of human exposure because they eat the same fish that we do and, in turn, humans eat them and their eggs.

APPENDIX

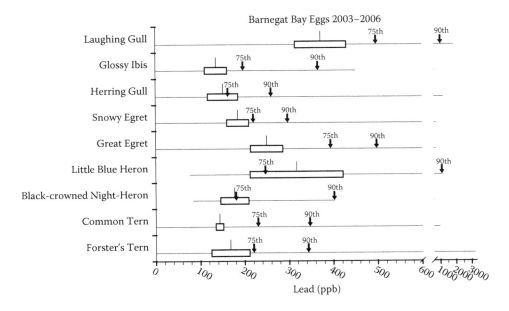

Figure 14A.1   Individual variation in lead levels (ppb, dry weight) in eggs of several species from Barnegat Bay. The graph shows mean (vertical bar), standard error (box), and range (horizontal bar), as well as 75% and 90% of exposure. This indicates what percent of the birds are exposed to higher levels.

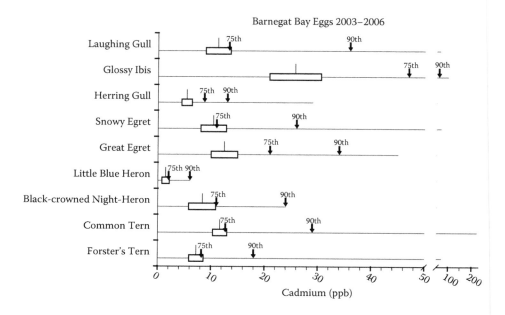

Figure 14A.2   Individual variation in cadmium levels (ppb, dry weight) in eggs of several species from Barnegat Bay. The graph shows mean (vertical bar), standard error (box), and range (horizontal bar), as well as 75% and 90% of exposure. This indicates what percent of the birds are exposed to higher levels.

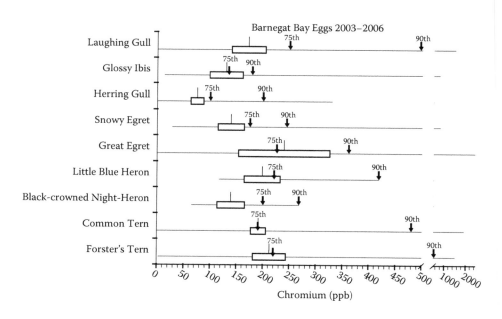

Figure 14A.3   Individual variation in chromium levels (ppb, dry weight) in eggs of several species from Barnegat Bay. The graph shows mean (vertical bar), standard error (box), and range (horizontal bar), as well as 75% and 90% of exposure. This indicates what percent of the birds are exposed to higher levels.

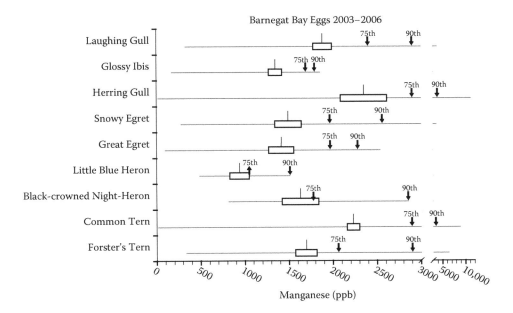

Figure 14A.4   Individual variation in manganese levels (ppb, dry weight) in eggs of several species from Barnegat Bay. The graph shows mean (vertical bar), standard error (box), and range (horizontal bar), as well as 75% and 90% of exposure. This indicates what percent of the birds are exposed to higher levels.

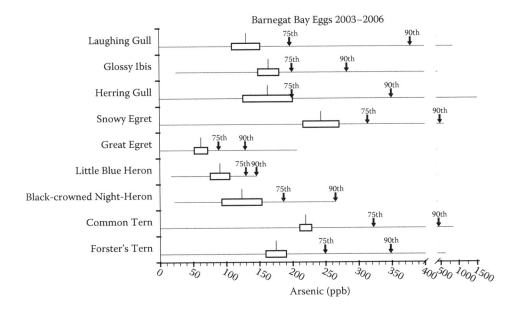

Figure 14A.5   Individual variation in arsenic levels (ppb, dry weight) in eggs of several species from Barnegat Bay. The graph shows mean (vertical bar), standard error (box), and range (horizontal bar), as well as 75% and 90% of exposure. This indicates what percent of the birds are exposed to higher levels.

CHAPTER 15

Colonial Waterbirds—The Future

INTRODUCTION

Our studies of colonial waterbirds in Barnegat Bay over a 40-year period provide a mixture of good news and bad news. The good news concerns the success of environmental concerns and regulation leading to widespread declines in lead and cadmium—and to a lesser extent, mercury—in the coastal environment of New Jersey and the United States. Lead and cadmium are great success stories, while the battle continues to control mercury emissions from coal-fired power plants—the major source of mercury in our estuaries. The bad news is that other environmental factors are worsening, particularly the spectra of climate change, which has current and palpable effects in Barnegat Bay.

We began our studies in Barnegat Bay in the 1970s. Year after year, we documented the shrinking and sinking of islands long before we became aware that we were seeing part of a global phenomenon called "sea level rise." The other piece of bad news is that the plight of some colonial nesting birds in some of the Northeast bays and estuaries considered in this book is dire, at least in some places. Maintaining healthy, viable populations is made difficult by increasing numbers of people moving to coasts, increased development, shifts in species (including Herring and Great Black-backed Gulls, and Double-crested Cormorants), invasive plant and predator species, and contaminants. Sustainability is a challenge because the Northeast region is a hot spot of accelerated sea level rise. Sea level rise is 3–4 times greater in this region than the global average (Sallenger et al. 2012). Furthermore, increases in the maximum high tide results in more frequent and catastrophic flooding of nests (van de Pol et al. 2010). A single flood tide or storm at a crucial period in a 4-month breeding season can nullify weeks of feeding, incubating, brooding, and feeding again, and can eliminate reproduction for that year. As observers of such failures, we can empathize with the birds facing repeated frustrations of failed nesting attempts. Faced with total failure, Black Skimmers have renested for a third and even fourth time in a season.

Regional studies, such as this one in Barnegat Bay, in Chesapeake Bay (Erwin et al. 2010), and on the salt marshes at Jamaica Bay (New York City; Hartig et al. 2002), indicate an overall decline or degradation of coastal salt marsh habitat. Although some species such as the Ardeids and Forster's Terns occur inland as well, others such as Laughing Gull and Black Skimmer are exclusively coastal and are therefore most vulnerable. Long-term studies of terns on Falkner Island and Bird Island document the need for maintenance to counter erosion and shrinkage of the islands (USFWS 1996), whereas Great Gull requires intense maintenance to remove invasive plants. Sustainability of our colonial waterbirds requires vigilance, monitoring, and investment in habitat restoration. Habitat loss and degradation and flooding may account for the overall decline in colonial nesting birds that has occurred, both in Barnegat Bay and in some of the other Northeast bays and estuaries. The colonial-nesting species are successful only on islands that are quite high, compared to their traditional low-lying islands that prevented predator access.

In this chapter, we briefly review the status of waterbirds in the Northeast bays and estuaries, including Barnegat Bay, consider future adaptations, discuss options, and discuss management techniques to preserve and conserve the diversity of waterbirds. It is our intent to look to the future and consider the options that both birds and people have to ensure stable populations of waterbirds in the Northeast bays and estuaries.

Waterbirds should be considered within a context of ecosystem health and well-being, and avian indicators (and endpoints) should be selected that provide information on many different aspects that are of interest not only to biologists managing populations, but to other scientists, public policymakers, and the public (Figure 15.1). Putting avian endpoints and ecology at the center allows managers and policymakers to integrate ecological concerns with the goods and services that ecosystems provide. Birds, their prey fish, and the ecosystems they are part of provide stability for coastal environments from severe storms and flood tides, food and fiber, and recreational and aesthetic values. Birds, especially colonial birds, also provide recreational opportunities for people, and seeing birds at the beach is part of the beach experience.

Managers, conservationists, and the public are increasingly interested in the health and well-being of both human and ecological communities, and want to see the development of indicators that address both (Burger and Gochfeld 1996d). Figure 15.1 shows some of the ways that indicators (species) and endpoints (attributes such as population size, reproductive success, metal levels) can be used to provide information on both human and ecological communities, and that will be useful for assessing both resiliency and sustainability in the face of global change. This figure also indicates the ways these measures can be used to provide early warning that require restoration or land management actions. Monitoring is really at the center of understanding ecological resiliency, and in turn suggesting management, engineering, and restoration actions.

One additional issue that managers and conservationists need to bear in mind is the extent of biological variation at all points in the continuum from species through communities to ecosystems. This variation is compounded when variation in a number of different variables are considered together. For example, understanding reproductive success is partly a function of variations in

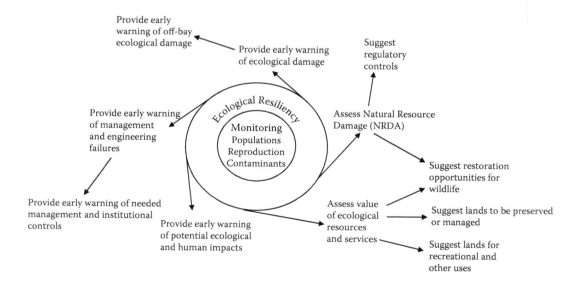

Figure 15.1 Sustainability and resiliency: schematic layout of integrating monitoring of populations, reproductive success, and contaminant levels to provide early warning of effects, assess value of resources, and Natural Resource Damage Assessment relevant to management and public policy.

timing of reproduction, nutrition and clutch size, incubation bouts, hatching success, feeding success, chick growth and development, size and type of prey, antipredator behavior, and chick defense.

At the species level, there are variations in niche width (e.g., habitat adaptations), species behavioral characteristics, reproductive potential (e.g., clutch size, longevity), and contaminants (e.g., exposure, effects) within a framework of species variation. Heinz et al. (2009) identified species vulnerabilities to mercury in the egg developmental phase (Table 15.1). Although this was discussed in Chapter 9, it bears repeating here. More species need to be examined to understand potential differences in how contaminants (in this case mercury) can differentially affect populations, and the implications of these differences for biomonitoring plans and managing populations. Furthermore, differences in vulnerabilities need to be examined for fledgling and adult stages.

Even for the species examined in this book, there were major differences: Laughing Gulls were the least sensitive, Great Egrets, Herring Gulls, and Common Terns were intermediate, and Snowy Egrets were most sensitive. From our perspective, this is interesting because we might have predicted that the more marine species (e.g., Common Terns, Herring Gulls) might be the least sensitive to mercury because they have spent more time in marine waters over their evolutionary history, and marine waters have higher natural (e.g., nonhuman caused) levels of mercury than freshwater. Marine birds, such as albatrosses and petrels, have higher tolerance for mercury and can apparently demethylate mercury more efficiently as suggested by Thompson and Furness (1989a,b), who reported very high mercury levels in albatross feathers.

The differences among species in this one endpoint (egg sensitivity to mercury) suggest similar interspecific differences in sensitivity or vulnerability to exposure for other metals, and such species differences have been reported for lead (Scheuhammer 1996). Similar differences may well occur for other life stages, such as the chick phase or older. This, however, requires laboratory testing similar to the experiments of Heinz et al. (2009), but with different endpoints than LC_{50}s (lethal concentration that kills 50% of test subjects). One measure of vulnerability of the birds examined in this book is the degree to which metal levels have increased or decreased over time. Presumably, if the major contaminants of concern (lead, mercury, cadmium) have declined, then the risk has decreased. But this is relative, and knowing the species-specific adverse effects levels is critical to determining risk.

Table 15.1 Differences in Vulnerability of the Eggs of Different Species to Mercury

Low Vulnerability or Sensitivity	Intermediate Vulnerability	High Vulnerability
$LC_{50} > 1$ ppm	$LC_{50} > 0.25$ to 1 ppm	$LC_{50} < 0.25$ ppm
Mallard	Clapper Rail	American Kestrel
Hooded Merganser	Sandhill Crane	Osprey
Lesser Scaup	Ring-necked Pheasant	White Ibis
Canada Goose	Chicken	Snowy Egret
Double-crested Cormorant	Common Grackle	Tricolored Heron
Laughing Gull	Tree Swallow	
	Herring Gull	
	Common Tern	
	Royal Tern	
	Caspian Tern	
	Great Egret	
	Brown Pelican	
	Anhinga	

Source: Heinz, G.H. et al., *Arch. Environ. Contam. Toxicol.*, 56, 129–138, 2009.
Note: Sensitivity was reported as ppm (wet weight). To convert egg wet weight results to dry weight, wet should be multiplied by 4 for gulls and terns, and multiplied by 5 for egrets and Night-Herons (see Table 3.4). Heinz et al. (2009) estimated the lethal concentration for 50% of the eggs (LC_{50}).

CURRENT STATUS OF WATERBIRD POPULATIONS IN NORTHEAST BAYS

In this book focusing on the breeding season, we described trends in populations of colonial birds in Barnegat Bay and in other Northeast estuaries and bays, and some of the major factors that affect them, including habitat (requirements, uses, losses), global change (human development and disturbance, sea level rise), and metals. Other factors affecting population levels (e.g., invasive species, human exploitation on the wintering grounds) are also important in affecting survival and reproduction in birds.

There were several sources of data on population trends that were presented in Chapters 5 and 6. In addition to our data on population trends, there are the *Birds of North America* (BNA) accounts and Nisbet et al.'s (2013) book on seabirds. The BNA accounts were not all completed at the same time, and some have not been updated for many years. Nisbet et al.'s (2013) book is up to date, but did not include the Ardeids. There are several major conclusions on which our data and these accounts generally agree: (1) Great Egrets are increasing; (2) Snowy Egrets are decreasing; (3) Herring Gulls and Black Skimmers are decreasing; (4) Great Black-backed Gulls are decreasing in the north and increasing in the south; and (5) the other species show variable patterns, increasing in a few places and decreasing in others (Black-crowned Night-Heron, Forster's Tern), or increasing in the north and decreasing in the south (e.g., Laughing Gull, Common Tern). Red Knots and Roseate Terns, the two federally listed species, are decreasing.

These data paint a grim picture for colonial birds. However, there are several reasons for optimism: (1) Some species seem to be moving north as habitat is lost in the south (Common Terns, Forster's Tern, Black Skimmer, Laughing Gull). (2) Some species show mixed trends, which means they might recover in some or most of their range (Black-crowned Night-Heron, Forster's Tern, Black Skimmer). (3) Some species' numbers were higher in 2015 than in previous years (e.g., Laughing Gull, Common Tern, Forster's Tern, Black Skimmer). (4) And lastly, managers and conservationists are taking a firmer hand in managing the public and predators, controlling access to colonies of the former and eliminating the latter to increase reproductive success, and in some places they are actively restoring habitat (see below). In the following section, we compare vulnerabilities and risks using data from Barnegat Bay.

INDICES OF VULNERABILITY AND RISKS OF SPECIES

There are many different factors that affect overall vulnerability of populations, including factors that affect thermal sensitivity, nesting success, foraging success, and survival. Others have examined some factors, such as the effect of temperature changes on habitat (e.g., vegetation changes) and on bird populations (National Audubon Society 2014). Analyses based primarily on climate variables do not take into account massive habitat changes due to sea level rise and subsidence. Thus, climate variables may predict range changes or expansions, but such changes or expansions cannot occur if suitable nesting islands (e.g., for colonial birds) are not available. Furthermore, any such analysis cannot replace having data on habitat lost, as we have presented for Barnegat Bay and Erwin et al. (2007a) have presented for Chesapeake Bay. In both cases, there has been massive loss of salt marsh nesting habitat for colonial species, and an increase in vulnerability of species nesting on marshes because of storm tides and flooding. Not only are some islands vulnerable themselves, but different habitats on them are vulnerable (e.g., low-lying sand banks, wrack on marsh) (Figure 15.2).

Although it is clearly impossible to address all the factors identified earlier, in the following discussion we attempt to summarize some of these factors from the data presented in this book (Table 15.2). We developed some indices of vulnerability, including habitat losses, population trends, mercury and lead trends (indicative of metal exposure), and risk to humans from consumption of eggs. We examined the relative "health" of the populations of colonial birds in Barnegat Bay.

(a)

(b)

(c)

(d)

(e)

(f)

Figure 15.2 Vulnerable habitats in Northeast bays and estuaries include wrack on salt marshes (a) used for nesting by terns and skimmers (b), muddy areas where wrack used to be (c), salt marshes where gulls nest (d), sand patches on salt marsh islands favored by Black Skimmers (e), and even low shrubs in salt marsh islands that may die with salt intrusion (f).

Although these relate only to Barnegat Bay, the same factors affect these species in the Chesapeake Bay (see Erwin et al. 2006, 2007a,b) and, to some extent, birds in the New York–New Jersey Harbor Estuary (Parsons et al. 2001; Craig 2013). Disturbingly, some of the nesting islands we assumed were most secure (e.g., rocky islands for nesting terns) are also being affected by sea level rise and erosion—notably, Falkner Island in Long Island Sound (USFWS 2011b) and Bird and Ram Islands in Buzzards Bay (Massachusetts 2015).

In Table 15.2, we consider habitat viability, population changes, trends in mercury and lead levels, percent of each species having levels above known adverse effects levels in birds (for fledgling

Table 15.2 Current Vulnerability of Primary Species Nesting in Barnegat Bay, New Jersey, Using Fledgling Feathers as an Indicator of Local Exposure and Potential Risk to Offspring, and Egg Levels as Female Exposure

Species	Habitat: % of 1975 Islands Still Suitable	Population Change, % Change (Fold Change)	Mercury: Local Exposure Trends (Feathers)	Lead: Local Exposure Trends (Feathers)	Mercury: Female Exposure (Eggs)	Lead: Female Exposure (Eggs)	Avian Effects: % above Mercury Threshold of 4500 ppb in Feathers	Avian Effects: % above Lead Threshold of 4000 ppb in Feathers	Human Risk: % above Human Mercury Threshold of 300 ppb in Eggs
Common Tern	46%	−84% (6.4 × decline)	Very slight decline	Decline	Very slight decline	Decline	6%	6%	39%
Forster's Tern	50%	−46% (1.8 × decline)	Decline	Increase[a]	No change	Decline	13%	30%	86%
Least Tern	0%[b]	Infinite	NA	NA	NA	NA	NA	NA	NA
Black Skimmer	32%	−92% (12.7 × decline)	No change	No change	NA	NA	13%	0%	83%
Great Black-backed Gull	61%	+1500% (16 × increase)	Slight decline	Decline	Increase	Decline	31%	8%	33%
Herring Gull	44%	−21% (1.3 × decline)	No change	Decline	Slight decline	Decline	10%	1%	7%
Laughing Gull	32%	−85% (6.5 × decline)	Decline	Decline	No change	Decline	0%	0%	2%
Great Egret	137%[c]	+818% (9 × increase)	Decline	Slight decline	No change	Decline	17%	0%	7%
Snowy Egret	75%	−52% (2.1 × decline)	No change	No change	No change	Decline	3%	0%	0%
Black-crowned Night-Heron	114%[c]	−15% (1.2 × decline)	No change	Decline	No change	Decline	15%	4%	10%
Number improved	2	2	5	6	2	8	2	8	
No change, NA			5	3	7	2	6	1	
Number worsened	8	8	0	1	1	0	1	0	

Note: The direct risk to humans from the birds considered is through the consumption of eggs (an illegal practice, but one we have observed). All data for metal levels in this table are from the 2000s. Under metal levels, decline = a higher correlation than 0.20, slight decline is less than 0.19. For avian effects, we give the percent above thresholds, but in the summary (last lines of table) we compare current levels (since 2012) to levels at the beginning of each temporal sequence. NA, data not available.

a The increase in lead in Forster's Tern fledgling feathers is an anomaly, but adults sometimes feed by Forsyth National Wildlife Refuge on polychaete worms, which differs from the diet of the other species.

b Least Terns normally nest on newly created spoil islands (there are none in Barnegat) or on barrier island beaches. The latter can be made suitable by addition of sand and removal of predators and people. In recent years, they have sometimes nested on Holgate (part of Forsythe National Wildlife Refuge) at the south end of Long Beach Island.

c More are now available because of growth of strong bushes (succession on spoil islands).

feathers), and percent of those having mercury levels above 300 ppb in eggs (indicative of potential harm to people eating bird eggs, if they were able to do so frequently). We calculated each factor as follows: (1) for habitat, we computed the percent of islands still usable based on sea level rise/subsidence (see Chapter 7, Table 7.5); (2) for population change, we divided the change by the original number (mean population size 1976–1978—mean population size 2014–2015)/(mean for 1976–1978); (3) for trends, we give the trend for fledgling feathers (indicative of local exposure), and eggs (indicative of female exposure); (4) for effects to the birds themselves, we give the percent of samples (= birds) with levels above the least conservative effects levels for lead and mercury; and (5) for human risk, we give the percent of eggs (eaten by some people) with mercury levels above 0.3 ppm (an effects level for fish; equivalent to 300 ppb; converted to wet weight for comparison). Although all of our waterbird species are "protected" by a variety of state and federal laws, and although the harvesting of eggs is prohibited, we have observed people collecting eggs, and they registered surprise when told that it was illegal, having made no effort to be surreptitious. We have been told, "At home we always collected eggs" with a variety of accents and languages. The effects levels in Table 15.2 are based on metal levels in 2003–2006. Effects would have been much greater in the 1970s and 1980s when levels of lead in particular were higher.

The data indicate that populations declined for 8 of 10 species, and these species had fewer viable colony sites (a bad sign). However, lead levels have clearly declined or remained similar for both eggs and fledgling feathers (19 of 20 possible trends, a good sign). Mercury levels have declined only slightly or remained the same in most species (a bad sign). Effects levels for embryos (an indication of female exposure) indicate potential problems for females. For many species, mercury remains a problem. Although it may have declined as a long-term trend (Cross et al. 2015), when the levels are compared from the 1970s and early 1980s there has been little change in recent years in most species.

Habitat, Populations, and Risk

The Roseate Tern and Piping Plover remain on the Federal Endangered and Threatened Species list, and Recovery Teams are actively involved with their management. Least Terns and Black Skimmers are on the New Jersey Endangered Species List. Table 15.2 indicates that 80% of the nesting species have lost nesting habitat, and 80% have experienced population declines since the mid-1970s in Barnegat Bay. Least Terns have largely disappeared as a colonial-nesting species on the Barnegat Bay islands, although they sometimes nest successfully at Holgate on Long Beach Island, but they require massive restoration of other barrier beaches and management to reduce human disturbance and predator losses. Red Knot, one of our primary species, is not included in Table 15.2 because it does not breed in the Northeast, but migrates through to arctic nesting grounds. However, Red Knot has recently (2015) been added to the U.S. Threatened List, a reflection of its high vulnerability. With active habitat restoration and management, there is hope for restoring and sustaining the populations of colonial birds in Barnegat Bay and those of shorebirds on Delaware Bay.

Coastal birds have had to evolve a certain level of resiliency, among which factors are longevity. Although these species do not breed until age 3 or older, they live for more than a decade and surviving adults may anticipate a dozen breeding seasons in their lifetime, in which each pair needs to raise only two young to recruitment age. In Barnegat Bay, populations of some species were higher in 2015 than in several recent years, perhaps because of a lack of severe storms during the 2014 and 2015 breeding season, and to habitat shifts to slightly higher places on the marshes. Black Skimmers are now nesting on *Spartina patens* (which is higher and slightly drier than *Spartina alterniflora*), and Laughing Gulls are nesting on higher areas closer to marsh creeks. Both gain some protection against flooding. High tides in winter during the past 2 years have brought wrack higher on some marshes, providing more secure nest sites for Common Terns and Black Skimmers, and more sand deposition on some islands providing higher nesting places (Common Terns on

Lavallette). Sometimes this just results in slightly higher marsh vegetation, but at other times sand deposits on the edge of islands, providing sand nesting habitat, which is preferred by terns. Over time, addition of more sand to Mordecai would increase habitat for Black Skimmers.

During our studies, there have been times when there were several years of bad spring storms (high tides, heavy rains) during the colony establishment and egg-laying period, which dampens the number of breeding adults on colony sites. Or later in the season, prolonged rain and high flood tides have washed out eggs or killed fledglings. Although weather (as opposed to climate) does not account for the overall population declines, it does so for some recent failed seasons. Several days of heavy rain and high tides that wash out the early nests can result in colony desertion. The lack of inclement weather and prolonged storms in 2015 may account for the increased breeding populations.

High tides, storms, and surges, particularly in winter, may bring wrack or wash it away. Additional management options, such as building wracks on higher places on the marsh for Common Terns and Black Skimmers, can help as short-term measures (Burger and Gochfeld 1990b, 1991b; Palestis 2009) in the event that the wrack is gone from traditional colony sites. This rather basic management technique in the early spring involves bringing eelgrass and dead grass stems from low-lying islands to the islands where terns and Skimmers have nested. In the month or two before the birds arrive, the wrack settles in place. It is not a permanent solution, lasting for only 1–3 years, but it compensates for the loss of wrack. Other habitat management options are discussed later.

Populations of most species have declined more than the available salt marsh island habitat has declined. For example, there has been an 84% decline in Common Tern populations, but a decline of only 54% in habitat suitability (only 46% if islands with Gulls are considered; refer to Table 15.2). Thus, there is ample room for the populations to rebound, which will likely occur if there are fewer severe storms or prolonged rains in the coming years. Black Skimmers have declined by 92%, but have lost only 68% of their habitat in the Bay, again indicating there is room for Skimmer populations to increase. The upturn in populations of some of these species in 2014 and 2015 is encouraging.

Metals and Risks

Metal levels, food chain effects, and relative risks were discussed more fully in Chapter 14. However, here we summarize and bring together the overall importance of habitats, population numbers, and metals to these colonial species. Lead levels have generally declined (often dramatically) in both fledgling feathers and egg. Regardless of the absolute levels, the declines indicate an overall lessening of contaminant risk to these populations, and a decrease in the percent of developing embryos and chicks with lead levels above an effects threshold. The average levels of lead in fledgling feathers are now well below the threshold for nonlethal adverse effects, except for Forster's Tern (which also had the higher levels in San Francisco Bay; Eagles-Smith and Ackerman 2010). This bears watching in the future, but they may be less sensitive to lead than other species. Unfortunately, Forster's Tern was not one of the species examined in the Heinz et al. (2009) study of mercury in eggs. The decline in lead tracks the removal of lead from gasoline and paint, whereas the declines in cadmium tracked various industrial ecology changes, including the removal of cadmium from batteries and the control of emissions from smelters (Boehme and Panero 2003; see Chapters 9, 11 through 13).

Mercury, on the other hand, showed only slight declines or no change for most fledgling feathers and eggs. Industrial ecology changes have reduced or eliminated many uses of mercury, and reduced emissions from waste incinerators and dental offices. Emissions from power plants continue only partially abated. It is disturbing that mercury levels have remained relatively constant over the past 20 years for many species. Mercury is the contaminant of utmost concern in marine and estuarine waters, and indeed in freshwater because of atmospheric deposition (Fitzgerald and

Mason 1996; Fitzgerald and O'Connor 2001; Davis and Fitzgerald 2004). Because mercury (especially methylmercury) accumulates and biomagnifies up the food chain, it potentially becomes more of a problem to top trophic level species, such as Great Egrets, Black-crowned Night-Herons, and Great Black-backed Gulls that eat larger fish or other vertebrates. This is also a potential threat to other predators, such as large predatory fish, other avian predators (Osprey, Bald Eagle), and people.

The last columns in Table 15.2 speak to the risks to the birds themselves (percent of fledgling feathers over the nonlethal effects threshold) and to people who might eat bird eggs. The risk from mercury in fledglings comes from the embryo (derived from their mother) and from local sources from food collected by their parents. In the present study, the percent of fledglings with levels above the nonlethal effects level (4500 ppb) varies for the different species.

Although many of the trends data suggest declines even in mercury, a closer examination of the data in Chapters 11 through 13 indicates that mercury levels since 2010 are similar to those in the 1970s and early 1980s. Many of the declining trends indicate declines since the late 1980s or 1990s. That current levels of mercury for many species are the same as those at the beginning of our studies indicates some cause for concern, especially given that China is adding new coal-fired power plants on line each month—which will surely increase atmospheric mercury deposition globally.

The direct risk to people from birds is only through consumption of bird eggs. We have observed people collecting gull and tern eggs (in buckets) in nesting colonies in the New York–New Jersey Harbor Estuary, in Barnegat Bay, and at Stone Harbor. Many of these observed collections occurred years ago, but the lack of recent observations may reflect that we are no longer sitting in blinds for days, where people might not be aware of our presence. Although collecting bird eggs for consumption is clearly illegal, some recent immigrants come from countries where collecting bird eggs is an accepted or traditional practice, and they may be unaware of our laws, much less of the potential health risks. The practice is still common in some places; the Aleuts in Alaska consider bird eggs a delicacy, and eat as many as they can during the egg-laying period (Burger et al. 2007d). The risk from such consumption is largely attributable to the short period when eggs are available and people have a peak exposure because they are only available for a short time as undeveloped embryos (if people like bird eggs, they eat them frequently during this period). Parenthetically, mercury is a large problem in many areas. For example, a quarter of adults living in New York City, where the primary source of mercury is from eating fish (and cultural uses in some cases; Rogers et al. 2008), have elevated blood mercury levels (McKelvey et al. 2007).

In this and most other studies, the risk from contaminants is considered singly, metal by metal, organic by organic. Yet fish, birds, and people live in a complex environment with many natural and anthropogenic chemicals. Exposures are complex mixtures, and it is challenging to conduct multi-element toxicity testing in birds. Evaluating mixtures is not always a matter of adding up the hazard values (or hazard quotients); this ignores a range of factors, such as synergisms, antagonisms, pathways, and endpoints or organs for effects (Burger et al. 2002c; Peakall and Burger 2003). The potential for synergisms and antagonisms between the metals or between metals and organics in birds (and other organisms) should be noted. For example, mercury and selenium interact, each to ameliorate the effects of the other (Ralston 2008, 2009).

Even when mixture studies are designed, interpretation can be challenging. Heinz et al. (2011) injected duck eggs with methylmercury and selenomethionine, and found that some doses of methylmercury counteracted the negative effect of selenium on hatching, while enhancing the teratogenic effect, which illustrates antagonism and synergism at the same time for different endpoints. Calcium interacts indirectly with lead in birds (Scheuhammer 1996); calcium deficiency enhances the uptake of lead and cadmium, and lead replaces cadmium in bones. Furthermore, the toxicodynamics of metals are influenced by age, gender, nutritional status, and trophic level (Burger 2008c; Burger and Gochfeld 2001b; Burger et al. 2012f). The risks of the other metals (and of interactions) to birds themselves, predators that eat them, and to people are described and discussed in Chapter 14, but there is no simple answer for ecoreceptors or for people.

In summary, populations for 8 of 10 species have declined, and suitable habitat has declined, but population losses have been proportionally greater than loss of suitable habitat, indicating that populations can rebound with fewer storms and flooding tides. Terns, skimmers, and Laughing Gulls have suffered greater losses than the larger gulls and egrets. Black-crowned Night-Herons have declined the least, and Great Egrets have increased considerably. Contaminant levels have declined for lead, or remained the same for mercury, the contaminants of greatest concern.

Management and Control of Risk

There are many options for local and regional management of chronic or acute exposure from pollutants. Industrial ecology begins with pollution prevention—eliminating the metal from commerce. This has been accomplished to a great extent with lead, mercury, and cadmium. Once in the environment, sources can be identified and removed or contained. Local and regional governments can institute engineering and legal controls, source containment, air emission controls, water effluent regulations and control technology, combined stormwater reduction controls, and improved recycling of metal. Great strides have been made in these areas, largely to reduce human exposure, but atmospheric deposition from other states, regions, and countries remains a major problem (Fitzgerald and O'Connor 2001).

In any case, biologists and conservationists have little control over these pressures on avian populations. Providing monitoring data on levels in tissues and eggs of birds, however, can provide the public, public policymakers, and government officials with the necessary tools to act. However, the provision of sound scientific data, especially trends data for metals that have declined (e.g., lead and cadmium, showing that regulations and laws work) and those that have not (mainly mercury), is an extremely important task, and vigilance is required to make sure that long-term data sets continue, such as mercury in feathers of young Great Egrets in the Everglades (Atkeson et al. 2003), mercury in eggs of Herring Gulls and other species in the Great Lakes (Weseloh et al. 2011), and metal levels in birds from Barnegat and Chesapeake Bays. These programs have provided a picture of changing metal levels in the Eastern United States.

RECOVERY, RESILIENCY, AND ADAPTATIONS

Recovery is the ability of a species to return to pre-stress levels, whether the stress is natural (high predation rates, normal high tides) or anthropogenic in nature (increased predation rates owing to introduced predators or human commensals). Recovery is limited by the reproductive potential of a species (age at first breeding, clutch size, longevity). Resiliency is the ability to respond to stressors, and to be able to withstand them or recover quickly. Both relate to the intrinsic ability of a species, at least for avian populations. Recovery—and to some extent, resiliency—can be aided by human interventions (e.g., removal of introduced or artificially high predator populations).

Species that are federally listed as Endangered or Threatened have recovery plans, and clear directives about monitoring populations, determining causes of declines, and assessing methods of aiding recovery of populations (Olive 2014). Similarly, some state-listed species also receive attention from state agencies. For the species we examined, the major causes of declines are habitat loss due to human development, human disturbance, invasive predators, and sea level rise. Being listed as Endangered is intended to confer some protection from the first two, but is powerless to hold back the sea. These problems will require innovative approaches, which will differ depending on the species, and the bays and estuaries involved.

The Northeast bays and estuaries considered in this book differ from south to north, partly because of geology and structure of the coast. The southern bays (Chesapeake, Delaware, Barnegat, and edging into New York–New Jersey Harbor) have large expanses of salt marshes between barrier islands and the mainland. Some of the salt marsh islands have sand on the edges, washed there by natural processes, but these habitats are very vulnerable to surging tides (Figure 15.3). Salt marsh islands are low, with some elevation, particularly on islands augmented over the years by dredge spoil. As a bird flies north, there are more rocky islands with more elevation, until one reaches Maine, where there are many high elevation rocky islands that will not be affected by sea level rise for centuries. This gradient in habitat type and elevation can provide colonially nesting waterbirds with the option to move north with climate change and sea level rise. If high-elevation, rocky islands are available, with suitable habitat, colonial birds can move to them.

Recently, Distler et al. (2015) evaluated the relative vulnerability of different species of birds based on a number of climate change variables (temperature, moisture, variations) using stacked species distribution models to estimate future biodiversity. That is, the projected future range of one species was overlaid on another, and for the 300+ whose future range they estimated, they were able to estimate future species diversity over the United States. They reported that many of the species we examined would have a wider geographical "temperature" habitat width than they currently have, whereas other species would not. However, they were examining species richness on a broad geographical scale, and were considering temperature variables and the suitability of landscapes to be used by birds—not the specific availability of habitats to support nesting and foraging. Specific studies that evaluated foraging habitat for birds have indicated a substantial loss of suitable intertidal mudflats, as well as other foraging areas for shorebirds, herons, egrets, and other coastal species. Although the future global temperature regime will be more favorable for many species to range further north than today, we predict that when they get north, they will find that available foraging habitats and nesting habitat have declined, and will continue to do so with sea level rise (Galbraith et al. 2002; Erwin et al. 2007a this book). The models developed by Galbraith et al. (2002, 2014) are more appropriate for the species considered in the book because they incorporated climate change and sea level rise to specific bays and estuaries (including Delaware Bay, Galbraith et al. 2002, 2014).

Figure 15.3 Narrow sandy beaches of some salt marsh islands in Barnegat Bay and other bays provide nesting for terns and skimmers, but are very vulnerable to tidal surges in the short run and sea level rise.

Avian Options

There are really only a few options for the colonially nesting birds discussed in this book: (1) they can be extirpated from a location or region (e.g., Least Terns from Barnegat Bay islands), (2) they can move elsewhere within a bay or estuary (egrets moving to more southern colonies in the Bay), (3) they can move elsewhere within the greater region (e.g., Chesapeake or Barnegat to Massachusetts or Maine), or (4) they can widen their habitat niche and behavior (e.g., Black Skimmers starting to nest on *Spartina patens*). There are several examples of the latter: Herring Gulls and Great Black-backed Gulls moving into salt marshes, Double-crested Cormorants nesting on pilings in the New York–New Jersey Harbor, and Black-crowned Night-Herons nesting on the ground in *Phragmites* (Figure 15.4). Many of the colonial species have already undergone habitat shifts, first from sandy beaches on barrier islands (terns, Skimmers) to sand strips on salt marshes, then to wrack strewn on the marsh by high winter tides (both species), and then to *Spartina alterniflora* and *Spartina patens*. Least Terns were unable to make the first switch, and have largely disappeared as a breeding bird on the Barnegat Bay islands, except for Holgate on Long Beach Island. Black-crowned Night-Heron's nesting on the ground in dense stands of *Phragmites* is interesting, because such colonies are usually monospecific and small, but successful (at least in Barnegat Bay). People usually think of *Phragmites* as sterile for nesting birds (although new data are changing this view), and perhaps predators avoid these areas as well. Another explanation for low predation rates, however, is that dense *Phragmites* are difficult to negotiate for some mammalian predators, including people.

Avian behavioral adaptations can partly stabilize some populations. For example, the ability to switch colony sites and switch habitats, even within one breeding season, allows birds to adjust to local conditions that year. When populations are low, younger birds can establish territories and reproduce successfully, thereby supplementing the normal breeding population. Both Herring Gulls and Great Black-backed Gulls established breeding colonies in New York and New Jersey with a higher-than-usual proportion of third- and fourth-year birds. With clutch sizes of more than one,

Figure 15.4 Black-crowned Night-Herons have adapted to nesting in dense stands of *Phragmites*, building nests on the ground and relying on the dense vegetation to hide themselves from predators.

birds can take advantage of good food years to raise all their chicks to fledging. Dividing parental feeding visits among chicks can compensate for a chick that is suboptimal (perhaps because of lead or mercury exposure; Burger and Gochfeld 1994a). That way, the behaviorally impaired chick does not have to compete with more active siblings. For herons and egrets, switching to ground nesting when small shrubs are no longer available allows for successful reproduction, especially when they nest on islands where there are no ground predators.

We, as citizens and managers, have only a few options: (1) we can let birds be extirpated from a region (not a good option); (2) we can provide (create, build, modify) suitable nesting habitat within a bay; (3) we can provide or continue to protect nesting habitat within the greater region (Chesapeake to Massachusetts); or (4) we can conduct massive, multidisciplinary restoration projects to enhance large areas to provide nesting or foraging habitat. All the old methods still apply—predator control, vegetation removal or addition, and wardening. All of our options require us to modify our views and behavior. For the purposes of this discussion, suitable refers not only to the physical and vegetative aspects of the habitat, but also to the social aspect, lack of predators, and lack of human disturbance.

Birds regularly relocate from one colony to another, often because of nest failures or predators (Austin 1940, 1949; Post and Gochfeld 1978; Burger 1979a, 1982a; Burger and Gochfeld 1990b, 1991b). This is a natural part of population dynamics (Ratcliffe et al. 2008; Spendelow et al. 2010; Breton et al. 2014; Palestis 2014). The exponential increase in Laughing Gulls in the breeding colony at Jamaica Bay (New York) (Buckley et al. 1978; Brown et al. 2001b) indicates how quickly this can happen. The colony grew much faster than could be accounted for by reproduction at the colony, and the growth was attributed to the immigration of birds from New Jersey (documented by wing-tagging; Burger, unpublished data).

Some species shift habitats when moving: both Herring and Great Black-backed Gulls shifted to nesting on *Spartina* and under *Iva* bushes on salt marsh islands as they moved south from Canada and Maine (Burger 1977a, 1978a). When they could no longer nest on the sand beaches of barrier islands, Black Skimmers moved into nesting on sandy patches or wrack on salt marshes (Burger and Gochfeld 1990b). Later, in the mid-2000s, when faced with no sand or wrack that was immune from flooding, skimmers began to nest on high areas of *Spartina patens* in marshes in Barnegat Bay (Burger, unpublished data). They successfully fledged young for a number of years, but increases in frequency and severity of floods have resulted in lower reproduction in these habitats, although at the writing of this book, they are still attempting to make these habitats work and are successfully raising young while nesting on *Spartina patens*.

Modification of human behavior and views should start first with biologists and conservationists. Humans created the mess that birds are in, and we should fix it. We developed most of the coastline in the Northeast (except for refuges or parks), preventing the natural succession and buildup of marshes, beaches, and sand dunes. Beach-nesting species, including Common and Least Terns, Black Skimmers, and even Laughing Gulls, nest on open beaches when they are not inundated with people, dogs, and predators. We need to "own" the fact that we already interfered with nature by destroying their habitat—fixing it is not interfering with natural habitats, but trying to return the developed coastline to some semblance of normalcy. There is a precedent for this in the removal of dams in the Northwest to return rivers to natural flow, allowing salmon to go upstream to spawn (Burger et al. 2015b,c). *Return to the River: Restoring Salmon to the Columbia River* (Williams 2006) is a blueprint to restore the Columbia River to its previous free-flowing state, with healthy and diverse populations of salmon. New Jersey and New York Rivers need their monographs that document the decline and restoration of water quality and wildlife. Some of these topics include *A Common Tragedy: History of an Urban River*, which is about the Passaic River (Iannuzzi et al. 2002), research books about the Hudson River (e.g., Smith 1988), and reports from the Harbor Estuary Programs, as well as *Life Along the Delaware River: Cape May, Gateway to a Million Shorebirds* (Niles et al. 2012). There are many popular books about the Chesapeake, as well as popular volumes that serve an important role to focus attention on specific bays and estuaries.

Management Options: Building an Ecological Ark

In 1990, we suggested that as seabird and waterbird biologists, we were "missing the Ark" by ignoring climate change and sea level rise (Burger 1990a). Now, we are suggesting that we need to provide colonial waterbirds with an "Ecological Ark" if they are to survive the continued increases in sea level rise. The "Ark" will vary with species, habitats, and threats, but will include building artificial wracks on salt marsh islands for safe nesting from flood tides (a temporary fix at best), increasing the height of islands where the addition of sand or dredge is possible (a longer-term fix), building artificial islands (a still longer fix), constructing islands that float with rising tides, and managing already-existing high islands so they are free from predators and people (which may involve evicting some people). Such actions will require major shifts in public views of wildlife management and habitat protection, as well as the development of well-thought-out adaptive management and conservation plans for marine and coastal birds. Such plans have been developed for shorebirds (Helmers 1992), herons and egrets (Hafner 2000; Kushlan 2000a,b, 2012; Kushlan and Hafner 2000b; Pineau 2000; Kushlan et al. 2002; Elbin and Tsipoura 2010), Least Terns (Burger 1989), marine birds (Nisbet et al. 2013), waterbirds and their habitats (Parsons et al. 2002; Parsons and Jedrey 2013), and even plans to reduce human disturbance generally (Goss-Custard 2014), and for waterbird colonies near Corps of Engineers projects (Valente and Fischer 2011). "Build it and they will come" works for reestablishing waterbird colonies, particularly with the aid of models and recordings (Burger 1984a, 1988b; Kotliar and Burger 1984). "Maintain and protect it, and they will stay" is a viable philosophy for existing colonies. There are also a number of "Recovery Plans" for endangered species. The protection and conservation of colonial-nesting waterbirds will require continued management; we cannot simply watch complacently, and continue to count birds. As long as there are massive global changes (human population shifts, temperature changes, sea level rise), there will be a need for massive management.

The physical and human dimension aspects of the habitats where colonial birds nest are changing rapidly. People continue to move to cities and to coastal areas, where there is also increased development. The physical environment is changing not only because of shifts in human development and activities along coasts, but because of global warming and sea level rise (IPCC 2007, 2014). Data from the Northeast (this chapter), from North America (BNA accounts), and from elsewhere in the world indicate that a great many colonial waterbirds, including some species of terns, gulls, egrets, and herons, are decreasing over most of their range. The environmental changes are happening too fast for them to adapt. Where they come into increasing contact with humans, they face direct exploitation for food, a problem that will only increase as human coastal populations grow. It is very clear that if these species are to survive and increase, it will require research, conservation, and management—all of which require will, money, time, and commitment. Merely chronicling the history of conservation of colonial waterbirds will not result in their conservation (Kushlan 2012). It will require maintaining inventories, data management, monitoring, predator control (Blodget and Henze 1992), habitat protection, organization building, and conservation planning (Erwin et al. 2002; Kushlan 2012).

Management and Conservation Plans have been developed, both for individual species, and for groups of species: Brown et al. (2001c) for shorebirds, Hafner and Kushlan (2002) for herons, Kushlan et al. (2002) for waterbirds, Parsons et al. (2002) for management of wetlands for waterbirds, and the North American Waterfowl Plan (2012), to name a few. Many other books have conservation woven throughout (e.g., Schreiber and Burger 2001). Even plans to manage birds around airports contain suggestions for protecting the birds while protecting aircraft (Brown et al. 2001a), for example, by creating favorable habitat and encouraging a colony to move to suitable alternative space. Inventories, monitoring, and developing management plans alone are necessary but insufficient to protect and conserve colonial nesting species. Managing the human dimension forces, as well as climate change and sea level rise, is a daunting task, and one that is made possible only with sufficient inventories, monitoring, and biological information about the species and their threats.

Although conservation actions must occur at all levels (local, regional, global), the actions themselves normally take place at a local scale, although they may be implemented over a region (Kushlan and Hafner 2000b). It is at the local scale that land is drained, developed, protected, enhanced, or otherwise rendered useful or useless. Without local involvement and protection, conservation measures will not succeed; without local people appreciating and valuing waterbirds, they will not be conserved or protected. Many of the management options listed in Table 15.3 can be undertaken by local conservation agencies or citizen scientists interested in preserving waterbirds. And we must remember that some of these species spend only part of their lives in our Northeast environment. In the remaining months, they are on wintering grounds where they face similar hazards, with different players.

Although wildlife managers have developed plans for enhancing waterfowl and other wildlife populations for decades, conservation efforts for colonial waterbirds were largely a state-by-state process, or were conducted by individual conservation organizations or scientists. Indeed, some of the waterbird management schemes focus on control of gulls and cormorants and relocation of Ardeid colonies. Some wildlife refuge managers were interested in enhancing colonial birds. In the 1970s, efforts at Agassiz National Wildlife Refuge included building nesting platforms for Double-crested Cormorants that were declining in Minnesota at the time. It worked rather well, and they fledged large numbers of chicks.

Intensive efforts to get rid of Phragmites have occupied the attention of many plant ecologists and managers. For example, Delaware Bay marshes, like the New Jersey Meadowlands, have been heavily impacted by the invasive Reed (*Phragmites australis*), which completely alters the appearance, topography, hydrology, and ecology to the detriment of many species such as killifish (Able and Hagan 2003). However, *Phragmites* do provide ecosystem services, although with a lower level of biodiversity (Kiviat 2013).

PSEG Nuclear (PSEG) operates the Salem Generating Station in Salem County, New Jersey. To address concerns about the large number of fish eggs and larvae killed by its cooling system, PSEG undertook a huge, multistate restoration project—the Estuary Enhancement Program (Teal and Peterson 2005; PSEG 2008). The program had several goals including "Provide long-term benefits to the environment, ecology, natural resources, economy, and people of the Delaware Estuary" (PSEG 2008). The restoration project had three components: the installation of 14 fish ladders to return access to historical river herring habitat, restoration of tidal flow to thousands of hectares of formerly diked salt hay farms, and restoration of *Phragmites*-choked wetlands back to fully functioning tidal marsh.

Achieving long-term control of *Phragmites* is difficult and requires a persistent effort over many years to eliminate regrowth from the well-established underground root system. Annual application of glyphosate-based herbicides is the most effective and least harmful of the control options evaluated by PSEG (Teal and Peterson 2005).

This is the largest privately funded wetland restoration program in the nation, beginning in 1994 and continuing today (PSEG 2015a). PSEG's restoration efforts have achieved unparalleled success. More than 1780 ha of "new" marsh have been added to the Delaware Estuary, and desirable vegetation coverage on the seven primary restoration sites has increased by 2110 ha. Invasive *Phragmites* coverage at the seven sites has been reduced by 1700 ha. The linear length of new tidal creeks, channels, and rivulets, an important measure of habitat quality, has increased by 2635 km as measured through annual monitoring. PSEG continues its efforts to control regrowth from underground *Phragmites* rhizomes and is pushing restoration bounds to get and keep *Phragmites* coverage down below 4% of the total marsh area (PSEG 2015b–d).

Table 15.3   Management Options for Conserving and Protecting Populations of Colonial Nesting Waterbirds

Time Frame	Options	Species	Reference
Short term	Creating impoundments	Wading birds and waterfowl	Parsons 2002
	Control water levels in impoundments FOR birds	Foraging areas for herons, egrets, ibises and shorebirds	Helmers 1992; Erwin et al. 1993; Jenkins and Gelvin-Innvaer 1995
	Add dredge material to dredge islands	Early successional species (Least Terns, Black Skimmers, Common Terns, Piping Plover)	Soots and Landin 1978; Jenkins and Gelvin-Innvaer 1995
	When reclaiming marsh from agricultural lands, ensure mudflats and foraging areas	Shorebirds and waders	Goss-Custard and Yates 1992
	Construct wracks for nesting species high on islands	Common Terns, Black Skimmers, Oystercatchers	Burger and Gochfeld 1990b, 1991b; Palestis 2009
	Vegetation control (invasive or noxious species). Manipulate native vegetation for optimal nesting.	Early successional stage species. Open space for nesting.	Jenkins and Gelvin-Innvaer 1995; Cook and Millenbah 2002
	Conduct sensitivity analyses for different species	All species	Heinz et al. 2008
	Regulating water levels to increase foraging space	All species, especially shorebirds	Collazo et al. 2002
	Create green roofs for nesting	Some shorebirds and ground nesting terns	Fernandez-Canero and Gonzalez-Redondo 2010
	Develop classifications for threats and actions so everyone uses the same language	All species	Salarsky et al. 2008
Long term	Reduce fragmentation of habitat (nesting and foraging)	All species	Fahrig 2003; Andred 2013
	Create new dredge spoil islands	Early successional species	Soots and Landin 1978; Erwin and Beck 2007
	Add small shrubs to new dredge spoil islands	Herons, egrets, ibises	Meredith and Saveikis 1987; Erwin et al. 1991
	Create salt marsh ponds with access to estuaries for fish populations	Foraging for Forster's Terns, herons and egrets	Merredith and Saveikis 1987
	Practice open-marsh water management where essential and with one objective the protection of birds and other wildlife	Marsh species	
	Reduction in contaminants and plastics in the environment	All species	Nisbet et al. 2013
	Develop partnerships to manage agricultural lands, salt ponds, and other habitats for waterbirds	All species	Colwell 2010

(Continued)

Table 15.3 (Continued) Management Options for Conserving and Protecting Populations of Colonial Nesting Waterbirds

Time Frame	Options	Species	Reference
Continuous (both short and long term)	Reduce human disturbance and exploitation	All species	Parsons and Burger 1982; Jenkins and Gelvin-Innvaer 1995; Nisbet et al. 2013; and many other authors
	Control predators (including human-enhanced predators)	All species where needed	Jenkins and Gelvin-Innvaer 1995; Magella and Brousseau 2001; Colwell 2010; Nisbet et al. 2013
	Use captive breeding populations to supplement species in trouble in the wild	Endangered and threatened species	Lyles 2000
	Regulatory protection of nesting habitat	All species	Erwin et al. 1993; Jenkins and Gelvin-Innvaer 1995
	Land use regulations	All species, depending on habitat under regulations	May et al. 2002
	Signage, wardening and education	Least Terns, Black Skimmers, Piping Plovers (as well as many other species)	Parsons 2013
	Develop public–private partnerships	All species	Andrew and Andres 2002
	Manage ecotourism	All species	Butler 2006; Parsons and Burger 1982; Boo 1990; Jenkins and Gelvin-Innvaer 1995
	Create buffers for human approach around nesting islands	All species, both colonial and solitary nesters	Erwin 1989; Erwin et al. 1993; Jenkins and Gelvin-Innvaer 1995
	Evaluate and regulate investigator activities	All species	Erwin 1989; Davis and Parsons 1991
	Monitoring and research	All species	Burger et al. 1994b; Dittman et al. 2010; Parsons 2013; Nisbet et al. 2013
	Evaluate and monitor to determine efficacy of restoration work; requires setting objectives	All species	Erwin and Beck 2007

Note: In many cases, the reference is to the technique that we are suggesting could aid nesting or foraging waterbirds. This is not a complete list of either techniques or references, but a starting point for consideration. They are all ones we endorse. The references listed are not inclusive, but are meant as suggestions for further reading.

All restoration sites provide increased fisheries and primary food source production; expanded habitat; nursery grounds, shelter, and foraging opportunities for fish and other aquatic species; and also provide increased habitat availability for endangered and threatened species, and resident and migrating birds. PSEG and Rutgers conducted extensive evaluation showing the enhancement of aquatic life. Numerous postrestoration studies showed increased fish populations of several species in the former salt hay farms and in the former *Phragmites* marsh (Able et al. 1998, 2003, 2008; PSEG 2015a–d).

Some scientists have taken management matters into their own hands. Roseate Terns prefer more cover over their nests than do Common Terns. Thus, in some rather open colonies, good nest sites are limited (Gochfeld and Burger 1987a). At Falkner Island, Jeff Spendelow solved this problem by providing old tires and making wooden boxes to increase nest site availability (Figure 15.5). The boxes and tires not only provided cover, but they provided shade from the sun, and some protection from predators. We, and later Brian Palestis, built wracks on salt marshes to provide nesting substrate for Common Terns and Black Skimmers (Burger and Gochfeld 1990b, 1991b; Palestis 2009). In some years in the 1980s and early 1990s, the only Skimmers that fledged in Barnegat Bay did so from these constructed wracks. These kinds of efforts clearly can increase the number of birds nesting, as well as overall reproductive success. But they require continuous effort, and large-scale restoration of habitat projects may succeed in providing more permanent solutions.

It is important, however, to step back and consider two things: (1) Is the supposed threat really a threat? (2) What unintended consequences of management might there be? For the first consideration, there are two factors that need to be addressed: How good are we at identifying threats (sensitivity; see Chapter 3)? If we identify something as a threat, is it real (specificity)? For example, Nisbet (2000) has suggested that there is little evidence that human disturbance "causes substantial harm to terns, gulls, or herons." We do not think this can be generalized to all species, colonies, or locations. For example, human activities such as nest checks may lead predators to nests (e.g., Piping Plover, Oystercatchers). Disturbance may keep adults off nests, exposing eggs and particularly chicks to heat stress, or to predators such as gulls (Kury and Gochfeld 1975). Human presence causes young birds to leave their nests, and in the case of Black Skimmer chicks, to run long distances until they reach cover. There is a substantial literature on human disturbance to colonial birds. Most likely, the impact of disturbance is site-specific, species-specific, and even activity-specific.

Figure 15.5 On Falkner Island, Jeff Spendelow provided tires and wooden boxes to increase the availability of suitable nest sites for Roseate Terns, which have a strong preference for nesting under cover.

Unintended consequences are obviously more difficult to consider. For example, vegetation removal to create optimal habitat for one species can bring in other unwanted species to nest. Culling of individual predatory gulls opens up opportunity for more individuals to move in, and predation rates can increase (Magella and Brousseau 2001). Creation of new sandy dredge spoil islands (or enhancement of other islands) can attract the large gulls to nest instead of the smaller terns and Skimmers (as happened in Barnegat Bay). Large wire enclosures are effective in keeping Crows away from Piping Plover nests, but the Crows patrol the enclosures, waiting for the hatchlings to exit. Restoration of habitat requires careful planning. In San Francisco Bay, for example, the restoration and management of salt ponds for migratory waterbirds resulted in low dissolved oxygen, fish die-offs, and increases in gulls to take advantage of the increased food supply. This, in turn, resulted in declines in survival of Forster's Tern chicks—surely an unintended consequence that affected trophic interactions in the system (Takekawa et al. 2015).

It is only in the past 30 years that the field of restoration ecology has emerged as a formal discipline. The first issue of the journal *Restoration Ecology* appeared in 1993. Restoration has become a buzzword for attempting to return degraded habitats to their former condition or to a pristine condition. Hobbs et al. (2013) argue in their book that a pristine condition, or even a former condition, is unrealistic. Years of agriculture, irrigation or drainage, invasive plants, animals, and pathogens, change both the ground and the neighborhood, often thwarting restoration efforts. They conclude that "it is simply impossible to recover historical ecosystems," and attempts to do so are fraught with disappointment, whereas sustainable restoration must account for many factors not usually incorporated in ecosystem descriptions (Hobbs et al. 2013). Restoration activities will produce novel ecosystems, intended to benefit target species, while hopefully doing no harm.

To some extent, restoration strategies or projects implies larger-scale management actions than just planting vegetation, removing vegetation, or controlling predators (Handel et al. 1994). Shrinking islands must be refurbished, dredged material added, and bulkheads or other containment maintained. There are a number of successful restoration projects, for example, the Merrimack River Watershed restoration strategy in New Hampshire (VHB 2009) for forests. One important aspect of these restoration strategies is the recognition that they are not short-term projects that one can complete, declare victory, and walk away. Most restorations require ongoing adaptive management plans that require frequent evaluation and adjustments and investments.

One important aspect of restoration work is the establishment of objectives, goals or endpoints, and indicators so that individual components of restoration projects can be evaluated (Ruiz-Jaen and Aide 2005). Such indicators are necessary to determine efficacy and sustainability of the project (Pastorok et al. 1979; Corvalan et al. 1999). This requires a framework with clear and concise decision points (who is doing what, when; who is evaluating what, when; what is success). Restoration ecology is widely practiced on closed landfills (Robinson and Handel 1993), urban industrial brownfields (Clemants and Handel 2005), closed military bases or nuclear facilities (Traynham et al. 2012), and depleted agricultural lands to create habitats from grasslands to forests (Montalvo et al. 1997). Restoration of beaches have targeted habitat enhancement for endangered Piping Plovers (Maslo et al. 2011), and restoration in estuaries offers the promise of enhancement for multiple species.

Restoration Examples for Northeast Estuaries

There are many large-scale restoration projects in Northeast Bays and estuaries, such as at Buzzards Bay and Jamaica Bay. Both of these had colonial birds and shorebirds as important components (Massachusetts 2015). There are also specific projects to restore bird nesting islands, such as at Bird Island, which just received funding ($5.5 million) to rebuild the protective wall around the island used by endangered Roseate Terns—it should increase the nesting habitat by 100% (I. Nisbet, personal communication). Here, we mention three—one ongoing in the Chesapeake, and two that are relatively new. Colonial nesting birds in the Chesapeake Bay are losing habitat not only because of development but also because of sea level rise (Erwin et al. 2007a,b). There have been a number of restoration projects, including restoration of submerged aquatic vegetation beds, tidal wetland restoration, and island creation/restoration (Erwin and Beck 2007). Poplar Island represents the latter. Poplar Island (Baltimore area) was on the verge of disappearing until the Poplar Island Restoration Project (officially the Paul S. Sarbanes Ecosystem Restoration Project) began more than 10 years ago (Figure 15.6). Like many large restoration projects, its goals were multidimensional, and included restoring and expanding wetland and terrestrial habitat for fish, shellfish, reptiles, amphibians, birds, and mammals (USACE 2015). It was made possible by the beneficial use of dredge material from channels to the Baltimore Harbor. The creation of sandy habitat from dredge material has supported the only colony of Common Terns in the Maryland portion of the Chesapeake for years. However, one of the disadvantages of coloniality is the conspicuousness of the birds, and sooner or later, predators are likely to find them. The major difficulty for the birds on Poplar Island is predation by Great Horned Owl, and more recently by foxes (M. Erwin, personal communication), which has hugely impacted tern reproductive success in recent years. This project illustrates that (1) such projects often have multiple goals or target species, (2) creating optimal habitat may draw in colonial nesters, but predators can succeed in reducing reproductive success (making it a sink, rather than a productive colony), and (3) the project is ongoing, and requires additional dredge material to sustain the ecosystem.

Two recent examples illustrate projects specifically designed for birds (with other secondary objectives). Superstorm *Sandy* devastated much of the coastline from Virginia to Massachusetts. New York and New Jersey were particularly hard-hit (Burger 2015b). The high-energy storm surge removed huge quantities of sand from Delaware Bay beaches, carving away some beaches almost completely, and undercutting many of the structures. The massive destruction of ecological systems and human communities resulted in a governmental response to foster recovery and to increase resiliency of both ecological system and human communities. A series of grants and contracts by different federal agencies aimed to improve what they considered soft infrastructures (sand dunes, salt marshes), engineered barriers (bulkheads, berms, and other structures), and water flow.

The immediacy of rebuilding the "shore" as an economic resource fostered federal and state governmental cooperation, as well as interagency cooperation, to act fast to protect communities

Figure 15.6 Mike Erwin and the Poplar Island Restoration Project in the Chesapeake. (Photo courtesy of Michael Erwin.)

and reopen facilities. One key outcome was that many projects were supported that had both an eco-logical component and a human community component. The projects were intended to improve the local ecology while preventing or reducing the impact of future storms and sea level rise on people. We discuss here two projects that directly benefited the birds considered in this book: (1) restora-tion of sand to several beaches along Delaware Bay to improve spawning habitat for Horseshoe Crabs, improve foraging space for shorebirds, and provide added protection to shore communities; and (2) restoration of Stone Harbor Point on the Atlantic coast to provide habitat for the federally endangered Piping Plover and state-endangered Least Terns and Black Skimmers. Both restoration projects substantially improved the recreational value of the beaches for the months when birds were not present (October through March).

The Delaware Bay Project was led by Larry Niles of Conserve Wildlife and Tim Dillingham of the American Littoral Society, but was a partnership of many different state and federal agencies, as well as conservation organizations. Its objective was to restore sand to Delaware Bay beaches that were used by spawning Horseshoe Crabs each spring. In turn, each spring thousands of shorebirds come to feed on the excess eggs of the crabs, and this spectacle is the largest concentration of spring migrant shorebirds in the United States (Figure 15.7) (Chapter 10). A complex food web in Delaware Bay depends on the preservation of the spawning beaches. The eggs are eaten by a variety of birds, as well as by some small mammals. The small, young Horseshoe Crabs are eaten by a variety of fish from killifish (*Fundulus* sp.) to Weakfish (*Cynoscion regalis*), and intermediate-sized crabs, still with soft shells, are eaten by larger fish as the crabs move out of Delaware Bay onto the Continental Shelf. The effects of *Sandy* on spawning beaches along the Delaware Bay were dramatic. Sand was removed from many beaches, exposing salt marshes to additional erosion. All the sand was washed away from the prime beaches, whereas huge quantities of rock and rubble were strewn ashore. Time was of the essence because the Horseshoe Crabs would not spawn if there was no sand on these beaches. Most participants agreed that there was no time for months of negotiation over responsibility and permits

Following *Sandy*, though, there was a 70% decline in optimal habitat for spawning Crabs and a 20% decline in suitable habitat (Niles et al. 2013). Thus, the shorebirds that require the Horseshoe

Figure 15.7 Foraging shorebirds on a restored beach on Delaware Bay.

Crab eggs to gain energy would not gain enough weight to be able to fly to their Arctic and sub-Arctic breeding grounds. Although there was a clear immediate objective, the long-term objective was to rebuild the beaches so they are part of the salt marsh ecosystem, and can protect the salt marsh ecosystem from sea level rise. This project involved moving truckloads of sand onto the beaches, and it required a whole new level of cooperation and collaboration among management agencies (state, federal), regulators (state, federal), funding agencies (governmental, nongovernmental, and private organizations), management personnel, and local landowners (many of whom lost their beach houses). About 3000 tons of asphalt and crushed concrete were used to repair three access roads to Delaware Bay beaches. More than 800 tons of rubble and other debris that could potentially trap Horseshoe Crabs were removed, and 40,000 tons of sand were placed on the spawning beaches (Niles et al. 2013). The project was completed by the end of April, in time for the spawning Crabs in early May, and more shorebirds used these newly restored beaches than in previous years. The numbers of Horseshoe Crabs and shorebirds using the restored beaches were comparable to numbers using beaches that were not damaged by *Sandy*, and were significantly higher than at damaged beaches that were not restored (Niles et al. 2013).

The Stone Harbor Project, led by David Mizrahi and Nellie Tsipoura of New Jersey Audubon Society and Larry Niles of Conserve Wildlife, has as its objective the restoration and rebuilding of Stone Harbor Point with natural dunes and low swales, with a natural sprinkling of shells on the sandy substrate. It is a partnership between nonprofit organizations, state and federal agencies, academic institutions, and the local municipality. Its stated objective was to restore the beach at Stone Harbor to increase coastal resiliency, improve wildlife habitat, and develop an adaptive strategy for protection from future storms. A stable dune system at Stone Harbor would protect local communities from storm surges and hurricanes. The project involved using sand from the end of the point to construct higher areas that would be above any storm tides, and that would provide safe nesting habitat for Piping Plover, American Oystercatchers, and colonial birds (Skimmers, Least Terns, Common Terns).

Long ago, in the 1970s, the beaches of New Jersey, and we suspect other Northeast bays, had old clam shells scattered about. These provided cover for young Piping Plover and Least Tern chicks, and Least Terns preferentially nested behind big shells. Shells provided protection from the scouring winds (Figure 15.8). With the increase in beachgoers who did not want to hurt their bare feet on broken shells, townships began scraping all the shells off the beach each morning before the people arrived. Nowadays, people do not remember what a natural beach looked like.

The Stone Harbor Project was a big success in drawing several pairs of nesting Piping Plover and American Oystercatchers. The latter successfully raised many young. Least Terns attempted to nest, but were discouraged by predators early in incubation. With the use of decoys, a thriving colony of Black Skimmers (more than 100 pairs), Royal Terns, and Common Terns nested on one of the highest sand "islands" in the midst of lower sections of beach (Figure 15.9). Although the habitat was less than a year old, Sea Rocket (*Cakile edentula*) colonized, providing some protection from the sun for nesting Skimmers. Eventually, however, the colony was destroyed by predators, and will require extensive predator control in future years. In the first year, the project succeeded in providing the habitat needs for colonial birds and nesting shorebirds, but did not deal with the predator problem. This project and the Poplar Island Project illustrate the importance of a total ecosystem approach to restoration. If the habitat is suitable and attractive, but there are too many predators, any colonies will still not be successful.

These projects, one for Horseshoe Crab spawning and shorebird foraging, and one for breeding birds were large-scale, expensive projects that involved moving tons of sand by truck or dredge, but they proved that such projects can be successful with the participation of various groups and agencies, including regulatory agencies. Establishing coalitions around restoration targets is important. These large-scale restorations relied on government funding that was responsive to the immediate needs of ecological and human systems that acted in concert. Both provided advantages to the birds, the coastal ecosystem, and the adjacent human communities.

Figure 15.8 Traditional beach physiognomy included relatively flat, sandy beach areas with abundant shells, providing optimal Least Tern nesting habitat. In recent decades, townships rake away beach debris including shells each morning before beachgoers arrive on the beaches. The shells, however, are important for nesting birds as they provide some protection from the scouring sand and enhance the camouflage effect.

We believe the wave of future conservation efforts will be large-scale restoration projects with multiple goals of protecting human health and well-being, as well as providing safe habitat for wildlife that is sustainable over long periods with minimal cost and effort. Single species projects will give way to multispecies, multiobjective projects aimed at restoring total ecosystems. As colonial waterbird biologists, we need to become team players in designing restoration projects that continue to protect and enhance populations of colonial waterbirds.

SUMMARY AND CONCLUSIONS

Habitat use, exposure to contaminants (specifically metals), and populations of colonial waterbirds are changing and shifting in Barnegat Bay and the other Northeast bays and estuaries. These changes are a direct result of human development along coasts, human disturbance of nesting, foraging and roosting sites, climate change, and sea level rise in addition to the pressures these species have always faced (competition, predation, inclement weather). Populations of many species are declining in the region, as well as in specific bays. Great Black-backed Gull and Great Egret are the only species that are clearly increasing, and the former may not be increasing currently. Some species are declining everywhere (e.g., Black Skimmer, Least Tern), whereas others are increasing in some places and declining in others (e.g., Common Tern, Laughing Gull). In the case of these species, they are declining in the southern areas (Chesapeake Bay, Barnegat Bay) and increasing to the north. This may be an adaptation to decreasing habitat for breeding (e.g., low-lying salt marsh island) in the south, caused by sea level rise and subsidence.

Overall, levels of lead and cadmium have declined sharply (or remained stable) in almost all species (eggs and fledgling feathers), whereas mercury is more variable and in some cases is similar to the levels in the 1970s and early 1980s. The number of eggs and fledgling feathers with levels

Figure 15.9   Restoration of a Black Skimmer colony at Stone Harbor Point, New Jersey, involved adding sand
to raise the colony level with a mixed shell-sand substrate and the use of decoys (a) to attract the
birds. Black Skimmers, Royal Terns, and Common Terns colonized the restored site, 3 months
after construction (b). Sea Rocket vegetation grew quickly on the barren sand.

above effects thresholds is variable, but specific threshold levels for each species and each endpoint have not been determined. However, the levels of some metals are clearly at levels that could result in adverse behavioral effects with adverse effects on reproduction and sustainable populations. Levels of mercury in eggs of some species are at levels that could pose a risk to human consumers.

Overall, the data and observations presented in the book indicate decreased availability of nesting islands and declining populations trends in 8 of 10 indicator species in Barnegat Bay and many other bays, but decreasing levels (and risk) of lead and cadmium, as well as some other metals. Mercury remains a problem. The future of these colonially nesting species, however, is improving because of movements northward to more suitable nesting areas, slight increases in populations in Barnegat in the past couple of years, and increased management of habitat for colonial nesting birds. Large restoration projects aimed at an ecosystem approach to improving coastal habitats for nesting birds and other wildlife are showing great potential to combat the effects of human development and sea levels rise.

In writing this book, we have had the pleasure to relive and reexamine 40 years of experiences with breeding birds, migrating shorebirds, Horseshoe Crabs, and beaches and marshes. Our work has focused around habitat use, population stability, and metals, and it is gratifying that some of the most serious metals have declined, so that significant toxic effects are probably local and infrequent. We recognize that there are many other important toxic contaminants including traditional ones such as chlorinated pesticides, polychlorinated biphenyls, and dioxins, which are persistent organic pollutants that have caused epidemics in the past. There are also new concerns such as fire retardants, pharmaceuticals, and personal care products that impact physiologic processes, particularly the endocrine system, and may take over in producing significant impacts on wildlife populations. Fishing, overfishing, and aquaculture constantly change prey availability, and wholesale starvation events, for example, experienced by penguins in the Falkland Islands, will likely be more conspicuous than the subtle impacts of toxic chemicals. Those chapters wait to be written. But it is by far human development, and sea level rise, that poses the greatest threat, and the greatest challenge for us to solve as we try to increase resiliency in avian populations, coastal ecosystems, and the human communities that live along the coasts.

Figure 1 A Common Tern standing over its nest, showing their preference for sandy beaches with some growing vegetation that will provide shade for their chicks.

Figure 2 Pair of Common Terns feeding small chicks; note that the nest has been placed under vegetation.

Figure 3 Once they moved onto salt marshes, Black Skimmers and Common Terns both preferred to nest on vegetation thrown up by winter high tides (wrack).

Figure 4 The other salt marsh habitat that Black Skimmers prefer is small patches of sand on salt marshes. Space is limited, and birds nest close together.

Figure 5 Black Skimmers nest very densely, and form a mass leaving a colony when disturbed by boaters.

Figure 6 As the tide rises in Delaware Bay, foraging Laughing Gulls, Red Knots, and other shorebirds are forced into smaller and smaller sandbars, increasing competition.

Figure 7 In the early spring, pairs of Laughing Gulls form and loaf along Delaware Bay. The male (right) is larger than the female (left).

Figure 8 Nest material is in short supply, and here an intruder is landing to steal nest material from the defending nest owner (on right).

Figure 9 Great and Snowy Egrets nest in bushes on salt marshes, and their colonies are often surrounded by colonies of Great Black-backed and Herring Gulls that nest in the *Spartina*.

Figure 10 Herring Gulls also prefer some cover to provide shade for their small chicks, and build substantial nests.

Figure 11 Herring Gulls have moved into the highest places on the marshes, forcing Laughing Gulls to nest lower.

Figure 12 As the tide moves into creeks, Snowy Egrets forage in groups at the edge, picking up small prey fish.

Figure 13 Great Egrets normally forage by standing in shallow water and waiting for fish to swim by, but have also adapted to feeding around docks.

Figure 14 Black-crowned Night-Herons also stand alone and wait for predators, mainly foraging at dawn and dusk.

(a)   (b)

Figure 15 Black-crowned Night-Herons (a) and Snowy Egrets (b) have asynchronous hatching, resulting in chicks of different sizes in the same nest. Black-crowned Night-Heron chicks are aggressive when disturbed.

Figure 16 When people or predators approach an egret and heron colony, the egrets come to the top of the vegetation, preparing to take flight. Night-Herons usually stay below until the egrets depart, or they depart out the back silently.

References

Abduljaleel, S. A., and M. Shuhaimi-Othman. 2013. Toxicity of cadmium and lead in *Gallus gallus domesticus* assessment of body weight and metal content in tissues after metal dietary supplements. *Pakistan Journal of Biological Sciences* 16:1551–1556.

Able, K. W. 2005. A re-examination of fish estuarine dependence: Evidence for connectivity between estuarine and ocean habitats. *Estuarine and Coastal Shelf Science* 64:5–17.

Able, K. W. 2015. *Station 119: From Lifesaving to Marine Research.* West Creek NJ: Down-the-Shore Publishing.

Able, K. W., and M. P. Fahay. 1998. *The First Year in the Life of Estuarine fishes in the Middle Atlantic Bight.* New Brunswick: Rutgers University Press.

Able, K. W., and M. P. Fahay. 2014. *Ecology of Estuarine Fishes: Temperate Waters of the Western North Atlantic.* Baltimore: John Hopkins Univ. Press.

Able, K. W., and S. M. Hagan. 2003. Impact of Common Reed, *Phragmites australis* on essential fish habitat: Influence on reproduction, embryological development, and larval abundance of Mummichog (*Fundulus heteroclitus*). *Estuaries* 26:40–50.

Able, K. W., J. H. Balletto, S. M. Hagen, P. R. Jivoff, and K. Strait. 2007. Linkages between salt marshes and other nekton habitats in Delaware Bay, USA. *Reviews in Fisheries Science* 15:1–60.

Able, K. W., T. M. Grothues, S. M. Hagan, M. E. Kimball, D. M. Nemerson, and G. L. Taghon. 2008. Long-term response of fishes and other fauna to restoration of former salt hay farms: Multiple measures of restoration success. *Reviews of Fishing Biology and Fisheries* 18:65–97.

Able, K. W., S. M. Hagan, and S. A. Brown. 2003. Mechanisms of marsh habitat alteration due to *Phragmites*: Response of young-of-the-year mummichog (*Fundulus heteroclitus*) to treatment for *Phragmites* removal. *Estuaries* 26:484–494.

Able, K. W., D. M. Nemerson, and T. M. Grothues. 1998. Evaluating salt marsh restoration in Delaware Bay: Analysis of fish response at former salt hay farms. *Estuaries* 27:58–69.

Achard, F., H. D. Eva, H. J. Stibig, P. Mayaux, J. Gallego, T. Richards, and J. Alingreau. 2002. Determination of deforestation rates of the world's humid tropical forests. *Science* 297:999–1002.

Ackerman, J. T., and C. A. Eagles-Smith. 2010. Accuracy of egg flotation throughout incubation to determine embryo age and incubation day in water bird nests. *Condor* 112:438–446.

Ackerman, J. T., C. A. Eagles-Smith, J. Y. Takekawa, and S. A. Iverson. 2008. Survival of postfledging Forster's terns in relation to mercury exposure in San Francisco Bay. *Ecotoxicology* 17:789–801.

Ackerman, J. T., C. A. Eagles-Smith, and M. P. Herzog. 2011. Bird mercury concentrations change rapidly as chicks age: Toxicological risk is highest at hatching and fledging. *Environmental Science Technology* 45:5418–5425.

Ackerman, J. T., T. Overton, M. L. Casazza, J. Y. Takekawa, C. A. Eagles-Smith, R. A. Keister, and M. P. Herzog. 2012. Does mercury contamination reduce body condition of endangered California Clapper Rails? *Environmental Pollution* 162:439–448.

Adams, D. H., R. H. McMichael, and G. E. Henderson. 2003. Mercury levels in marine estuarine fishes of Florida 1989–2001. Florida Marine Research Institute Technical Reports. Florida Fish and Wildlife Commission.

Agency for Toxic Substances and Disease Registry (ATSDR). 1999. *Toxicological Profile for Mercury.* Agency for Toxic Substances and Disease Registry, Atlanta, GA.

Agency for Toxic Substances and Disease Registry (ATSDR). 2007a. *Toxicological Profile for Lead.* Agency for Toxic Substances and Disease Registry, Atlanta, GA. http://www.atsdr.cdc.gov/toxprofiles/tp.asp?id=96&tid=22 (accessed April 19, 2015).

Agency for Toxic Substances and Disease Registry (ATSDR). 2007b. *Toxicological Profile for Arsenic.* Agency for Toxic Substances and Disease Registry, Atlanta, GA. http://www.atsdr.cdc.gov/toxprofiles/tp.asp?id=22&tid=3 (accessed April 19, 2015).

Agency for Toxic Substances and Disease Registry (ATSDR). 2012a. *Toxicological Profile for Cadmium.* Agency for Toxic Substances and Disease Registry, Atlanta, GA.

Agency for Toxic Substances and Disease Registry (ATSDR). 2012b. *Toxicological Profile for Manganese.* Agency for Toxic Substances and Disease Registry, Atlanta, GA.

Agency for Toxic Substances and Disease Registry (ATSDR). 2012c. *Toxicological Profile for Chromium.* Agency for Toxic Substances and Disease Registry, Atlanta, GA.

Agency for Toxic Substances and Disease Registry (ATSDR) 2013. *Addendum to the Toxicological Profile for Mercury: (Alkyl and Dialkyl Compounds).* Agency for Toxic Substances and Disease Registry, Atlanta GA. http://www.atsdr.cdc.gov/toxprofiles/mercury_organic_addendum.pdf (accessed August 9, 2015).

Agency for Toxic Substances and Disease Registry (ATSDR). 2015. Minimal Risk Levels Agency for Toxic Substances and Disease Registry, Atlanta, GA.

Akearok, J. A., C. E. Hebert, B. M. Braune, and M. L. Mallory. 2010. Inter- and intraclutch variation in egg mercury levels in marine bird species from the Canadian Arctic. *Science of the Total Environment* 408:836–840.

Alber, M., E. M. Swenson, S. C. Adamowicz, and I. A. Mendelssohn. 2008. Salt marsh diback: An overview of recent events in the US. *Estuarine, Coastal and Shelf Science* 80:1–11.

Albers P. H., M. T. Koterba, R. Rossmann, W. A. Link, J. B. French, R. S. Bennett, and W. C. Bauer. 2007. Effects of methylmercury on reproduction in American Kestrels. *Environmental Toxicology and Chemistry* 26:1856–1866.

Aldenberg, T., and W. Slob. 1993. Confidence-limits for hazardous concentrations based on logistically distributed NOEC toxicity data. *Ecotoxicology and Environmental Safety* 25:48–63.

Alford, J. B., M. S. Peterson, and C. C. Green. 2015. *Impacts of Oil Spill Disasters on Marine Habitats and Fisheries in North America.* Boca Raton: CRC Press.

Alexander, K. S., A. Ryan, and T. G. Measham. 2012. Managed retreat of coastal communities: Understanding responses to projected sea level rise. *Journal of Environmental Planning and Management* 55:409–433.

Alsobayel, A. A., M. A. Almarshade, and M. A. Albadry. 2013. Effect of breed, age and storage period on egg weight, egg weight loss and chick weight of commercial broiler breeders raised in Saudi Arabia. *Journal of the Saudi Society of Agricultural Sciences* 12:53–57.

American Association for the Advancement of Science (AAAS). 2014. What we know: The reality, risks, and response to climate change. AAAS Climate Science Panel. http://whatweknow.aaas.org/get-the-facts/ (accessed February 13, 2014).

American Bird Conservancy (ABC). 2007 The United States Watch List of Birds of Conservation Concern. American Bird Conservancy. http://www.abcbirds.org/abcprograms/science/watchlist/index.html (accessed September 9, 2015).

American Oystercatcher Working Group, E. Nol, and R. C. Humphrey. 2012. American Oystercatcher (*Haematopus palliatus*). In: *The Birds of North America Online*, ed. A. Poole. Ithaca, NY: Cornell Laboratory of Ornithology. http://bna.birds.cornell.edu.bnaproxy.birds.cornell.edu/bna/species/082 (accessed February 9. 2015).

Anderson, J. G., and S. Garthe. 2013. Gulls in two worlds: The rise and fall of gull populations in the North Atlantic. Abstract only. Waterbird Society 37th Annual Meeting. September 2013.

Anderson, D. W., and J. J. Hickey. 1970. Oological data on egg and breeding characteristics of brown pelicans. *The Wilson Bulletin* 82:14–28.

Anderson, D. W., J. J. Hickey, R. W. Risebrough, D. F. Hughes, and R. E. Christensen. 1969. Significance of chlorinated hydrocarbon residues to breeding pelicans and cormorants. *Canadian Field Naturalist* 83:91–112.

Anderson, J. G., G. Mittelhauser, and J. Ellis. 2013. Rise and fall of Herring and Great black-backed gulls in the Northeastern Unites States: An overview. Abstract only. Waterbird Society 37th Annual Meeting. September 2013.

Andren, H. 1994. Effects of habitat fragmentation on birds and mammals in landscapes with different proportions of suitable habitat: A review. *Oikos* 71:355–366.

Andres, B. A., P. A. Smith, R. G. Morrison, C. L. Gratto-Trevor, S. C. Brown, and C. A. Friis. 2013. Population estimates of North American shorebirds, 2012. *Wader Study Group Bulletin* 119:178–194.

Annest, J. L., J. L. Pirkle, D. Makuc, J. W. Neese, D. D. Bayse, and M. G. Kovar. 1983. Chronological trend in blood lead levels between 1976 and 1980. *New England Journal of Medicine* 308:1373–1377.

Apostolakis, G. A., and S. E. Pickett. 1998. Deliberation: Integrating analytical results into environmental decisions involving multiple stakeholders. *Risk Analysis* 18:621–634.

Armstrong, F. A. J. 1979. Effects of mercury compounds on fish. In: *The Biogeochemistry of Mercury in the Environment,* ed. J. O. Nriagu, 657–670. New York: Elsevier.

Arnold, J. M., J. J. Hatch, and I. C. T. Nisbet. 2004. Seasonal declines in reproductive success of the common tern *Sterna hirundo*. *Journal of Avian Biology* 35:33–45.

Arnold, J. M., D. J. Tyerman, and S. Oswald. 2015. Does contemporary premature feather loss among common tern chicks in Lake Ontario reflect persistent pollutants, enigmatic diseases or novel pathogens? *PeerJ Preprint #1196*. https://peerj.com/preprints/1196/ (accessed September 1, 2015).

Ashmole, N. P. 1971. Seabird ecology and the marine environment. In: *Avian Biology*, ed. D. S. Farner and J. R. King, 223–286. New York: Academic Press.

Atkinson, P. W. 2003. Can we recreate or restore intertidal habitats for shorebirds? *Bulletin-Wader Study Group* 100:67–72.

Atkinson, P. W., A. J. Baker, K. A. Bennett, N. A. Clark, J. A. Clark, K. B. Cole, A. Dekinga et al. 2007. Rates of mass gain and energy deposition in Red Knot on their final spring staging site is both time- and condition-dependent. *Journal of Applied Ecology* 44:885–895.

Atkeson, T., D. Axelrad, C. Pollman, and G. Keeler. 2003. Integrating Atmospheric Mercury Deposition and Aquatic Cycling in the Florida Everglades. Final Report to U.S. Environmental Protection Agency. http://www.dep.state.fl.us/water/sas/mercury/docs/everglades_hg_tmdl_oct03.pdf (accessed July 29, 2015).

Atkinson, P. W., N. A. Clark, M. C. Bell, P. J. Dare, J. A. Clark, and P. L. Ireland. 2003. Changes in commercially fished shellfish stocks and shorebird populations in the Wash, England. *Biological Conservation* 114:127–141.

Atlantic States Marine Fisheries Commission (ASMFC). 1998. *Interstate Fishery Management Plan for Horseshoe Crab*. Washington, D.C.: ASMFC.

Atlantic States Marine Fisheries Commission (ASMFC). 1999. *Horseshoe Crab Stock Assessment for Peer Review (Stock Assessment Report No. 98-01 (suppl.))*. Washington, D.C.: ASMFC.

Austin, O. L. 1940. Some aspects of individual distribution in the Cape Cod tern colonies. *Bird Banding* 11:155–168.

Austin, O. L. 1949. Site tenacity, a behaviour trait of the Common Tern (*Sterna hirundo* Linn). *Bird-Banding* 20:1–39.

Austin, O. L., and O. L. Austin Jr. 1956. Some demographic aspects of the Cape Cod population of Common Terns. *Bird-Banding* 27:55–56.

Austin, G. E., and M. M. Rehfisch. 2005. Shifting nonbreeding distributions of migratory fauna in relation to climatic change. *Global Change Biology* 11:31–38.

Axelrad, D. A., D. C. Bellinger, L. M. Ryan, and T. J. Woodruff. 2007. Dose–response relationship of prenatal mercury exposure and IQ: An integrative analysis of epidemiologic data. *Environmental Health Perspectives* 115:609–615.

Baker, A., P. Gonzalez, R. I. G. Morrison, and B. A. Harrington. 2013. Red Knot (*Calidris canutus*). In: *The Birds of North America Online*, ed. A. Poole. Ithaca, NY: Cornell Laboratory of Ornithology. http://bna.birds.cornell.edu.bnaproxy.birds.cornell.edu/bna/species/563 (accessed January 3, 2015).

Baker, A. J., P. M. Gonzalez, T. Piersma, and L. J. Niles. 2004. Rapid population decline in Red Knots: Fitness consequences of refueling rates and late arrival in Delaware Bay. *Proceedings of the Royal Society of London* 271:875–882.

Ballachey, B. E., J. L. Bodkin, D. Esler, and S. D. Rice. 2015. Lessons from the 1989 *Exxon Valdez* oil spill: A biological perspective. In: *Impacts of Oil Spill Disasters on Marine Habitats and Fisheries in North America*, ed. J. B. Alford, M. S. Peterson, and C. C. Green, 181–198. CRC Press: Boca Raton.

Bancroft, G. T., A. M. Strong, R. J. Sawicki, W. Hoffman, and J. D. Jewell. 1994. Relationships among wading bird foraging patterns, colony locations, and hydrology in the Everglades. In: *Everglades, the Ecosystem and Its Restoration*, ed. S. M. Davis and J. C. Ogden, 615–657. Florida: St. Lucie Press.

Barata, C., M. C. Fabregat, J. Cotínb, D. Huertas, M. Solé, L. Quiró, C. Sanpera et al. 2010. Blood biomarkers and contaminant levels in feathers and eggs to assess environmental hazards in heron nestlings from impacted sites in Ebro basin (NE Spain). *Environmental Pollution* 158:704–710.

Barbraud, C., and H. Weimerskirch. 2006. Antarctic birds breed later in response to climate change. *Proceedings of the National Academy of Sciences* 103:6248–6251.

Barbraud, C., V. Rolland, S. Jenouvrier, M. Nevoux, K. Delord, and H. Weimerskirch. 2012. Effects of climate change and fisheries bycatch on Southern Ocean seabirds: A review. *Marine Ecology Press Series:* 454:285–307.

Barnegat Bay Beat (BBB). 2012. Special Report: *Sandy*—A record setting storm. Barnegat Bay Partnership Quart. Publ. http://bbp.ocean.edu/PDFFiles/BarnegatBay/BBP%20Newsletters%202013/BBP-newsletter1 _Apr2013_spring_forWeb%20Final.pdf (accessed September 9, 2015).

Barnegat Bay Shellfish (BBS). 2015. Barnegat Bay Shellfish. http://barnegatshellfish.org/crab01.htm (accessed January 31, 2015).

Barnegat Bay National Estuary Program (BBNEP). 2002. Final Comprehensive Conservation and Management Plan. http://bbp.ocean.edu/pdffiles/barnegatbay/chapter_1.pdf (accessed July 22, 2015).

Barnegat Bay Partnership (BBP). 2010. Watershed Land Use: Population and Land Use Trends 1974–2001. Ocean County College. http://bbp.ocean.edu/pages/147.asp (accessed July 10, 2015).

Barrett, R. T., J. U. Skaare, G. Norheim, W. Vader, and A. Froslie. 1985. Persistent organochlorines and mercury in eggs of Norwegian seabirds 1983. *Environmental Pollution* 39:79–93.

Bartell, S. M. 2006. Biomarkers, bioindicators, and ecological risk assessment—A brief review and evaluation. *Environmental Bioindicators* 1:39–50.

Bartell, S. M., R. H. Gardner, and R. V. O'Neill. 1992. *Ecological Risk Estimation*. Boca Raton: Lewis Publishing.

Battin, J. 2004. When good animals love bad habitats: Ecological traps and the conservation of animal populations. *Conservation Biology* 18:1482–1491.

Bears, H., J. G. Richards, and P. M. Schulte. 2006. Arsenic exposure alters hepatic arsenic species composition and stress-mediated gene expression in the Common Killifish (*Fundulus heteroclitus*). *Aquatic Toxicology* 1:257–266.

Becker, P. H. 1992. Egg mercury levels decline with the laying sequence in *Charadriiformes*. *Archives of Environmental Contamination and Toxicology* 48:762–767.

Becker, P. H. 1995. Effects of coloniality on gull predation on Common Tern (*Sterna hirundo*) Chicks. *Colonial Waterbirds* 18:11–22.

Becker, P. H. 2003. Biomonitoring with birds. In: *Bioindicators and Biomonitors: Principles, Assessment, Concepts*, ed. B. A. Markert, A. M. Breure, and H. G. Zechmeister, 677–736. Oxford: Elsevier.

Becker, P. H., and J. S. Bradley. 2007. The role of intrinsic factors for the recruitment process in long-lived birds. *Journal of Ornithology* 148:S377–S384.

Becker, P. H., and T. Dittmann. 2009. Contaminants in bird eggs. In: *Quality Status Report 2009*, ed. H. Marencic and J. de Vlas, 3–12. Wilhelmshaven, Germany: Common Waden Sea Secretariat, Trilateral Monitoring and Assessment Group.

Becker, P. H., P. Finck, and A. Anlauf. 1985. Rainfall preceding egg-laying—A factor of breeding success in Common Terns (*Sterna hirundo*). *Oecologia* 65:431–446.

Becker, P. H., R. W. Furness, and D. Henning. 1993. Mercury dynamics in young common tern (*Sterna hirundo*) chicks from polluted environment. *Ecotoxicology* 2:33–40.

Becker, P. H., J. Gonzalez-Solis, B. Behrends, and J. Croxall. 2002. Feather mercury levels in seabirds at South Georgia: Influence of trophic position, sex, and age. *Marine Ecology Progress Series* 243:261–269.

Becker, P. H., D. Henning, and R. W. Furness. 1994. Differences in mercury contamination and elimination during feather development in gull and tern broods. *Archives of Environmental Contamination and Toxicology* 27:162–167.

Behmke, S., J. Fallon, A. E. Duerr, A. Lehner, J. Buchweitz, and T. Katzner. 2015. Chronic lead exposure is epidemic in obligate scavenger populations in eastern North America. *Environmental Indicators* 79:51–55.

Belant, J. L., T. W. Seamans, S. W. Gabrey, and S. K. Ickes. 1993. The importance of landfills to nesting Herring Gulls. *Condor* 95:817–830.

Bellrose, F. C. 1959. Lead poisoning as a mortality factor in waterfowl populations. *Illinois Natural History Survey Bulletin* 27:235–288.

Belton, T. J., B. E. Ruppel, K. Lockwood, and M. Boried. 1982. *PCB's (Aroclor 1254) in the Fish Tissues Throughout the State of New Jersey: A Comprehensive Survey*. Trenton, NJ: NJ Department of Environmental Protection, Office of Cancer and Toxic Substances Research.

Berg, J. A., M. G. Felton, J. L. Gecy, A. D. Laderman, C. R. Mayhew, J. L. Mengler, W. H. Meredith et al. 2010. Mosquito control and wetlands. *Wetlands Science Practice* 27:24–34.

Berg, W., A. G. Johnels, B. Sjöstrand, and T. Westermark. 1966. Mercury content in feathers of Swedish birds from the past 100 years. *Oikos* 17:71–83.

Bertilsson, L., and H. Y. Neujahr. 1971. Methylation of mercury compounds by methylcobalamin. *Biochemistry* 10:2805–2808.

Beyer, W. N., and J. P. Meador. 2011. *Environmental Contaminants in Wildlife: Interpreting Tissue Concentrations.* Boca Raton, FL: CRC Press.

Beyer, W. N., G. H. Heinz, and A. Redmon-Norwood. 1996. *Environmental Contaminants in Wildlife: Interpreting Tissue Concentrations.* SETAC Special Publications Series. Florida: Lewis Publishers.

Bidone, E. D., Z. C. Castilhos, T. J. S. Santos, T. M. C. Souza, and L. D. Lacerda. 1997. Fish contamination and human exposure to mercury in Tartarugalzinho River, Northern Amazon, Brazil: A screening approach. *Water Air and Soil Pollution* 97:9–15.

Bjorgo, K. A., J. J. Isley, and C. S. Thomason. 2000. Seasonal movement and habitat use of striped bass in the Cobbahee River, South Carolina. *Transactions of American Fisheries Society* 129:1281–1287.

Blodget, B. G., and L. Henze. 1991. Use of DRC-1339 to eliminate gulls and re-establish a tern nesting colony in Buzzards Bay, Massachusetts. *Fifth Eastern Wildlife Damage Control Conference* 5:212–215.

Blumsetin, D. T., L. L. Anthony, R. Harcourt, and G. Ross. 2003. Testing a key assumption of wildlife buffer zones: Is flight initiation distance a species-specific trait? *Biological Conservation* 110:97–100.

Bodkin, J. I., D. Eisler, S. D. Rice, C. O. Matkin, and B. E. Ballachey. 2014. The effect of spilled oil on coastal ecosystems: Lessons from the *Exxon Valdez* spill. In: *Coastal Conservation*, ed. B. Maslo and J. L. Lockwood, 311–346. Cambridge: Cambridge Univ. Press.

Boehme, S. E., and M. A. Panero. 2003. *Pollution Prevention and Management Strategies for Cadmium in the New York/New Jersey Harbor.* New York: New York Academy of Sciences for the New York/New Jersey Harbor Consortium.

Boesch, D. F. 1973. Classification and community structure of macrobenthos in the Hampton Roads area, Virginia. *Marine Biology* 21:226–244.

Boesch, D., J. Field, and D. Scavia. 2000. The potential consequences of climate variability and change on coastal areas and marine resources. NOAA Coastal Ocean Program, Decision Analysis Series Number # 21, Silver Spring, Maryland. USA.

Boersma, P. D., J. A. Clarke, and N. Hillgarth. 2001. Seabird conservation. In: *Biology of Marine Birds*, ed. E. A. Schreiber and J. Burger, 559–579. Boca Raton: CRC Press.

Bohnee, G., J. P. Mathews, J. Pinkham, A. Smith, and J. Stanfill. 2011. Nez Perce involvement with solving environmental problems: History, perspectives, tribal rights, and obligations. In: *Stakeholders and Scientists*, ed. J Burger, 149–184. New York: Springer.

Bologna, P. A. 2002. Growth of juvenile Winter Flounder (*Pseudopleuronectes americanus*) among eelgrass (*Zostera marina*) habitats in Little Egg Harbor, New Jersey. *Bulletin of the New Jersey Academy of Science* 47:7–12.

Boncompagni, E., A. Muhammad, R. Jabeen, E. Orvini, C. Gandini, C. Sanpera, X. Ruiz, and M. Fasola. 2003. Egrets as monitors of trace-metal contamination in wetlands of Pakistan. *Archives of Environmental Contamination and Toxicology* 45:399–406.

Bond, A. L., and A. W. Diamond. 2009. Total and methylmercury concentrations in seabird feathers and eggs. *Archives of Environmental Contamination and Toxicology* 56:286–291.

Bond, A. L., and J. L. Lavers. 2011. Trace element concentrations in feathers of Flesh-footed Shearwaters (*Puffinus carneipes*) from across their breeding range. *Archives of Environmental Contamination and Toxicology* 61:318–326.

Boo, E. 1990. *Ecotourism. The Potentials and Pitfalls. Volume 1.* Maryland: World Wildlife Fund Publishing. *Archives of Environmental Contamination and Toxicology* 45:399–406.

Borg, K., H. Wanntorp, K. Erne, and E. Hanko. 1969. Alkyl mercury poisoning in Swedish wildlife. *Viltrevy* 6:301–379.

Bortolotti, G. R., 2010. Flaws and pitfalls in the chemical analysis of feathers: Bad news–good news for avian chemoecology and toxicology. *Ecological Applications* 20:1766–1774.

Botton, M. L. 1984. The importance of predation by Horseshoe Crabs, *Limulus polyphemus*, to an intertidal sand flat community. *Journal of Marine Research* 42:139–161.

Botton, M. L. 2000. Toxicity of cadmium and mercury to Horseshoe Crab (*Limulus polyphemus*) embryos and larvae. *Bulletin of Environmental Contamination and Toxicology* 64:137–143.

Botton, M. L., and R. E. Loveland. 2001. Updating the life history of the Atlantic Horseshoe Crab, *Limulus polyphemus. Jersey Shoreline* 20:6–9.

Botton, M. L., R. E. Loveland, and T. R. Jacobsen. 1994. Site selection by migratory shorebirds in Delaware Bay, and its relationship to beach characteristics and abundance of Horseshoe Crab (*Limulus polyphemus*) eggs. *Auk* 111:605–616.

Botton, M. L., C. N. Shuster Jr., and J. A. Keinath. 2003. Horseshoe Crabs in a food web: Who eats whom? In: *The American Horseshoe Crab,* ed. C. N. Shuster Jr., R. B. Barlow, and H. J. Brockmann, 133–150. Cambridge, MA: Harvard University Press.

Bouton, S. N., P. Frederick, M. G. Spalding, and H. McGill. 1999. Effects of chronic, low concentrations of dietary methylmercury on the behavior of juvenile Great Egrets. *Environmental Toxicology and Chemistry* 18:1934–1939.

Bouton, C. F., S. Gorlach, J. P. Candelone, M. A. Bolshov, and R. J. Delmas. 1991. Decrease in anthropogenic lead, cadmium, and zinc in Greenland snows since the late 1960s. *Nature* 353:153–156.

Bowes, G. W., B. R. Simoneit, A. L. Burlingame, B. W. deLappe, and R. W. RiseBrough. 1973. The search for chlorinated dibenzofurans and chlorinated dibenzodioxins in wildlife populations showing elevated levels of embryonic death. *Environmental Health Perspectives* 5:191–198.

Boyle, D., K. V. Brix, H. Amlund, A. K. Lundebye, C. Hogstrand, and N. R. Bury. 2008. Natural arsenic contaminated diets perturb reproduction in fish. *Environmental Science and Technology* 42:5354–5360.

Bradshaw, W. E., and C. M. Holzapfel. 2006. Evolutionary response to rapid climate change. *Science* 312:1477–1478.

Brasso, R. L., M. K. Abdel Latif, and D. A. Cristol. 2010. Relationship between laying sequence and mercury concentration in tree swallow eggs. *Environmental Toxicology and Chemistry* 29:1155–1159.

Braune, B. W., and D. E. Gaskin. 1987. Mercury levels in Bonaparte's Gull (*Larus philadelphia*) during autumn molt in the Quoddy region, New Brunswick, Canada. *Archives of Environmental Contamination and Toxicology* 16:539–549.

Breed, G. A., S. Stichter, and E. E. Crone. 2013. Climate-driven changes in northeastern US butterfly communities. *Nature Climate Change* 3:142–145.

Breeding Bird Atlas Explorer (BBAE). 2015a Birds of Delaware. Online resource for 1983–1987. U.S. Geological Survey Patuxent Wildlife Research Center. http://www.pwrc.usgs.gov/bba (accessed April 14, 2015).

Breeding Bird Atlas Explorer (BBAE). 2015b. Delaware Breeding Bird Atlas 2008–2012. Online resource for 1983–1987. U.S. Geological Survey Patuxent Wildlife Research Center. http://www.pwrc.usgs.gov/bba (accessed April 14, 2015).

Breton, A. R., G. A. Fox, and J. W. Chardine. 2008. Survival of adult Herring Gulls (*Larus argentatus*) from a Lake Ontario colony over two decades of environmental change. *Waterbirds* 31:15–23.

Breton, A. R., I. C. Nisbet, C. S. Mostello, and J. J. Hatch. 2014. Age-dependent breeding dispersal and adult survival within a metapopulation of Common Terns *Sterna hirundo. Ibis* 156:534–547.

Bried, J., and P. Jouventin. 2001. Site and mate choice in seabirds: An evolutionary approach. In: *Biology of Marine Birds*, ed. E. A. Schreiber and J. Burger, 263–305. Boca Raton: CRC Press.

Brinker, D. F., J. M. McCann, B. Williams, and B. D. Watts, 2007. Colonial-nesting seabirds in the Chesapeake Bay Region: Where have we been and where are we going? *Waterbirds* 30:93–104.

Broccoli, A. J., N. C. Lau, and M. J. Nath. 1998. The cold ocean–warm land pattern: Model simulation and relevance to climate change detection. *Journal of Climate* 11:2743–2763.

Brown, R. G. B. 1980. Seabirds as marine animals. In: *Behavior of Marine Animals. Vol.4. Marine Birds,* ed. J. Burger, B. L. Olla, and H. E. Winn, 1–40. New York: Plenum.

Brown, J., and S. Erdle. 2009. Amphibians, reptiles, birds, and mammals of the York River. *Journal of Coastal Research* 57:111–117.

Brown, K. M., R. M. Erwin, M. E. Richmond, P. A. Buckley, J. T. Tanacredi, and D. Avrin. 2001a. Managing birds and controlling aircraft in the Kennedy airport–Jamaica Bay Wildlife Refuge Complex: The need for hard data and soft opinions. *Environmental Management* 28:207–224.

Brown, K. M., J. L. Tims, R. M. Erwin, and M. E. Richmond. 2001b. Changes in the nesting populations of colonial waterbirds in Jamaica Bay Wildlife Refuge, New York, 1974–1998. *Northeastern Naturalist* 8:275–292.

Brown, S., C. Hickey, B. Harrington, and R. Gill (Eds.). 2001c. *The U.S. Shorebird Conservation Plan,* 2nd ed. Manomet, MA: Manomet Center for Conservation Sciences.

Bryan, A. L., H. A. Brant Jr., C. H. Jagoe, C. S. Romanek, and I. L. Brisbin Jr., 2012. Mercury concentrations in nestling wading birds relative to diet in the southeastern United States: A stable isotope analysis. *Archives of Environmental Contamination and Toxicology* 63:144–152.

Bryan, G. W., P. E. Gibbs, L. G. Hummerstone, and G. R. Burt. 1986. The decline of the gastropod *Nucella paillus* around southwest England: Evidence for the effect of tributyltin from antifouling paints. *Journal of Marine Biology Association* 66:611–640.

Brzorad, J. N., A. D. Maccarone, and H. M. Stone. 2015. A telemetry-based study of Great Egret (*Ardea alba*) nest-attendance patterns, food-provisioning rates, and foraging activity in Kansas, USA. *Waterbirds* 38:162–172.

Buckley, P. A., and F. G. Buckley. 1977. Hexagonal packing of Royal Tern nests. *Auk* 94:36–43.

Buckley, F. G., and P. A. Buckley. 1980. Habitat selection and marine birds. In: *Behavior of Marine Animals. Vol. 4. Marine birds*, ed. J. Burger, B. L. Olla, and H. E. Winn, 69–112. New York: Plenum.

Buckley, F. G., M. Gochfeld, and P. A. Buckley. 1978. Breeding Laughing Gulls return to Long Island. *Kingbird* 28:202–207.

Bugoni, L., T. D. Cormons, A. W. Boyne, and H. Hays. 2005. Feeding grounds, daily foraging activities, and movements of Common Terns in Southern Brazil, determined by radio-telemetry. *Waterbirds* 28:468–477.

Bull, J. 1964. *Birds of the New York Area*. New York: Harper and Row.

Bullard, R. D. 1990. *Dumping in Dixie: Race, Class, and Environmental Quality*. Boulder, CO: Westview.

Burger, J. 1974. Breeding adaptations of the Franklin's Gull (*Larus pipixcan*) to a marsh habitat. *Animal Behavior* 22:521–567.

Burger, J. 1976. Daily and seasonal activity patterns in breeding Laughing Gulls, *Larus atricilla. Auk* 93:308–323.

Burger, J. 1977a. Nesting behavior of Herring Gulls: Invasion into *Spartina* salt marsh areas of New Jersey. *Condor* 79:162–179.

Burger, J. 1977b. The role of visibility in nesting behavior of *Larus* gulls. *Journal of Comparative Physiology and Psychology* 91:1347–1358.

Burger, J. 1978a. Great Black-backed Gulls breeding in salt marsh in New Jersey. *Wilson Bulletin* 90:304–305.

Burger, J. 1978b. Competition between Cattle Egrets and native North America Ardeids. *Condor* 80:15–23.

Burger, J. 1978c. The pattern and mechanism of nesting in mixed-species heronries. *Wading Birds*. Research Rep. #7, 45–58. New York: National Audubon Society.

Burger, J. 1978d. Determinants of nest repair in Laughing Gulls. *Animals Behavior* 26:856–861.

Burger, J. 1979a. Competition and predation: Herring Gulls versus Laughing Gulls. *Condor* 81:269–277.

Burger, J. 1979b. Nest repair behavior in birds nesting in salt marshes. *Journal of Comparative Physiology and Psychology* 11:189–199.

Burger, J. 1979c. Resource partitioning: Nest site selection in mixed-species colonies of herons, egrets and ibises. *American Midland Naturalist* 101:191–210.

Burger, J. 1979d. Colony Size: A test for breeding synchrony in Herring Gulls (*Larus argentatus*). *Auk* 96:694–703.

Burger, J. 1980a. Behavioral adaptations of Herring Gulls (*Larus argentatus*) to salt marshes and storm tides. *Biology Behavior* 5:147–162.

Burger, J. 1980b. Territory size differences in relation to reproductive stage and type of intruder in Herring Gulls (*Larus argentatus*). *Auk* 91:733–741.

Burger, J. 1980c. The transition from dependence to independence and post-fledging parental care in marine birds. In: *Behavior of Marine Organisms: Perspectives in research, Vol. 4: Marine Birds*, ed. J. Burger, B. Olla, and H. Winn, 367–447. New York: Plenum Press.

Burger, J. 1981a. A model for the evolution of mixed species of colonies in *Ciconiiformes. Annual Review of Biology* 56:1443–1467.

Burger, J. 1981b. Super-territories: A comment. *American Naturalist* 118:578–580.

Burger, J. 1981c. Feeding competition between Laughing Gulls and Herring Gulls at a Sanitary Landfill. *Condor* 83:328–335.

Burger, J. 1981d. On becoming independent in Herring Gulls: Parent–young conflict. *American Naturalist* 117:444–456.

Burger, J. 1982a. The role of reproductive success in colony-site selection and abandonment in Black Skimmers (*Rynchops niger*). *Auk* 99:109–125.

Burger, J. 1982b. An overview of proximate factors affecting reproductive success in colonial birds: Concluding remarks and summary of panel discussion. *Colonial Waterbirds* 5:58–65.

Burger, J. 1982c. On the nesting of location of Cattle Egrets, *Bubulcus ibis* in South African heronries. *Ibis* 124:523–529.

Burger, J. 1983. Competition between two species of nesting gulls: On the importance of timing. *Behavioral Neuroscience* 97:492–501.

Burger, J. 1984a. Grebes nesting in gull colonies: Protective associations and early warning. *American Naturalist* 123:327–337.

Burger, J. 1984b. Colony stability in Least Terns. *Condor* 86:61–67.

Burger, J. 1984c. Pattern, mechanism, and adaptive significance of territoriality in Herring Gulls (*Larus argentatus*). *American Ornithologists Union Monograph* 34:1–92.

Burger, J. 1984d. Advantages and disadvantages of mixed-species colonies of seabirds. *Proceedings of the 18th International Ornithological Congress*, 905–918.

Burger, J. 1984e. Environmental influences on the abundance and distribution of migrating shorebirds. In: *Behavior of Marine Animals, Vol. 6: Shorebirds: Migration and Foraging Behavior*, ed. J. Burger and B. Olla, 1–72. New York: Plenum Press.

Burger, J. 1985a. Habitat selection in temperate marsh-nesting birds. In: *Habitat Selection in Birds*, ed. M. Cody, 253–281. Waltham: Academic Press.

Burger, J. 1985b. Factors affecting bird strikes on aircraft at a coastal airport. *Biological Conservation* 33:1–28.

Burger, J. 1986. The effect of human activities on shorebirds in two coastal bays in the Northeastern United States. *Environmental Conservation* 13:123–130.

Burger, J. 1987a. Physical and social determinants of nest-site selection in Piping Plover in New Jersey. *The Condor* 89:811–818.

Burger, J. 1987b. Foraging efficiency in Gulls: A congeneric comparison of age differences in efficiency and age of maturity. *Studies in Avian Biology* 10:83–89.

Burger, J. 1988a. *Seabirds and Other Marine Vertebrates: Competition, Predation, and Other Interactions*. New York: Columbia University Press.

Burger, J. 1988b. Social Attraction in Least Terns (*Sterna antillarum*): Effects of numbers, placement and social status. *Condor* 90:575–582.

Burger, J. 1988c. Jamaica Bay Studies: VIII. An overview of abiotic factors affecting several avian groups. *Journal of Coastal Research* 4:193–205.

Burger, J. 1988d. Foraging behavior of gulls: Differences in habitats, prey items and foraging methods. *Colonial Waterbirds* 11:9–23.

Burger, J. 1988e. Effects of age on foraging on birds. *Proceedings of the 19th International Ornithologist Congress*, Ottawa, Canada, 1127–1140.

Burger, J. 1989. Least Tern populations in coastal New Jersey: Monitoring and management of a regionally-endangered species. *Journal of Coastal Research* 5:801–811.

Burger, J. 1990a. Seabirds, tropical biology and global warning: Are we missing the Ark. *Colonial Waterbirds* 13:81–156.

Burger, J. 1990b. Behavioral effects of early postnatal lead exposure in Herring Gull (*Larus argentatus*) chicks. *Pharmacology, Biochemistry and Behavior* 35:7–13.

Burger, J. 1991a. Foraging behavior and the effect of human disturbance on the Piping Plover (*Charadrius melodus*). *Journal of Coastal Research* 7:39–52.

Burger, J. 1991b. Coastal landscapes, coastal colonies and seabirds. *Aquatic Reviews* 4:23–43.

Burger, J. 1993. Metals in avian feathers: Bioindicators of environmental pollution. *Reviews in Environmental Toxicology* 5:197–306.

Burger, J. 1994a. The effect of human disturbance on foraging behavior and habitat use in Piping Plover (*Charadrius melodus*). *Estuaries* 17:695–701.

Burger, J. 1994b. *Before and After an Oil Spill: The Arthur Kill*. New Brunswick: Rutgers University Press.

Burger, J. 1994c. Heavy metals in avian eggshells: Another excretion method. *Journal of Toxicology and Environmental Health* 41:207–220.

Burger, J. 1994d. Introduction. In: *Before and After an Oil Spill: The Arthur Kill*, ed. J. Burger, 1–22. New Brunswick: Rutgers University Press.

Burger, J. 1994e. Effects of oil on vegetation. In: *Before and After an Oil Spill: The Arthur Kill*, ed. J. Burger, 130–141. New Brunswick: Rutgers University Press.

Burger, J. 1994f. From the past to the future: Conclusions from the Arthur Kill. In: *Before and After an Oil Spill: The Arthur Kill*, ed. J. Burger, 283–290. New Brunswick: Rutgers University Press.

Burger, J. 1995a. A risk assessment for lead in birds. *Journal of Toxicology and Environmental Health* 45:369–396.

Burger, J. 1995b. Heavy metal and selenium levels in feathers of Herring Gulls (*Larus argentatus*): Differences due to year, gender, and age at Captree, Long Island. *Environmental Monitoring and Assessment* 38:37–50.

Burger, J. 1996a. *A Naturalist Along the Jersey Shore.* New Brunswick: Rutgers University Press.

Burger, J. 1996b. Laughing Gull (*Leucophaeus atricilla*). In: *Birds of North America Online*, ed. A. Poole. Ithaca, NY: Cornell Laboratory of Ornithology. http://bna.birds.cornell.edu.bnaproxy.birds.cornell.edu/bna/species/225 (accessed February 9, 2015).

Burger, J. 1997a. *Oil Spills.* New Brunswick: Rutgers University Press.

Burger, J. 1997b. Heavy metals in the eggs and muscle of Horseshoe Crabs (*Limulus Polyphemus*) from Delaware Bay. *Environmental Monitoring and Assessment* 46:279–287.

Burger, J. 1997c. Heavy metals and selenium in Herring Gulls (*Larus argentatus*) nesting in colonies from Eastern Long Island to Virginia. *Environmental Monitoring and Assessment* 48:285–296.

Burger, J. 1998a. Effects of motorboats and personal watercraft on flight behavior over a colony of Common Terns. *Condor* 100:528–534.

Burger, J. 1998b. Personal watercraft and Common Terns on Barnegat Bay: Taking a Turn for the Worse. Barnegat Bay Watershed Association, Newsletter. http://www.bbwa.org/summer988/watercraft_s98.htm (accessed March 7, 2015)

Burger, J. 1998c. Effects of lead on sibling recognition in young Herring Gulls. *Toxicological Sciences* 43:155–160.

Burger, J. 2001a. Introduction. In: *Protecting the Commons: A Framework for Resource Management in the Americas*, ed. J. Burger, E. Ostrom, R. Norgaard, and B. Goldstein, 1–16. Washington D.C.: Island Press.

Burger J. 2001b. Multiuse coastal commons: Personal watercraft. In: *Protecting the Commons: A Framework for Resource Management in the Americas*, ed. Burger, J., R. Norgaard, E. Ostrom, D. Policansky, and B. D. Goldstein, 195–214. Washington, D.C.: Island Press.

Burger, J. 2002a. Consumption patterns and why people fish. *Environmental Research* 90:125–135.

Burger, J. 2002b. Effects of motorboats and personal watercraft on nesting terns: Conflict Resolution and the Need for Vigilance. *Journal of Coastal Research* 37:7–17.

Burger, J. 2002c. Food chain differences affect heavy metals in bird eggs in Barnegat Bay, New Jersey. *Environmental Research* 90:33–39.

Burger, J. 2003a. Perceptions about environmental use and future restoration of an urban estuary. *Journal of Environmental Planning and Management* 46:399–416.

Burger, J. 2003b. Assessing perceptions about ecosystems health and restoration options in three East Coast estuaries. *Environmental Monitoring and Assessment* 83:145–162.

Burger, J. 2003c. Personal watercraft and boats: Coastal conflicts with Common Terns. *Lake and Reservoir Management* 19:26–34.

Burger, J. 2006a. Bioindicators: A review of their use in the environmental literature 1970–2005. *Environmental Bioindicators* 1:136–144.

Burger, J. 2006b. Bioindicators: Types, development, and use in ecological assessment and research. *Environmental Bioindicators* 1:22–39.

Burger, J. 2008a. Environmental management: Integrating ecological evaluations, remediation, restoration, Natural Resource Damage Assessment, and long-term stewardship on contaminated lands. *Science of the Total Environment* 400:6–19.

Burger, J. 2008b. Perceptions as indicators of potential risk from fish consumption and health of fish populations. *Environmental Bioindicators* 3:90–105.

Burger, J. 2008c. Assessment and management of risk to wildlife from cadmium. *Science of the Total Environment* 389:37–45.

Burger, J. 2009a. Stakeholder involvement in indicator selection: Case studies and levels of participation. *Environmental Bioindicators* 4:170–190.

Burger, J. 2009b. Perceptions of the risks and benefits of fish consumption: Individual choices to reduce risk and increase health benefits. *Environmental Research* 109:343–349.

Burger, J. 2009c. Risk to consumers from mercury in bluefish (*Pomatomus saltatrix*) from New Jersey: Size, season, and geographical effects. *Environmental Research* 109:803–811.

Burger, J. 2012. Perceptions of goods, services, and eco-cultural attributes of Native Americans and Caucasians in Idaho. *Remediation* 22:105–121.

Burger, J. 2013a. Temporal trends (1989–2011) in levels of mercury and other heavy metals in feathers of fledgling Great Egrets nesting in Barnegat Bay, NJ. *Environmental Research* 122:11–17.

Burger, J. 2013b. Role of self-caught fish in total fish consumption rates for recreational fisherman: Average consumption rates exceed allowable intake. *Journal of Risk Research* 16:1057–1075.

Burger, J. 2015a. Laughing Gull (*Leucophaeus atricilla*), Revised. *Birds of North America Online*, ed. A. Poole. Ithaca, NY: Cornell Laboratory of Ornithology. http://bna.birds.cornell.edu.bnaproxy.birds.cornell.edu/bna/species/225 (accessed February 9, 2015).

Burger, J. 2015b. Ecological damages and responses to Hurricane *Sandy*: Physical damage, avian and food web responses, and anthropogenic attempts to aid ecosystem recovery. *Urban Ecosystems* 18:553–575.

Burger, J., and S. Elbin. 2015a. Metal levels in eggs of waterbirds in the New York Harbor (USA): Trophic relationships and possible risk to human consumers. *Journal of Toxicology and Environmental Health* 78:78–91.

Burger, J., and S. Elbin. 2015b. Contaminant levels in Herring (*Larus argentatus*) and Great Black-backed Gull (*Larus marinus*) eggs from colonies in the New York Harbor complex between 2012–2013. *Ecotoxicology* 24:445–452.

Burger, J., and M. Gochfeld. 1981a. Age-related differences in piracy behavior of four species of gulls, *Larus*. *Behaviour* 77:242–267.

Burger, J., and M. Gochfeld. 1981b. Discrimination of the threat of direct versus tangential approach to the nest by incubating Herring and Great Black-backed Gulls. *Journal of Comparative and Physiological Psychology* 95:676–684.

Burger, J., and M. Gochfeld. 1983a. Behavioural responses to human intruders of Herring Gulls (*Larus argentatus*) and great black-backed gulls (*Larus marinus*) with varying exposure to human disturbance. *Behavioral Processes* 8:327–344.

Burger, J., and M. Gochfeld. 1983b. Jamaica Bay studies: V. Flocking associations and behavior of shorebirds at an Atlantic coastal estuary. *Biology of Behavior* 8:289–318.

Burger, J., and M. Gochfeld. 1983c. Feeding behavior in Laughing Gulls: Compensatory site-selection by young. *Condor* 85:467–473.

Burger, J., and M. Gochfeld. 1984a. Great black-backed Gull predation on Kittiwake fledglings in Norway. *Bird Study* 31:149–151.

Burger, J., and M. Gochfeld. 1984b. The effects of relative numbers on aggressive interactions and foraging efficiency in gulls: The cost of being outnumbered. *Bird Behavior* 5:81–89.

Burger, J., and M. Gochfeld. 1985a. Nest site selection by Laughing Gulls: Comparison of tropical colonies (Culebra, Puerto Rico) with temperate colonies (New Jersey). *Condor* 87:364–373.

Burger, J., and M. Gochfeld. 1985b. Early postnatal lead exposure: Behavioral effects on Common Tern chicks (*Sterna hirundo*). *Journal of Toxicology and Environmental Health* 16:869–886.

Burger, J., and M. Gochfeld. 1985c. Comparisons of nine heavy metals in salt gland and liver of Greater Scaup (*Aythya marila*), Black Duck (*Anas rubripes*) and Mallard (*A. platyrhynchos*). *Comparative Biochemistry Physiology* C81:287–292.

Burger, J., and M. Gochfeld. 1987a. Nest-site selection: Comparison of Roseate and Common Terns (*Sterna dougallii* and *S. hirundo*) in a Long Island, New York colony. *Bird Behavior* 7:58–66.

Burger, J., and M. Gochfeld. 1988a. Defensive aggression in terns: Effect of species, density and isolation. *Aggressive Behavior* 14:169–178.

Burger, J., and M. Gochfeld. 1988b. Effect of lead on growth in young Herring Gulls (*Larus argentatus*). *Journal of Toxicology and Environmental Health* 25:227–236.

Burger, J., and M. Gochfeld. 1988c. Lead and behavioral development: Effects of varying dosage and schedule on survival and performance of young Common Terns (*Sterna hirundo*). *Journal of Toxicology and Environmental Health* 24:173–182.

Burger, J., and M. Gochfeld. 1988d. Metals in tern eggs in a New Jersey estuary: A decade of change. *Environmental Monitoring and Assessment* 11:127–135.

Burger, J., and M. Gochfeld. 1990a. Tissue levels of lead in experimentally-exposed Herring Gull (*Larus argentatus*) chicks. *Journal of Toxicology and Environmental Health* 29:219–233.

Burger, J., and M. Gochfeld. 1990b. *The Black Skimmer: Social Dynamics of a Colonial Species*. New York: Columbia University Press.

Burger, J., and M. Gochfeld. 1990c. Nest site selection in Least Terns (*Sterna antillarum*) in New Jersey and New York. *Colonial Waterbirds* 13:31–40.

Burger, J., and M. Gochfeld. 1990d. Vertical nest stratification in a heronry in Madagascar. *Colonial Waterbirds* 13:143–146.

Burger, J., and M. Gochfeld. 1991a. Cadmium and lead in Common Terns (Aves: *Sterna hirundo*): Relationship between levels in parents and eggs. *Environmental Monitoring and Assessment* 16:253–258.

Burger, J., and M. Gochfeld. 1991b. *The Common Tern: Its Breeding Biology and Behavior*. New York: Columbia University Press.

Burger, J., and M. Gochfeld. 1991c. Nest site selection in Least Terns (*Sterna antillarum*) in New Jersey and New York. *Colonial Waterbirds* 13:31–40.

Burger, J., and M. Gochfeld. 1991d. Vigilance and feeding behaviour in large feeding flocks of Laughing Gulls, *Larus atricilla*, on Delaware Bay. *Estuarine, Coastal and Shelf* Science 32:207–212.

Burger, J., and M. Gochfeld. 1991e. Lead, mercury, and cadmium in feathers of tropical terns in Puerto Rico and Australia. *Archives of Environmental Contamination and Toxicology* 21:311–315.

Burger, J., and M. Gochfeld. 1992a. Trace element distribution in growing feathers: Additional excretion in feather sheaths. *Archives of Environmental Contamination and Toxicology* 23:105–108.

Burger, J., and M. Gochfeld. 1992b. Experimental evidence for aggressive antipredator behavior in Black Skimmers (*Rynchops niger*). *Aggressive Behavior* 18:241–248.

Burger, J., and M. Gochfeld. 1992c. Heavy metal and selenium concentrations in Black Skimmers (*Rynchops niger*): Gender differences. *Archives of Environmental Contamination and Toxicology* 23:431–434.

Burger, J., and M. Gochfeld. 1993a. Lead and cadmium accumulation in eggs and fledgling seabirds in the New York bight. *Environmental Toxicology and Chemistry* 12:261–267.

Burger, J., and M. Gochfeld. 1993b. When is a heronry crowded: A case study of Huckleberry Island, New York, U.S.A. *Journal of Coastal Research* 9:221–228.

Burger, J., and M. Gochfeld. 1993c. Lead and behavioral development in young Herring Gulls: Effects of timing of exposure on individual recognition. *Fundamental and Applied Toxicology* 21:187–195.

Burger, J., and M. Gochfeld. 1993d. Heavy metal and selenium levels in feathers of young egrets and herons from Hong Kong and Szechuan, China. *Archives of Environmental Contamination and Toxicology* 25:322–327.

Burger, J., and M. Gochfeld. 1994. Behavioral impairments of lead-injected young Herring Gulls in nature. *Fundamental & Applied Toxicology* 23:553–561.

Burger, J., and M. Gochfeld. 1995a. Effects of varying temporal exposure to lead on behavioral development in Herring Gull (*Larus argentatus*) chicks. *Pharmacology, Biochemistry, and Behavior* 52:601–608.

Burger, J., and M. Gochfeld. 1995b. Growth and behavioral effects of early postnatal chromium and manganese exposure in Herring Gull (*Larus argentatus*) chicks. *Pharmacology, Biochemistry and Behavior* 50:607–612.

Burger, J., and M. Gochfeld. 1995c. Behavior effects of lead exposure on different days for gull (*Larus argentatus*) chick. *Pharmacology Biochemistry and Behavior* 50:97–105.

Burger, J., and M. Gochfeld. 1995d. Heavy metal and selenium concentrations in eggs of Herring Gulls (*Larus argentatus*): Temporal differences from 1989 to 1994. *Archives of Environmental Contamination and Toxicology* 29:192–197.

Burger, J., and M. Gochfeld. 1996a. Heavy metal and selenium levels in Franklin's Gulls (*Larus pipixcan*): Parents and their eggs. *Archives of Environmental Contamination and Toxicology* 30:487–491.

Burger, J., and M. Gochfeld. 1996b. Heavy metal and selenium levels in birds at Agassiz National Wildlife Refuge, Minnesota: Food chain differences. *Environmental Monitoring and Assessment* 43:267–282.

Burger, J., and M. Gochfeld. 1996c. Family *Laridae* (gulls). In: *Handbook of the Birds of the World, Vol. 3*, ed. J. del Hoyo, A. Elliot, and J. Sargatal, 572–623. Barcelona: Lynx Editions.

Burger, J., and M. Gochfeld. 1996d. Ecological and human health risk assessment: A comparison. In: *Interconnections between Human and Ecosystem Health*, ed. R. T. DiGuilio and E. Monosson, 127–148. London: Chapman and Hall.

Burger, J., and M. Gochfeld. 1996e. Lead and behavioral development: Parental compensation for behaviorally impaired chicks. *Pharmacology Biochemistry and Behavior* 55:339–349.

Burger, J., and M. Gochfeld. 1997a. Risk, mercury levels, and birds: Relating adverse laboratory effects to field biomonitoring. *Environmental Research* 75:160–172.

Burger, J., and M. Gochfeld. 1997b. Lead and neurobehavioral development in gulls: A model for understanding effects in the laboratory and the field. *Neurotoxicology* 18:279–287.

Burger, J., and M. Gochfeld. 2000a. Effects of lead on birds (*Laridae*): A review of laboratory and field studies. *Journal of Toxicology and Environmental Health* 3:59–78.

Burger, J., and M. Gochfeld. 2000b. Metal levels in feathers of 12 species of seabirds from Midway Atoll in the Northern Pacific Ocean. *Science of Total Environment* 257:37–52.

Burger, J., and M. Gochfeld. 2000c. Metals in albatross feathers from Midway Atoll: Influence of species, age, and nest location. *Environmental Research* 82:207–221.

Burger, J., and M. Gochfeld. 2001a. On developing bioindicators for human and ecological health. *Environmental Monitoring and Assessment* 66:23–46.

Burger, J., and M. Gochfeld. 2001b. Effects of chemicals and pollution on seabirds. In: *Biology of Marine Birds*, ed. E. A. Schreiber and J. Burger, 485–526. Boca Raton: CRC Press.

Burger, J., and M. Gochfeld. 2003. Spatial and temporal patterns in metal levels in eggs of Common Terns (*Sterna hirundo*) in New Jersey. *Science of the Total Environment* 311:91–100.

Burger, J., and M. Gochfeld. 2004a. Metal levels in eggs of Common Terns (*Sterna hirundo*) in New Jersey: Temporal trends from 1971 to 2002. *Environmental Research* 94:336–343.

Burger, J., and M. Gochfeld. 2004b. Effects of lead and exercise on endurance and learning in young Herring Gulls. *Ecotoxicology and Environment Safety* 57:136–144.

Burger, J., and M. Gochfeld. 2005a. Heavy metals in commercial fish in New Jersey. *Environmental Research* 99:403–412.

Burger, J., and M. Gochfeld. 2005b. The Peconic River: Concerns associated with different risk evaluations for fish consumption. *Journal of Environmental Planning & Management* 48:789–808.

Burger, J., and M. Gochfeld. 2011a. Conceptual environmental justice model: Evaluation of chemical pathways of exposure in low-income, minority, Native American, and other unique exposure populations. *American Journal of Public Health* 101 Suppl. 1:S64–S73.

Burger, J., and M. Gochfeld. 2011b. Mercury and selenium in 19 species of saltwater fish from New Jersey as a function of species, size, and season. *Science of the Total Environment* 409:1418–1429.

Burger, J., and M. Gochfeld. 2013a. Selenium: Mercury molar ratios in freshwater, marine and commercial fish from the United States: Variation, risk, and health management. *Reviews on Environmental Health* 28:129–143.

Burger, J., and M. Gochfeld. 2013b. Selenium and mercury molar ratios in commercial fish from New Jersey and Illinois: Variation within species and relevance to risk communication. *Food and Chemical Toxicology* 57:234–345.

Burger, J., and M. Gochfeld. 2014a. Health concerns and perceptions of central and coastal New Jersey residents in the 100 days following Superstorm *Sandy*. *Science of the Total Environment* 481:611–618.

Burger, J., and M. Gochfeld. 2014b. Perceptions of personal and governmental actions to improve responses to disasters such as Superstorm *Sandy*. *Environmental Hazards* 13:200–210.

Burger, J., and M. Gochfeld. 2015. Concerns and perceptions immediately following Superstorm *Sandy*: Ratings for property damage were higher than for health issues. *Journal of Risk Research* 18:249–265.

Burger, J., and F. Lesser. 1978. Selection of colony sites and nest site selection by Common Terns *Sterna hirundo* in Ocean County, New Jersey. *Ibis* 120:443–449.

Burger, J., and F. Lesser. 1979. Breeding behavior and success in salt marsh Common Tern *Sterna hirundo* colonies. *Bird Banding* 50:322–337.

Burger, J., and F. Lesser. 1980. Nest site selection in an expanding population of Herring Gulls. *Journal of Field Ornithology* 5:270–280.

Burger, J., and L. M. Miller. 1977. Factors determining colony and nest site selection in White-faced and Glossy Ibis (*Plegadis*). *Auk* 94:664–678.

Burger J., and L. J. Niles. 2013a. Shorebirds and stakeholders: Effects of beach closure and human activities on shorebirds at a New Jersey coastal beach. *Urban Ecosystems* 16:657–673.

Burger, J., and L. J. Niles. 2013b. Closure versus voluntary avoidance as a method of protecting migrating shorebirds on beaches in New Jersey. *Wader Study Group Bulletin* 120:20–25.

Burger, J., and L. J. Niles. 2014. Effects on five species of shorebirds of experimental closures of a beach in New Jersey: Implications for severe storms and sea-level rise. *Journal of Toxicology and Environmental Health* 77:1102–1113.

Burger, J., and J. Shisler. 1978a. The effects of ditching a salt marsh on colony and nest site selection by Herring Gulls (*Larus argentatus*). *American Midland Naturalist* 100:54–63.

Burger, J., and J. Shisler. 1978b. Nest site selection of Willets in New Jersey salt marsh. *Wilson Bulletin* 90:599–607.

Burger, J., and J. Shisler. 1978c. Nest site selection and competitive interactions of Herring and Laughing gulls in New Jersey. *Auk* 95:252–266.

Burger, J., and J. Shisler. 1979. Immediate effects of ditching a marsh on breeding Herring Gulls. *Biological Conservation* 15:85–104.

Burger, J., and J. Shisler. 1980a. Colony site and nest selection in Laughing Gulls in response to tidal flooding. *Condor* 82:251–258.

Burger, J., and J. Shisler. 1980b. The process of colony formation among Herring Gull (*Larus argentatus*) nesting in New Jersey. *Ibis* 122:15–26.

Burger, J., and R. Trout. 1979. Additional data on body size as a difference related to niche. *Condor* 81:305–307.

Burger, J., and N. Tsipoura. 2014. Metals in Horseshoe Crab eggs from Delaware Bay, USA: Temporal patterns from 1993–2012. *Environmental Monitoring and Assessment* 186:6947–6958.

Burger, J., D. Gladstone, L. M. Miller, and D. C. Hahn. 1977a. Intra and inter-specific interactions at a mixed-species roost of *Ciconiiformes* in San Blas, Mexico. *Biology of Behavior* 2:309–327.

Burger, J., M. A. Howe, C. Hahn, and J. Chase. 1977b. Effects of tide cycles on habitat selection and habitat partitioning by migrating shorebirds. *Auk* 94:743–758.

Burger, J., J. Shisler, and F. Lesser. 1982a. Avian utilization of six salt marshes. *Biological Conservation* 23:187–212.

Burger, J., M. Gochfeld, and W. I. Boarman. 1988a. Experimental evidence for sibling recognition in Common Terns (*Sterna hirundo*). *The Auk* 105:142–148.

Burger, J., I. Nisbet, and M. Gochfeld. 1992a. Metal levels in regrown feathers: Assessment of contamination on the wintering and breeding grounds in the same individuals. *Toxicology and Environmental Health* 36:327–338.

Burger, J., Parsons, K., T. Benson, T. Shukla, D. Rothstein, and M. Gochfeld. 1992b. Heavy metals and selenium levels in young Cattle Egrets from nesting colonies in the northeastern United States, Puerto Rico, and Egypt. *Archives of Environmental Contamination and Toxicology* 23:435–439.

Burger, J., D. Shealer, and M. Gochfeld. 1993a. Defensive aggression in terns: Discrimination and response to individual researchers. *Aggressive Behavior* 19:303–311.

Burger, J., F. Lesser, and M. Gochfeld. 1993b. Brown pelicans attempt nesting in New Jersey. *Records of New Jersey Birds* 18:78–80.

Burger, J., M. Pokras, R. Chafel, and M. Gochfeld. 1994a. Heavy metal concentrations in feathers of Common Loons (*Gavia immer*) in the Northeastern Unites States and age differences in mercury levels. *Environmental Monitoring and Assessment* 30:1–7.

Burger, J., K. Parsons, D. Wartenberg, C. Safina, J. O'Conner, and M. Gochfeld. 1994b. Biomonitoring using Least Terns and Black Skimmers in the northeastern United States. *Journal of Coastal Research* 10:39–47.

Burger, J., J. N. Bryzorad, and M. Gochfeld. 1994c. Fiddler Crabs (*Uca spp*) as bioindicators for oil spills. In *Before and After an Oil Spill: The Arthur Kill*, ed. J. Burger, 160–177. New Brunswick: Rutgers Univ. Press.

Burger, J., I. Nisbet, and M. Gochfeld. 1994d. Heavy metal and selenium levels in feathers of known-aged Common Terns (*Sterna hirundo*). *Archives of Environmental Contamination and Toxicology* 26:351–355.

Burger, J., M. H. Lavery, and M. Gochfeld. 1994e. Temporal changes in lead levels in Common Tern feathers in New York and relationship of field levels to adverse effects in the laboratory. *Environmental Toxicology and Chemistry* 13:581–586.

Burger, J., K. L. Clark, and L. Niles. 1997a. Importance of beach, mudflat and marsh for migrant shorebirds on Delaware Bay. *Biological Conservation* 79:283–292.

Burger, J., T. Shukla, T. Benson, and M. Gochfeld. 1997b. Lead levels in exposed Herring Gulls: Differences in the field and laboratory. *Toxicology and Industrial Health* 13:193–202.

Burger, J., C. D. Trivedi, and M. Gochfeld. 2000. Metals in Herring and Great Black-backed Gulls from the New York Blight: *Environmental Monitoring and Assessment* 64:569–581.

Burger, J., M. Gochfeld, C. W. Powers, L. Waishwell, C. Warren, and B. D. Goldstein. 2001a. Science, policy, stakeholders, and fish consumption advisories: Developing a fish fact sheet for the Savannah River. *Environmental Management* 27:4:501–514.

Burger, J., C. D. Jenkins, Jr., F. Lesser, and M. Gochfeld. 2001b. Status and trends of colonially-nesting birds in barnegat bay. *Journal of Coastal Research* 32:197–211.

Burger, J., R. Norgaard, E. Ostrom, D. Policansky, and B. D. Goldstein. 2001c. *Protecting the Commons A Framework for Resource Management in the Americas*. Washington DC: Island Press.

Burger, J., K. F. Gaines, J. D. Peles, W. L. Stephens Jr., C. S. Boring, I. L. Brisbin Jr., J. Snodgrass, A. L. Bryan Jr., M. H. Smith, and M. Gochfeld. 2001d. Radiocesium in Fish from the Savannah River and Steel Creek: Potential food chain exposure to the public. *Risk Analysis* 21:545–559.

Burger, J., C. Dixon, T. Shukla, N. Tsipoura, and M. Gochfeld. 2002a. Metal Levels in Horseshoe Crabs (*Limulus polyphemus*) from Maine to Florida. *Environmental Research*. 90:227–236.

Burger, J., K. F. Gaines, C. S. Boring, W. L. Stephens, J. Snodgrass, C. Dixon, M. McMahon, S. Shukla, T. Shukla, and M. Gochfeld. 2002b. Metal Levels in Fish from the Savannah River: Potential hazards to fish and other receptors. *Environmental Research*. 89:85–97.

Burger, J., K. Kannan, J. P. Giesy, C. Grue, and M. Gochfeld. 2002c. Effects of environmental pollutants on avian behaviour. In *Behavioural Ecotoxicology*, ed. G. Dell'Omo, 337–375. New York: John Wiley & Sons.

Burger, J., F. Diaz-Barriga, E. Marafante, J. Pounds, M. Robson. 2003a. Methodologies to examine the importance of host factors in bioavailability of metals. *Ecotoxicology and Environmental Safety* 56:1:20–31.

Burger, J., C. Dixon, T. Shukla, N. Tsipoura, H. Jensen, M. Fitzgerald, R. Ramos, and M. Gochfeld. 2003b. Metals in Horseshoe Crabs from Delaware Bay. *Archives of Environmental Contamination and Toxicology* 44:36–42.

Burger, J., C. Jeitner, K. Clark, and L. Niles. 2004. The effect of human activities on migrant shorebirds: Successful adaptive management. *Environmental Conservation* 31:4:283–288.

Burger, J., M. Gochfeld, D. Kosson, C. W. Powers, B. Friedlander, J. Eichelberger, D. Barnes, L. K. Duffy, S. C. Jewett, and C. D. Volz. 2005. Science, policy, and stakeholders: Developing a consensus science plan for Amchitka Island, Aleutians, Alaska. *Environmental Management* 35:557–568.

Burger, J., H. Mayer, M. Greenberg, C. S. Powers, C. D. Volz, and M. Gochfeld. 2006. Conceptual site models as a tool in evaluating ecological health: The case of the Department of Energy's Amchitka Island nuclear site. *Journal of Toxicology and Environmental Health* 69:1217–1238.

Burger, J., S. A. Carlucci, C. W. Jeitner, and L. Niles. 2007a. Habitat choice, disturbance, and management of foraging shorebirds and gulls at a migratory stopover. *Journal of Coastal Research* 23:1159–1166.

Burger, J., M. Gochfeld, D. S. Kosson, and C. W. Powers. 2007b. A biomonitoring plan for assessing potential radionuclide exposure using Amchitka Island in the Aleutian chain of Alaska as a case study. *Journal of Environmental Radioactivity* 98:315–328.

Burger, J., M. Gochfeld, C. W. Powers, D. Kosson, J. Halverson, G. Siekaniec, A. Morkill, R. Patrick, L. K. Duffy, and D. Barnes. 2007c. Scientific research, stakeholders, and policy: Continuing dialogue during research on radionuclides on Amchitka Island, Alaska. *Journal of Environmental Management* 85:232–244.

Burger, J., M. Gochfeld, C. Jeitner, S. Burke, T. Stamm, R. Snigaroff, D. Snigaroff, R. Patrick, and J. Weston. 2007d. Mercury levels and potential risk from subsistence foods from the Aleutians. *Science Total Environment*. 384:93–105.

Burger, J., M. Gochfeld, K. Pletnikoff, R. Snigaroff, D. Snigaroff, and T. Stamm. 2008a. Ecocultural attributes: Evaluating, ecological degradation in terms of ecological goods and services versus subsistence and tribal values. *Risk Analysis* 28:1261–1271.

Burger, J., M. Gochfeld, S. Shukla, C. Jeitner, R. Ramos, N. Tsipoura, and M. Donio. 2008b. Pollution, contamination, and future land use at Brookhaven National Laboratory. *Archive of Environmental Contamination and Toxicology* 55:341–347.

Burger, J., M. Gochfeld, C. Jeitner, D. Snigaroff, R. Snigaroff, T. Stamm, C. Volz. 2008c. Assessment of metals in down feathers of female common eiders and their eggs from the Aleutians: Arsenic, cadmium, chromium, lead, manganese, mercury, and selenium. *Environmental Monitoring Assessment* 143:247–256.

Burger, J., M. Gochfeld, and K. Pletnikoff. 2009a. Collaboration versus communication: The Department of Energy's Amchitka Island and the Aleut Community. *Environmental Research* 109:503–510.

Burger, J., C. Jeitner, M. Donio, S. Shukla, and M.Gochfeld. 2009b. Factors affecting mercury and selenium levels in New Jersey flatfish: Low risk to human consumers. *Journal of Toxicology and Environmental Health* 72:853–860.

Burger, J., M. Gochfeld, C. Jeitner, S. Burke, C. Volz, R. Snigaroff, D. Snigaroff, T. Shukla, and S. Shukla. 2009c. Mercury and other metals in eggs and feathers of Glaucous-winged Gulls (*Larus glaucescens*) in the Aleutians. *Environmental Monitoring and Assessment*. 152:179–194.

Burger, J., M. Gochfeld, C. D. Jenkins, and F. Lesser. 2010a. Effects of approaching boats on nesting Black Skimmers: Distances to establish protective buffer zones. *Journal of Wildlife Management* 74:102–108.

Burger, J., S. Harris, B. Harper, and M. Gochfeld. 2010b. Ecological information needs for environmental justice. *Risk Analysis* 30:893–905.

Burger, J., C. Gordon, J. Lawrence, J. Newman, G. Forcey, and L. Vlietstra. 2010c. Risk evaluation for federally listed (Roseate Tern, Piping Plover) or candidate (Red Knot) bird species in offshore waters: A first step for managing the potential impacts of wind facility development on the Atlantic Outer Continental Shelf. *Renewable Energy* 36:338–351.

Burger, J., M. Gochfeld, C. Jeitner, and T. Pittfield. 2011a. Comparing perceptions of the important environmental characteristics of the places people engage in consumptive, non-consumptive, and spiritual activities. *Journal of Risk Research* 14:1219–1236.

Burger, J., M. Gochfeld, and C. Jeitner. 2011b. Locational differences in mercury and selenium levels in 19 species of saltwater fish from New Jersey. *Journal of Toxicology and Environmental Health* 74:863–874.

Burger J., M. Gochfeld, C. Jeitner, M. Donio, and T. Pittfield. 2012a. Activity patterns and perceptions of goods, services, and eco-cultural attributes by ethnicity and gender for Native Americans and Caucasians. *International Journal of Sports ManageFment Recreation and Tourism* 9:34–51.

Burger, J., L. Niles, R. Porter, A. Dey, S. Koch, and C. Gordon. 2012b. Migration and overwintering of Red Knots (*Calidris canutus rufa*) along the Atlantic coast of the United States. *Condor* 114:302–313

Burger, J., L. J. Niles, R. R. Porter, and A. D. Dey. 2012c. Using geolocators to reveal incubation periods and breeding biology in Red Knots. *Wader Study Group Bulletin* 119:26–36.

Burger, J., L. Niles, R. Porter, A. Dey, S. Koch, and C. Gordon. 2012d. Using a shorebird (Red Knot) fitted with geolocators to evaluate a conceptual risk model focusing on off shore wind. *Renewable Energy* 43:370–377.

Burger, J., M. Gochfeld, C. Jeitner, M. Donio, and T. Pittfield. 2012e. Lead (Pb) in biota and perceptions of lead exposure at a recently-designated superfund beach site in New Jersey. *Journal of Toxicology and Environmental Health* 75:272–287.

Burger, J., M. Gochfeld, F. Lesser, C. Jeitner, M. Donio, and T. Pittfield. 2012f. Bioindicators for environmental assessment and monitoring: Metals in wading birds from New Jersey. *International Journal of Environmental Science and Engineering Research* 3:147–160.

Burger, J., M. Gochfeld, C. W. Powers, J. Clarke, and D. Kosson. 2013a. Types and integration of environmental assessment and monitoring plans. *International Journal of Environmental Science and Engineering Research* 4:31–51.

Burger, J., M. Gochfeld, C. W. Powers, L. Niles, R. Zappalorti, J. Feinberg, and J. Clarke. 2013b. Habitat protection for sensitive species: Balancing species requirements and human constraints using bioindicators as examples. *Natural Science* 5:50–62.

Burger, J., M. Gochfeld, C. W. Powers, J. H. Clarke, K. Brown, D. Kosson, L. Niles, A. Dey, C. Jeitner, and T. Pittfield. 2013c. Determining environmental impacts for sensitive species: Using iconic species as bioindicators for management and policy. *Journal of Environmental Protection* 4:87–95.

Burger, J., M. Gochfeld, J. Clarke, C. W. Powers, and D. Kosson. 2013d. An ecological multidisciplinary approad to protecting society, human health, and the environment at nuclear facilities. *Remediation* 23:123–148.

Burger, J., M. Gochfeld, and T. Fote. 2013e. Stakeholder participation in research design and decisions: Scientists, fishers, and mercury in saltwater fish. *EcoHealth* 10:21–30.

Burger, J., C. Jeitner, M. Donio, T. Pittfield, and M. Gochfeld. 2013f. Mercury and selenium levels, and selenium: Mercury molar ratios of brain, muscle and other tissues in Bluefish (*Pomatomus saltatrix*) from New Jersey, USA. *Science of the Total Environment* 443:278–286.

Burger, J., J. R. Jehl, Jr., and M. Gochfeld. 2013g. Selenium: Mercury molar ratio in Eared Grebes (*Podiceps nigricollis*) as a possible biomarker of exposure. *Ecological Indicators* 34:60–68.

Burger, J., M. Gochfeld, L. Niles, A. Dey, C. Jeitner, T. Pittfield, and N. Tsiourpa. 2014a. Metals in tissues of migrant Semipalmated Sandpipers (*Calidris pusilla*) from Delaware Bay, New Jersey. *Environmental Research* 133:362–370.

Burger, J., N. Tsipoura, L. J. Niles, M. Gochfeld, A. Dey, and D. Mizrahi. 2015a. Mercury, lead, cadmium, arsenic, chromium and selenium in feathers of Shorebirds during migrating through Delaware Bay, New Jersey: Comparing the 1990s and 2011/2012. *Toxics* 3:63–74.

Burger, J., M. Gochfeld, L. Niles, C. W. Powers, K. Brown, J. Clarke, and A. Dey. 2015b. Complexity of bioindicator selection for ecological, human, and cultural health: Chinook Salmon and Red Knot as case studies. *Environmental Monitoring and Assessment* 187:1–18.

Burger, J., M. Gochfeld, C. W. Powers, K. G. Brown, and J. H. Clarke. 2015c. Salmon exposure to chromium in the Hanford reach of the Columbia River: Potential effects on the life history and population biology. Consortium for Risk Evaluation with Stakeholder Participation III Tennessee: Vanderbilt University.

Burgess, N. M., and M. W. Meyer. 2008. Methylmercury exposure associated with reduced productivity in common loons. *Ecotoxicology* 17:83–91.

Burgess, N. M., A. L. Bond, C. E. Hebert, E. Neugebauer, and L. Champoux. 2013. Mercury trends in Herring Gull (*Larus argentatus*) eggs from Atlantic Canada, 1972–2008: Temporal change or dietary shift? *Environmental Pollution* 172:216–222.

Burgess, N. M., D. C. Evers, and J. D. Kaplan. 2005. Mercury and other contaminants in Common Loons breeding in Atlantic Canada. *Ecotoxicology* 14:241–252.

Bustnes, J. O., T. Anker-Nilssen, K. E. Erikstad, S. H. Lorentsen, and G. H. Systad. 2013a. Changes in the Norwegian breeding population of European Shag correlate with forage fish and climate. *Marine ecology Progress series* 489:235–244.

Butchart, S. H., M. Walpole, B. Collen, A. van Strien, J. P. W. Scharlemann, R. E. A. Almond, J. E. M. Baillie et al. 2010. Global biodiversity: Indicators of recent declines. *Science* 328:1164–1168.

Butler, C. J. 2003. The disproportionate effect of global warming on the arrival dates of short-distance migratory birds in North America. *Ibis* 145:484–495.

Butler, R. W. 2006. The concept of a tourist area cycle of evolution: Implications for management of resources. *The Tourism Area Life Cycle* 1:3–12.

Butler, V. L., and J. E. O'Connor. 2004. 9000 years of salmon fishing on the Columbia River, North America. *Quaternary Research* 62:1–8.

Buzzards Bay National Estuary Program. 2015. Buzzards Bay Quick Facts. http://buzzardsbay.org/bayshed .htm (accessed July 22, 2015).

Cahoon, D. R., P. F. Hensel, T. Spencer, D. J. Reed, K. L. McKee, and N. Saintilan. 2006. Coastal wetland vulnerability to relative sea-level rise: Wetland elevation trends and process controls. In: *Wetlands and Natural Resource Management*, ed. J. T. A. Verheven, B. Beltman, R. Bobbink, and D. F. Whigham, 271–292. Berlin: Springer.

Cairns, J. Jr., and B. R. Niederlehner. 1996. Developing a field of landscape ecotoxicology. *Ecological Applications* 6:780–796.

Carey, C. 2009. The impacts of climate change on the annual cycles of birds. *Philosophical Transactions of the Royal Society B: Biological Sciences* 364:3321–3330.

Caribbean Business Online Staff (CBOS). 2012. MD's push DC to clean Vieques, Culebra. http://www.caribbean businesspr.com/news/mds-push-dc-to-clean-vieques-culebra-90415.html (accessed August 31, 2015).

Carignan, V., and M. A. Villard. 2001. Selecting indicator species to monitor ecological integrity: A review. *Environmental Monitoring and Assessment* 78:45–61.

Carneiro, M., B. Colaço, R. Brandão, B. Azorín, O. Nicolas, J. Colaço, M. J. Pires et al. 2015. Assessment of the exposure to heavy metals in Griffon vultures (*Gyps fulvus*) from the Iberian Peninsula. *Ecotoxicology and Environmental Safety* 113:295–301.

Caro, T. M., and J. M. Eadie. 2005. Animal behavior and conservation. In: *The Behavior of Animals: Mechanism, Function and Evolution*, ed. J. J. Bolhuis and L. A. Giraldeau, 367–392. Oxford: Blackwell Publishers Ltd.

Caro, T., and P. W. Sherman. 2013. Eighteen reasons animal behaviourists avoid involvement in conservation. *Animal Behaviour* 85:305–312.

Carvalho A. R., V. V. Cardoso, A. Rodrigues, E. Ferreira, M. J. Benoliel, and E. A. Duarte. 2015. Occurrence and analysis of endocrine-disrupting compounds in a water supply system. *Environmental Monitoring and Assessment* 187:139.

Castro, G., and J. P. Myers. 1993. Shorebird predation on eggs of Horseshoe Crabs during spring stopover on Delaware Bay. *Auk* 110:927–930.

Chaine, A. S., and J. Colbert. 2012. Dispersal. In: *Behavioral Responses to a Changing World: Mechanisms and Consequences*, ed. U. Candolin and B. B. M Wong, 63–79. Oxford: Oxford Univ. Press.

Champoux, L., J. Rodrigue, J. L. DesGranges, S. Trudeau, A. Hontela, M. Boily, and P. Spear. 2002. Assessment of contamination and biomarker responses in two species of herons on the St. Lawrence River. *Environmental Monitoring and Assessment* 79:193–215.

Chapman, F. M. 1899. The passing of the tern. *Bird-Lore* 1(6):2.

Chen, L., G. Qu, X. Sun, S. Zhang, L. Wang, N. Sang, and S. Liu. 2013. Characterization of the interaction between cadmium and chlorpyrifos with integrative techniques in incurring synergistic hepatoxicity. *PLoS One* 8(3).

Chesapeake Bay Foundation (CBF). 2010. On the brink: Chesapeake's Native Oysters. What will it take to bring them back. http://www.cbf.org/document.doc?id=523 (accessed March 12, 2015).

Chesapeake Bay Foundation (CBF). 2015. Habitats of the Chesapeake. Chesapeake Bay Foundation. http://www.cbf.org/about-the-bay/more-than-just-the-bay/habitats-of-the-chesapeake (accessed March 12, 2015).

Chesapeake Bay Program (CBP). 2015. Fact and figures—Chesapeake Bay Program. http://www.chesapeake bay.net/discover/bay101/facts (accessed March 12, 2015).

Chess, C., J. Burger, and M. H. McDermott. 2005. Policy review and essays—Speaking like a state: Environmental justice and fish consumption advisories. *Society and Natural Resources* 18:267–278.

Chu, S., J. Wang, G. Leong, L. A. Woodward, R. J. Letcher, and Q. X. Li. 2015. Perfluoroalkyl sulfonates and carboxylic acids in liver, muscle and adipose tissues of Black-footed Albatross (*Phoebastria nigripes*) from Midway Island, North Pacific Ocean. *Chemosphere* 138:60–66.

Cifuentes, J. M., P. H. Becker, U. Sommer, P. Pacheco, and R. Schlatter. 2003. Seabird eggs as bioindicators of chemical contamination in Chili. *Environmental Pollution* 126:123–137.

Claggett, P. R. 2007. Human population growth and land-use change. In: *Synthesis of U.S. Geological Survey Science for the Chesapeake Ecosystem and Implications for Environmental Management,* Circular 1316, ed. S. W. Phillips. U. S. Geological Survey, Reston Virginia, 10–13.

Clark, K., and B. Wurst. 2013. The 2013 Osprey Project in New Jersey. http://www.conservewildlifenj.org/downloads/cwnj_478.pdf (accessed February 28, 2015).

Clark, K. E., L. J. Niles, and J. Burger. 1993. Abundance and distribution of migrant shorebirds in Delaware Bay. *Condor* 95:694–705.

Clemants, S. E., and S. N. Handel. 2005. Restoring urban ecology: The New York–New Jersey metropolitan area experience. In: *The Humane Metropolis: People and Nature in the 21st Century City*, ed. R. H. Platt, 127–140. Amherst: University of Massachusetts Press.

Climate Central. 2014. Jersey Retreating From Rivers, But Not Coast, After Sandy. Climate Central website Nov 4, 2014. http://www.climatecentral.org/news/nj-hurricane-sandy-blue-acres-18275 (accessed August 23, 2015).

Colborn, T., and C. Clement. 1992. Chemically-induced alterations in sexual and functional development: The wildlife/human connection. In: *Advances in Modern Environmental Toxicology*, ed. M. A. Mehlman. Princeton, NJ: Princeton Scientific Publ. Co, Inc.

Collins, C. T. 1970. The Black-crowned Night Heron as a predator of tern chicks. *Auk* 87:584–586.

Colonial Waterbird Working Group (CWWG). 2007. Harbor Herons, Cormorants, and more current research and future planning. Staten Island, New York. Proceedings of the Greater New York/New Jersey Harbor Colonial Waterbirds Working Group.

Columbia River Inter-Tribal Fish Commission (CRITFC). 2013. We Are Salmon People. CRITFC. http://critfc.org/salmon-culture/columbia-river-salmon/columbia-river-salmon-species (accessed March 12, 2015).

Colwell, M. A. 2010. *Shorebird Ecology: Conservation and Management*. Berkeley, CA: University of California Press.

Connell, D. W., B. S. F. Wong, P. K. S. Lam, K. F. Poon, M. H. W. Lam, R. S. S. Wu, B. J. Richardson, and Y. F. Yen. 2002. Risk to breeding success of Ardeids by contaminants in Hong Kong: Evidence from trace metals in feathers. *Ecotoxicology* 11:49–59.

Connors, P. G., V. C. Anderlini, R. W. Risebrough, M. Gilbertson, and H. Hays.1975. Investigations of heavy metals in common tern populations. *Canadian Field-Naturalist* 89:157–162.

Conserve Wildlife. 2014. Conservation status overview. http://www.conservewildlifenj.org/species/status/ (accessed February 28, 2015).

Conti, D. B. 2008. *Biological Monitoring: Theory and Applications—Bioindicators and Biomarkers for Environmental Quality and Human Exposure Assessment*. Boston: WIT Press.

Convertino, M., A. Bockelie, G. A. Kiker, R. Muñoz-Carpena, and I. Linkov. 2012. Shorebird patches as fingerprints of fractal coastline fluctuations due to climate change. *Ecological Processes* 1:1–17.

Conway, T. M., and R. G. Lathrop. 2005. Alternative land use regulations and environmental impacts: Assessing future land use in an urbanizing watershed. *Landscape and Urban Planning* 71:1–15.

Cook-Haley, B. S., and K. F. Millenbah. 2002. Impacts of vegetative manipulations on Common Tern nest success at Lime Island, Michigan. *Journal of Field Ornithology* 73:174–179.

Cornell Laboratory of Ornithology. 2014. All About Birds: Herring Gull. http://www.allaboutbirds.org/guide/herring_gull/lifehistory (accessed March 4, 2015).

Corsi, I., M. Mariottini, C. Sensini, L. Lancini, and S. Focardi. 2003. Fish as bioindicators of brackish eco-system health: Integrating biomarker responses and target pollutant concentrations. *Oceanologica Acta* 26:129–138.

Corvalan, C. F., T. Kjellstrom, and K. R. Smith. 1999. Health, environment, and sustainable development. Identifying links and indicators to promote action. *Epidemiology* 10:656–660.

Costanza, R., R. d'Arge, R. de Groot, S. Farber, M. Grasso, B. Hannon, K. Limburg et al. 1997. The value of the world's ecosystem services and natural capital. *Nature* 387:253–260.

Costanza, R., R. de Groot, P. Sutton, S. van der Ploeg, S. J. Anderson, I. Kubiszewski, S. Farber, and R. K. Turner. 2014. Changes in the global value of ecosystem services. *Global Environmental Change* 26:152–158.

Cotin, J., M. Garcia-Tarrason, C. Sanpera, L. Jover, and X. Ruiz. 2011. Sea, freshwater, or saltpans? Foraging ecology of terns to assess mercury inputs in a wetland landscape: The Ebro Delta. *Estuarine, Coatsal, and Shelf Science* 92:188–194.

Cotter, R. C., J. F. Rail, A. W. Boyne, D. V. C Weseloh, K. G. Chaulk, and A. W. Diamond. 2013. Recent trends in gull numbers nesting in eastern Canada. Abstract only. Waterbird Society 37th Annual Meeting, September 2013.

Coulson, J. C. 2002. Colonial breeding in seabirds. In: *Biology of Marine Birds*, ed. E. A. Schreiber and J. Burger, 87–113. Boca Raton, FL: CRC Press.

Coulson, J. C. 2011. *The Kittiwake*. London: Poyser Monographs.

Craig, E. 2009. *New York City Audubon's Harbor Herons Project: 2009 Interim Nesting Survey Report*. New York: New York City Audubon Society.

Craig, E. 2013. New York City Audubon's Harbor Herons Project: 2013 Nesting Survey Report. *New York City Audubon*. New York, NY. http://www.nycaudubon.org/images/pdf/2013_HH_Survey_Report_-12-23-13 .compressed.pdf (accessed March 29, 2015).

Craig, E., S. Elbin, J. Danoff-Burg, and M. Palmer. 2007. Impacts of double-crested cormorants on New York Harbor: Vegetation and arthropods (Abstract). Colonial Waterbird Working Group. Harbor Herons, Cormorants, and more current research and future planning. *Proceedings of the Greater New York/New Jersey Harbor Colonial Waterbirds Working Group*, Staten Island, New York.

Craig, E. C., S. B. Elbin, J. A. Danoff-Burg, and M. I. Palmer. 2012. Impacts of Double-Crested Cormorants (*Phalacrocorax auritus*) and other colonial waterbirds on plant and arthropod communities on islands in an urban estuary. *Waterbirds* 35:4–12.

Craig, E. H., J. Pasher, C. Weseloh, T. Dobbie, S. Dobbyn, D. Moore, V. Minelga, and J. Duffe. 2014. Nesting cormorants and temporal changes in Island habitat. *Journal of Wildlife Management* 78:307–313.

Crick, H. Q. 2004. The impact of climate change on birds. *Ibis* 146:48–56.

Cristol, D. A., R. L. Brasso, A. M. Condone, R. E. Fovargue, S. L. Friedman, K. K. Hallinger, A. P. Monroe, and A. E. White. 2008. The movement of aquatic mercury through terrestrial food webs. *Science* 320:335.

Crosby, G. T. 1972. Spread of the cattle egret in the Western Hemisphere. *Bird-Banding* 43:205–212.

Cross, F. A., D. W. Evans, and R. T. Barber. 2015. Decadal declines of mercury in adult bluefish (1972–2011) from the mid-Atlantic coast of the U.S.A. Environmental and Scientific Technology. 2015 Jul 21. [Epub ahead of print]

Cronin, W. B. 2005. *The Disappearing Islands of the Chesapeake*. Baltimore: Johns Hopkins University Press.

Cross, F. A., D. W. Evans, and R. T. Barber. 2015. Decadal declines of mercury in adult Bluefish (1972–2011) from the mid-Atlantic coast of the U.S.A. *Environmental Science and Technology* 49:9064–9072.

Crosset, K., T. Cultiton, P. Wiley, and T. Goodspeed. 2013. Population trends along the coastal United States 1980–2008. http://oceanservice.noaa.gov/programs/mb/pdfs/coastal.pop.trends.complete.pdf (accessed February 28, 2015).

Croxall, J. P., P. N. Trathan, and E. J. Murphy. 2002. Environmental change and Antarctic seabird populations. *Science* 297:1510–1514.

Crump, K. L., and V. L. Trudeau. 2009. Mercury-induced reproductive impairment in fish. *Environmental Toxicology and Chemistry* 28:895–907.

Custer, T. W. 2000. Environmental contaminants. In: *Heron Conservation*, ed. J. A. Kushlan and H. Hafner, 251–267. New York: Academic Press.

Custer, T. W., and W. L. Hoffman. 1994. Trace elements in canvasbacks (*Aythya valisineria*) wintering in Louisiana, USA, 1987–1988. *Environmental Pollution* 84:253–259.

Custer, C. M., T. W. Custer, and P. M. Dummer. 2010. Patterns of organic contaminates in eggs of an insectivorous, an omnivorous, and piscivorous bird nesting on the Hudson River, New York, USA. *Environmental Toxicology and Chemistry* 29:2286–2296.

Custer, T. W., C. M. Custer, B. A. Eichorst, and D. Warburn. 2007a. Selenium and metal concentrations in waterbird eggs and chicks at Agassiz National Wildlife Refuge, Minnesota. *Archives of Environmental Contamination and Toxicology* 53:103–109.

Custer, C. M., T. W. Custer, and E. F. Hill. 2007b. Mercury exposure and effects on cavity-nesting birds from the Carson River, Nevada. *Archives of Environmental Contamination and Toxicology* 52:129–136.

Custer, T. W., N. H. Golden, and B. A. Rattner. 2008. Element patterns in feathers of nestling Black-crowned Night herons, *Nycticorax nycticorax* L., from four colonies in Delaware, Maryland, and Minnesota. *Bulletin of Environmental Contamination and Toxicology* 81:147–151.

Cuthbert, F. J., L. R. Wires, and J. E. McKearnan. 2002. Potential impacts of nesting Double-crested Cormorants on Great Blue Herons and Black-crowned Night-Herons in the U.S. Great Lakes region. *Journal of Great Lakes Research* 28:145–154.

Danchin, E., and R. H. Wagner. 1997. The evolution of coloniality: The emergence of new perspectives. *Trends in Ecology and Evolution* 12:342–347.

Darling, F. F. 1938. *Bird Flocks and the Breeding Cycle.* Cambridge: Cambridge University Press.

Darling, F. F., and J. P. Milton. 1966. *Future Environments of North America: Transformation of a Continent.* Garden City, NY: Natural History Press.

Dauwe, T., L. Bervoets, R. Pinxten, R. Blust, and M. Eens. 2003. Variation of heavy metals within and among feathers of birds of prey: Effects of molt and external contamination. *Environmental Pollution* 124:429–436.

Davidson, P. W., G. J. Myers, B. Weiss, C. F. Shamlaye, and C. Cox. 2006. Prenatal methyl mercury exposure from fish consumption and child development: A review of evidence and perspectives from the Seychelles Child Development Study. *Neurotoxicology* 27:1106–1109.

Davis, W. E., Jr. 1993. Black-crowned Night-Heron (*Nycticorax nycticorax*). In: *The Birds of North America, No. 74*, ed. A. Poole and F. Gill. Philadelphia: Academy of Natural Sciences; Washington, D.C.: American Ornithologists' Union.

Davis, R. A., and D. M. Fitzgerald. 2004. *Beaches and Coasts.* Malden: Blackwell Science.

Davis, W. E. Jr., and J. Kricher. 2000. Glossy Ibis (*Plegadis falcinellus*). In: *The Birds of North America Online*, ed. A. Poole. Ithaca, NY: Cornell Laboratory of Ornithology. http://bna.birds.cornell.edu.bnaproxy.birds .cornell.edu/bna/species/545 (accessed February 28, 2015).

Davis, W. E., and K. C. Parsons. 1991. Effects of investigator disturbance on the survival of snowy egret nestlings. *Journal of Field Ornithology* 62:432–435.

de Camp, W. Jr. 2012. Declare Barnegat Bay as impaired. *Inside the Pinelands* 19:1–5.

de Cerreño, A. L. C., M. Panero, and S. Boehme. 2002. *Pollution Prevention and Management Strategies for Mercury in the New York/New Jersey Harbor.* New York: New York Academy of Sciences for the New York/New Jersey Harbor Consortium.

Delaware Audubon Society. 2002. Important Bird Areas in Delaware. http://www.delawareaudubon.org/bird ing/globaliba.html (accessed February 9, 2015).

De Luca-Abbott, S. B., B. S. Wong, D. B. Peakall, P. K. Lam, L. Young, M. H. Lam, and B. J. Richardson. 2001. Review of effects of water pollution on the breeding success of waterbirds, with particular reference to Ardeids in Hong Kong. *Ecotoxicology* 10:327–349.

Delaware Birding Trail. 2015. Site #6: Fort Delaware State Park http://www.delawarebirdingtrail.org/drc2.html (accessed February 9, 2015).

Delistraty, D., and J. Yokel. 2014. Ecotoxicological study of arsenic and lead contaminated soils in former orchards at the Hanford Site, USA. *Environmental Toxicology* 29:10–20.

Dell'Omo, G. 2002. *Behavioural Ecotoxicology.* New York: John Wiley & Sons.

Depew, D. C., N. M. Burgess, and L. M. Campbell. 2013. Spatial patterns of methylmercury risks to Common Loons and piscivorous fish in Canada. *Environmental Science Technology* 47:13093–13103.

DesGranges, J., J. Rodrigue, B. Tardif, and M. Laperle. 1998. Mercury accumulation and biomagnification in ospreys (*Pandion haliaetus*) in the James Bay and Hudson Bay regions of Québec. *Archive of Environmental Contamination and Toxicology* 35:330–341.

Deshpande, A. D., A. F. Draxler, V. S. Zdanowicz, M. E. Schrock, and A. J. Paulson. 2002. Contaminant levels in the muscle of four species of fish important to the recreational fishery of the New York Bight Apex. *Marine Pollution bulletin* 44:164–171.

Dey, P. M., J. Burger, M. Gochfeld, and K. R. Reuhl. 2000. Development lead exposure disturbs expression synaptic neural cell adhesion molecules in Herring Gull brains. *Toxicology* 146:137–147.

Dey, A. D., L. J. Niles, H. P. Sitters, K. Kalasz, and R. I. G. Morrison. 2015. Update to the status of the Red Knot Calidris canutus in the Western Hemisphere, August 2014. Draft update to Status of the Red Knot (*Calidris canutus rufa*) in the Western Hemisphere. New Jersey Environmental Protections April 2014. Unpublished Report.

Distler, T., J. G. Schuetz, J. Velasquez-Tibata, and G. M. Langham. 2015. Stacked species distribution models and macroecological models provide congruent projections of avian species richness under climate change. *Journal of Biogeography* 42:976–988.

Dittmann, T., P. H. Becker, J. Bakker, A. Bignert, E. Nyberg, M. G. Pereira, U. Pijanowska et al. 2011. The EcoQO on mercury and organohalogens in coastal bird eggs: Report on the pilot study 2008–2010. INBO.R.2011.43. Research Institute for Nature and Forest, Brussels.

Dittmann, T., P. H. Becker, J. Bakker, A. Bignert, E. Nyberg, M. G. Pereira, U. Pijanowska et al. 2012. Large-scale spatial pollution patterns around the North Sea indicated by coastal bird eggs within an EcoQO programme. *Environmental Science and Pollution Research* 19:4060–4072.

Dittmann, T., D. Zinsmeister, and P. H. Becker. 2005. Dispersal decisions: Common Terns, *Sterna hirundo*, choose between colonies during prospecting. *Animal Behaviour* 70:13–20.

Dolbeer, R. A., J. L. Belant, and J. L. Sillings. 1993. Shooting gulls reduces strikes with aircraft at John F. Kennedy International Airport. *Wildlife Society Bulletin* 21:442–450.

Dorr, B. S., J. J. Hatch, and D. V. Weseloh. 2014. Double-crested Cormorant (*Phalacrocorax auritus*). In: *The Birds of North America Online*, ed. A. Poole. Ithaca, NY: Cornell Laboratory of Ornithology. http://bna .birds.cornell.edu.bnaproxy.birds.cornell.edu/bna/species/441 (accessed February 28, 2015).

Dosch, J. J. 2003. Movement patterns of adult Laughing Gulls *Larus atricilla* during the nesting season. *Acta Ornithologica* 38:15–25.

Doughty, R. W. 1974. *Feather Fashions and Bird Preservation: A Study in Nature Protection*. Oakland, CA: Univ. California Press.

Downs, S. G., C. L. Macleod, and J. M. Lester. 1998. Mercury in precipitation and its relation to bioaccumulation in fish: A literature review. *Water Air and Soil Pollution* 108:149–187.

Drilling, N., R. Titman, and F. Mckinney. 2002. *Mallard (Anas platyrhynchos)*. In: *Birds of North America Online*, ed. A. Poole. Ithaca, NY: Cornell Laboratory of Ornithology. http://bna.birds.cornell.edu .bnaproxy.birds.cornell.edu/bna/species/658 (accessed August 19, 2015).

Drown, D. B., S. G. Oberg, and R. P. Sharma. 1986. Pulmonary clearance of soluble and insoluble forms of manganese. *Journal of Toxicology and Environmental Health* 17:201–212.

Duarte, C. M. 2009. Introduction. In: *Global Loss of Coastal Habitats: Rates, Causes, and Consequences*. Madrid, Spain: Fundacion BBVA.

Dubos, R. J. 1972. *A God Within*. New York: MacMillan.

Duke, N. C., J. O. Meynecke, S. Ditmann, A. M. Ellison, K. Anger, U. Berger, S. Cannicci et al. 2013. A world without mangroves. *Science* 317:41–43.

Dunne, P., D. Sibley, C. Sutton, and W. Wander. 1982. Aerial surveys in Delaware Bay: Confirming an enormous spring staging area for shorebirds. *Wader Study Group Bulletin* 35:32–33.

Duquesne, S., L. C. Newton, L. Giusti, S. B. Marriott, H. J. Stärk, and D. J. Bird. 2006. Evidence for declining levels of heavy-metals in the Severn Estuary and Bristol Channel, U.K., and their spatial distribution in sediments. *Environmental Pollution* 143:187–196.

Dżugan, M., M. Lis, M. Droba, and J. W. Niedziółka. 2011. Effect of cadmium injected in ovo on hatching results and the activity of plasma hydrolytic enzymes in newly hatched chicks. *Acta Veterinaria Hungarica* 59:337–347.

Eagles-Smith, C. A., and J. T. Ackerman. 2010. Developing impairment thresholds for the effects of mercury on Forster's Tern reproduction in San Francisco Bay: Data Summary. U.S. Geological Survey. Western Ecological Research Center, Davis. CA, 21 pp.

Eagles-Smith, C. A., J. T. Ackerman, S. E. De La Cruz, and J. Y. Takekawa. 2009b. Mercury bioaccumulation and risk to three waterbird foraging guilds is influenced by foraging ecology and breeding stage. *Environmental Pollution* 157:1993–2002.

Eagles-Smith, C. A., J. T. Ackerman, J. Yee, and T. L. Adelsbach. 2009a. Mercury demethylation in water-bird livers: Dose–response thresholds and differences among species. *Environmental Toxicology and Chemistry* 28:568–577.

Ehrlich, P. R., D. S. Dobkin, and D. Wheye. 1988. The Plume Trade. https://web.stanford.edu/group/stanford birds/text/essays/Plume_Trade.html (accessed March 12, 2015).

Eisler, R. 1985a. Cadmium—Hazards to fish, wildlife and invertebrates: A synoptic review. U.S. Fish and Wildlife Service Reports, No. 85(1–2). Washington, DC.

Eisler, R. 1985b. Selenium hazards to fish, wildlife, and invertebrates: A synoptic review. U.S. Fish and Wildlife Service. Biological Report 85 (1.5). Report No. 5.

Eisler, R. 1986a. Polychlorinated biphenyl hazards to fish, wildlife, and invertebrates: A synoptic review. U.S. Fish and Wildlife Service. Biological Report 85(1.7).

Eisler, R. 1986b. Dioxin hazards for fish, wildlife and invertebrates. US Fish and Wildlife Service Biological Report, No. 85, 36. Washington, DC.

Eisler, R. 1986c. Chromium hazards to fish, wildlife, and invertebrates: A synoptic review. US Fish and Wildlife Service Report, No. 85, 1.6. Washington, DC.

Eisler, R. 1987a. Mercury Hazards to Fish, Wildlife, and Invertebrates: A Synoptic Review. US Fish and Wildlife Service Report, No. 85(1.10). Washington, DC.

Eisler, R. 1988a. Lead Hazards to Fish, Wildlife, and Invertebrates: A Synoptic Review. US Fish and Wildlife Service. Washington, DC.

Eisler, R. 1988b. Arsenic hazards to fish, wildlife, and invertebrates: A synoptic review. U. S. Fish and Wildlife Service. Biological Report 85 (1.12).

Eisler, R. 1994. A review of arsenic hazards to plants and animals with emphasis on fishery and wildlife resources. In: *Arsenic in the Environment, Part II*, ed. J. O. Nriagu, 374–399. New York: Wiley.

Eisler, R. 2000a. *Handbook of chemical Risk Assessment: Health Hazards to Humans, Plants and Animals.* Boca Raton, FL: Lewis Publishers.

Eisler, R. 2000b. Selenium. In: *Handbook of Chemical Risk Assessment: Health Hazards to Humans, Plants and Animals*, 141. Boca Raton, FL: CRC Press.

Eisner, T., and J. Meinwald. 1995. *Chemical Ecology: The Chemistry of Biotic Interaction.* Washington, DC: National Academy Press.

El-Bayoumy, K. 2001. The protective role of selenium on genetic change and cancer. *Mutation Research* 475:123–139.

El-Begearmi, M. M., M. L. Sunde, and H. E. Ganther. 1977. A mutual protective effect of mercury and sele-nium in Japanese Quail. *Poultry Science* 56:313–322.

Elbin, S. B., and E. Craig. 2007. Status of Double-crested Cormorants in New York Harbor, 2007. In: *Proceedings of the Greater NY/NJ Harbor Colonial Waterbirds Working Group*, 12. Staten Island, NY.

Elbin, S. B., and N. K. Tsipoura. 2010. Harbor Herons Conservation Plan—NY/NJ Harbor Region. NY–NJ Harbor Estuary Program. http://www.harborestuary.org/reports/harborheron/ConservationPlan_for _HarborHerons0610.pdf (accessed June 16, 2015).

Elliott-Smith, E., and S. M. Haig. 2004. Piping Plover (*Charadrius melodus*). In: *The Birds of North America Online*, ed. A. Poole. Ithaca, NY: Cornell Laboratory of Ornithology. http://bna.birds.cornell.edu .bnaproxy.birds.cornell.edu/bna/species/002 (accessed March 12, 2015).

Elliot-Smith, E., S. M. Haig, and B. M. Powers. 2009. Data from the 2006 International Piping Plover Census. US Department of the Interior, US Geological Survey Data Series 426, 332.

Elliot-Smith, E., M. M. Machmer, L. K. Wilson, and C. J. Henny. 2000. Contaminants in Ospreys from the Pacific Northwest: II. Organochorline pesticides, polychlorinated biphenyls, and mercury, 1191–1997. *Archives of Environmental Contamination and Toxicology* 38:93–106.

Ellis, H. I., and G. W. Gabrielsen. 2001. Energetics of free-ranging seabirds. In: *Biology of Marine Birds*, ed. E. A. Schreiber, and J. Burger, 359–407. Boca Raton, FL: CRC Press.

Ellwood, E. R., R. B. Primack, and M. L. Talmadge. 2010. Effects of climate change on spring arrival times of birds in Thoreau's Concord from 1851 to 2007. *Condor* 112:754–762.

Ellwood, E. R., S. A. Temple, R. B. Primack, N. L. Bradley, and C. C. Davis. 2013. Record-breaking early flowering in the Eastern United States. *PLoS ONE* 8: e53788.

Environmental Protection Agency. 1983. Chesapeake Bay: A Framework for Action. http://www.chesapeake bay.net/content/publications/cbp_12405.pdf (accessed August 23, 2015).

Environmental Protection Agency (EPA). 1997a. *Ecological Indicators: Evaluation Criteria.* Washington, DC: Environmental Protection Agency.

Environmental Protection Agency (EPA). 1997b. Mercury Study: Report to Congress. Volume III: Fate and Transport of Mercury in the Environment U.S. Environmental Protection Agency EPA-452/R-97-005 http://www.epa.gov/ttn/oarpg/t3/reports/volume3.pdf (accessed February 2, 2015).

Environmental Protection Agency (EPA). 1997c. About the National Estuary Program. http://water.epa.gov /type/oceb/nep/about2.cfm (accessed July 6, 2015).

Environmental Protection Agency (EPA). 1997d. Mercury Study: Report to Congress. Volume VII: Characterization of Human Health and Wildlife Risks from Mercury Exposure in the United States U.S. Environmental Protection Agency. EPA-452/R-97-005. http://www.epa.gov/ttn/caaa/t3/reports/volume7 .pdf (accessed August 31, 2015).

Environmental Protection Agency (EPA). 2001. Water Quality Criterion for the protection of human health: Methylmercury. EPA-823-R-01-001. http://www.epa.gov/waterscience/criteria/methylmercury/pdf/mer cury-criterion.pdf. (accessed January 15, 2015).

Environmental Protection Agency (EPA). 2002. Methods for Measuring the Acute Toxicity of Effluents and Receiving Waters to Freshwater and Marine Organisms, Fifth Edition. http://www.epa.gov/Region6 /water/npdes/wet/wet_methods_manuals/atx.pdf (accessed May 1, 2015).

Environmental Protection Agency (EPA). 2007a. *Northeast National Estuary Program Coastal Condition, Massachusetts Bays Program.* Washington, DC: Environmental Protection Agency. http://ww.epa.gov /owow/oceans/nepccr/index.html (accessed January 15, 2015).

Environmental Protection Agency (EPA). 2007b. *Northeast National Estuary Program Coastal Condition, New York/New Jersey Harbor Estuary Program.* Washington, DC: Environmental Protection Agency. http:// ww.epa.gov/owow/oceans/nepccr/index.html (accessed January 15, 2015).

Environmental Protection Agency (EPA). 2007c. *Northeast National Estuary Program Coastal Condition, Barnegat Bay National Estuary Program.* Washington, DC: Environmental Protection Agency. http:// ww.epa.gov/owow/oceans/nepccr/index.html (accessed January 15, 2015).

Environmental Protection Agency (EPA). 2007d. *Northeast National Estuary Program Coastal Condition, Partnership for the Delaware Estuary.* Washington, DC: Environmental Protection Agency. http:// ww.epa.gov/owow/oceans/nepccr/index.html (accessed January 15, 2015).

Environmental Protection Agency (EPA). 2007e. Chapter 3: Northeast National Estuary Program Coastal Condition Report. Chapter 3: Northeast National Estuary Program Coastal Condition Report, Buzzards Bay National Estuary Program. http://water.epa.gov/type/oceb/nep/upload/2007_05_09_oceans_nepccr _pdf_nepccr_nepccr_ne_parte.pdf (accessed March 7, 2015).

Environmental Protection Agency (EPA). 2007f. National Estuary Program Coastal Condition Report. Chapter 3: Northeast National Estuary Program Coastal Condition Report, Northeast NEPs. http://water.epa.gov /type/oceb/nep/upload/2007_05_09_oceans_nepccr_pdf_large_section2.pdf (accessed March 7, 2015).

Environmental Protection Agency (EPA). 2007g. National Estuary Program Coastal Condition Report. Chapter 3: Northeast National Estuary Program Coastal Condition Report, Long Island Sound Study. http://water .epa.gov/type/oceb/nep/upload/2007_05_09_oceans_nepccr_pdf_nepccr_nepccr_ne_partg.pdf (accessed March 7, 2015).

Environmental Protection Agency (EPA). 2009. *Environmental Justice: Compliance and Environment.* Washington, DC: EPA.

Environmental Protection Agency (EPA). 2011. New Bedford Harbor: Harbor Cleanup. http://www2.epa.gov /new-bedford-harbor/harbor-cleanup#Why (accessed March 5, 2015).

Environmental Protection Agency (EPA). 2012a. Toxic Contaminants in the Chesapeake Bay and Its Watershed: Extent and Severity of Occurrence and Potential Biological Effects. USEPA Chesapeake Bay Program Office, Annapolis, MD. http://executiveorder.chesapeakebay.net/ChesBayToxics_finaldraft_11513b.pdf (accessed February 7, 2015).

Environmental Protection Agency (EPA). 2012b. Advanced Analysis: Species Sensitivity Distributions. http:// www.epa.gov/caddis/da_advanced_2.html.

Environmental Protection Agency (EPA). 2012c. The State of the Estuary—2012. NY-NJ Harbor Estuary. http://water.epa.gov/type/oceb/nep/upload/New-York-New-Jersey-SOE_Rprt.pdf (accessed August 23, 2015).

Environmental Protection Agency (EPA). 2012c. Ecoregions of New Jersey. http://www.epa.gov/wed/pages /ecoregions/nj_eco.htm (accessed June 30, 2015).

Environmental Protection Agency (EPA). 2013a. Region 2 Superfund: Vineland Chemical Co., Inc. Environmental Protection Agency, National Priorities List. www.epa.gov/region02/superfund/npl/vineland/ (accessed February 7, 2015).

Environmental Protection Agency (EPA). 2013b. Chesapeake Bay Program. Report to Congress. http://www .chesapeakebay.net/documents/cbp_rtc_2013.pdf (accessed July 23, 2015).

Environmental Protection Agency (EPA). 2014a. EPA Credits States for Making Progress in Bay Cleanup; Says More Effort Needed to Get Back on Track for a Restored Bay. http://yosemite.epa.gov/opa/adm press.nsf/90829d899627a1d98525735900400c2b/d6fc92040ef8cc4585257d03005d7199!OpenDocu ment (accessed February 7, 2015).

Environmental Protection Agency (EPA). 2014b. Emerging Contaminants—Perfluorooctane Sulfonate (PFOS) and Perfluorooctanoic Acid (PFOA). http://www2.epa.gov/sites/production/files/2014-04/documents /factsheet_contaminant_pfos_pfoa_march2014.pdf (accessed May 13, 2015).

Environmental Protection Agency (EPA). 2014c. Cleaner Powerplants http://www.epa.gov/mats/powerplants .html.

Environmental Protection Agency (EPA). 2015a. Mid-Atlantic Superfund: Aberdeen Proving Ground (Edgewood). http://www.epa.gov/reg3hwmd/npl/MD2210020036.htm (accessed March 12, 2015).

Environmental Protection Agency (EPA). 2015b. The Great Waters program: Chesapeake Bay. http://www.epa .gov/oaqps001/gr8water/xbrochure/chesapea.html (accessed March 12, 2015).

Erwin, R. M. 1989. Responses to human intruders by birds nesting in colonies: Experimental results and management guidelines. *Colonial Waterbirds* 12:104–108.

Erwin, R. M., and R. A. Beck. 2007. Restoration of waterbird habitat in Chesapeake Bay: Great expectation or *Sisyphus* revisited? *Colonial Waterbirds* 30:163–176.

Erwin, R. M., and T. Custer. 2000. Herons as indictors. In: *Heron Conservation*, ed. J. A. Kushlan and H. Hafner, 311–330. New York: Academic.

Erwin, R. M., D. F. Brinker, B. D. Watts, G. R. Costanzo, and D. D. Morton. 2010. Islands at bay: Rising seas, eroding islands, and waterbird habitat loss in Chesapeake Bay (USA). *Journal of Coastal Conservation* 15:51–60.

Erwin, R. M., C. J. Conway, S. W. Hadden, J. S. Hatfield, and S. M. Melvin. 2002. Waterbird Monitoring Protocol for Cape Cod National Seashore and other Coastal Parks, Refuges, and Protected Areas. USGS report to the National Park Service.

Erwin, R. M., D. K. Dawson, D. B. Stotts, L. S. McAllister, and P. H. Geissler. 1991. Open marsh water management in the mid-Atlantic region: Aerial surveys of waterbird use. *Wetlands* 11:209–228.

Erwin, R. M., J. Galli, and J. Burger. 1981. Colony site dynamics and habitat use in Atlantic coast seabirds. *The Auk* 98:550–561.

Erwin, R. M., G. M. Haramis, D. G. Krementz, and S. L. Funderburk. 1993. Resource protection for waterbirds in Chesapeake Bay. *Environmental Management* 17(5):613–619.

Erwin, R. M., G. M. Haramis, M. C. Perry, and B. D. Watts. 2007b. Waterbirds of the Chesapeake region: An introduction. *Waterbirds* 30:1–3.

Erwin, R. M., G. M. Sanders, and D. J. Prosser. 2004. Changes in lagoonal marsh morphology at selected Northeastern Atlantic coast sites of significance to migratory waterbirds. *Wetlands* 24:891–903.

Erwin, R. M., G. M. Sanders, D. J. Prosser, and D. R. Cahoon. 2006. High tides and rising seas: Potential effects on estuarine waterbirds. *Studies in Avian Biology* 32:214–228.

Erwin, R. M., B. D. Watts, G. M. Haramis, M. C. Perry, and K. A. Hobson. 2007a. Waterbirds of the Chesapeake Bay and vicinity: Harbingers of change. *Waterbirds* 30(Special Publication):1.

Espín, S., E. Martínez-López, P. Jiménez, P. María-Mojica, and A. J. García-Fernández. 2014. Effects of heavy metals on biomarkers for oxidative stress in Griffon vulture (*Gyps fulvus*). *Environmental Research* 129:59–68.

European Environmental Agency (EEA). 2003. *Environmental Indicators: Typology and Use in Reporting*. Copenhagen, Denmark: EEU.

Evers, D. C., A. K. Jackson, T. H. Tear, and C. E. Osborne. 2012. *Hidden Risk: Mercury in Terrestrial Ecosystems of the Northeast*. Gorham, ME: Biodiversity Research Institute.

Evers, D. C., J. D. Paruk, J. W. Mcintyre, and J. F. Barr. 2010. Common Loon (*Gavia immer*). In: *Birds of North America Online*, ed. A. Poole. Ithaca, NY: Cornell Laboratory of Ornithology. http://bna.birds.cornell .edu/bna/species/313 (accessed February 14, 2016).

Evers, D. C., L. J. Savoy, C. R. DeSorba, D. E. Yates, W. Hanson, K. M. Taylor, and J. Fair. 2008. Adverse effects from environmental mercury loads on breeding Common Loons. *Ecotoxicology* 17:69–81.

Evers, D. C., J. G. Wiener, N. Basu, R. A. Bodaly, H. A. Morrison, and K. A. Williams. 2011. Mercury in the Great Lakes region: Bioaccumulation, spatiotemporal patterns, ecological risks, and policy. *Ecotoxicology* 20:1487–1499.

Ezard, T. H., P. H. Becker, and T. Coulson. 2007. Correlations between age, phenotype, and individual contribution to population growth in Common Terns. *Ecology* 88:2496–2504.

Fahrig, L. 2003. Effects of habitat fragmentation on biodiversity. *Annual Review of Ecology and Evolutionary Systematics* 34:487–515.

Falfushynska, H. I., L. L. Gnatyshyna, and O. B. Stoliar. 2015. Effect of in situ exposure history on the molecular responses of freshwater bivalve *Anodonta anatina* (Unionidae) to trace metals. *Ecotoxicology and Environmental Safety* 89:73–83.

Fanelli, D. 2012. Negative results are disappearing from most disciplines and countries. *Scientometrics* 90:891–904.

Faria, M. L., M. A. López, M. Fernández-Sanjuan, S. Lacorte, and C. Barata. 2010. Comparative toxicity of single and combined mixtures of selected pollutants among larval stages of the native freshwater mussels (*Unio elongatulus*) and the invasive zebra mussel (*Dreissena polymorpha*). *Science of the Total Environment* 408:2452–2458.

Farness, S. 1966. Resources planning versus regional planning. In: *Future Environments of North America*, ed. F. F. Darling and J. P. Milton, 494–502. Garden City, NY: Natural History Press.

Fasola, M., P. A. Movalli, and C. Gandini. 1998. Heavy metal, organochlorine pesticide, and PCB residues in eggs and feathers of herons breeding northern Italy. *Archives of Environmental Contamination* 34:87–93.

Faurby, S., T. L. King, M. Obst, E. M. Hallerman, C. Pertoldi, and P. Funch. 2010. Population dynamics of American Horseshoe Crabs—Historic climatic events and recent anthropogenic pressures. *Molecular Ecology* 19:3088–3100.

Favero, M., R. Mariano-Jelicich, M. P. Silva, M. S. Bó, and C. Garcia-Mata. 2001. Food and feeding biology of the Black Skimmer in Argentina: Evidence supporting offshore feeding in nonbreeding areas. *Waterbirds* 24:413–418.

Feagin, R. A., S. M. Lozeda-Bernard, T. M. Ravens, I. Moller, K. M. Yeager, and A. H. Baird. 2009. Does vegetation prevent wave erosion of salt marsh edges. *Proceedings of the National Academy of Sciences of the United States of America* 106:10109–10113.

Feagin, R. A., D. J. Sherman, and W. E. Grant. 2005. Coastal erosion, global sea-level rise, and the loss of sand dune plant habitats. *Frontiers in Ecology and the Environment* 3:359–364.

Fernandez-Canero, R., and P. Gonzalez-Redondo. 2010. Green roofs as a habitat for birds. *Journal of Animal and Veterinary Advances* 9:2041–2052.

Fertig, B., M. J. Kennish, and G. P. Sakowicz. 2013. Changing eelgrass (*Zostera marina* L.) characteristics in a highly eutrophic temperate coastal lagoon. *Aquatic Botany* 14:70–79.

Field, D. W., C. E. Alexander, and M. Broutman. 1988. Toward developing an inventory of U.S. coastal wetlands. *Marine Fisheries Review* 50:40–46.

Fimreite, N. 1971. Effects of dietary methylmercury on Ring-necked Pheasants. *Canadian Wildlife Service Occasions Paper* 9:1–39.

Fimreite, N. 1974. Mercury contamination of aquatic birds in northwestern Ontario. *Journal of Wildlife Management* 38:120–131.

Fimreite, N. 1979. Accumulation and Effects of mercury in birds, In: *The Biogeochemistry of mercury in the environment*, ed. J. O. Nriagu, 601–628. New York: Elsevier.

Finkelstein, M. E., D. F. Doak, D. George, J. Burnett, J. Brandt, M. Church, J. Grantham, and D. R. Smith. 2012. Lead poisoning and the deceptive recovery of the critically endangered California Condor. *Proceedings of the National Academy of Science* 109:11449–11454.

Finkelstein, M. E., D. George, S. Scherbinski, R. Gwiarzda, M. Johnson, J. Burnett, J. Brandt et al. 2010. Feather lead concentrations and (207) Pb/(206) Pb ratios reveal lead exposure history of California Condors (*Gymnogyps californianus*). *Environmental Science and Technology* 44:2639–2647.

Finkelstein, M. E., Z. E. Kuspa, A. Welch, C. Eng, M. Clark, J. Burnett, and D. R. Smith. 2014. Linking cases of illegal shootings of the endangered California Condor using stable lead isotope analysis. *Environmental Research* 134:270–279.

Fish and Wildlife Service (FWS). 2014a. Species Profile for Piping Plover (*Charadrius melodus*). http://ecos .fws.gov/speciesProfile/profile/speciesProfile.action?spcode=B079 (accessed February 28, 2015).

Fish and Wildlife Service (FWS). 2014b. Piping Plover *Charadrius melodus* [Threatened]. http://www.fws .gov/northeast/njfieldoffice/endangered/plover.html (accessed February 28, 2015).

Fitzgerald, W. F. 1989. Atmospheric and oceanic cycling of mercury. *Chemical Oceanography* 10:151–186.

Fitzgerald, W. F., and R. P. Mason. 1996. The Global Mercury Cycle: Oceanic and Anthropogenic Aspects. In: *Global and Regional Mercury Cycles: Sources, Fluxes, and Mass Balances*, ed. W. Baeyens, R. Ebinghaus, and O. Vasiliev, 185–208. NATO-ASI Series 2. Dordrecht, The Netherlands: Kluwer.

Fitzgerald, W. F., and J. S. O'Connor. 2001. *Mercury Cycling in the Hudson/Raritan River Basin. Industrial Ecology, Pollution Prevention and the NY/NJ Harbor Project.* New York: New York Academy of Sciences.

Fleischer, R. C. 1983. Relationships between tidal oscillations and ruddy turnstone flocking, foraging, and vigilance behavior. *Condor* 85:22–29.

Fleming, W. J., B. P. Pullin, and D. M. Swineford. 1984. Population trends and environmental contaminants in herons in the Tennessee Valley, 1980–81. *Colonial Waterbirds* 7:63–73.

Forman, R. T. T. 1996. *Pine Barrens: Ecosystem and Landscape.* New Brunswick, NJ: Rutgers University Press.

Forman, R. T. T., and M. Godron. 1986. *Landscape Ecology.* New York: Wiley.

Fox, G. A.1991. Practical causal inference for ecoepidemiologists. *Journal of Toxicology and Environmental Health* 33:359–373.

Fox, G. A. 2001. Wildlife as sentinels of human health effects in the Great Lakes–St. Lawrence Basin. *Environmental Health Perspectives* 109:853–861.

Fox, G. A., M. Gilbertson, A. P. Gilman, and T. J. Kubiak. 1991. A rationale for the use of colonial fish-eating birds to monitor the presence of development toxicants in Great Lakes fish. *Journal of Great Lakes Research* 17:151–162.

Frank, C. A. 2010. *Important Bird Areas of New Jersey.* Bernardsville, NJ: New Jersey Audubon Society.

Franson, J. C., and D. J. Pain. 2011. Lead in birds. In: *Environmental Contaminants in Wildlife: Interpreting Tissue Concentrations*, ed. W. N. Beyer and J. P. Meador, 563–594. Boca Raton, FL: CRC Press.

Fraser, D. 2015. Red Knots in decline and threatened. *Cape Cod Times.* http://capecodtimes.com/article /20150111/NEWS/150119913/101100/NEWS06 (accessed April 17, 2015).

Frederick, P. C. 2013. Tricolored Heron (*Egretta tricolor*). In: *The Birds of North America Online*, ed. A. Poole. Ithaca, NY: Cornell Laboratory of Ornithology. http://bna.birds.cornell.edu/bna/species/306 (accessed March 4, 2015).

Frederick, P., and N. Jayasena. 2010. Altered pairing behaviour and reproductive success in White Ibises exposed to environmentally relevant concentrations of methylmercury. *Proceedings of the Royal Society Biological Sciences* 282:1851–1857.

Frederick, P. C., and M. G. Spalding. 1994. Factors affecting reproductive success of wading birds (*Ciconiiformes*) in the Everglades. In: *Everglades: The ecosystem and its restoration,* ed. S. Davis, and J. C. Ogden, 659–691. Del Ray Beach, FL: St. Lucie Press.

Frederick, P. C., R. B. Bjork, G. T. Bancroft, and G. V. N. Powell. 1992. Reproductive success of three species of herons relative to habitat in southern Florida. *Colonial Waterbirds* 15:192–201.

Frederick, P., A. Campbell, N. Jayasena, and R. Borkhataria. 2011. Survival of White Ibises (*Eudocimus albus*) in response to chronic experimental methylmercury exposure. *Ecotoxicology* 20:358–364.

Frederick, P. C., B. Hylton, J. A. Heath, and M. G. Spalding. 2004. A historical record of mercury contamination in southern Florida (USA) as inferred from avian feather tissue. *Environmental Toxicology and Chemistry* 23:1474–1478.

Frederick, P. C., M. G. Spalding, and R. Dusek. 2002. Wading birds as bioindicators of mercury contamination in Florida, USA: Annual and geographic variation. *Environmental Toxicology and Chemistry* 21:163–167.

Frederick, P. C., M. G. Spalding, M. S. Sepulveda, G. E. Williams Jr., S. Bouton, H. Lynch, J. Arrecis, S. Loerzel, and D. Hoffman. 1997. Effects of environmental mercury exposure on reproduction, health and survival of wading birds in Florida Everglades. Report to Florida Department of Environmental Protection, Gainesville, FL.

Frederick, P. C., M. G. Spalding, M. S. Sepulveda, G. E. Williams, L. Nico, and R. Robins. 1999. Exposure of Great Egret (*Ardea alba*) nestlings to mercury through diet in the Evergreens ecosystem. *Environmental Toxicology and Chemistry* 18:1940–1947.

Freedman, A. 2013. Heeding *Sandy*'s lessons, before the next big storm. Climate Central. http://www.climatecentral .org/news/four-lay-lessons-learned-from-hurricane-sandu-15928 (accessed March 4, 2015).

French, H. 2000. *Vanishing Borders: Protecting the Planet in the Age of Globalization.* New York: W. W. Norton.

Friberg, L., and N. K. Mottet.1989. Accumulation of methylmercury and inorganic mercury in the brain. *Biological Trace Elements Research* 21:201–206.

Friend, M. 2006. Evolving changes in disease of waterbirds. In: *Waterbirds Around the World*, ed. G. C. Boere, C. A. Galbraith, and D. A. Strout, 412–417. Edinburgh: The Stationery Office.

Fry, D. M., and C. K. Toone. 1981. DDT-induced feminization of gull embryos. *Science* 213:922–924.

Fulford, R. S., R. J. Griffin, N. J. Brown-Peterson, H. Perry, and G. Sanchez-Rubio. 2015. Impacts of the *Deepwater Horizon* Oil Spill on Blue Crab, *Callinectes sapidus,* Larval Settlement in Mississippi. In: *Impacts of Oil Spill Disasters on Marine Habitats and Fisheries in North America*, ed. J. B. Alford, M. S. Peterson, and C. C. Green, 253–268. Boca Raton, FL: CRC Press.

Fusco, P. 2012. The wide-ranging, yet threatened—Roseate Tern. *Connecticut Wildlife* (May/June):12–13.

Furness, R. W.1993. Birds and monitors of pollutants. In: *Birds as Monitors of Environmental Change,* ed. R. W. Furness and J. J. D. Greenwood, 86–143. London: Chapman and Hall.

Furness, R. W., and T. R. Birkhead. 1984. Seabird colony distributions suggest competition for food supplies during the breeding season. *Nature* 311:655–656.

Furness, R. W., S. J. Muirhead, and M. Woodburn. 1986. Using bird feathers to measure mercury in the environment: Relationship between mercury content and moult. *Marine Pollution Bulletin* 17:27–37.

Furness, R. W., D. R. Thompson, and P. H. Becker. 1995. Spatial and temporal variation in mercury contamination of seabirds in the North Sea. *Helgolander Meeresuntersuchungen* 49:605–615.

Galbraith, H., D. W. DesRochers, S. Brown, and J. M. Reed. 2014. Predicting vulnerabilities of North American shorebirds to climate change. *PLoS ONE* 9:e108899.

Galbraith, H., J. J. Hatch, I. Nisbet, and T. H. Kunz. 1999. Age-related changes in efficiency among breeding Common Terns *Sterna hirundo:* Measurement of energy expenditure using doubly-labelled water. *Journal of Avian Biology* 30:85–96.

Galbraith, H., R. Jones, R. Park, J. Clough, S. Herrod-Julius, B. Harrington, and G. Page. 2002. Global climate change and sea level rise: Potential losses of intertidal habitat for shorebirds. *Colonial Waterbirds* 25:173–183.

Garshelis, D. L. 2000. Delusions in habitat evaluation: Measuring use, selection, and importance. In: *Research Techniques in Animal Ecology: Controversies and Consequences*, ed. L. Boitani and T. K. Fuller, 111–164. New York: Columbia University Press.

Garthe, S. 2013. Population trends of gulls in Germany: Continuous changes at the southern coasts of North and Baltic Seas. Abstract only. Waterbird Society 37th Annual Meeting. September 2013.

Gaston, A. J., H. G. Gilchrist, and M. Hipfner. 2005. Climate change, ice conditions and reproduction in an Arctic nesting marine bird: Brunnich's guillemot (*Uria lomvia* L.). *Journal of Animal Ecology* 74:832–841.

Gedan, K. B., B. R. Silliman, and M. D. Bertness. 2009. Centuries of human-driven change in salt marsh ecosystems. *Annual Review of Marine Science* 1:117–141.

Genovese, E., and V. Przyluski. 2013. Storm surge disaster risk management: The *Xynthia* case study in France. *Journal of Risk Research* 16:825–841.

Gilbertson M., T. Kubiak, J. Ludwig, and G. Fox. 1991. Great Lakes embryo mortality, edema, and deformities syndrome (GLEMEDS) in colonial fish-eating birds: Similarity to chick-edema disease. *Journal of Toxicology and Environmental Health* 33:455–520.

Giroux, J. 2013. The rise and fall of Ring-billed Gulls in eastern North America. 2013. Abstract only. Waterbird Society 37th Annual Meeting. September 2013.

Giudice, J. H., and J. T. Ratti. 2001. Ring-necked Pheasant (*Phasianus colchicus*), *Birds of North America Online*, ed. A. Poole. Ithaca, NY: Cornell Laboratory of Ornithology. http://bna.birds.cornell.edu .bnaproxy.birds.cornell.edu/bna/species/572 (accessed August, 18 2015).

Gobeille, A. K., K. V. Morland, R. F. Bopp, J. H. Gobold, and P. J. Landrigan. 2006. Body burdens of mercury in lower Hudson River area anglers. *Environmental Research* 101:205–212.

Gochfeld, M. 1975a. Developmental defects in Common Terns of western Long Island, New York. *Auk* 92:58–65.

Gochfeld, M. 1975b. Hazards of tail-first fish-swallowing by young terns. *Condor* 77:345–346.

Gochfeld, M. 1980a. Mechanisms and adaptive value of reproductive synchrony in colonial seabirds. In: *Behavior of Marine Animals*, ed. J. Burger, B. L. Olla, and H. E. Winn, 207–270. New York: Plenum.

Gochfeld, M. 1980b. Tissue distribution of mercury in normal and abnormal young Common Terns. *Marine Pollution Bulletin* 11:362–377.

Gochfeld, M. 1983. Colony site selection by Least Terns: Physical attributes of sites. *Colonial Waterbirds* 6:205–213.

Gochfeld, M. 1997. Spatial patterns in a bioindicator: Heavy metal and selenium concentration in eggs of Herring Gulls (*Larus argentatus*) in the New York Bight. *Archives of Environmental Contamination and Toxicology* 33:63–70.

Gochfeld, M., and J. Burger. 1982a. Feeding enhancement by social attraction in Sandwich Tern. *Behavioral Ecology and Sociobiology* 10:15–17.

Gochfeld, M., and J. Burger. 1982b. Biological concentration of cadmium in estuarine birds of the New York Bight. *Colonial Waterbirds* 5:116–123.

Gochfeld, M., and J. Burger. 1987a. Nest-site selection: Comparison of Roseate and Common Terns (*Sterna dougallii* and *S. hirundo*) in a Long Island New York colony. *Bird Behavior* 7:58–66.

Gochfeld, M., and J. Burger. 1988. Effects of lead on growth and feeding behavior of young Common Terns (*Sterna hirundo*). *Archives of Environmental Contamination* 17:513–517.

Gochfeld, M., and J. Burger. 1989. Tissue distribution of lead in young Common Terns: Influence of the time since exposure. *Environmental Research* 50:262–268.

Gochfeld, M., and J. Burger. 1994. Black Skimmer (*Rynchops niger*), The Birds of North America Online (A. Poole, Ed.). Ithaca, NY: Cornell Laboratory of Ornithology. http://bna.birds.cornell.edu/bna/species/108 (accessed July 8, 2015).

Gochfeld, M., and J. Burger. 1996. Family Sternidae (terns). In: *Handbook of the Birds of the World*, ed. J. del Hoyo, A. Elliot, and J. Sargatal, 624–643. Barcelona: Lynx Editions.

Gochfeld, M., and J. Burger. 1997. *Butterflies of New Jersey*. New Brunswick, NJ: Rutgers University Press.

Gochfeld, M., and J. Burger. 1998. Temporal trends in metal levels in eggs of the Endangered Roseate Tern (*Sterna dougallii*) in New York. *Environmental Research* 77:36–42.

Gochfeld, M., and J. Burger. 2011. Disproportionate exposures in environmental justice and other populations: Outliers Matter. *American Journal of Public Health* 1:S53–S63.

Gochfeld, M., J. L. Belant, T. Shukla, T. Benson, and J. Burger. 1996. Heavy metals in laughing gulls: Gender, age and tissue differences. *Environmental Toxicology and Chemistry* 15:2275–2283.

Gochfeld, M., J. Burger, C. Jeitner, M. Donio, and T. Pittfield. 2012. Seasonal, locational and size variations in mercury and selenium levels in Striped Bass (*Morone saxatilis*) from New Jersey. *Environmental Research* 112:8–19.

Gochfeld, M., J. Saliva, F. Lesser, T. Shukla, D. Bertrand, and J. Burger, J. 1991. Effects of color on cadmium and lead levels in avian contour feathers. Archives of environmental contamination and toxicology, 20(4), 523–526.

Golden, N. H., and B. A. Rattner. 2003. Ranking terrestrial vertebrate species for utility in biomonitoring and vulnerability to environmental contaminants. *Review of Environmental Contamination and Toxicology* 176:67–136.

Golden, N. H., B. A. Rattner, J. B. Cohen, D. J. Hoffman, E. Russek-Cohen, and M. A. Ottinger. 2003a. Lead accumulation in feathers of nestling Black-crowned Night Herons (*Nycticorax nycticorax*) experimentally treated in the field. *Environmental Toxicology and Chemistry* 22:1517–1524.

Golden, N. H., B. A. Rattner, P. C. McGowan, K. C. Parsons, and M. A. Ottinger. 2003b. Concentrations of metals in feathers and blood of nestling Black-Crowned Night Herons (*Nycticorax nycticorax*) in Chesapeake and Delaware Bays. *Bulletin of Environmental Contamination and Toxicology* 70:385–393.

Good, T. P. 1998. Great Black-backed Gull (*Larus marinus*). In: *The Birds of North America Online*, ed. A. Poole. Ithaca, NY: Cornell Laboratory of Ornithology. http://bna.birds.cornell.edu.bnaproxy.birds.cornell.edu/bna/species/330doi:10.2173/bna.330 (accessed February 21, 2015).

Gornitz, V., S. Couch, and E. L. Hartig. 2001. Impacts of sea level rise in the New York City metropolitan area. *Global and Planetary Change* 32:61–88.

Goss-Custard, J. D. 2014. Birds and people: Resolving the conflict on estuaries. *Ardea* 102:225–226.

Goss-Custard, J. 2015. *Birds and People: Resolving the Conflict on Estua*ries. Kindle Direct Publishing E-Book.

Goss-Custard, J. D., and M. G. Yates. 1992. Towards predicting the effect of salt-marsh reclamation on feeding bird numbers on the Wash. *Journal of Applied Ecology* 29:330–340.

Goss-Custard, J. D., R. W. Caldow, R. T. Clarke, S. E. L. V. Durell, J. Urfi, and Y. D. West. 1995. Consequences of habitat loss and change to populations of wintering migratory birds: Predicting the local and global effects from studies of individuals. *Ibis* 137:S56–S66.

Goss-Custard, J. D., P. Triplet, F. Sueur, and A. D. West. 2006. Critical thresholds of disturbance by people and raptors in foraging wading birds. *Biological Conservation* 127:88–97.

Goto, D., and W. G. Wallace. 2009. Relevance of intracellular partitioning of metals in prey to differential metal bioaccumulation among populations of Mummichogs (*Fundulus heteroclitus*). *Marine Environmental Research* 68:257–267.

Gottschalk, T. K., F. Huettmann, and M. Ehlers. 2005. Thirty years of analyzing and modelling avian habitat relationships using satellite imagery data: A review. *International Journal of Remote Sensing* 26:2631–2656.

Goutner, V., and R. W. Furness. 1997. Mercury in feathers of little egret *Egretta garzetta* and night heron *Nycticorax nycticorax* chicks and in their prey in the Axios Delta, Greece. *Archives of Environmental Contamination and Toxicology* 32:211–216.

Goutner, V., I. Papagiannis, and V. Kalfakakou. 2001. Lead and cadmium in eggs of colonially nesting waterbirds of different position in the food chain of Greek wetlands of international importance. *The Science of the Total Environment* 267:169–176.

Goyert, H. F. 2014. Relationship among prey availability, habitat, and the foraging behavior, distribution, and abundance of Common Terns *Sterna hirundo* and roseate terns *S. dougallii*. *Marine Ecology Press Series* 506:291–609.

Goyert, H. F., L. L. Manne, and R. R. Veit. 2014. Facilitative interactions among the pelagic community of temperate migratory terns, tunas, and dolphins. *Oikos* 123:1400–1408.

Grajewska, A., L. Falkowska, E. Szumito-Pilarska, J. Hajdrych, M. Szubska, T. Frączek, W. Meissner et al. 2015. Mercury in the eggs of aquatic birds from the Gulf of Gdansk and Wloclawek Dam (Poland). *Environmental Science and Pollution Research* 22:1–10.

Grandjean, P., P. Weihe, F. Debes, A. L. Choi, and E. Budtz-Jørgensen. 2014. Neurotoxicity from prenatal and postnatal exposure to methylmercury. *Neurotoxicology and Teratology* 43:39–44

Grant, C., D. Flaten, M. Tenuta, X. Gapo, S. Malhi, and E. Gowalko. 2010. Impact of long-term application of phosphate fertilizer on cadmium accumulation in crops. *Proceedings of the 19th World Congress of Soil Science*, Soil Solutions for a Changing World, 2010. http://iuss.org/19th%20WCSS/Symposium /pdf/1643.pdf (accessed July 27, 2015).

Grasman, K. A., K. R. Echols, T. M. May, P. H. Peterman, R. W. Gale, and C. E. Orazio.2013. Immunological and reproductive health assessment in Herring Gulls and black-crowned night herons in the Hudson–Raritan Estuary. *Environmental Toxicology and Chemistry* 32:548–561.

Green, N. W., and J. Knutzen. 2003. Organohalogens and metals in marine fish and mussels and some relationships to biological variables at reference localities in Norway. *Marine Pollution Bulletin* 46:362–377.

Green, E. P., and E. T. Short. 2003. *World Atlas of Seagrasses: Present Status and Future Conservation*. Berkeley, CA: University of California Press.

Greenberg, R., J. Maldonado, S. Droege, and M. V. McDonal. 2006. Tidal marshes: A global perspective on the evolution and conservation of their terrestrial vertebrates. *Bioscience* 56:675–685.

Greenlaw, J. S., and J. D. Rising. 1994. Saltmarsh Sparrow (*Ammodramus caudacutus*). In: *The Birds of North America Online*, ed. A. Poole. Ithaca, NY: Cornell Laboratory of Ornithology. http://bna.birds.cornell .edu.bnaproxy.birds.cornell.edu/bna/species/112 (accessed February 21, 2015).

Greig, R. A., and R. A. McGrath. 1977. Trace metals in sediments of *Raritan Bay. Marine Pollution Bulletin* 8:188–192.

Grémillet, D., and T. Boulinier. 2009. Spatial ecology and conservation of seabirds facing global climate change: A review. *Marine Ecology Progress Series* 391:121–137.

Grieb, T. M., C. T. Driscoll, S. P. Gloss, C. L. Schofield, G. L. Bowie, and D. B. Porcella. 1990. Factors affecting mercury accumulaton in fish in the upper Michigan peninsula. *Environmental Toxicology and Chemistry* 9:919–930.

Gross, A. 1951. The Herring Gull–Cormorant control project. *Proceedings of the International Ornithological Congress* 10:532–536.

Guillemette, M., and P. Brousseau. 2001. Does culling of predatory gulls enhance the productivity of breeding Common Terns? *Journal of Applied Ecology* 38:1–8.

Guitart, R., R. Mateo, C. Sanpera, A. Hernández-Matías, and X. Ruiz. 2003. Mercury and selenium levels in eggs of Common Terns (*Sterna hirundo*) from two breeding colonies in the Ebro Delta, Spain. *Bulletin of Environmental Contamination and Toxicology* 70:71–77.

Gundlach, E. R. 2006. Oil spills: Impacts, recovery and remediation. *Journal of Coastal Resources* 39:39–42.

Hackl, E., and J. Burger. 1988. Factors affecting kleptoparasitism in Herring Gulls at a New Jersey landfill. *Wilson Bulletin* 100:424–430.

Hackensack Riverkeeper (HR). 2014. People's climate march—United for climate justice. Hackensack Riverkeeper 25.

Hafner, H. 2000. Heron nest site conservation. In: *Heron Conservation*, ed. J. A. Kushlan and H. Hafner, 201–218. New York: Academic Press.

Hafner, J., and J. A. Kushlan. 2002. Action plan for conservation of the herons of the world. Heron Specialist Group and Station Biologique de la Tour du Valat, Arles, France.

Hall, C. S., and S. W. Kress. 2008. Diet of nestling Black-crowned Night–herons in a mixed-species colony: Implications for tern conservation. *The Wilson Journal of Ornithology* 120:637–640.

Hamede, R. K., A. M. Pearse, K. Swift, L. A. Barmuta, E. P. Murchison, and M. E. Jones. 2015. Transmissible cancer in Tasmanian devils: Localized lineage replacement and host population response. *Proceedings Biological Sciences (Royal Society)*: 282(1814) DOI: 10.1098/rspb.2015.1468.

Hamer, K. C., E. A. Schreiber, and J. Burger. 2001. Breeding biology, life histories, and life history–environment interactions. In: *Biology of Marine Birds*, ed. E. A. Schreiber and J. Burger, 217–261. Boca Raton, FL: CRC Press.

Hammerschmidt, C. R., and W. F. Fitzgerald. 2006. Bioaccumulation and trophic transfer of methylmercury in Long Island Sound. *Archives of Environmental Contamination and Toxicology* 51:416–424.

Hammerschmidt, C. R., W. F. Fitzgerald, C. H. Lamborg, P. H. Balcom, and C. M. Tseng. 2006. Biogeochemical cycling of methylmercury in lakes and tundra watersheds of Arctic Alaska. *Environmental Science and Technology* 40:1204–1211.

Hancock, J. A., and J. A. Kushlan.1984. *The Heron Handbook*. 1st edition, New York: Harper and Rowe.

Hancock, J. A., and J. A. Kushlan. 2010. *The Heron Handbook*. 2nd edition, New York: Harper and Rowe.

Handel, S. N., G. R. Robinson, and A. J. Beattie. 1994. Biodiversity resources for restoration ecology. *Restoration Ecology* 2:230–241.

Hanowski, J. M., G. J. Niemi, A. R. Lima, and R. R. Regal. 1997. Response of breeding birds to mosquito control treatments of wetlands. *Wetlands* 17:485–492.

Hardin, G. 1968. The tragedy of the commons. *Science* 162:1243–1248.

Harris, R., D. P. Krabbenhoft, R. Mason, M. W. Murray, R. Reash, and T. Saltman. 2006. *Ecosystem Responses to Mercury Contamination: Indicators of Change*. Boca Raton, FL: CRC Press.

Hart, M. 1999. *Guide to Sustainable Community Indicators*. North Andover, MA: Hart Environmental Data.

Hartig, E. K., V. Gornnitz, A. Kolker, F. Mushacke, and D. Fallon. 2002. Anthropogenic and climate-change impacts on salt marshes of Jamaica bay, New York City. *Wetlands* 22:71–89.

Hartwell, S. I., and J. Hameedi. 2007. Magnitude and Extent of Contaminated Sediment and Toxicity in Chesapeake Bay. NOAA Technical Memorandum NOS NCCOS 47. http://ccma.nos.noaa.gov/publica tions/NCCOSTM47.pdf (accessed July 22, 2015).

Hartwell, S. I., J. Hameedi, and M. Harmon. 2001. Magnitude and Extent of Contaminated Sediment and Toxicity in Delaware Bay. NOAA Technical Memorandum NOS ORCA 148. http://ccma.nos.noaa.gov /publications/TechMemo148.pdf (accessed July 22, 2015).

Hatch, J. J., and D. V. C. Weseloh. 1999. Double-crested Cormorant (*Phalacrocorax auritus*). In: *The Birds of North America*, ed. A. Poole and F. Gill. Ithaca, NY: Cornell Laboratory of Ornithology.

Haymes, G. T., and H. Blokpoel. 1978. Seasonal distribution and site tenacity of the Great Lakes Common Tern. *Bird-Banding* 49:142–151.

Hays, H. 1970. Common Terns pirating fish on Great Gull Island. *The Wilson Bulletin* 99–100.

Hays, H., and M. LeCroy. 1971. Field criteria for determining incubation stage in eggs of the Common Tern. *Wilson Bulletin* 83:425–429.

Hays, H., and R. W. Risebrough. 1972. Pollutant concentrations in abnormal young terns from Long Island Sound. *Auk* 89:19–35.

Hazen, E. L., S. Jorgensen, R. R. Rykaczewski, S. J. Bograd, D. G. Foley, I. D. Jonsen, and B. A. Block. 2013. Predicted habitat shifts of Pacific top predators in a changing climate. *Nature and Climate Change* 3:234–238.

Head, J. A., A. Debofsky, J. Hinshaw, and N. Basu. 2011. Retrospective analysis of mercury content in feathers of birds collected from the state of Michigan (1895–2007). *Ecotoxicology* 20:1636–1643.

Heath, J. A., and P. C. Frederick. 2005. Relationships among mercury concentrations, hormones, and nesting effort of White Ibises (*Eudocimus albus*) in the Florida Everglades. *The Auk* 122:255–267.

Heink, U., and I. Kowarik. 2010. What are indicators? On the definition of indicators in ecology and environmental planning. *Ecological Indicators* 1:584–593.

Heinz, G. H. 1979. Methylmercury: Reproductive and behavioral effects on three generations of Mallard Ducks. *Journal Wildlife Management* 43:394–401.

Heinz, G. H. 1996. Selenium in birds. In: *Environmental Contaminants in Wildlife: Interpreting Tissue Concentrations*, ed. W. N. Beyer, G. H. Heinz, and A. W. Redmon-Norwood, 447–458. Boca Raton, FL: Lewis Publishers.

Heinz, G. H., and D. J. Hoffman. 1998. Methylmercury chloride and selenomethionine interactions on health and reproduction in mallards. *Environmental Toxicology and Chemistry* 17:139–145.

Heinz, G. H., S. D. Haseltine, and L. Sileo. 1983. Altered avoidance behavior of young Black Ducks fed cadmium. *Environmental Toxicology and Chemistry* 2:419–421.

Heinz, G. H., D. J. Hoffman, J. D. Klimstra, K. R. Stebbins, S. L. Kondrad, and C. A. Erwin. 2009. Species differences in the sensitivity of avian embryos to methylmercury. *Archives of Environmental Contamination and Toxicology* 56:129–138.

Heinz, G. H., D. J. Hoffman, J. D. Klimstra, K. R. Stebbins, S. L. Kondrad, and C. A. Erwin. 2011. Teratogenic effects of injected methylmercury on avian embryos. *Environmental Toxicology and Chemistry* 30:1593–1598.

Heinz, G. H., D. J. Hoffman, A. J. Krynitsky, and D. M. G. Weller. 1987. Reproduction in mallards fed selenium. *Environmental Toxicology and Chemistry* 6:423–433.

Helmers, D. L. 1992. *Shorebird Management Manual*. Manomet, MA: Western Hemisphere Shorebird Reserve Network.

Helgason, L. B., R. Barrett, E. Lie, A. Polder, J. U. Skaare, and G. W. Gabrielsen. 2008. Levels and temporal trends (1983–2003) of persistent organic pollutants (POPS) and mercury (Hg) in seabirds eggs from northern Norway. *Environmental Pollution* 155:190–198.

Henny, C. J., T. W. Anderson, and J. J. Crayon. 2008. Organochlorine pesticides, polychlorinated biphenyl, metals, and trace elements in waterbird eggs, Salton Sea, California, 2004. *Hydrobiologia* 604:137–149.

Henny, C. J., E. F. Hill, R. A. Grove, and J. L. Kaiser. 2007. Mercury and drought along the lower Carson River, Nevada: I. Snowy Egret and Black-crowned Night-Heron annual exposure to mercury, 1997–2006. *Archive of Environmental Contamination and Toxicology* 53:269–280.

Henny, C. J., E. F. Hill, D. J. Hoffman, M. Spalding, and R. A. Grove. 2002. Nineteenth century mercury: Hazard to wading birds and cormorants of the Carson River, Nevada. *Ecotoxicology* 11:213–231.

Herbert, C. E., D. V. C. Weseloh, S. MacMillan, D. Campbell, and W. Nordstrom. 2011. Metals and polycyclic aromatic hydrocarbons in colonial waterbird eggs from Lake Athabasca and the Peace–Athabasca Delta, Canada. *Environmental Toxicology and Chemistry* 30:1178–1183. (accessed February 9, 2015).

Herring G., J. T. Ackerman, and C. A. Eagles-Smith. 2010. Embryo malposition as a potential mechanism for mercury-induced hatching failure in bird eggs. *Environmental Toxicology and Chemistry* 29:1788–1794.

Herring G., D. E. Gawlik, and D. G. Rumbold. 2009. Feather mercury concentrations and physiological condition of Great Egret and White Ibis nestlings in the Florida Everglades. *Science of the Total Environment* 407:2641–2649.

Hettick, B. E., J. E. Cañas-Carrell, A. D. French, and D. M. Klein. 2015. Arsenic: A review of the element's toxicity, plant interactions, and potential methods of remediation. *Journal of Agricultural and Food Chemistry* 63:7097–7107.

Hill, A. B.1965. The Environment and Disease: Association or Causation. *Proceedings of the Royal Society of Medicine* 58:295–300.

Hill, C. H. 1975. Interrelationships of selenium with other trace elements. *Federation Proceedings* 34:2096–2100.

Hill, E. F., C. J. Henny, and R. A. Grove. 2008. Mercury and drought along the lower Carson River, Nevada: II. Snowy Egret and Black-Crowned Night-Heron reproduction on Lahontan Reservoir, 1997–2006. *Ecotoxicology* 17:117–131.

Hinzman, L. D., N. Bettez, F. S. Chapin III, M. Dyurgerov, C. Fastie, D. B. Griffith, A. Hope et al. 2005 Evidence and implications of recent climate change in terrestrial regions of the Arctic. *Climatic Change* 72:251–298.

Hirata, S. H., Y. Yasuda, S. Urakami, T. Isobe, T. K. Yamada, Y. Tajima, M. Amamo, N. Miyazaki, S. Takahashi, and S. Tanabe. 2010. Environmental monitoring of trace elements using marine mammals as bioindicators—Species-specific accumulations and temporal trends. In: *Environmental Specimen Bank*, ed. R. J. Hobbs, E. S. Higgs, and C. Hall, 75–79. 2013. *Novel Ecosystems: Intervening in the New Ecological World Order*. New York: Wiley-Blackwell.

Hobbs, R. J., E. S. Higgs, and C. Hall. 2013. *Novel Ecosystems: Intervening in the New Ecological World Order*. New York: Wiley-Blackwell.

Hobson, K. A., J. F. Piatt, and J. Pitocchelli. 1994. Using stable isotopes to determine seabird trophic relationships. *Journal of Animal Ecology* 63:786–798.

Hobson, K. A., J. Sirois, and M. L. Gloutney. 2000. Tracing nutrient allocation to reproduction with stable isotopes: A preliminary investigation using colonial waterbirds of Great Slave Lake. *Auk* 117:760–774.

Hoegh-Guldberg, O., and J. F. Bruno. 2010 The impact of climate change on the world's marine ecosystems. *Science* 328:1523–1528.

Hoffman, R. D., and R. D. Curnow. 1979. Mercury in heron, egrets, and their foods. *Journal of Wildlife Management* 43:85–93.

Hoffman, D. J., C. A. Eagles-Smith, J. T. Ackerman, T. L. Adelsbach, and K. R. Stebbins. 2011. Oxidative stress response of Forster's terns (*Sterna forsteri*) and Caspian terns (*Hydroprogne caspia*) to mercury and selenium bioaccumulation in liver, kidney, and brain. *Environmental Toxicology* 30:920–929.

Hoffman D. J., C. J. Henny, E. F. Hill, R. A. Grove, J. L. Kaiser, and K. R. Stebbins. 2009. Mercury and drought along the lower Carson River, Nevada: III. Effects on blood and organ biochemistry and histopathology of Snowy Egrets and Black-crowned Night-Herons on Lahontan reservoir, 2002–2006. *Journal of Toxicology and Environmental Health Part A* 72:1223–1241.

Hoffman, D. J., B. A. Rattner, G. A. Burton Jr., and J. Cairns Jr. 2002. *Handbook of Ecotoxicology*, 2nd Edition. Boca Raton: CRC Press.

Hoffman D. J., M. G. Spalding, and P. C. Frederick. 2005. Subchronic effects of methylmercury on plasma and organ biochemistries in Great Egret nestlings. *Environmental Toxicology and Chemistry* 24:3078–3084.

Honda, K., D. P. Lee, and R. Tatsukawa. 1990. Lead poisoning in swans in Japan. *Environmental Pollution* 65:209–218.

Honda, K., B. Y. Min, and R. Tatsukawa. 1985. Heavy metal distribution in organs and tissues of the eastern Great White Egret *Egretta alba modesta*. *Archives of Environmental Contamination and Toxicology* 35:781–789.

Honda, K., B. Y. Min, and R. Tatsukawa.1986. Distribution of heavy metals and their age-related changes in the Eastern Great White Egret, *Egretta alba modesta*. *Archives of Environmental Contamination and Toxicology* 15:185–197.

Hood, E. 2006. The apple bites back: Claiming old orchards for residential development. *Environmental Health Perspectives* 114:A471–A475.

Hornberger, M. I., S. N. Luoma, A. van Geen, C. Fuller, and R. Anima. 1999. Historical trends of metals in the sediments of San Francisco Bay, California. *Marine Chemistry* 64:39–55.

Hosono, T., S. Chih-Chieh, R. Delinomc, Y. Umezawa, T. Toyota, S. Kaneko, and M. Taniguichi. 2011. Decline in heavy metal contamination in marine sediments in Jakarta Bay, Indonesia due to increasing environmental regulations. *Estuarine, Coastal and Shelf Science* 92:297–306.

Hothem, R. L., B. E. Brussee, and W. E. Davis Jr. 2010. Black-crowned Night-Heron (*Nycticorax nycticorax*). In: *The Birds of North America Online*, ed. A. Poole. Ithaca, NY: Cornell Laboratory of Ornithology. http://bna .birds.cornell.edu.bnaproxy.birds.cornell.edu/bna/species/074doi:10.2173/bna.74 (accessed July 16, 2015).

Hothem, R. L., J. J. Crayon, and M. A. Law. 2006. Effects of contaminants on reproductive success of aquatic birds nesting at Edwards Air Force Base, California. *Archives of Environmental Contamination and Toxicology* 51:711–719.

Houtman C. J., J. Kroesbergen, K. Lekkerkerker-Teunissen, and J. P. van der Hoek. 2014. Human health risk assessment of the mixture of pharmaceuticals in Dutch drinking water and its sources based on frequent monitoring data. *Science of the Total Environment* 496:54–62.

Hu, J., H. Hu, and Z. Jiang. 2010. The impacts of climate change on the wintering distribution of an endangered migratory bird. *Oecologia* 164:555–565.

Hudson, R. H., R. K. Tucker, and M. A. Haegele. 1984. *Handbook of Toxicity of Pesticides to Wildlife*. U.S. Fish Wildl. Serv. Resour. Publ. 153. 90 pp.

Hughs, M. F. 2002. Arsenic toxicity and potential mechanisms of action. *Toxicological Letters* 133:1–16.

Hulse, M., J. S. Mahoney, G. D. Schroder, C. S. Hacker, and S. M. Pier. 1980. Environmentally acquired lead, cadmium, and manganese in the Cattle Egret, *Bubulcus ibis*, and the Laughing gull *Larus atricilla*. *Archives of Environmental Contamination and Toxicology* 9:65–78.

Hume, R. 1993. *The Common Tern*. London: Hamlyn, 30 pp.

Hunt, S., and S. Hedgecott. 1992. Revised Environmental Quality Standards for chromium in water. WRC report to the Department of the Environment, 2858 (1).

Hunter, D. 1955. *Diseases of Occupation*, 1st ed. London: English Universities Press.

Hunter, K. L., M. G. Fox, and K. W. Able. 2007. Habitat influences on reproductive allocation and growth of the mummichog (*Fundulus heteroclitus*) in a coastal salt marsh. *Marine Biology* 151:617–627.

Huntley, B., Y. C. Collingham, R. E. Green, G. M. Hilton, C. Rahbek, and S. G. Willis. 2006. Potential impacts of climatic change upon geographical distributions of birds. *Ibis* 148:8–28.

Hurton, L., and J. Berkson. 2006. Potential causes of mortality for Horseshoe Crabs (*Limulus polyphemus*) during the biomedical bleeding process. *Fish Bulletin* 104:293–298.

Hutchinson, T. C., and K. M. Meema. 1987. *Lead, Mercury, Cadmium, and Arsenic in the Environment*. New York: John Wiley & Sons, p. 349.

Hyrenbach, K. D., and R. R. Veit. 2003. Ocean warming and seabird communities of the southern California Current System (1987–98): Response at multiple temporal scales. *Deep Sea Research Part II: Topical Studies in Oceanography* 50:2537–2565.

Iannuzzi, T. J., D. F. Ludwig, J. C. Kinnell, J. M. Wallin, W. H. Desvousges, and R. W. Dunford. 2002. *A Common Tragedy: History of an Urban River (Passaic R)*. Amherst, MA: Amherst Scientific Publ., 200 pp.

Ingersoll, R. T., E. B. Montgomery, and H. V. Aposhain. 1995. Central system toxicity if manganese. *Fundamental and Applied Toxicology* 27:106–113.

Institute of Medicine (IOM). 1991. *Seafood Safety*. Washington, DC: National Academy Press.

Institute of Medicine (IOM). 2006. *Seafood Safety: Balancing Benefits and Risks*. Washington, DC: National Academy Press.

Integrated Risk Information System (IRIS). 2001. Methylmercury (MeHg) (CASRN 22967-92-6) Integrated Risk Information System, Environmental Protection Agency. http://www.epa.gov/iris/subst/0073.htm (accessed September 2, 2015).

Integrated Risk Information System (IRIS). 2004. Lead and compounds (inorganic) (CASRN 7439-92-1). Integrated Risk Information System, Environmental Protection Agency. http://www.epa.gov/iris/subst /0277.htm (accessed August 31, 2015)

Intergovernmental Panel on Climate Change (IPCC). 2007. *Climate change 2007: Synthesis report. Contribution of working groups I, II and III to the fourth assessment report of the intergovernmental panel on climate change*. Geneva, Switzerland: IPPC.

Intergovernmental Panel on Climate Change (IPCC). 2014. *Climate Change 2014: Impacts, Adaptation, and Vulnerability*. Copenhagen, Denmark: IPCC. http://www.ipcc.ch/report/ar5/wg2/ (accessed December 29, 2014).

Jackson, A. K., D. C. Evers, E. M. Adams, D. A. Cristol, C. Eagles-Smith, S. T. Edmonds, C. E. Gray et al. 2015 Songbirds as sentinels of mercury in terrestrial habitats of eastern North America. *Ecotoxicology* 24:453–67.

Jackson, A. K., D. C. Evers, M. A. Etterson, A. M. Condon, S. B. Folsom, J. Detweiler, J. Schmerfeld, and D. A. Cristol. 2011. Mercury exposure affects the reproductive success of a free-living terrestrial songbird, the Carolina Wren (*Thryothorus ludovicianus*). *Auk* 128:759–769.

Jacobs, R. M., A. O. Jones, M. R. Fox, and B. E. J. Fry. 1978. Retention of dietary cadmium and the ameliorative effect of zinc, copper, and manganese in Japanese quail. *Journal of Nutrition* 108:22–32.

Jaspers, V., T. Dauwe, R. Pinxten, L Bervoets, R. Blust, and M. Eens. 2004. The importance of exogenous contamination on heavy metal levels in bird feathers. A field experiment with free-living great tits, *Parus major*. *Journal of Environmental Monitoring* 6:356–360.

Jehl, J. R. 1973. Studies of declining population of Brown Pelicans in northwestern Baja California. *Condor* 75:69–79.

Jenkins, D., and L. A. Gelvin-Innvaer.1995. Living resources of the Delaware Bay: Colonial Wading birds. New Jersey Department of Environmental Protection Division of Fish, Game & Wildlife Endangered & Nongame Species Program. http://www.pdcbank.state.nj.us/dep/fgw/ensp/pdf/literature/colonial_wading _birds.pdf (accessed June 15, 2015).

Jenouvrier, S., C. Barbraud, and H. Weimerskirch. 2003. Effects of climate variability on the temporal population dynamics of southern fulmars. *Journal of Animal Ecology* 72:576–587.

Jenouvrier, S., H. Weimerskirch, C. Barbraud, Y. H. Park, and B. Cazelles. 2005. Evidence of a shift in the cyclicity of Antarctic seabird dynamics linked to climate. *Proceedings of the Royal Society B: Biological Sciences* 272:887–895.

Jensen, S., and A. Jernelov. 1969. Biological methylation of mercury in aquatic organisms. *Nature* 223:753–754.

Jerez, S., M. Motas, M. J. Palacios, F. Valera, J. J. Cuervo, and A. Barbosa. 2011. Concentration of trace elements in feathers of three Antarctic penguins: Geographical and interspecific differences. *Environmental Pollution* 159:2412–2419.

Jivoff, P., and K. W. Able. 2001. Characterization of the fish and selected decapods in Little Egg Harbor. *Journal of Coastal Research* 32:178–196.

Joint Nature Conservation Committee (JNCC). 2002. Herring Gull *Larus argentatus*. http://jncc.defra.gov.uk /page-2887 (accessed March 7, 2015).

Jones, J. 2001. Habitat selection studies in avian ecology: A critical review. *The Auk* 118:557–562.

Jones, R. L., and T. E. Marquardt. 1987. Monitoring of aldicarb residues in Long Island, New York potable wells. *Archives of Environmental Contamination and Toxicology* 16:643–647.

Jones, H. P., B. R. Tershy, E. S. Zavaleta, D. A. Croll, B. S. Keitt, M. E. Finkelstein, and G. R. Howald. 2008. Severity of the effects of invasive rats on seabirds: A global review. *Conservation Biology* 22:16–26.

Kadlec, J. A. 1971. Effects of introducing foxes and raccoons on Herring gull colonies. *Journal of Wildlife Management* 35:625–636.

Karpanty, S. M., J. D. Fraser, J. Berkson, L. J. Niles, A. Dey, and E. P. Smith. 2006. Horseshoe Crab eggs determine Red Knot distribution in Delaware Bay. *Journal of Wildlife Management* 70:1704–1710.

Kartman, L. 1968. Future Environments of North America: Book Review: 1966. *American Journal of Public Health* 58:601.

Kearney, M. S., A. S. Rogers, J. R. G. Townshend, E. Rizzo, D. Stutzer, J. C. Stevenson, and K. Sundborg. 2002. Landsat imagery shows decline of coastal marshes in Chesapeake and Delaware Bays. *Transactions of the American Geophysical Union* 83:173.

Keegan, G. C. Jr. 2011. The dredging crisis in New York Harbor—Present and future problems, present and future solutions. *Fordham Environmental Law Reviews* 8:351–388.

Kench, J. E., and P. J. Gubb. 1970. The activity of certain enzymes in cadmium-poisoned chicks. *Biochemistry Journal* 120(4):27.

Kennamer, R. A., J. R. Stout, B. P. Jackson, S. V. Colwell, I. L. Brisbin Jr., and J. Burger. 2005. Mercury patterns in wood duck eggs from a contaminated reservoir in South Carolina, USA. *Environmental Toxicology and Chemistry* 24:1793–1800.

Kennish, M. J. 2001a. Barnegat Bay–Little Egg Harbor, New Jersey: Estuary and watershed assessment. *Journal of Coastal Research* Special 32.

Kennish, M. J. 2001b. Characterization of the Barnegat Bay–Little Egg Harbor estuary and watershed. *Journal of Coastal Research* 81:3–12.

Kennish, M. J. 2001c. Physical description of the Barnegat Bay–Little Egg Harbor estuarine system. *Journal of Coastal Research* 81:13–27.

Kennish, M. J. 2002. Environmental threats and environmental future of estuaries. *Environmental Conservation* 29:78–109.

Kent, D. M. 1986. Behavior, habitat use, and food of three egrets in a marine habitat. *Colonial Waterbirds* 9:25–30.

Kerlinger, P., and D. Weidner. 1991. The economics of birding at Cape May, New Jersey. In: *Ecotourism and Resource Conservation, a Collection of Papers*, ed. J. Kassler, 324–334. New York: Holt, Rinehart & Winston.

Kharin, V. V., F. W. Zwiers, X. Zhang, and G. C. Hegerl. 2007. Changes in temperature and precipitation extremes in the IPCC ensemble of global coupled simulations. *Journal of Climate* 20:1419–1444.

Kim, J., and T. Koo. 2007. Heavy metal concentrations in diet and livers of Black-crowned Night Heron *Nycticorax nycticorax* and Grey Heron *Ardea cinerea* chicks from Pyeongtaek, Korea. *Ecotoxicology* 16:411–416.

Kim, J., and T. Koo. 2008. Heavy metal distribution in chicks of two heron species from Korea. *Archives of Environmental Contamination and Toxicology* 54:740–747.

Kim, J., and J. Oh. 2014. Lead and cadmium contaminations in feathers of heron and egret chicks. *Environmental Monitoring and Assessment* 186:2321–2327.

Kim, E. Y., T. Murakami, D. Saeki, and R. Tatsukawa. 1996. Mercury levels and its chemical form in tissues and organs of seabirds. *Archives of Environmental Contamination and Toxicology* 30:259–266.

Kim, J., J. Shin, and T. Koo. 2009. Heavy metal distribution in some wild birds from Korea. *Archives of Environmental Contamination and Toxicology* 56:317–324.

King, K. A., T. W. Custer, and J. S. Quinn. 1991. Effects of mercury, selenium, and organochlorine contaminants on reproduction of Forster's terns and black skimmers nesting in a contaminated Texas Bay. *Archives of Environmental Contamination and Toxicology* 20:32–40.

Kiviat, E. 2013. Ecosystem services of *Phragmites* in North America with emphasis on habitat functions AoB Plants. Published online. http://aobpla.oxfordjournals.org/content/5/plt008.full.pdf (accessed August 23, 2015).

Kiviat, E., and E. Hamilton. 2001. *Phragmites* use by native North Americans. *Aquatic Botany* 69:341–357.

Klimstra, J. D., J. L. Yee, G. H. Heinz, D. J. Hoffman, and K. R. Stebbins. 2012. Interactions between methylmercury and selenomethionine injected into mallard eggs. *Environmental Toxicology and Chemistry* 31:579–584.

Knight, R. L., W. E. Walton, G. F. O'Meara, W. K. Reisen, and R. Wass. 2003. Strategies for effective mosquito control in constructed treatment wetlands. *Ecological Engineering* 21:211–232.

Kobell, R. 2014. Botulism, virus, down birds on Poplar Island. Bay Journal. Sept. 15, 2014. http://www.bayjournal.com/article/botulism_virus_down_birds_on_poplar_island (accessed January 19, 2016).

Koepp, S. J., E. D. Santoro, and G. DiNardo. 1988. Heavy metals in finfish and selected macroinvertebrates of the Lower Hudson River estuary. In: *Fisheries Research in the Hudson River*, ed. C. L. Smith, 273–286. Albany, NY: State University of New York Press.

Koffijberg, K., L. Dijksen, B. Hälterlein, K. Laursen, P. Potel, and P. Südbeck. 2006. Breeding Birds in the Wadden Sea in 2001: Results of the total survey in 2001 and trends in numbers between 1991 and 2001. Common Wadden Sea Secretariat, trilateral Monitoring and Assessment Group, Joint Monitoring Group.

Kolbert, E. 2014. *The Sixth Extinction: An Unnatural History.* New York: Henry Holt, 319 pp.

Kolluru, R. V., S. M. Bartell, R. M. Pitblado, and R. S. Stricoff. 1995. *Risk Assessment and Management Handbook: For Environmental Health, and Safety Profession.* New York: McGraw-Hill.

Konvicka, M., M. Maradova, J. Benes, Z. Fric, and P. Kepka. 2003. Uphill shifts in distribution of butterflies in the Czech Republic: Effects of changing climate detected on a regional scale. *Global Ecology and Biogeography* 12:403–410.

Koster, M. D., D. P. Ryckman, D. V. C. Weseloh, and J. Struger. 1996. Mercury levels in great lakes Herring gull (*Larus argentatus*) eggs, 1972–1992. *Environmental Pollution* 93:261–270.

Kostich, M. S., A. L. Batt, and J. M. Lazorchak. 2014. Concentrations of prioritized pharmaceuticals in effluents from 50 large wastewater treatment plants in the US and implications for risk estimation. *Environmental Pollution* 184:354–359.

Kotliar, N., and J. Burger. 1984. The use of decoys to attract Least Terns (*Sterna antillarum*) to abandoned colony sites in New Jersey. *Colonial Waterbirds* 7:134–138.

Kotliar, N. B., and J. Burger. 1986. Colony selection and abandonment by Least Terns (*Sterna Antillarum*) in New Jersey. *Biological Conservation* 37:1–22.

Kozicky, E. L., and F. V. Schmidt. 1949. Nesting habits of the Clapper Rail in New Jersey. *Auk* 66:355–364.

Krabbenhoft, D. P., and E. M. Sunderland. 2013. Global change and mercury. *Science* 341:1457–1458.

Krabbenhoft, D. P., R. Mason, M. W. Murray, R. Reash, and T. Saltman. 2007. *Ecosystem Responses to Mercury Contamination: Indicators of Change.* Boca Raton, FL: CRC Press.

Kreamer, G., and S. Michels. 2009. History of Horseshoe Crab harvest in Delaware Bay. In: *Biology and Conservation of Horseshoe Crabs*, ed. J. T. Tanacredi, M. L. Botton, and D. R. Smith, 299–313. New York: Springer.

Kury, C. R., and M. Gochfeld. 1975. Human interference and gull predation in cormorant colonies. *Biological Conservation* 8:23–34.

Kushlan, J. A. 1976. Feeding behavior of North American herons. *Auk* 93:86–94.

Kushlan, J. A. 1978. Feeding ecology of wading birds. In: *Wading Birds*, ed. A. Sprunt, J. C. Ogden, and S. Winckler, 249–297. New York: National Audubon Society.

Kushlan, J. A. 2000a. Heron nest site conservation. In: *Heron Conservation*, ed. J. A. Kushlan and H. Hafner, 331–342. New York: Academic.

Kushlan, J. A. 2000b. Research and information needs for heron conservation. In: *Heron Conservation*, ed. J. A. Kushlan and H. Hafner, 377–380. New York: Academic.

Kushlan, J. A. 2012. A history of conserving colonial waterbirds in the United States. *Waterbirds* 35:608–625.

Kushlan, J. A., and H. Haffner. 2000a. *Heron Conservation*. New York: Academic.

Kushlan, J. A., and H. Hafner. 2000b. Reflections on heron conservation. In: *Heron Conservation*, ed. J. A. Kushlan and H. Hafner, 377–380. New York: Academic.

Kushlan, J. A., M. J. Steinkamp, K. C. Parsons, J. Capp, M. A. Cruz, M. Coulter, I. J. Davidson et al. 2002. Waterbird conservation for the Americas: The North American waterbird conservation plan, version 1. Waterbird Conservation for the Americas, Washington, DC, USA. http://www.waterbirdconservation.org/nawcp.html (accessed June 13, 2015).

La Brea, 2015. La Brea Tar Pits and Museum Bird Faunal List. http://www.tarpits.org/research-collections/collections/bird-faunal-list (accessed March 14, 2016).

La Sorte, F. A., and W. Jetz. 2012. Tracking of climatic niche boundaries under recent climate change. *Journal of Animal Ecology* 81:914–925.

La Sorte, F. A., and F. R. Thompson. 2007. Poleward shifts in winter ranges of North American birds. *Ecology* 88:1803–1812.

Lack, D. 1968. *Ecological Adaptations for Breeding in Birds*. London: Methuen.

Lam, J. C. W., S. Tanabe, M. H. W. Lam, and P. K. S. Lam. 2005. Risk to breeding success of waterbirds by contaminants in Hong Kong: Evidence from trace elements in eggs. *Environmental Pollution* 135:481–490.

Lane, L. K., K. Charles-Guzman, Z. Wheeler, N. Abid, N. Graber, and T. Matte. 2013. Health effects of coastal storms and flooding in urban areas: A review and vulnerability assessment. *Journal of Environmental and Public Health* 2013:1–13.

Lange, T. R., H. E. Royals, and L. L. Connor. 1994. Mercury accumulation in largemouth bass (*Micropterus salmoides*) in a Florida Lake. *Archives of Environmental Contamination* 27:466–471.

Lantz, S. M., D. E Gawlik, and M. I. Cook. 2010. The effects of water depth and submerged aquatic vegetation on foraging habitat selection and foraging success of wading birds. *Condor* 112:460–469.

Lantz, S. M., D. E. Gawlik, and M. I. Cook. 2011. The effects of water depth and emergent vegetation on foraging success and habitat selection of wading birds in the everglades. *Waterbirds* 34:439–447.

Laskey, J. W., G. L. Rehnberg, J. F. Hein, and S. D. Carter. 1982. Effects of chronic manganese (Mg^3O^4) exposure on selected reproductive parameters in rats. *Journal of Toxicology and Environmental Health Part B* 9:677–687.

Lasky, T., W. Sun, A. Kadry, and M. K. Hoffman. 2004. Mean total arsenic concentrations in chicken 1989–2000 and estimated exposures for consumers of chicken. *Environmental Health Perspectives* 112:18–21.

Lathrop, R. G., Jr., and J. A. Bognar. 2001. Habitat loss and alteration in the Barnegat Bay region. *Journal of Coastal Research* 32:212–228.

Lauro, B., and J. Burger. 1989. Nest-site selection in the American Oystercatcher in salt marshes. *Auk* 106:185–192.

Lavers, J. L., A. L. Bond, and I. Hutton. 2014. Plastic ingestion by Flesh-footed Shearwaters (*Puffinus carneipes*): Implications for fledgling body condition and the accumulation of plastic-derived chemicals. *Environmental Pollution* 187:124–129.

Lavoie, R. A., C. J. Baird, L. E. King, T. K. Kyser, V. L. Friesen, and L. M. Campbell. 2014. Contamination of mercury during the wintering period influences concentrations at breeding sites in two migratory piscivorous birds. *Environmental Science & Technology* 48(23):13694–13702.

Lavoie, R. A., T. K. Kyser, V. L. Friesen, and L. M. Campbell. 2015. Tracking overwintering areas of fish-eating bids to identify mercury exposure. *Environmental Science and Technology* 49:863–872.

Le V. Dit Durell, S. E. 2000. Individual feeding specialisation in shorebirds: Population consequences and conservation implications. *Biological Reviews of the Cambridge Philosophical Society* 75:503–518.

Leafe, M., M. Irigoyen, C. DeLago, A. Hassan, and L. Braitman. 2015. Change in childhood lead exposure prevalence with new reference level. *Journal of Environmental Health*. 77:14–16.

Leatherman, S., R. Chalfont, E. Pendleton, T. McCandless, and S. Funderburk. 1995. Vanishing lands: Sea level, society, and Chesapeake Bay. Chesapeake Bay Field Office, U.S. Fish and Wildlife Service, Annapolis, Maryland USA.

Leberg, P. L., P. Deshotels, S. Pius, and M. Carloss. 1995. Nest sites of seabirds on dredge islands in coastal Louisiana. *Proceedings of the Southeastern Association of Fish and Wildlife Agencies* 49:356–366.

Lee, S. Y., R. J. K. Dunn, R. A. Young, R. M. Connolly, P. E. R. Dale, R. Dehayr, C. J. Lemckert et al. 2006. Impact of urbanization on coastal wetland structure and function. *Austral Ecology* 31:149–163.

Leiserowitz, A. A. 2005. American risk perceptions: Is climate change dangerous? *Risk analysis* 25:1433–1442.

Leiserowitz, A., E. Maibach, C. Roser-Renouf, N. Smith, and J. D. Hmielowski. 2010. Climate change in the American mind: Public support for climate and energy policies in June 2010. Yale Project on Climate Change Communication. http://environment.yale.edu/climate/files/PolicySupportJune2010.pdf (accessed March 1, 2015).

Leitao, A. G., and J. Ahern. 2002. Applying landscape ecological concepts and metrics in sustainable landscape planning. *Landscape and Urban Planning* 59:65–93.

Lellis-Dibble, K. A., K. E. McGlynn, and T. E. Bigford. 2008. Estuarine Fish and Shellfish Species in U.S. Commercial and Recreational Fisheries: Economic Value as an Incentive to Protect and Restore Estuarine Habitat. U.S. Dep. Commerce, NOAA Tech. Memo. http://www.habitat.noaa.gov/pdf/publications_gen eral_estuarinefishshellfish.pdf (accessed March 1, 2015).

Lemly, D. A. 1993a. Guidelines for evaluating selenium data from aquatic monitoring and assessment studies. *Environmental Monitoring and Assessment* 28:83–100.

Lemly, D. A. 1993b. Metabolic stress during winter increases the toxicity of selenium to fish. *Aquatic Toxicology* 27:133–158.

Leschen, A. S., and S. J. Correia. 2010. Mortality in female Horseshoe Crabs (*Limulus polyphemus*) from biomedical bleeding and handling: Implications for fisheries management. *Marine and Freshwater Behaviour and Physiology* 43:135–147.

Lester, L. A., H. W. Avery, A. S. Harrison, and E. A. Standora. 2013. Recreational boats and turtles: Behavioral mismatches result in high rates of injury. *PLoS One* 8:1–8.

Lewis, H. F. 1929. The natural history of the double-crested Cormorant (*Phalacrocorax auritus auritus* (Lesson)). Ottawa, Ontario: Ru-Mi-Lou Books, p. 94.

Lewis, S. A., and R. W. Furness. 1991. Mercury accumulation and excretion by laboratory reared Black-headed Gulls (*Larus ridibundus*) chicks. *Archives of Environmental Contamination and Toxicology* 21:316–320.

Lewis, J. C., and R. L. Garrison. 1983. Habitat suitability index models: Clapper Rail. U.S. Fish & Wildlife Service. FWS/OBS-82/l0.51. 15 pp.

Li, C., J. Cornett, and K. Ungar. 2003. Long-term decrease of cadmium concentration in the Canadian Arctic. *Geophysical Research Letters* 30:1–4.

Li, J. L., C. Y. Jiang, S. Li, and S. W. Xu. 2013. Cadmium induced hepatotoxicity in chickens (*Gallus domesticus*) and ameliorative effect by selenium. *Ecotoxicology and Environmental Safety* 96:103–109.

Li, L., B. Zheng, and L. Liu. 2010. Biomonitoring and bioindicators used for river ecosystems: Definitions, approaches and trends. *Proceedings of Environmental Sciences* 2:1510–1524.

Lieske, D. J., T. Wade, and L. A. Roness. 2014. Climate change awareness and strategies for communicating the risk of coastal flooding: A Canadian Maritime case example. *Estuarine, Coastal and Shelf Science* 140:83–94.

Lin, L., W. Zhou, H. Dai, F. Cao, G. Zhang, and F. Wu. 2012. Selenium reduces cadmium uptake and mitigates cadmium toxicity in rice. *Journal of Hazardous Materials* 15:343–351.

Lindström, Å., and J. Agrell. 1999. Global change and possible effects on the migration and reproduction of arctic-breeding waders. *Ecological Bulletins* 145–159.

Linthurst, R. A., P. Bourdeau, and R. G. Tardiff. 1995. *Methods to Assess the Effects of Chemicals on Ecosystems. SCOPE Monograph 53.* Scientific Committee on Problems of the Environment. New York: John Wiley.

Lioy, P. J., and P. G. Georgopoulos. 2011. New Jersey: A case study of the reduction in urban and suburban air pollution from the 1950s to 2010. *Environmental Health Perspectives* 119:1351–1355.

Liu, X. F., Z. P. Li, F. Tie, N. Liu, Z. W. Zhang, and S. W. Xu. 2012. Effects of manganese-toxicity on immune-related organs of cocks. *Chemosphere* 90:2085–2100.

Liu, L. L., C. M. Li, Z. W. Zhang, J. L. Zhang, H. D. Yao, and S. W. Xu. 2014. Protective effects of selenium on cadmium-induced brain damage in chickens. *Biological Trace Elements and Research* 158:176–185.

Liu, X. F., L. M. Zhang, H. N. Guan, Z. W. Zhang, and S. W. Xu. 2013. Effects of oxidative stress on apoptosis in manganese-induced testicular toxicity in cocks. *Food and Chemical Toxicology* 60:168–776.

Llanso, R. J., L. C. Scott, D. M. Dauer, J. L. Hyland, and D. E. Russell. 2002. An estuarine benthic index of biotic integrity for the mid-Atlantic region of the United States: I. Classification of assemblages and habitat definition. *Estuaries* 25:1219–1230.

Long Island Sound Study (LISS) 2010. Status and Trends: LISS Environmental Indicators. Wading Birds. Long Island Sound Study. http://longislandsoundstudy.net/2010/06/colonial-waterbirds/ (accessed September 6, 2015).

Long Island Sound Study (LISS). 2015a. Status and Trends: LISS Environmental Indicators. http://longisland soundstudy.net/indicator/chlorophyll-a-concentration/ (accessed March 7, 2015).

Long Island Sound Study (LISS). 2015b. Habitats. http://longislandsoundstudy.net/issues-actions/habitat-quality/ (accessed March 25, 2015).

Lockwood, J., and B. Maslo. 2014. The conservation of coastal biodiversity. In: *Coastal Conservation*, ed. B. Maslo and J. L. Lockwood, 382. Cambridge UK: Cambridge Univ. Press.

Lockwood, J. L., K. H. Fenn, J. M. Caudill, D. Okines, O. L. Bass Jr., J. R. Duncan, and S. L. Pimm. 2001. The implications of Cape Sable seaside sparrow demography for Everglades restoration. *Animal Conservation* 4:275–281.

Lodenius, M., and T. Solonen. 2013. The use of feathers of birds of prey as indicators of metal pollution. *Ecotoxicology* 22:1319–1334.

Lomba, A., P. Alves, and J. Honrado. 2008. Endemic sand dune vegetation of the Northwest Iberian Peninsula: Diversity, dynamics, and significance for bioindication and monitoring of coastal landscapes. *Journal of Coastal Research* 24:113–121.

Lord, C. G., K. F. Gaines, C. S. Boring, I. L. Brisbin Jr., M. Gochfeld, and J. Burger. 2002. Raccoon (*Procyon lotor*) as a bioindicator of mercury contamination at the U.S. Department of Energy's Savannah River Site. *Archives of Environmental Contamination and Toxicology* 43:356–363.

Love, S. E., and D. B. Carter. 2001. Pea Patch Island Heronry Region: Special Area Management Plan Progress Report—June 2001. Delaware Coastal Management Program, Dover DE. http://www.dnrec.delaware .gov/coastal/Documents/PPISAMP/ProgReport2001.pdf (accessed August 26, 2015).

Lowther, P. E., H. D. Douglas III, and C. L. Gratto-Trevor. 2001. Willet (*Tringa semipalmata*). In: *The Birds of North America Online Cornel*, ed. A. Poole. Ithaca, NY: Laboratory of Ornithology. http://bna.birds .cornell.edu.bnaproxy.birds.cornell.edu/bna/species/579 (accessed February 7, 2015).

Lucia, M., J. M. André, P. Gonzalez, M. Baudrimont, M. D. Bernadet, K. Gontier, R. Maury-Brachet, G. Guy, and S. Davail. 2010. Effect of dietary cadmium on lipid metabolism and storage of aquatic bird *Cairina moschata*. *Ecotoxicology* 19:163–170.

Ludwig, J. P., H. J. Auman, D. V. Weseloh, G. A. Fox, J. P. Giesy, and M. E. Ludwig. 1995. Evaluation of the effects of toxic chemicals in Great Lakes Cormorants: Has causality been established? *Colonial Waterbirds* 18:60–69.

Lucia, M., P. Bocher, R. P. Cosson, C. Churlaud, and P. Bustamante. 2012. Evidence of species-specific detoxification processes for trace elements in shorebirds. *Ecotoxicology* 21:2349–2362.

Lyles, A. M. 2000. Captive populations. In: *Heron Conservation*, ed. J. A. Kushlan and H. Hafner, 293–310. New York: Academic.

Maccarone, A. D., and J. N. Brzorad. 2005. Foraging microhabitat selection by wading birds in a tidal estuary, with implications for conservation. *Waterbirds* 28; 383–391.

Maccarone, A. D., J. N. Brzorad, and H. M. Stone. 2012. A telemetry-based study of Snowy Egret (*Egretta thula*) nest-activity patterns, food-provisioning rates and foraging energetics. *Waterbirds* 35:394–401.

MacCoun, R. J. 1998. Biases in the interpretation and use of research results. *Annual Review of Psychology* 49:259–287.

Mackay, G. H. 1899. The Terns of Muskeget and Penikese Islands, Massachusetts. *Auk* 16:259–266.

MacKenzie, C. L., Jr. 1992. *The Fisheries of Raritan Bay*. New Brunswick, NJ: Rutgers University Press.

Mackie, J. A., S. M. Natali, J. S. Levinton, and S. A. Sañudo-Wilhelmy. 2007. Declining metal levels at Foundry Cove (Hudson River, New York): Response to localized dredging of contaminated sediments. *Environmental Pollution* 149:141–148.

Maclean, I. M. D. 2014. Global change and conservation of waders. In: *Coastal Conservation*, ed. B. Maslo and J. L. Lockwood, 265–286. Cambridge UK: Cambridge Univ. Press.

Maclean, I. M. D., G. E. Austin, M. M. Rehfisch, J. Blew, O. Crowe, S. Delany, K. Devos et al. 2008. Climate change causes rapid changes in the distribution and site abundance of birds in winter. *Global Change Biology* 14:2489–2500.

Magella, G., and P. Brousseau. 2001. Does culling predatory gulls enhance the productivity of breeding Common Terns? *Journal of Applied Ecology* 38:1–8.

Mailman, R. B. 1980. Heavy metals. In: *Introduction to Environmental Toxicology*, ed. F. E. Gunthrie and J. J. Perry, 34–43. New York: Elsevier.

Malik, R. N., and N. Zeb. 2009. Assessment of environmental contamination using feathers of *Bubulcus ibis* L., as a biomonitor of heavy metal pollution, Pakistan. *Ecotoxicology* 18:522–536.

Mangold, R. E. 1974. Research on shore and upland migratory birds in New Jersey: Clapper rail studies. Division of Fish, Game, and Shellfisheries, Trenton, NJ.

Marcarenhas, A., P. Coelho, E. Subtil, and T. B. Ramos. 2010. The role of common local indicators in regional sustainability assessment. *Ecological Indicators* 10:644–656.

Marettová, E., M. Maretta, and J. Legáth. 2015. Toxic effects of cadmium on testis of birds and mammals: A review. *Animal Reproductive Science* 155:1–10.

Mariette, M. M., and S. C. Griffith. 2013. Does coloniality improve foraging efficiency and nestling provisioning? A field experiment in the wild Zebra Finch. *Ecology* 94:325–335.

Markert, B., O. Wappelhorst, V. Weckert, U. Herpin, U. Siewers, K. Friese, and G. Breulmann. 1999. The use of bioindicators for monitoring the heavy-metal status of the environment. *Journal of Radioanalytical and Nuclear Chemistry* 240:425–429.

Marples, G., and Marples. 1934. *Sea Terns or Sea Swallows*. London: Country Life Ltd.

Martín-Doimeadios, R. C. R., J. J. Berzas Nevado, F. J. Guzmán Bernardo, M. Jiménez Moreno, G. P. F. Arrifano, A. M. Herculano, J. L. M. do Nascimento, and M. E. Crespo-López. 2014. Comparative study of mercury speciation in commercial fishes of the Brazilian Amazon. *Environmental Science and Pollution Research* 21:7466–7479.

Maslo, B., and J. L. Lockwood. 2014. *Coastal Conservation*, 382. Cambridge, UK: Cambridge University Press.

Maslo, B., J. Burger, and S. N. Handel. 2012. Modeling foraging behavior of piping plovers to evaluate habitat restoration success. *The Journal of Wildlife Management* 76:181–188.

Maslo, B., S. N. Handel, and T. Pover. 2011. Restoring beaches for Atlantic Coast Piping Plovers (*Charadrius melodus*): A classification and regression tree analysis of nest-site selection. *Restoration Ecology* 19:194–203.

Mason, R. P. 2004. Mercury Concentrations in Fish from Tidal Waters of the Chesapeake Bay. Final Report to Maryland Department of Natural Resources http://www.dnr.state.md.us/irc/docs/00006644.pdf (accessed July 24, 2015).

Massachusetts. 2015. Buzzards Bay Tern Restoration Project. Massachusetts Department of Energy and Environmental Affairs. http://www.mass.gov/eea/agencies/dfg/dfw/natural-heritage/species-information -and-conservation/rare-birds/buzzards-bay-tern-restoration-project.html (accessed 7/16/2015).

Master, T. L., J. K. Leiser, K. A. Bennett, and J. K. Bretsch. 2005. Patch selection by snowy egrets. *Waterbirds* 28:220–224.

Matthiopoulos, J., J. Harwood, and L. E. N. Thomas. 2005. Metapopulation consequences of site fidelity for colonially breeding mammals and birds. *Journal of Animal Ecology* 74:716–727.

Mauco, L., and M. Favero. 2005. The food and feeding biology of Common Terns wintering in Argentina: Influence of environmental conditions. *Waterbirds* 28:450–457.

Mauco, L., M. Favero, and M. S. Bó. 2001. Food and feeding biology of the Common Tern during the non-breeding season in Samborombon Bay, Buenos Aires, Argentina. *Waterbirds* 24:89–96.

Maxted, A. M., M. P. Luttrell, V. H. Goekjian, J. D. Brown, L. J. Niles, A. D. Dey, K. S. Kalasz et al. 2012. Avian influenza virus infection dynamics in shorebird hosts. *Journal of Wildlife Diseases Apr* 48(2): 322–334.

McCarty, J. P. 2001. Ecological consequences of recent climate change. *Conservation Biology* 15:320–331.

McCauley, D. J., M. L. Pinsky, S. R. Palumbi, J. A. Estes, F. H. Joyce, and R. R. Warner. 2015. Marine defaunation: Animal loss in the global ocean. *Science* 347:247–269.

McCool, S. F., and G. H. Stankey. 2004. Indicators of sustainability: Challenges and opportunities at the interface of science and policy. *Environmental Management* 33:294–305.

McCrimmon, D. A., Jr., J. C. Ogden, and G. T. Bancroft. 2011. Great Egret (*Ardea alba*). In: *The Birds of North America Online Cornell*, ed. A. Poole. Ithaca, NY: Cornell Laboratory of Ornithology. http://bna.birds .cornell.edu/bna/species/570 (accessed February 7, 2015).

McKelvey, W., R. C. Gwynn, N. Jeffery, D. Kass, L. E. Thorpe, R. K. Garg, C. D. Palmer, and P. J. Parsons. 2007. A biomonitoring study of lead, cadmium, and mercury in the blood of New York city adults. *Environmental Health Perspectives* 115:1435–1441.

McKim, J. M., G. F. Holson, G. W. Holcome, and E. P. Hunt. 1976. Long-term effects of methylmercuric chloride on three generations of Brook Trout (*Salvelinus fontinalis*): Toxicity, accumulation, distribution and elimination. *Journal Fisheries Research Board Canada* 33:2726–2739.

McManus, M. C., P. Licandro, and S. H. Coombs. 2015. Is the Russell Cycle a true cycle? Multidecadal zooplankton and climate trends in the western English Channel. *Journal of Marine Systems* 77:296–311.

McNicholl, M. K. 1975. *Larid* site tendency and group adherence in relation to habitat. *Auk* 92:98–104.

McNicholl, M. K., P. E. Lowther, and J. A. Hall. 2001. Forster's Tern (*Sterna forsteri*). In: *The Birds of North America Online*, ed. A. Poole. Ithaca, NY: Cornell Laboratory of Ornithology. http://bna.birds.cornell .edu.bnaproxy.birds.cornell.edu/bna/species/595 (accessed February 8, 2015).

McNutt, M. K., R. Camilli, T. J. Crone, G. D. Guthrie, P. A. Hsieh, T. B. Ryerson, O. Savas, and F. Shaffer. 2012. Review of the flow estimates of the *Deepwater Horizon* oil spill. *Proceedings of the National Academy of Science* 109:20260–20267.

Melvin, S. M. 2010. Survey of coastal nesting colonies of cormorants, gulls, night-herons, egrets, and ibises in Massachusetts, 2006–08. Final Report. National heritage and Endangered Species Program, Massachusetts Division of Fisheries and Wildlife.

Mendonça, V. M., D. G. Raffaelli, and P. R. Boyle. 2007. Interactions between shorebirds and benthic invertebrates at Culbin Sands lagoon, NE Scotland: Effects of avian predation on their prey community density and structure. *Scientia Marina* 71:579–591.

Meredith, W. H., and D. E. Saveikis. 1987. Effects of open marsh water management (OMWM) on bird populations of a Delaware tidal marsh, and OMWM's use in waterbird habitat restoration and enhancement. In: *Waterfowl and wetlands symposium: Proceedings of Symposium on Waterfowl and Wetlands Management in the Coastal Zone of the Atlantic Flyway*. Delaware Coastal Management Program, 298–321. Delaware Department of Natural Resources and Environmental Control.

Meyer, T. 2013. Four ways the government subsidizes risky coastal rebuilding. Propublic. New York: Journalism in the Public Interest. Available at https://www.propublica.org/article/four-ways-the-government-subsidizes -risky-coastal-rebuilding (accessed March 6, 2016).

Michel, J., E. H. Owens, S. Zengel, A. Graham, Z. Nixon, T. Allard, W. Holton et al. 2013. Extent and degree of shoreline oiling: *Deepwater Horizon* oil spill, Gulf of Mexico, USA. *PLOS One* 8:e6587.

Michener, W. K., E. R. Blood, K. L. Bildstein, M. M. Brinson, and L. R. Gardner. 1997. Climate change, hurricanes and tropical storms and rising sea level in coastal wetlands. *Ecological Applications* 7:770–801.

Mierzykowski, S. E. 2008. Environmental contaminants in tern eggs from Monomoy NWR and Seal Island NWR. USFWS Spec. Proj. Rep. FY07-MEFO-6-EC. Maine Field Office. Old Town, ME. 27 pp.

Mierzykowski, S. E., L. J. Welch, W. Goodale, D. C. Evers, C. S. Hall, S. W. Kress, and R. B. Allen. 2005. Mercury in bird eggs from coastal Maine. USFWS Special Project Report FY05-MEFO-1-EC. Old Town, ME.

Mierzykowski, S. E., L. J. Welch, C. S. Hall, S. W. Kress, and R. B. Allen. 2008. Contaminant assessment of Common Terns in the Gulf of Maine. USFWS. Spec. Proj. Rep. FY07-MEFO-2-EC. Maine Field Office. Old Town, ME.

Mikula, G., and M. Wenzel. 2000. Justice and social conflict. *International Journal of Psychology* 35: 126–135.

Miller, K. G., P. J. Sugarman, and J. V. Browning. 2014. Sea level and climate change should I sell my shore house? http://climatechange.rutgers.edu/docman-list/events/symposia-past-events/may-25-2011/79-sea -level-change-in-new-jersey-should-i-sell-my-shore-house/file (accessed January 12, 2015).

Mizrahi, D. S., K. A. Peters, and Hodgetts, P. A. 2012. Energetic conditions of Semipalmated and Least Sandpipers during northbound migration staging periods in Delaware Bay. *Waterbirds* 35:135–145.

Møller, A. P. 1980. Breeding cycle of the Gull-billed Tern *Gelochelidon nilotica* Gmel., especially in relation to colony size. *Ardea* 69:193–198.

Møller, A. P., E. Flensted-Jensen, and W. Mardal. 2006. Dispersal and climate change: A case study of the Arctic Tern *Sterna paradisaea*. *Global Change Biology* 12:2005–2013.

Montalvo, A. M., S. L. Williams, K. J. Rice, S. L. Buchmann, C. Cory, S. N. Handel, G. P. Nabhan, R. Primack, and R. H. Robichaux. 1997. Restoration biology: A population biology perspective. *Restoration Ecology* 5:277–290.

Monteiro, L. R. 1996. Seabirds as monitors of mercury in the marine environment. *Water, Air, Soil Poll.* 80:851–870.

Monteiro, L. R., and R. W. Furness. 1995. Seabirds as monitors of mercury in the environment. *Water, Air, Soil Pollution.* 80:831–870.

Monteiro, L. R., V. Costa, R. W. Furness, and R. S. Santos. 1996. Mercury concentrations in prey fish indicate enhanced bioaccumulation in mesopelagic environments. *Marine Ecology Progress Series* 141:21–25.

Montevecchi, W. A., and R. A. Myers. 1997. Centurial and decadal oceanographic influences on changes in northern gannet populations and diets in the north-west Atlantic: Implications for climate change. *ICES Journal of Marine Science: Journal du Conseil* 54:608–614.

Montevecchi, W. A., A. Hedd, L. M. Tranquilla, D. A. Fifield, C. M. Burke, P. M. Regular, and R. A. Phillips. 2012. Tracking seabirds to identify ecologically important and high risk marine areas in the western North Atlantic. *Biological Conservation* 156:62–71.

Moore, S. A., T. J. Wallington, R. J. Hobbs, P. R. Ehrlich, C. S. Holling, S. Levin, D. Lindenmayer et al. 2009. Diversity in current ecological thinking: Implications for environmental management. *Environmental Management* 43:17–27.

Moore-Colyer, R. J. 2000. Feathered women and persecuted birds: The struggle against the plumage trade, c. 1860–1922. *Rural History* 11:57–73.

Mora, M. A. 1996. Organochlorines and trace elements in four colonial waterbird species nesting in the Lower Laguna Madre, Texas. 31:533–537.

Moriarty, F. 1983. *Ecotoxicology: The Study of Pollutants in Ecosystems.* New York: Academic Press.

Morris, R. D. 1986. Seasonal differences in courtship feeding rates of male Common Terns. *Canadian Journal of Zoology* 64:501–507.

Morris, R. D., C. Pekarik, and D. J. Moore. 2012. Current status and abundance trends of Common Terns breeding at known coastal and inland nesting regions in Canada. *Waterbirds* 35:194–207.

Morrison, R. I. G., Y. Aubry, R. W. Butler, G. W. Beyersbergen, G. M. Donaldson, C. L. Gratto-Trevor, P. W. Hicklin, V. H. Johnston, and R. K. Ross. 2001. Declines in North American shorebird populations. *Wader Group Study Bulletin* 94:34–38.

Morrison, R. I. G., N. C. Davidson, and J. R. Wilson. 2007. Survival of the fittest: Body stores on migration and survival in Red Knots, *Calidris canutus islandica. Journal of Field Ornithology* 38:479–487.

Morrison, R. I. G., R. K. Ross, and L. J. Niles. 2004. Declines in wintering populations of Red Knots in southern South America. *Condor* 106:60–70.

Mostello, C. 2014. Inventory of terns, laughing gulls, and black skimmers nesting in Massachusetts in 2013. Natural Heritage & Endangered Species Program. http://www.mass.gov/eea/docs/dfg/nhesp/species -and-conservation/matern12report-040313.pdf (accessed April 6, 2015).

Muñoz, G., and M. A. Panero. 2006. *Pollution Prevention and Management Strategies for Dioxins in the New York/New Jersey Harbor.* New York: New York Academy of Sciences for the New York/New Jersey Harbor Consortium.

Muñoz, G., and M. A. Panero. 2008. Sources of suspended solids to the New York/New Jersey Harbor watershed. New York Academy of Sciences for the New York/New Jersey Harbor Consortium. https://www .researchgate.net/profile/Gabriela_Munoz5/publication/268326365_Sources_of_suspended_solids_to _the_New_YorkNew_Jersey_harbor_watershed/links/54dab5fd0cf2ba88a68daed9.pdf (accessed March 14, 2016).

National Audubon Society (NAS). 2010. The Christmas Bird Count Historical Results. http://www.christmas birdcount.org (accessed February 8, 2015).

National Audubon Society. 2014. Audubon's Birds and Climate Change Report: A Primer for Practitioners. National Audubon Society, New York. Version 1.2. http://climate.audubon.org/sites/default/files/Audubon -Birds-Climate-Report-v1.2.pdf (accessed September 1, 2015).

National Park Service (NPS). 2015a. Upper Delaware: Scenic and Recreational River. http://www.nps.gov /upde/index.htm (accessed August 23, 2015).

National Park Service (NPS). 2015b. The Maurice Wild & Scenic River. http://www.nps.gov/ncrc/programs /pwsr/maurice_pwsr_sub.html (accessed August 23, 2015).

National Oceanic and Atmospheric Administration (NOAA). 2011. Mussel Watch: Indicators of Successful Restoration. http://stateofthecoast.noaa.gov/musselwatch/musselwatch.html (accessed March 9, 2015).

National Oceanographic and Atmospheric Administration (NOAA). 2012a. Communities: The U.S. population living in coastal watershed counties. http://www.noaanews.noaa.gov/stories2013/20130325_coastalpopulation .html. (accessed March 14, 2016).

National Oceanic and Atmospheric Administration (NOAA). 2012b. U.S. Seafood Landings Remain Near High 2011 levels. http://www.noaanews.noaa.gov/stories2013/20131030_2012usseafoodlandings.html (accessed March 7, 2015).

National Oceanographic and Atmospheric Administration (NOAA). 2013. NOAA's State of the Coast: Rebuilding the Striped Bass Fishery on the Atlantic Coast. http://stateofthecoast.noaa.gov/rec_fishing /striped_bass.html (accessed August 4, 2015).

National Oceanographic and Atmospheric Administration (NOAA). 2014. Mean Sea Level Trend. http://tides andcurrents.noaa.gov/sltrends/sltrends_station.shtml?stnid=8534720 (accessed September 8, 2015).

National Research Council (NRC). 1977. Mercury MONOGRAPH. Washington D.C.: National Academy Press.

National Research Council (NRC). 1983. *Risk Assessment in the Federal Government: Managing the Process*, 92. Washington D.C.: National Academy Press.

National Research Council (NRC). 1991. *Animals as Sentinels of Environmental Health Hazards*. Washington D.C.: National Academy Press.

National Research Council (NRC). 1993. *Issues in Risk Assessment*. Washington D.C.: National Academy Press.

National Research Council (NRC). 2000. *Ecological Indicators for the Nation*. Washington, D.C.: National Academy Press.

National Research Council (NRC). 2008. *Public Participation in Environmental Assessment Decision Making*, 180. Washington D.C.: National Academy Press.

Nemerson, D. M., and K. W. Able. 2003. Spatial and temporal patterns in the distribution and feeding habits of *Morone saxatilis* in marsh creeks of Delaware Bay, USA. *Fisheries Management and Ecology* 10:337–348.

Nemerson, D. M., and K. W. Able. 2004. Spatial patterns in diet and distribution of juveniles of four fish species in Delaware Bay marsh creeks: Factors influencing fish abundance. *Marine Ecology Progress Series* 276:249–262.

Nettleship, D. N., J. Burger, and M. Gochfeld. 1994. *Threats to Seabirds on Islands*. Cambridge, England: International Council for Bird Preservation.

Neuman, M. J., G. Ruess, and K. W. Able. 2004. Species composition and food habits of dominant fish predators in salt marshes of an urbanized estuary, the Hackensack Meadowlands, New Jersey. *Urban Habitats* 2:3–22.

New Jersey Audubon Society (NJAS). 1999. *Birds of New Jersey*, ed. J. Walsh, V. Elia, R. Kane, and T. Halliwell. Cape May, NJ: New Jersey Audubon Society.

New Jersey. (NJCOM) 2015. Businesses struggle to stay afloat in winter on remote Long Beach Island. http:// www.nj.com/news/index.ssf/2011/02/businesses_struggle_to_stay_af.html (accessed February 7, 2015).

New Jersey Endangered and Nongame Species Program (NJDEP). 2012. NJ Endangered and Nongame Species Program: Status Definitions. http://www.state.nj.us/dep/fgw/ensp/pdf/spclspp.pdf (accessed February 7, 2015).

New Jersey Endangered and Nongame Species Program (NJDEP). 2015. New Jersey's endangered and threatened wildlife. New Jersey Department of Environmental Protection, Trenton, NJ. http://www.nj.gov/dep /fgw/tandespp.htm (accessed February 7, 2015).

NJ Mercury Task Force. 2001. Vol. 1 Executive Summary and Recommendations. http://www.state.nj.us/dep /dsr/mercury_task_force.htm (accessed August 9, 2015).

New York City Panel on Climate Change (NPCC2). 2013. *Climate Risk Information 2013: Observations, Climate Change Projections, and Maps*. New York City: New York City Mayor's Office.

New York Environmental Department of Conservation (NYDEC). 1996. New York–New Jersey Harbor & Estuary Program. http://www.dec.ny.gov/lands/31849.html (accessed February 7, 2015).

New York Environmental Department of Conservation (NYDEC). 2015. Long Island Sound Study http://www .dec.ny.gov/lands/31851.html (accessed February 7, 2015).

Niemeijer, D., and R. S. deGroot. 2008. A conceptual framework for selecting environmental indicator sets. *Ecological Indicators* 8:14–25.

Niemi, G. J., A. E. Hershey, L. Shannon, J. M. Hanowski, A. Lima, R. P. Axler, and R. R. Regal. 1999. Ecological effects of mosquito control on zooplankton, insects, and birds. *Environmental Toxicology and Chemistry* 18:549–559.

Nijhuis, M. 2014. The Audubon report: A gathering storm for North American birds. Audubon Society. *Audubon* 116:24–30.

Niles, L. J., J. Bart, H. P. Sitters, A. D. Dey, K. E. Clark, P. W. Atkinston, A. J. Baker et al. 2009. Effects of Horseshoe Crab harvest in Delaware Bay on Red Knots: Are harvest restrictions working. *BioScience* 59:153–164.

Niles, L., J. Burger, R. R. Porter, A. D. Dey, C. D. T. Minton, P. M. Gonzalez, A. J. Baker, J. W. Fox, and C. Gordon. 2010. First results using light level geolocators to track Red Knots in the Western Hemisphere show rapid and long intercontinental flights and new details of migration pathways. *Wader Study Group Bulletin* 117:1–8.

Niles, L. J., J. Burger, and A. Dey, 2012. *Life along the Delaware Bay, Cape May: Gateway to a Million Shorebirds.* New Brunswick, NJ: Rutgers University Press.

Niles, L. J., A. D. Dey, and B. Maslo. 2014. Overexploitation of marine species and its consequences for terrestrial biodiversity along coasts. In: *Coastal conservation*, ed. B. Maslo and J. L. Lockwood, 347–368. Cambridge UK: Cambridge University Press.

Niles, L. J., J. Smith, D. Daly, T. Dillingham, W. Shadel, A. Dey, S. Hafner, and D. Wheeler. 2013. Restoration of Horseshoe Crab and migratory shorebird habitat on five Delaware Bay beaches damaged by Superstorm Sandy. Published electronically November 22, 2013. http://arubewithaview.com/wordpress/w//p-content /uploads/2012/12/RestorationReport_112213.pdf.

Niles, L. J., H. P. Sitters, A. D. Dey, P. W. Atkinson, A. J. Baker, K. A. Bennett, R. Carmona et al. 2008. Status of the Red Knot, *Calidris canutus rufa*, in the Western Hemisphere. *Studies in Avian Biology* 36:1–185.

Nisbet, I. C. T. 1975. Selective effects of predation in a tern colony. *Condor* 77:221–226.

Nisbet, I. C. T. 1978. Dependence of fledging success on egg-size, parental performance and egg-composition among Common and Roseate Terns *Sterna hirundo* and *S. dougallii. Ibis* 120:207–215.

Nisbet, I. C. T. 1994. Effects of pollution on marine birds. In: *Seabirds on Islands: Threats, Case Studies and Action Plans*, ed. D. N. Nettleship, J. Burger, and M. Gochfeld, 8–25. Cambridge, UK: BirdLife International.

Nisbet, I. C. T. 2000. Disturbance, habituation, and management of waterbird colonies. *Waterbirds* 312–332.

Nisbet, I. C. T. 2002. Common Tern (*Sterna hirundo*). In: *The Birds of North America Online*, ed. A. Poole. Ithaca, NY, USA: Cornell Laboratory of Ornithology. http://bna.birds.cornell.edu.bnaproxy.birds.cornell .edu/bna/species/618 (accessed February 9, 2015).

Nisbet, I. C. T., and E. Paul. 2004. Ethical issues concerning animal research outside the laboratory. *Institute for Laboratory Animal Research (ILAR) Journal* 45:375–377.

Nisbet, I. C. T., and L. M. Reynolds. 1984. Organochlorine residues in Common Terns and associated estuarine organisms, Massachusetts, USA, 1971–81. *Marine Environmental Research* 11:33–66.

Nisbet, I. C. T., and M. J. Welton. 1984. Seasonal variations in breeding success of Common Terns: Consequences of predation. *Condor* 53–60.

Nisbet, I. C. T., D. M. Fry, J. J. Hatch, and B. Lynn. 1996. Feminization of male Common Tern embryos is not correlated with exposure to specific PCB congeners. *Bulletin of Environmental Contamination and Toxicology* 57:895–901.

Nisbet, I. C. T., M. Gochfeld, and J. Burger. 2014. Roseate Tern (*Sterna dougallii*). In: *The Birds of North America Online*, ed. A. Poole. Ithaca, NY, USA: Cornell Laboratory of Ornithology. http://bna.birds .cornell.edu.bnaproxy.birds.cornell.edu/bna/species/370 (accessed February 7, 2015).

Nisbet, I. C. T., J. P. Montoya, J. Burger, and J. J. Hatch. 2002. Use of stable isotopes to investigate individual differences in diets and mercury exposures among Common Terns *Sterna hirundo* in breeding and wintering grounds. *Marine Ecology Progress Series* 242:267–274.

Nisbet, I. C. T., C. S. Mostello, R. R. Veit, J. W. Fox, and V. Afanasyev. 2011. Migrations and winter quarters of five Common Terns tracked using geolocators. *Waterbirds* 34:32–39.

Nisbet, I. C. T., R. R. Veit, S. A. Auer, and T. P. White. 2013. *Marine Birds of the Eastern United States and the Bay of Fundy.* Ornithological Monograph No. 29. Nuttall Ornithological Club, Cambridge.

Nixon, S. W. 1982. The ecology of New England high marshes: A community profile. *Fish and Wildlife Service, Office of Biological Services*, Vol. RWS/OBS-81/55. Washington, DC: US. Department of Interior.

Nogales, M., A. Martini, B. R. Tershy, C. J. Donlan, D. Veitch, N. Puertal, B. Woods, and J. Alonso. 2004. A review of feral cat eradication on islands. *Conservation Biology* 18:310–319.

Nol, E., and R. C. Humphrey. 1994. American Oystercatcher (*Haematopus palliatus*). *The Birds of North America Online*, ed. A. Poole. Ithaca, NY: Cornell Labortatory of Ornithology. http://bna.birds.cornell .edu/bna/species/082 (accessed March 14, 2016).

Nordstrom, K. F. 2008. *Beach and Dune Restoration.* Cambridge, UK: Cambridge Univ. Press.

Nordstrom, K. F., and E. L. Lotstein. 1989. Perspectives on resource use of dynamic coastal dunes. *Geographical Review* 79:1–12.

Nordstrom, K. F., and W. A. Mitteager. 2001. Perceptions of the value of natural and restored beach and dune characteristics by high school students in New Jersey, USA. *Ocean and Coastal Management* 44:545–559.

Norris, K., P. W. Atkinson, and J. A. Gill. 2004. Climate change and coastal waterbird populations—Past declines and future impacts. *Ibis* 146:82–89.

North American Waterfowl Management Plan. 2012. North American Waterfowl Management Plan 2012—People Conserving Waterfowl and Wetlands. U.S. Fish and Wildlife Service and Canadian Wildlife Service. Washington, DC and Ottawa, Canada. http://static.nawmprevision.org/sites/default/files/NAWMP-Plan-EN -may23.pdf (accessed June 13, 2015).

Nriagu, J. O. 1979. *The Biogeochemistry of Mercury in the Environment*. New York: Elsevier.

Nye, J. A., D. D. Davis, and T. J. Miller. 2007. The effect of maternal exposure to contaminated sediment on the growth and condition of larval *Fundulus heteroclitus*. *Aquatic Toxicology* 82:242–250.

Nygård, T., E. Lie, N. Røv, and E. Steinnes. 2001. Metal dynamics in an Antarctic food chain. *Marine Pollution Bulletin* 42:598–602.

Nyman, J. A., and C. G. Green. 2014. A brief review of the effects of oil and dispersed oil on coastal wetlands including suggestions for future research. In: *Impacts of Oil Spill Disasters on Marine Habitats and Fisheries in North America*, ed. J. B. Alford, M. S. Peterson, and C. C. Green, 97–112. Boca Raton, FL: CRC Press.

NYNJ-HEP. 1996. *Comprehensive Conservation Management Plan*. New York–New Jersey Harbor Estuary Program. http://www.harborestuary.org/pdf/ccmpintro.pdf (accessed August 23, 2015).

Nystrom, R. R. 1984. Cytological changes occurring in the liver of *Coturnix* quail with an acute arsenic exposure. *Drug and Chemical Toxicology* 7:587–594.

Ocean County. 2010. Census data for Ocean County, NJ. http://www.planning.co.ocean.nj.us/census.htm (accessed March 5, 2015).

Ocean County. 2012. About our region: Barnegat Bay Watershed. Ocean County Soil Conservation District. http://www.soildistrict.org/about-us/about-our-region/ (accessed March 8, 2015).

O'Connell, T. J., and R. A. Beck. 2003. Gull predation limits nesting success of terns and skimmers on the Virginia barrier islands. *Journal of Field Ornithology* 74:66–73.

Odum, E. P. 1959. *Fundamentals of Ecology*, 2nd ed. Philadelphia: W.B. Saunders.

Ohlendorf, H. M. 2000. Ecotoxicology of selenium. In: *Handbook of ecotoxicology*, ed. D. J. Hoffmanm, B. A. Rattner, G. S. Burton, and J. Cairns, 465–500. Boca Raton: CRC Press.

Ohlendorf, H. M. 2011. Selenium, salty water, and deformed birds. In: *Wildlife Ecotoxicology: Emerging Topics in Ecotoxicology*, ed. J. E. Elliot, C. A. Bishop, and C. A. Morrissey, 325–357. New York: Springer.

Ohlendorf, H. M., and G. H. Heinz. 2011. Selenium in birds. In: *Environmental Contaminants in Wildlife: Interpreting Tissue Concentrations*, ed. W. N. Beyer and J. P. Meador, 669–702. Boca Raton, FL: CRC Press.

Ohlendorf, H. M., T. W. Custer, R. W. Lowe, M. Rigney, and E. Cromartie. 1988. Organochlorines and mercury in eggs of coastal terns and herons in California, USA. *Colonial Waterbirds* 11:85–94.

Ohlendorf, H. M., D. J. Hoffman, M. K. Salki, and T. W. Aldrich. 1986b. Embryonic mortality and abnormalities of aquatic birds: Apparent impacts of selenium from irrigation drain water. *Science of the Total Environment* 52:49–63.

Ohlendorf, H. M., R. L. Hothem, C. M. Bunck, T. W. Aldrich, and J. R. Moore. 1986a. Relationship between selenium concentrations and avian reproduction. *Transactions of the 51st North American Wildlife Research Conference* 51:330–342.

Ohlendorf, H. M., R. L. Hothem, C. M. Bunck, and K. C. Marois. 1990. Bioaccumulation of selenium in birds at Kesterson Reservoic, California. *Archives of environmental Contamination and Toxicology* 19:495–507.

Ohlendorf, H. M., R. L. Hothem, and D. Walsh. 1989. Nest success, cause-specific nest failures and hatchability of aquatic birds at selenium contaminated Kesterson Reservoir and a reference site. *Condor* 91:787–796.

Olafsdottir, K., E. Petersen, E. V. Magnusdottir, T. Bjornsson, and T. Johannesson. 2005. Temporal trends of organochlorine contamination in Black Guillemots in Iceland from 1976 to 1996. *Environmental Pollution* 133:509–515.

Olive, A. 2014. The road to recovery: Comparing Canada and US recovery strategies for shared endangered species. *The Canadian Geographer* 58:263–275.

Olivero-Verbel, J., D. Agudelo-Frias, and K. Caballero-Gallardo. 2013. Morphometirc parameters and total mercury in eggs of snowy egret (*Egretta thula*) from Cartagena Bay and Totumo Marsh, north of Colombia. *Marine Pollution Bulletin* 69:105–109.

Omae, I. 2006. Chemistry and fate of organotin antifouling biocides in the environment. *Handbook of Environmental Chemistry* 5:17–50.

Omernik, J. M. 1987. Ecoregions of the conterminous United States. *Annals of the Association of American geographers* 77:118–125.

Omernik, J. M. 2004. Perspectives on the nature and definition of ecological regions. *Environmental Management* 34:527–538.

Osachoff, H. L., M. Mohammadali, R. C. Skirrow, E. R. Hall, L. L Brown, G. C. van Aggelen, C. J. Kennedy, and C. C. Helbing. 2014. Evaluating the treatment of a synthetic wastewater containing a pharmaceutical and personal care product chemical cocktail: Compound removal efficiency and effects on juvenile rainbow trout. *Water Research* 62:271–280.

Osland, M. J., R. H. Day, A. S. From, M. L. McCoy, J. L. McLeod, and J. J. Kelleway. 2015. Life stage influences the resistance and resilience of black mangrove forests to winter climate extremes. *Ecosphere* 6:1–15.

Oswald, S. A., and J. M. Arnold. 2012. Direct impacts of climatic warming on heat stress in endothermic species: Seabirds as bioindicators of changing thermoregulatory constraints. *Integrative Zoology* 7:121–136.

Otorowski, C. I. 2006. Mercury in gulls of the Bay of Fundy. *Masters Abstracts International* 47:150 pp.

Outridge, P. M., and A. M. Scheuhammer.1993. Bioaccumulation and toxicology of chromium: Implications for wildlife. *Reviews of Environmental Contamination and Toxicology* 130:31–77.

Pacyna, E. G., J. M. Pacyna, F. Steenhuisen, and S. Wilson. 2006. Global anthropogenic mercury emission inventory for 2000. *Atmospheric Environment* 40:4048–4063.

Padula, V., J. Burger, S. H. Newman, S. Elbin, and C. Jeitner. 2010. Metals in feathers of Black-crowned Night-Heron (*Nycticorax nycticorax*) chicks from the New York Harbor Estuary. *Archives Environmental Contamination and Toxicology* 59:157–165.

Palestis, B. G. 2009. Use of artificial eelgrass mats by saltmarsh-nesting Common Terns. *Animal Zoology* 37:31A.

Palestis, B. G. 2014. The role of behavior in tern conservation. *Current Zoology* 60:500–551.

Palestis, B. G., and J. Burger. 1999. Individual sibling recognition in experimental broods of Common Tern chicks. *Animal Behaviour* 58:375–381.

Palestis, B. G., and J. Burger. 2001a. The effect of siblings on nest site homing by Common Tern chicks: A benefit of kin recognition. *Waterbirds* 24:175–181.

Palestis, B. G., and J. Burger. 2001b. Development of Common Tern (*Sterna hirundo*) sibling recognition in the field. *Bird Behavior* 14:75–80.

Palestis, B. G., and J. E. Hines. 2015. Adult survival and breeding dispersal of Common Terns (*Sterna hirundo*) in a declining population. *Waterbirds* 38:221–320.

Palestis, B. G., and M. D. Stanton. 2013. Responses of common tern chicks to feather sample removal. *The Wilson Journal of Ornithology* 125:646–650.

Panero, M., S. Boehme, and G. Muñoz. 2005. *Pollution Prevention and Management Strategies for Polychlorinated Biphenyls in the New York/New Jersey Harbor*. New York: New York Academy of Sciences for the New York/New Jersey Harbor Consortium.

Parmesan, C. 1996. Climate and species' range. *Nature* 382:756–766.

Parmesan, C. 2006. Ecological and evolutionary responses to recent climate change. *Annual Review of Ecology, Evolution, and Systematics.* 37:637–669.

Parsons, K. 1987. *The Harbor Herons Project. New York City.* New York: Audubon Society.

Parsons, K. C., 1994. The Arthur Kill oil spills: Biological effects in birds. In: *Before and After an Oil Spill: The Arthur Kill*, ed. J. Burger, 215–236. New Brunswick: Rutgers University Press.

Parsons, K. C., 1995. Heron nesting at pea Patch Island, upper Delaware Bay, USA: Abundance and reproductive success. *Colonial Waterbirds* 18:69–78.

Parsons, K. C., and J. Burger. 1982. Human disturbance and nestling behavior in black crowned night herons. *Condor* 84:184–187.

Parsons, K. C., and E. Jedrey. 2013. Boston Harbor Islands national park area: Coastal waterbird program. Massachusetts Audubon. Available directly online (accessed April 17, 2015).

Parsons, K. C., and T. L. Master. 2000. Snowy Egret (*Egretta thula*). In: *The Birds of North America Online*, ed. A. Poole. Ithaca, NY, USA: Cornell Laboratory of Ornithology. http://bna.birds.cornell.edu.bnaproxy .birds.cornell.edu/bna/species/489 (accessed February 9, 2015).

Parsons, K. C., S. C. Brown, R. M. Erwin, H. A. Czech, and J. C. Coulson. 2002. Managing wetlands for water-birds: Integrated approaches. *Waterbirds* 25(Special Publication 2):1–4.

Parsons, K. C., A. D. Maccarone, and J. Brzorad. 1991. First breeding record of Double-crested Cormorant in New Jersey. *Records of NJ Birds* 17:51–52.

Parsons, K. C., A. C. Matz, and S. R. Schmidt. 1998. Wading birds and cholinesterase-inhibiting insecti-cides: An examination of exposure and effects in free-living populations. Interim Report to Delaware Department of Natural Resources & Environmental Control.

Parsons, K. C., S. R. Schmidt, and A. C. Matz. 2001. Regional patterns of wading bird productivity in north-eastern US estuaries. *Waterbirds*: 24:323–330.

Pastorok, R. A., A. MacDonald, J. R. Sampson, P. Wilber, D. J. Yozzo, and J. P. Titre. 1979. An ecological deci-sion framework for environmental restoration projects. *Ecological Engineering* 9:89–107.

Paton, P. W., R. J. Harris, and C. L. Trocki. 2005. Distribution and abundance of breeding birds in Boston Harbor. *Northeastern Naturalist* 12:145–168.

Patton, S. R. 1988. Abundance of Gulls at Tampa Bay landfills. *Wilson Bulletin* 100:431–442.

Paul, J. F., K. J. Scott, D. E. Campbell, J. H. Gentile, C. S. Strobel, R. M. Valente, S. B. Weisberg, A. F. Holland, and J. A. Ranasinghe. 2001. Developing and applying a benthic index of estuarine condition for the Virginian Biogeographic Province Ecological Indicators. *United States Environmental Protection Agency* 1:83–99.

Pauley, D., V. Christensen, J. Dalsgaard, R. Froese, and F. Torres Jr. 1998. Fishing down marine food webs. *Science* 418:860–863.

Pauley, D., V. Christensen, S. Guenette, T. J. Pitcher, R. Sumaila, C. J. Walters, R. Watson, and D. Zeller. 2002. Towards sustainability in world fisheries. *Nature* 418:689–695.

Peakall, D. 1992. *Animal Biomarkers as Pollution Indicators*. London: Chapman and Hall.

Peakall, D. G. 1993. DDE-induced eggshell thinning: An environmental detective story. *Environmental Reviews* 1:13–20.

Peakall, D., and J. Burger. 2003. Methodologies for assessing exposure to metals: Speciation, bioavailability of metals, and ecological host factors. *Ecotoxicology and Environmental Safety* 56:110–121.

Peconic. 2015. Peconic Estuary Program. http://www.peconicestuary.org/about.php (accessed March 7, 2015).

Pesch, C. E., R. A. Voyer, J. S. Latimer, J. Copeland, G. Morrison, and D. McGovern. 2011. Imprint of the Past: Ecological History of New Bedford Harbor Environmental Protection Agency: Region 1. http://www2 .epa.gov/sites/production/files/documents/imprintofthepast.pdf (accessed March 5, 2015).

Peterson, M. J., R. A. Efroymson, and S. M. Adams. 2011. Long-term biological monitoring of an impaired stream: Synthesis and environmental management implications. *Environmental Management* 47:1125–1140.

Pffaf, L. G. 2012. Sea change: The post-*Sandy* rebuilding is about to begin. *New Jersey Monthly*, Jan. 2012 45–51.

Pflugh, K. K., A. H. Stern, L. Nesposudny, L. Lurig, B. Ruppel, and G. A. Buchanan. 2011. Consumption pat-terns and risk assessment of crab consumers from the Newark Bay complex, New Jersey, USA. *Science of the Total Environment* 409:4536–4544.

Phillips, S. W. 2007. Synthesis of U.S. Geological Survey Science for the Chesapeake Ecosystem and Implications for Environmental Management. *U.S. Geological Survey Circular* 1316:63.

Pidgeon, N. 2012. Climate change risk perception and communication: Addressing a critical moment? *Risk Analysis* 32:951–956.

Piehler, G. R. 2000. *Exit Here for Fish: Enjoying and Conserving New Jersey's Recreational Fisheries*, 229. New Brunswick, NJ: Rutgers University Press.

Pierce, R. L., and D. E. Gawlik. 2010. Wading bird foraging habitat selection in the Florida Everglades. *Waterbirds* 33:494–503.

Pierotti, R. J., and T. P. Good. 1994. Herring Gull (*Larus argentatus*). In: *The Birds of North America Online*, ed. A. Poole. Ithaca, NY, USA: Cornell Laboratory of Ornithology. http://bna.birds.cornell.edu.bnaproxy .birds.cornell.edu/bna/species/124 (accessed February 7, 2015).

Piersma, T., and Å Lindström. 2004. Migrating shorebirds as integrative sentinels of global environmental change. *Ibis* 146:61–69.

Pilastro, A., L. Congiu, L. Tallandini, and M. Turchetto. 1993. The use of bird feathers for the monitoring of cadmium, pollution. *Archives of Environmental Contamination and Toxicology* 24:355–358.

Pimm, S. L., G. J. Russell, J. L. Gittleman, and T. M. Brooks. 1995. The future of biodiversity, *Science* 269:347–350.

Pineau, T. 2000. Conservation of wintering and migratory habitats. In: *Heron Conservation*, ed. J. A. Kushlan, and H. Hafner, 201–218. New York: Academic.

Pius, S. M., and P. L. Leberg. 1998. The protector species hypothesis: Do Black Skimmers find refuge from predators in Gull-billed Tern colonies? *Ethology* 104:273–284.

Pol, M. V. D., Y. Vindenes, B. E. Sæther, S. Engen, B. J. Ens, K. Oosterbeek, and J. M. Tinbergen. 2010. Effects of climate change and variability on population dynamics in a long-lived shorebird. *Ecology* 91:1192–1204.

Poole, A. F., R. O. Bierregaard, and M. S. Martell. 2002. Osprey (*Pandion haliaetus*). In: *The Birds of North America Online*, ed. A. Poole. Ithaca, NY, USA: Cornell Laboratory of Ornithology. http://bna.birds .cornell.edu.bnaproxy.birds.cornell.edu/bna/species/683 (accessed February 7, 2015).

Post, G. B., P. D. Cohn, and K. R. Cooper. 2012. Perfluorooctanoic acid (PFOA), an emerging drinking water contaminant: A critical review of recent literature. *Environmental Research* 116:93–117.

Post, P. W., and M. Gochfeld. 1978. Recolonization by Common Terns at Breezy Point, New York. *Proceedings Colonial Waterbird Group* 2:128–136.

Post, W., and J. S. Greenlaw. 2009. Seaside Sparrow (*Ammodramus maritimus*). In: *The Birds of North America Online*, ed. A. Poole. Ithaca, NY, USA: Cornell Laboratory of Ornithology. http://bna.birds.cornell.edu .bnaproxy.birds.cornell.edu/bna/species/127 (accessed February 9, 2015).

Post, P. W., and G. S. Raynor. 1964. Range expansion of the American Oystercatcher into New York. *Wilson Bulletin* 76:339–346.

Post, P. W., and D. Reipe. 1980. Laughing gulls colonize Jamaica Bay. *Kingbird* 30:11–13.

Pottern, G. B., M. T. Huish, and J. H. Kerby. 1989. Species profiles: Life histories and environmental requirements of coastal fishes and invertebrates (mid-Atlantic): Bluefish. *U.S. Fish & Wildlife Service, Biological Report* 82:1–21.

Powell, G. N. 1987. Dynamics of habitat use by wading birds in a subtropical estuary: Implications of hydrography. *Auk* 104:740–749.

Pratt J. R., and J. Cairns Jr. 1996. Ecotoxicology and the redundancy problem: Understanding effects of community structure and function. In: *Ecotoxicology: A Hierarchical Treatment*, ed. M. C. Newman and C. H. Jagoe, 347–369. New York: Lewis Publishers.

Pries, A. J., D. L. Miller, and L. C. Branch. 2008. Identification of structural features that influence storm-related dune erosion along a barrier island ecosystem in the Gulf of Mexico. *Journal of Coastal Research* 24:168–176.

Prince, R., and K. R. Cooper. 2009. Comparisons of the effects of 2,3,7,8-tetrachlorodibenzo-*p*-dioxin on chemically impacted and nonimpacted subpopulations of *Fundulus heteroclitus*: I. TCDD toxicity. *Environmental Toxicology and Chemistry* 14:579–587.

Prosser, D. 2014. Chesapeake Bay Forage Base: Waterbird Predation and Population Trends. Chesapeake Bay Biological Laboratory, Solomons, Maryland. http://www.chesapeake.org/stac/presentations/234_20141112 _Prosser_Waterbirds.pdf (accessed August 8, 2015).

Psuty, N. P., and D. D. O' Fiara. 2002. *Coastal Hazard Management*. New Brunswick, NJ: Rutgers University Press.

Public Service Enterprise Group (PSEG). 2008. Estuary Enhancement Program. Public Service Electric and Gas, Newark, NJ. https://www.pseg.com/info/environment/pdf/program_overview.pdf (accessed September 3, 2015).

Public Service Enterprise Group (PSEG) Nuclear, LLC (PSEG). 2015a. Biological Monitoring Program 2014 Annual Report. Hancocks Bridge, NJ 549 pp.

Public Service Enterprise Group (PSEG). 2015b. Commercial Township Salt Hay Farm—2014 Site Status Report. Hancocks Bridge, NJ. June 22, 2015. 52 pp.

Public Service Enterprise Group (PSEG). 2015c. Alloway Creek Watershed Phragmites-Dominated Wetland Restoration Site—2014 Site Status Report. Hancocks Bridge, NJ. June 22, 2015. 47 pp.

Public Service Enterprise Group (PSEG). 2015d. Cedar Swamp and the Rocks Phragmites-Dominated Wetland Restoration Sites—2014 Site Status Report. Hancocks Bridge, NJ. June 22, 2015. 65 pp.

Pyke, G. H., H. R. Pulliam, and E. L. Charnov. 1977. Optimal foraging: A selective review of theory and tests. *Quarterly Review of Biology* 137–154.

Quinn, J. R. 1997. *Fields of Sun and Grass: An Artist's Journal of the New Jersey Meadowlands.* New Brunswick, NJ: Rutgers Univ. Press.

Rabinowitz, P., and L. Cont. 2013. Links among human health, animal health, and ecosystem health. *Annual Reviews Public Health* 34:189–204.

Raichel, D. L., K. W. Able, and J. M. Hartman. 2003. The influence of Phragmites (common reed) on the distribution, abundance, and potential prey of a resident marsh fish in the Hackensack Meadowlands, New Jersey. *Estuaries* 26:511–521.

Ralston, N. V. C. 2008. Selenium health benefit values as seafood safety criteria. *Eco-Health* 5:442–455.

Ralston, N. V. C. 2009. Introduction to 2nd issue on special topic: Selenium and mercury as interactive environmental indicators. *Environmental Bioindicators* 4:286–290.

Ralston, N. V. C., and L. J. Raymond. 2010. Dietary selenium's protective effects against methylmercury toxicity. *Toxicology* 278:112–123.

Ramel, C. 1969. Genetic effects of organic mercury compounds: I. Cytological investigations on Allium roots. *Hereditas* 61:208–230.

Ramsar. 1987. Chesapeake Bay Estuarine Complex. http://www.ramsar.org/chesapeake-bay-estuarine -complex (accessed August 23, 2015).

Ramsar. 2015. Delaware Bay Estuary. http://www.ramsar.org/delaware-bay-estuary (accessed August 23, 2015).

Ratcliffe, D. A. 1970. Changes attributable to pesticides in egg breakage, frequency, and eggshell Thickness in some British Birds. *Journal of Applied Ecology* 7:67–115.

Ratcliffe, N., S. Newton, P. Morrison, O. Merne, T. Cadwallender, and M. Frederiksen. 2008. Adult survival and breeding dispersal of Roseate Terns within the northwest European metapopulation. *Waterbirds* 31:320–329.

Rattner, B. 2000. Environmental contaminants and colonial waterbirds. USGS Patuxent Wildlife Research Center, Laurel, MD, USA. http://www.waterbirdconservation.org/plan/rpt-contaminants.pdf (accessed March 14, 2016).

Rattner, B. A., and P. C. McGowan. 2007. Potential hazards of environmental contaminants in avifauna residing in the Chesapeake Bay estuary. *Waterbirds* 30:63–81.

Rattner, B. A., D. J. Hoffman, M. J. Melancon, G. H. Olsen, S. R. Schmidt, and K. C. Parsons. 2000. Organochlorine and metal contaminant exposure and effects in hatching black-crowned night herons (*Nycticorax nycticorax*) in Delaware Bay. *Archives Environmental Contamination and Toxicology* 39:38–45.

Ratter, B. M., K. H. Philipp, and H. von Storch. 2012. Between hype and decline: Recent trends in public perception of climate change. *Environmental Science and Policy* 18:3–8.

Rattner, B. A., A. M. Scheuhammer, and J. E. Elliott. 2011. History of wildlife toxicology and the interpretation of contaminant concentrations in tissues. In: *Environmental Contaminants in Wildlife: Interpreting Tissue Concentrations*, ed. W. N. Beyer and J. P. Meador, 9–44. Boca Raton, FL: CRC Press.

Raye, S. S. C., and J. Burger. 1979. Behavioral determinants of nestling success of Snowy Egrets (*Leucophoyr thula*). *The American Midland Naturalist* 102:76–85.

Raymond, L. J., and N. V. C. Ralston. 2004. Mercury: Selenium interactions and health implications. *Seychelles Medical and Dental Journal* 17:72–77.

Recher, H. F., and J. A. Recher. 1980. Why are there different kinds of herons? *Transactions of the Linnaean Society New York* 9:135–158.

Rehfisch, M. M., and H. Q. P. Crick. 2003. Predicting the impact of climate change on Arctic breeding waders. *Wader Study Group Bulletin* 100:86–95.

Richardson, M. E., M. R. S. Fox, and B. E. Fry Jr. 1974. Pathological changes produced in Japanese Quail by ingestion of cadmium. *Journal of Nutrition* 104:323–338.

Ricklefs, R. E. 1977. On the evolution of reproductive strategies in birds: Reproductive effort. *American Naturalist* 111:453–478.

Robards, M. D., M. F. Willson, R. H. Armstrong, and J. F. Piatt. 1999. Sand lance: A review of biology and predator relations and annotated bibliography. *Exxon Valdez* Oil Spill Restoration Project. U.S. Department of Agriculture PNW-RP-521.

Robbins, C. S., B. Bruun, and H. S. Zim. 1966. *Birds of North America.* New York: Golden Press.

Robichaud, A., and Y. Bégin. 1997. The effects of storms and sea-level rise on a coastal forest margin in New Brunswick, eastern Canada. *Journal of Coastal Research* 429–439.

Robinson, G. R., and S. N. Handel. 1993. Forest restoration on a closed landfill: Rapid addition of new species by bird dispersal. *Conservation Biology* 7:271–278.

Robinson, R. A., N. A. Clark, R. Lanctot, S. Nebel, B. Harrington, J. A. Clark, J. A. Gill et al. 2005. Long term demographic monitoring of wader populations in non-breeding areas. *Wader Study Group Bulletin* 106:17–29.

Rodgers, J. A., and S. T. Schwikert. 2002. Buffer-zone distance to protect foraging and loafing waterbirds from disturbance by personal watercraft and outboard-powered boats. *Conservation Biology* 16:216–224.

Rodgers, J. A. Jr., and H. T. Smith. 2012. Little Blue Heron (*Egretta caerulea*). In: *The Birds of North America Online*, ed. A. Poole. Ithaca, NY, USA: Cornell Laboratory of Ornithology. http://bna.birds.cornell.edu .bnaproxy.birds.cornell.edu/bna/species/145 (accessed February 7, 2015).

Rodgers, J. A., P. S. Kubilis, and S. A. Nesbit. 2005. Accuracy of aerial surveys of waterbird colonies. *Waterbirds* 28:230–237.

Roemmich, D., and J. McGowan. 1995. Climatic warming and the decline of zooplankton in the California Current. *Science* 267:1324–1326.

Rogers, H. S., N. Jeffery, S. Kieszak, P. Fritz, H. Spliethoff, C. D. Palmer, P. J. Parsons et al. 2008. Mercury exposure in young children living in New York City. *Journal Urban Health* 85:39–51.

Root, T. L., J. T. Price, K. R. Hall, S. H. Schneider, C. Rosenzweig, and J. A. Pounds. 2003. Fingerprints of global warming on wild animals and plants. *Nature* 421:57–60.

Roth, T., M. Plattner, and V. Amrhein. 2014. Plants, birds and butterflies: Short-term responses of species communities to climate warming vary by taxon and with altitude. *PLOS One* 9:e82490.

Rothschild, B. J., J. S. Ault, P. Goulletquer, and M. Heral. 1994. Decline of the Chesapeake Bay oyster population: A century of habitat destruction and overfishing. *Marine Ecology Progress Series* 111: 29–39.

Rountree, R. A., and K. W. Able. 1992. Foraging habits, growth, and temporal patterns of salt-marsh creek habitat use by young-of-year summer flounder in New Jersey. *Transactions of the American Fisheries Society* 121:765–776.

Rowcliffe, J. M., A. R. Watkinson, W. J. Sutherland, and J. A. Vickery. 2001. The depletion of algal beds by geese: A predictive model and test. *Oecologia* 127:361–371.

Ruiz-Jaen, M. C., and T. M. Aide. 2005. Restoration success: How is it being measured? *Restoration Ecology* 13:596–597.

Rumbold, D. G. 2005. A probabilistic risk assessment of the effects of methylmercury on Great Egrets and Bald Eagles foraging at a constructed wetland in South Florida relative to the Everglades. *Human and Ecological Risk Assessment* 11:365–388.

Rumbold, D. G., S. L. Niemczyk, L. E. Fink, T. Chandrasekhar, B. Harkanson, and K. A. Laine. 2001. Mercury in eggs and feathers of Great Egrets (*Ardea albus*) from the Florida Everglades. *Archives of Environmental Contamination and Toxicology* 41:501–507.

Rush, S. A., K. F. Gaines, W. R. Eddleman, and C. J. Conway. 2012. Clapper Rail (*Rallus longirostris*). In: *The Birds of North America Online*, ed. A. Poole. Ithaca, NY, USA: Cornell Laboratory of Ornithology. http:// bna.birds.cornell.edu.bnaproxy.birds.cornell.edu/bna/species/340 (accessed February 9, 2015).

Russo, S., and A. Sterl. 2012. Global changes in seasonal means and extremes of precipitation from daily climate model data. *Journal of Geophysical Research* 117:1984–2012.

Ryckman, D. P., D. V. Chip Weseloh, and C. A. Bishop. 2005. Contaminants in Herring Gull eggs from the Great Lakes: 25 years of monitoring levels and effects. Great Lakes Fact Sheet 10. October 2005. Canadian Wildlife Service, Environment Canada, Ottawa, Ontario, Canada.

Safina, C. 1990. Bluefish mediation of foraging competition between Roseate and Common Terns. *Ecology* 71:1804–1809.

Safina, C. 1998. *Song for the Blue Ocean*. New York: Henry Holt.

Safina, C. 2002. *Eye of the Albatross: Visions of Hope and Survival*. New York: Henry Holt.

Safina, C. 2014a. The Elements: Sea. How climate change is sinking seabirds. *Audubon Magazine*, September. http://www.audubon.org/magazine/september-october-2014/how-climate-change-sinking-seabirds (accessed March 14, 2016).

Safina, C. 2014b. The elements: Crash diet. In warming oceans, seabirds' food chains are headed straight to the bottom. *Audubon Magazine* 116:20–23.

Safina, C., and J. Burger. 1985. Common Terns (*Sterna hirundo*) foraging: Seasonal trends in prey fish densities, and competition with Bluefish (*Pomatomus saltatrix*). *Ecology* 66:1457–1463.

Safina, C., and J. Burger. 1988a. Prey dynamics and the breeding phenology of Common Terns. *Auk* 105:720–726.

Safina, C., and J. Burger. 1988b. Use of sonar for studying foraging ecology of seabirds from a small boat. *Colonial Waterbirds* 11:234–244.

Safina, C., and J. Burger. 1988c. Ecological dynamics among prey fish, Bluefish and Common Terns in an Atlantic Coastal System. In: *Seabirds and Other Marine Vertebrates*, ed. J. Burger, 93–173. New York: Columbia University Press.

Safina, C., and J. Burger. 1989a. Inter-annual variation in prey available for Common Terns at different stages in their reproductive cycle. *Colonial Waterbirds* 12:37–42.

Safina, C., and J. Burger. 1989b. Population interactions among free-living Bluefish and prey fish in an ocean environment. *Oecologia* 79:91–95.

Safina, C., J. Burger, M. Gochfeld, and R. H. Wagner. 1988. Evidence for food limitation of Common and Roseate tern reproduction. *Condor* 90:852–859.

Salafsky, N., D. A. J. Salzer, C. Stattersfield, C. Hilton-Taylor, R. Neugarten, S. H. Butchart, B. Collen et al. 2008. A standard lexicon for biodiversity conservation: Unified classifications of threats and actions. *Conservation Biology* 22:897–911.

Sallenger, A. H., K. S. Doran, and P. A. Howd. 2012. Hotspot of accelerated sea-level rise on the Atlantic coast of North America. *Nature* 2:884–888.

Saha, S., and S. Ray. 2014. Sublethal effect of arsenic on oxidative stress and antioxidant status in *Scylla serrata*. *Soil, Air, Water* 42:1216–1222.

Saha, S. M., M. Ray, and S. Ray. 2010. Screening of phagocytosis and intrahemocytotoxicity in arsenic exposed crab as innate immune response. *Asian Journal of Experimental Biology Science* 1:47–54.

Sánchez-Virosta, P., S. Espín, A. J. García-Fernández, and T. Eeva. 2015. A review on exposure and effects of arsenic in passerine birds. *Science of the Total Environment* 512–513:506–525.

Sarewitz, D. 2004. How science makes environmental controversy worse. *Environmental Science and Policy* 7:385–403.

Sanpera, C., M. Morera, X. Ruiz, and L. Jover. 2000. Variability of mercury and selenium levels in clutches of Audouin's gulls (*Larus audouiniii*) breeding at the Chafarinas Islands, southwest Mediterranean. *Archives of Environmental Contamination and Toxicology* 39:119–123.

Sather, M. E., S. Mukerjee, K. L. Allen, L. Smith, J. Mathew, C. Jackson, R. Callison et al. 2014. Gaseous oxidized mercury dry deposition measurements in the Southwestern USA: A comparison between Texas, Eastern Oklahoma, and the Four Corners Area. *Scientific World Journal* (Article 580723:1–14).

Scarton, F. 2010. Long term decline of a Common Tern (*Sterna hirundo*) population nesting in salt marshes in Venice Lagoon, Italy. *Wetlands* 30:1153–1159.

Schaefer, H. C., W. Jetz, and K. Böhning-Gaese. 2007. Impact of climate change on migratory birds: Community reassembly versus adaptation. *Global Ecology and Biogeography* 17:38–49.

Scheifler, R., M. Gauthier-Clerc, C. Le Bohec, N. Crini, M. Coeurdassier, P. M. Badot, P. Giraudoux, and Y. Le Maho. 2005. Mercury concentrations in King Penguin (*Aptenodytes patagonicus*) feathers at Crozet Islands (sub-Antarctic): Temporal trend between 1966–1974 and 2000–2001. *Environmental Toxicology Chemistry*. 24:125–128.

Scheuhammer, A. M. 1987. Erythrocyte delta-aminolevulinic acid dehydratase in birds: II. The effects of lead exposure in vivo. *Toxicology* 45:165–175.

Scheuhammer, A. M. 1996. Influence of reduced dietary calcium on the accumulation and effects of lead, cadmium, and aluminum in birds. *Environmental Pollution* 94:337–343.

Schreiber, B. A. 2001. Climate and weather effects on seabirds. In: *Biology of Marine Birds*, ed. E. A. Schreiber and J. Burger, 179–215. Boca Raton, FL: CRC Press.

Schreiber, B. A., and J. Burger. 2001. *The Biology of Seabirds*. Boca Raton, FL: CRC Press.

Schubel, J. 1986. The life and death of the Chesapeake Bay. University of Maryland, College Park, Maryland USA, Maryland Sea Grant Office.

Schwarzbach, S. E., J. D. Albertson, and C. M. Thomas. 2006. Effects of predation, flooding, and contamination on reproductive success of California clapper rails (*Rallus longirostris obsoletus*) in San Francisco Bay. The *Auk* 123:45–60.

Sealy, S. G. 1973. Interspecific feeding assemblages of marine birds off British Columbia. *Auk* 90:796–802.

Seewagen, C. L. 2010. Threats of environmental mercury to birds: Knowledge gaps and priorities for future research. *Bird Conservation International* 20:112–123.

Seitz, S., and S. Miller. 1996. *The Other Islands of New York City*. Vermont: The Countrymen Press.

Sell, J. L. 1977. Comparative effects of selenium on metabolism on methylmercury by chickens and quail: Tissue distribution and transfer into eggs. *Poultry Science* 56:939–948.

Sell, J. L., and W. Magat. 1979. Distribution of mercury and selenium ion egg components and egg-white proteins. *Proceedings of Social and Experimental Biological Medicine* 161:458–463.

Senturk, U. K., and G. Oner. 1996. The effect of manganese-induced hypercholesterolemia on learning in rats. *Biological Trace Element Research* 51:249–257.

Sepúlveda, M. S., G. E. Jr. Williams, P. C. Frederick, and M. G. Spalding.1999a. Effects of mercury on health and first-year survival of free-ranging Great Egrets (*Ardea albus*) from southern Florida. *Archives of Environmental Contamination and Toxicology* 37:369–376.

Sepúlveda, M. S., P. C. Frederick, M. G. Spalding, and G. E. Williams. 1999b. Mercury contamination in free-ranging Great Egret nestlings (*Ardea albus*) from southern Florida, USA. *Environmental Contamination and Toxicology* 18:985–992.

Shahbaz, M., M. Z. Hashmi, R. N. Malik, and A. Yasmin. 2013. Relationship between heavy metals concentrations in egret species, their environment and food chain differences from two Headworks of Pakistan. *Chemosphere* 93:274–282.

Shaw-Allen, P. L., C. S. Romanek, A. L. Bryan Jr., H. Brandt, and C. H. Jagoe. 2005. Shifts in relative tissue N^{15} values in Snowy Egret nestlings with dietary mercury exposure: A marker for increased protein degradation. *Environmental Science and Technology* 39:4226–4233.

Shealer, D. A. 2001. Foraging behavior and foods of seabirds. In: *Biology of Marine Birds*, ed. E. A. Schreiber and J. Burger, 137–178. Boca Raton, FL: CRC Press.

Shealer, D. A., and J. A. Haverland. 2000. Effects of investigator disturbance on the reproductive behavior and success of Black Terns. *Waterbirds* 15–23.

Shealer, D. A., T. Floyd, and J. Burger. 1997. Host choice and success of gulls and terns kleptoparasitizing brown pelicans. *Animal Behavior* 53:655–665.

Shear, M. K., K. A. McLaughlin, A. Ghesquiere, M. J. Gruber, N. A. Sampson, and R. C. Kessler. 2011. Complicated grief associated with hurricane Katrina. *Depression and Anxiety* 28:648–657.

Shields, M. 2002. Brown Pelican (*Pelecanus occidentalis*). In: *The Birds of North America Online* ed. A. Poole. Ithaca, NY, USA: Cornell Laboratory of Ornithology. http://bna.birds.cornell.edu/bna/species/609 (accessed February 1, 2012).

Shields, M. 2014. Brown Pelican (*Pelecanus occidentalis*). In: *The Birds of North America Online*, ed. A. Poole. Ithaca, NY: Cornell Laboratory of Ornithology. http://bna.birds.cornell.edu/bna/species/609 (accessed March 14, 2016).

Shore, R. F., M. G. Pereira, L. A. Walker, and D. R. Thompson. 2011. Mercury in nonmarine birds and mammals. In: *Environmental Contaminants in Wildlife: Interpreting Tissue Concentrations,* ed. W. N. Beyer and J. P. Meador, 609–626. Boca Raton, FL: CRC Press.

Shrestha, R. K., T. V. Stein, and J. Clark. 2007. Valuing nature-based recreation in public natural areas of the Apalachicola River region, Florida. *Journal of Environmental Management* 85:977–985.

Shuster, C. N., Jr. 1982. A pictorial review of the natural history and ecology of the Horseshoe Crab (*Limulus polyphemus*), with references to other *Limulidae*. In: *Physiology and Biology of Horseshoe Crabs*, ed. J. Bonaventura, C. Bonaventura, and S. Tesh, 1–52. New York: Alan R. Liss.

Shuster, C. N., Jr., and M. L. Botton. 1985. A contribution to the population biology of Horseshoe Crabs (*Limulus polyphemus*) in Delaware Bay. *Estuaries* 8:363–372.

Siegel, S. 1956. *Nonparametric Statistics for the Behavioral Sciences*. New York: McGraw-Hill.

Sih, A. 2013. Understanding variation in behavioural responses to human-induced rapid environmental change: A conceptual overview. *Animal Behaviour* 85:1077–1088.

Sih, A., D. I. Bolnick, B. Luttbeg, J. L. Orrock, S. D. Peacor, L. M. Pintor, and J. R. Vonesh. 2010. Predator–prey naïveté, antipredator behavior, and the ecology of predator invasions. *Oikos* 119:610–621.

Sileo, L., and S. I. Fefer. 1987. Paint chip poisoning of Laysan albatross at Midway Atoll. *Journal of Wildlife Diseases* 23:432–437.

Sims, S. A., J. R. Seavey, and C. G. Curtin. 2013. Room to move? Threatened shorebird habitat in the path of sea level rise—Dynamic beaches, multiple users, and mixed ownership: A case study from Rhode Island, USA. *Journal of Coastal Conservation* 17:339–350.

Simmons, K. E. L. 1972. Some adaptive features of seabird plumage types. *Birds* 65:465–521.

Skinner, L. C., M. W. Kane, K. Gottschall, and D. A. Simpson. 2009. Chemical Residue Concentrations in Four Species of Fish and the American Lobster from Long Island Sound, Connecticut and New York: 2006 and 2007. New York State Department of Environmental Conservation and Connecticut Department of Environmental Protection. http://www.dec.ny.gov/docs/fish_marine_pdf/lis2009rep.pdf (accessed July 27, 2015).

Small, M. J., U. Guvene, and M. L. DeKay. 2014. When can scientific studies promote consensus among conflicting stakeholders? *Risk Analysis* 14:1978–1994.

Smith, C. L. (ed.) 1988. *Fisheries Research in the Hudson River*. New York: State University of New York Press.

Smith, G., and J. S. Weis. 1997. Predator/prey interactions in *Fundulus heteroclitus*: Effects of living in a polluted environment. *Journal of Experimental Marine Biology and Ecology* 209:75–87.

Sokal, R. R., and F. J. Rohlf. 1995. *Biometry: The Principles and Practices of Statistics*. New York: MacMillan.

Somers, C. M., M. N. Lozer, and J. S. Quinn. 2007. Interactions between Double-crested Cormorants and Herring Gulls at a shared breeding site. *Waterbirds* 30:241–250.

Soots, R. F. Jr., and M. C. Landin. 1978. Development and management of avian habitat on dredged material islands. U.S. Army Corps of Engineers Stn. Tech. Rep. DS-78-18.

South Florida Water Management District (SFWMD). 2014. *South Florida Environmental Report*. Ft. Lauderdale, FL: SFWMD.

Spahn, S. A., and T. W. Sherry. 1999. Cadmium and lead in exposure associated with reduced growth rates, poor fledging success of Little Blue Heron chicks (*Egretta caerulea*) in South Louisiana wetlands. *Archives of Environmental Contamination and toxicology* 37:377–384.

Spalding, M. G., R. D. Bjork, G. V. N. Powell, and S. F. Sundlof. 1994. Mercury and cause of death in Great White Herons. *Journal of Wildlife Management* 58:735–739.

Spalding, M. G., P. C. Frederick, H. C. McGill, S. N. Bouton, and L. R. McDowell. 2000a. Methylmercury accumulation in tissues and its effects on growth and appetite in captive Great Egrets. *Journal of Wildlife Diseases* 36:411–422.

Spalding, M. G., P. C. Frederick, H. C. McGill, S. N. Bouton, L. J. Richey, I. M. Schumacher, C. G. M. Blackmore, and J. Harrison. 2000b. Histologic, neurologic, and immunologic effects of methylmercury on appetite and hunting behavior in juvenile Great Egrets (*Ardea albus*). *Environmental Toxicology and Chemistry* 18:1934–1939.

Spallholz, J. E., and D. J. Hoffman. 2002. Selenium toxicity: Cause and effects in aquatic birds. *Aquatic Toxicology* 57:27–37.

Spann, J. W., J. F. Kreitzer, R. G. Heath, and L. N. Locke. 1972. Ethyl mercury para-toluene sulfonanilide— Lethal and reproductive effects on pheasants. *Science* 175:328–331.

Sparks, T. H., and C. F. Mason. 2004. Can we detect change in the phenology of winter migrant birds in the UK? *Ibis* 146:57–60.

Spellerberg, I. F., and P. J. Fedor. 2003. A tribute to Claude Shannon (1916–2001) and a plea for more rigorous use of species richness, species diversity and the 'Shannon–Wiener' Index. *Global Ecology and Biogeography* 12:177–179.

Spendelow, J. A., J. E. Hines, J. D. Nichols, I. C. T Nisbet, G. Cormons, H. Hays, J. J. Hatch, and C. S. Mostello. 2008. Temporal variation in adult survival rates of Roseate Terns during periods of increasing and declining populations. *Waterbirds* 31:309–319.

Spendelow, J. A., C. S. Mostello, I. C. Nisbet, C. S. Hall, and L. Welch. 2010. Interregional breeding dispersal of adult roseate terns. *Waterbirds* 33:242–245.

Spendelow, J. A., J. D. Nichols, I. C. T. Nisbet, H. Hays, G. D. Cormons, J. Burger, C. Safina, J. E. Hines, and M. Gochfeld. 1995. Estimating annual survival and movement rates of adults within a metapopulation of Roseate Terns. *Ecology* 76:2415–2428.

Spry, D. J., and J. G. Wiener. 1991. Metal bioavailability and toxicity to fish in low-alkalinity lakes: A critical review. *Environmental Pollution* 71:243–304.

Statistical Analysis System (SAS). 2005. *SAS Users' Guide*. Cary, NC: SAS Institute.

Steers, J. A. 1966. Coastal changes. In: *Future Environments of North America*, ed. F. F. Darling and J. P. Milton, 538–551. Garden City, NY: Natural History Press.

Steinberg, N., D. J. Suszkowski, L. Clark, and J. Way. 2004. Health of the Harbor: The first comprehensive look at the state of the NY/NJ Harbor Estuary. Report to the NY/NJ Harbor Estuary Program. Hudson River Foundation, New York.

Steinnes, E. 1987. Impact of long-range atmospheric transport of heavy metals to the terrestrial environment in Norway. In: *Lead, Mercury, Cadmium, and Arsenic in the Environment*, ed. T. C. Hutchinson and K. M. Meema, 107–117. New York: John Wiley & Sons.

Stern, A. H., and A. E. Smith. 2003. An assessment of the cord blood: Maternal blood methylmercury ratio: Implications for risk assessment. *Environmental Health Perspectives* 111:1465–1470.

Stern, A. H., M. Gochfeld, and P. J. Lioy. 2013. Two decades of exposure assessment studies on chromate production waste in Jersey City, New Jersey—What we have learned about exposure characterization and its value to public health and remediation. *Journal of Exposure Science and Environmental Epidemiology* 23:2–12.

Stewart, F. M., D. R. Thompson, R. W. Furness, and N. Harrison.1994. Seasonal variation in heavy metal levels in tissues of Common Buillemots, *Uria aalge* from northwest Scotland. *Archives of Environmental Contamination and Toxicology* 27:168–175.

Stillman, R. A. 2008. Predicted effect of shellfishing on the oystercatcher and knot population of the Solway Firth—Final report. In: *Solway Shellfish Management Association*. Poole, UK: Bournemouth University, p. 23.

Stillman, R. A., and J. D. Goss-Custard. 2002. Seasonal changes in the response of oystercatchers Haematopus ostralegus to human disturbance. *Journal of Avian Biology* 33:358–365.

Stillman, R. A., and J. D. Goss-Custard. 2010. Individual-based ecology of coastal birds. *Biological Reviews* 85:413–434.

Stillman, R. A., A. D. West, R. W. Caldow, and S. E. L. V. Durell. 2007. Predicting the effect of disturbance on coastal birds. *Ibis* 149:73–81.

Stirling, I., N. J. Lunn, and J. Iacozza. 1999. Long-term trends in the population ecology of Polar Bears in Western Hudson Bay in relation to climatic change. *Arctic* 52:294–306.

Stone, W. B., S. R. Overmann, and J. C. Okoniewski. 1984. Intentional poisoning of birds with parathion. *Condor* 86:333–336.

Strom, S. 2013. F.D.A. bans three arsenic drugs used in poultry and pig feeds. *New York Times.* http://www.nytimes.com/2013/10/02/business/fda-bans-three-arsenic-drugs-used-in-poultry-and-pig-feeds.html?_r=1 (accessed June 15, 2015).

Strong, C. M., L. B. Spear, T. P. Ryan, and R. E. Dakin. 2004. Forster's tern, Caspian tern, and California gull colonies in San Francisco Bay: Habitat use, numbers and trends, 1982–2003. *Waterbirds* 27:411–423.

Struger, J., J. E. Elliot, and D. V. Weseloh. 1987. Metals and essential elements in Herring Gulls from the Great Lakes, 1983. *Journal Great Lakes Research* 13:43–45.

Sun, H. J., H. B. Li, P. Xiang, X. Zhang, and L. Q. Ma. 2015b. Short-term exposure of arsenite disrupted thyroid endocrine system and altered gene transcription in the HPT axis in zebrafish. *Environmental Pollution* 205:145–152.

Sun J., Q. Luo, D. Wang, and Z. Wang. 2015a. Occurrences of pharmaceuticals in drinking water sources of major river watersheds, China. *Ecotoxicology and Environmental Safety* 117:132–140.

Surgeon General. 1964. *Report on Smoking and Health*. Washington DC: U.S. Government Printing Office.

Sutherland, W. J. 2004. Climate change and coastal birds: Research questions and policy responses. *Ibis* 146:120–124.

Sutton, C. C., J. C. O'Heron II, and R. T. Zappalorti. 1996. The Scientific Characterization of the Delaware Estuary. Delaware Estuary Program. https://s3.amazonaws.com/delawareestuary/pdf/ScienceReports byPDEandDELEP/PDE-DELEP-Report-96-02-SciChar.pdf (accessed August 19, 2015).

Swan, B. 2005. Migrations of adult Horseshoe Crabs, *Limulus polyphemus*, in the mid Atlantic bight: A 17-year tagging study. *Estuaries* 28:28–40.

Sydeman, W., S. Thompson, and A. Kitaysky. 2012 Seabirds and climate change: Roadmap for the future. *Marine Ecology Progress Series* 454:1–203.

Syers, J. K., and M. Gochfeld. 2001. Environmental cadmium in the food chain: Sources, pathways, and risks. *Proceedings of the SCOPE Workshop. Scientific Committee on Problems of the Environment*, Belgian Academy of Sciences, 2000.

Szostek, K. L., and P. H. Becker. 2012. Terns in trouble: Demographic consequences of low breeding success and recruitment on a common tern population in the German Wadden Sea. *Journal of Ornithology* 153:313–326.

Szumilo, E., M. Szubska, W. Meissner. M. Bełdowska, and L. Falkowska. 2013. Mercury in immature and adults Herring Gulls (*Larus argentatus*) wintering on the Gulf of Gdansk area. *Oceanological and Hydrobiological Studies* 42:260–267.

Takekawa, J. Y., J. T. Ackerman, L. A. Brand, T. R. Graham, C. A. Eagles-Smith, M. P. Herzog, and N. D. Athearn. 2015. Unintended consequences of management actions in salt pond restoration: Cascading effects in trophic interactions. *PloS One* 10: e0119345.

Talbot, C. W., K. W. Able, and J. K. Shisler. 1986. Fish species composition in New Jersey salt marshes: Effects of marsh alterations for mosquito control. *Transactions of the American Fisheries Society* 115:269–278.

Tardiff, R. G. 1992. *Methods to Assess Effects of Pesticides on Non-target Organisms.* United Kingdom: Wiley, on behalf of SCOPE/IPCS.

Tarr, N. M., T. R. Simons, and K. H. Pollock. 2010. An experimental assessment of vehicle disturbance effects on migratory shorebirds. *Journal of Wildlife Management* 74:1776–1783.

Taylor, D. L., R. S. Nichols, and K. W. Able, K. W. 2007. Habitat selection and quality for multiple cohorts of young-of-the-year bluefish (*Pomatomus saltatrix*): Comparisons between estuarine and ocean beaches in southern New Jersey. *Estuarine, Coastal and Shelf Science* 73:667–679.

Terborgh, J. 2009. Preservation of natural diversity. *BioScience* 24:715–722.

Teal, J. M., and S. Peterson. 2005. The interaction between science and policy in the control of Phragmites in oligohaline marshes of Delaware Bay. *Restoration Ecology* 13:223–227.

Teal, J., and M. Teal. 1969. *Life and Death of the Salt Marsh.* New York: Ballantine Books.

Teal, J. M., and L. Weishar. 2005. Ecological engineering: Adaptive management, and restoration management in Delaware Bay salt marsh restoration. *Ecological Engineering* 25:304–314.

Tejning, S. 1967. Biological effects of methyl mercury dicyandiamide-treated grain in the domestic fowl *Gallus gallus* L. *Oikos* Suppl 8:1–116.

Thom, R. M., G. W. Williams, and H. L. Diefenderfer. 2005. Balancing the need to develop coastal areas with the desire for an ecologically functioning coastal environment: Is net ecosystem improvement possible? *Restoration Ecology* 13:193–203.

Thomas, K., R. G. Kvitek, and C. Bretz. 2003. Effects of human activity on the foraging behavior of Sanderlings *Calidris alba. Biological Conservation* 109:67–71.

Thompson, D. R. 1996. Mercury in birds and terrestrial mammals. In: *Environmental Contaminants in Wildlife: Interpreting Tissue Concentrations*, ed. W. N. Beyer, G. H. Heinz, and A. W. Redmon-Norwood, 341–356. Boca Raton, FL: SETAC, Lewis Publishers.

Thompson, D. R., and R. W. Furness. 1989a. The chemical form of mercury stored in South Atlantic seabirds. *Environmental Pollution* 60:305–317.

Thompson, D. R., and R. W. Furness. 1989b. Comparison of the levels of total and organic mercury in seabird feathers. *Marine Pollution Bulletin* 20:577–579.

Thompson, D. R., and R. W. Furness. 1995. Stable isotope ratios of carbon and nitrogen in feathers indicate seasonal dietary shifts in northern fulmars. *Auk* 112:493–498.

Thompson, P. M., and J. C. Ollason. 2001. Lagged effects of ocean climate change on fulmar population dynamics. *Nature* 413:417–420.

Thompson, D. R., S. Bearhop, J. R. Speakman, and R. W. Furness. 1998a. Feathers as a means of monitoring mercury in seabirds: Insights from stable isotope analysis. *Environmental Pollution* 101:193–200.

Thompson, D. R., P. H. Becker, and R. W. Furness. 1993. Long-term changes in mercury concentrations in Herring Gulls *Larus argentatus* and Common Terns *Sterna hirundo* from the German North Sea coast. *Journal of Applied Ecology* 30:316–320.

Thompson, D. R., R. W. Furness, and L. R. Monteiro. 1998b. Seabirds as biomonitors of mercury inputs to epipelagic and mesopelagic marine food chains. *Science of the Total Environment* 213:299–305.

Thompson, D. R., K. C. Hamer, and R. W. Furness. 1991. Comparison of the levels of total and organic mercury in seabird feathers. *Marine Pollution Bulletin* 20:577–579.

Thompson, B. C., J. A. Jackson, J. Burger, L. A. Hill, E. M. Kirsch, and J. L. Atwood. 1997. Least Tern (*Sterna antillarum*). In: *The Birds of North America*, ed. A. Poole. Ithaca, NY, USA: Cornell Laboratory of Ornithology. http://bna.birds.cornell.edu.bnaproxy.birds.cornell.edu/bna/species/290 (accessed February 6, 2015).

Tims, J., I. C. T. Nisbet. M. S. Friar, C. Mostello, and J. J. Hatch. 2004. Characteristics and performance of Common Terns in old and newly-established colonies. *Waterbirds* 27:321–332.

Tinbergen, N. 1953. *The Herring Gull's World.* London: Collins New Naturalist Series #9.

Titus, J. G. 1990. Greenhouse effect, sea level rise, and barrier islands: Case study of Long Beach Island, New Jersey. *Coastal Management* 18:65–90.

Travers, S. E., B. Marquardt, N. J. Zerr, J. B. Finch, M. J. Boche, R. Wilk, and S. C. Burdick. 2015. Climate change and shifting arrival date of migratory birds over a century in the northern Great Plains. *The Wilson Journal of Ornithology* 127:43–51.

Traynham, B., J. Clarke, J. Burger, and J. Waugh. 2012. Engineered contaminant systems: Identification of dominant ecological processes for long term performance assessment and monitoring. *Remediation* 22:93–103.

Triplet, P., R. A. Stillman, and J. D. Goss-Custard. 1999. Prey abundance and the strength of interference in a foraging shorebird. *Journal of Animal Ecology* 68:254–265.

Trocki, C. L. 2014. Boston Harbor Islands national recreation area coastal breeding bird monitoring. 2013 field season summary. Natural Resource Technical Report NPS/NETN/NRTR-2014/863.

Tsao, D. C., A. K. Miles, J. Y. Takekawa, and I. Woo. 2009. Potential effects of mercury on threatened California Black Rails. *Archive of Environmental Contamination and Toxicology* 56:292–301.

Tsipoura, N., and J. Burger. 1999. Shorebird diet during spring migration stop-over on Delaware Bay. *Condor* 101:635–644.

Tsipoura, N., J. Burger, R. Feltes, J. Yacabucci, D. Mizrahi, C. Jeitner, and M. Gochfeld. 2008. Metal concentrations in three species of passerine birds breeding in the Hackensack Meadowlands of New Jersey. *Environmental Research* 107:218–228.

Tsipoura, N., J. Burger, D. Mizrahi, L. Niles, A. Dey, C. Jeitner, M. Peck, and T. Pittfield. 2015. Metals in blood of shorebirds. *Environmental Research* (manuscript).

Uchupi, E., N. Driscoll, R. D. Ballard, and S. T. Bolmer. 2001. Drainage of late Wisconsin glacial lakes and the morphology and late quaternary stratigraphy of the New Jersey–southern New England continental shelf and slope. *Marine Geology* 172:117145.

United Nations Environmental Programme (UNEP). 2006. Protection of the oceans, all kinds of seas, including enclosed and semi-enclosed seas, and coastal areas, and the protection, rational use, and development of their living resources. Agenda 21, Chapter 17. New York: United National Division for Sustainable Development.

United States Army Corps of Engineers (USACE). 2015. Poplar Island Overview. Baltimore District. http://www.nab.usace.army.mil/Missions/Environmental/PoplarIsland.aspx (accessed June 29, 2015).

United States Climate Change Science Program (USCCSP). 2009. Coastal sensitivity to sea-level rise: A focus on the mid-Atlantic region. U.S. Environmental Protection Program, Washington DC (USA) http://www.epa.gov/climatechange/effects/coastal/sap4-1.html (accessed June 11, 2015).

United States Fish and Wildlife Service (USFWS). 1996. Significant Habitats and Habitat Complexes of the New York Bight Watershed. Technical Report, U.S. Fish and Wildlife Service, Southern New England–New York Bight Coastal Ecosystems Program. Charlestown, Rhode Island. http://nctc.fws.gov/resources/knowledge-resources/pubs5/begin.htm (accessed March 1, 2015).

United States Fish and Wildlife Service (USFWS). 2010. Endangered and Threatened Wildlife and Plants; 5-Year Status Review of Roseate Tern [FWS-R4-ES-2010-N037; 40120-1113-0000-C4] Federal Register April 5, 2010:75(64):17153–17154. http://www.gpo.gov/fdsys/pkg/FR-2010-04-05/pdf/2010-7709.pdf#page=1 (accessed August 17, 2015).

United States Fish and Wildlife Services (USFWS). 2011a. 2011 National Survey of Fishing, Hunting, and Wildlife-Associated Recreation. https://www.census.gov/prod/2012pubs/fhw11-nat.pdf (accessed March 1, 2015).

United States Fish and Wildlife Services (USFWS). 2011b. Connecticut: Shoring up a Shrinking Island for Endangered Roseate Terns: May 30, 2011. http://www.fws.gov/news/blog/index.cfm/2011/5/30/Connecticut-Shoring-up-a-Shrinking-Island-for-Endangered-Roseate-Terns (accessed July 16, 2015).

United States Fish and Wildlife Services (USFWS). 2013. List of Migratory Bird Species Protected by the Migratory Bird Treaty Act as of December 2, 2013. U.S. Fish and Wildlife Service. http://www.fws.gov/migratorybirds/RegulationsPolicies/mbta/MBTANDX.HTML (accessed February 6, 2015).

United States Food and Drug Administration (USFDA). 2001. *FDA Consumer Advisory*. Washington, DC: USFDA. http://www.fda.gov/Food/RecallsOutbreaksEmergencies/SafetyAlertsAdvisories/default.htm (accessed March 14, 2016).

United States Food and Drug Administration (USFDA). 2005. *Mercury Levels in Commercial Fish and Shellfish*. Washington, DC: USFDA. http://www.fda.gov/food/foodborneillnesscontaminants/metals/ucm115644.htm (accessed March 14, 2016).

United States Food and Drug Administration (USFDA). 2011. 1991–2004 Total Diet Study—Analytical Results. http://www.fda.gov/food/foodscienceresearch/totaldietstudy/ucm184293.htm (accessed March 14, 2016).

United States Geological Survey (USGS). 2010. Cadmium Risks to Freshwater Life: Derivation and Validation of Low-Effect Criteria Values Using Laboratory and Field Studies. http://pubs.usgs.gov/sir/2006/5245 /pdf/sir20065245.pdf (accessed May 13, 2015).

United States Geological Survey (USGS). 2012. Technical Report on Toxic Contaminants in the Chesapeake Bay and its Watershed: Extent and Severity of Occurrence and Potential Biological Effect. http://executive order.chesapeakebay.net/ChesBayToxics_finaldraft_11513b.pdf (accessed March 8, 2015).

United States Geological Survey (USGS). 2013. Hurricane *Sandy*: Updated Assessment of Potential Coastal-Change Impacts. http://coastal.er.usgs.gov/hurricanes/*Sandy*/coastal-change/initialassessment.php (accessed February 9, 2015).

United States Geological Survey (USGS). 2014. Concerns Rise over Known and Potential Impacts of Lead on Wildlife. http://www.nwhc.usgs.gov/disease_information/lead_poisoning/ (accessed August 8, 2015).

United States Maritime Administration (USMA). 2015. 2002–2012 Total vessel calls in U.S. ports, terminals, and lightering areas report. http://www.marad.dot.gov/resources/data-statistics/ (accessed July 21, 2016).

Vahl, W. K., and S. A. Kingma. 2007. Food divisibility and interference competition among captive ruddy turnstones, *Arenaria interpres*. *Animal Behaviour* 74:1391–1401.

Vahl, W. K., T. Lok, J. Van der Meer, T. Piersma, and F. J. Weissing. 2005a. Spatial clumping of food and social dominance affect interference competition among ruddy turnstones. *Behavioral Ecology* 16:834–844.

Vahl, W. K., J. Van der Meer, F. J. Weissing, D. Van Dullemen, and T. Piersma. 2005b. The mechanisms of interference competition: Two experiments on foraging waders. *Behavioral Ecology* 16:845–855.

Valente, J. J., and R. A. Fischer. 2011. Reducing human disturbance to waterbird communities near Corps of Engineers projects. ERDC TN-DOER-E29 http://el.erdc.usace.army.mil/elpubs/pdf/doere29.pdf (accessed August 18, 2015).

Valiela, I., E. Kinney, J. Culbertson, E. Peacock, and S. Smith. 2009. Global losses of mangroves and salt marshes. In: *Global loss of coastal habitats: Rates causes and consequences*, ed. C. Duarte, 175. Madrid: Fundacion BBVA.

Valle, S., M. A. Panero, and L. Shore. 2007. *Pollution Prevention and Management Strategies for Polycyclic Aromatic Hydrocarbons in the New York/New Jersey Harbor*. New York: New York Academy of Sciences for the New York/New Jersey Harbor Consortium.

Vallius, H. 2013. Heavy metal concentrations in sediment cores from the northern Baltic Sea: Declines over the last two decades. *Marine Pollution Bulletin* 79:359–364.

Vallverdú-Coll, N., A. López-Antia, M. Martinez-Haro, M. E. Ortiz-Santaliestra, and R. Mateo. 2015. Altered immune response in mallard ducklings exposed to lead through maternal transfer in the wild. *Environmental Pollution* 205:350–356.

van de Pol, M., B. J. Ens, D. Heg, L. Brouwer, J. Krol, M. Maier, K.-M. Exo et al. 2010. Do changes in the frequency, magnitude and timing of extreme climatic events threaten the population viability of coastal birds? *Journal of Applied Ecology* 47:720–730.

Van der Schalie, W. H., H. S. Gardner Jr., J. A. Bantle, C. T. De Rosa, R. A. Finch, J. S. Reif, R. H. Reuter et al. 1999. Animals as sentinels of human health hazards of environmental chemicals. *Environmental Health Perspectives* 107:309.

Vanasse Hangen Brustlin, Inc. (VHB). 2009. Merrimack River Watershed Wetland Restoration Society. Technical report for New Hampshire Department of Environmental Services, March. http://www.restore nhwetlands.com/pdf/finalreport/WatershedReport_final.pdf (accessed August 5, 2015).

Vane, C. H., I. Harrison, A. W. Kim, V. Moss-Hayes, B. P. Vickers, and B. P. Horton. 2008. Status of organic pollutants in surface sediments of Barnegat Bay–Little Egg Harbor, New Jersey, USA. *Marine Pollution Bulletin* 56:1802–1814.

Vermeer, K. 1973. Comparison of food habits and mercury residues of Caspian and Common Terns. *Canadian Field-Naturalist* 87:305–308.

Visser, G. H. 2001. Chick growth and development in seabirds. In: *Biology of Marine Birds*, ed. E. A. Schreiber and J. Burger, 439–465. Boca Raton, FL: CRC Press.

Visser, M. E., and C. Both. 2005. Shifts in phenology due to global climate change: The need for a yardstick. *Proceedings of the Royal Society of London B: Biological Sciences* 272(1581):2561–2569.

Vitousek, P. M., H. A. Mooney, J. Lubchenco, and J. M. Melillo. 1997. Human domination of earth's ecosystems. *Science* 277:494–499.

Viverette, C. B., G. C. Garman, S. P. McIninch, A. C. Markham, B. D. Watts, and S. A. Macko. 2007. Finfish—Waterbird trophic interactions in tidal freshwater tributaries of the Chesapeake Bay. *Waterbirds* 30:50–62.

Vo, A. E., M. S. Bank, J. P. Shine, and S. V. Edwards. 2011. Temporal increase in organic mercury in an endangered pelagic seabird assessed by century-old museum specimens. *Proceedings of the National Academy of Sciences* 108:7466–7471.

Vogt, W. 1938. Preliminary notes on the behavior and ecology of the Eastern Willet. *Proceedings of the Linnaean Society New York* 49:8–42.

Walker, C. H., S. P. Hopkin, R. M. Sibly, and D. B. Peakall.1996. *Principles of Ecotoxicology*. London: CRC Press/Taylor & Francis.

Walsh, P. M. 1990. The use of seabirds as monitors of heavy metals in the marine environment. In: *Heavy Metals in the Marine Environment*, ed. R. W. Furness and P. S. Rainbow, 183–204. Boca Raton: CRC Press.

Ward, L., and J. Burger. 1980. Survival of Herring Gull and domestic chicken embryos after simulated flooding. *Condor* 82:142–148.

Warnock, N., C. Elphick, and M. A. Rubega. 2001. Shorebirds in the marine environment. In: *Biology of Marine Birds*, ed. E. A. Schreiber and J. Burger, 582–655. Boca Raton: CRC Press.

Warren, R. S., P. E. Fell, R. Rozsa, A. H. Brawley, A. C. Orsted, E. T. Olson, V. Swamy, and W. A. Niering. 2002. Salt marsh restoration in Connecticut: 20 years of science and management. *Restoration Ecology* 10:497–513.

Washburn, B. E., M. S. Lowney, and A. L. Gosser. 2012. Historical and current status of Laughing Gulls breeding in New York State. *The Wilson Journal of Ornithology* 124:525–530.

Washburn, B. E., S. B. Elbin, and C. Davis. 2016. Historical and current population trends of Herring Gulls (*Larus argentatus*) and Great Black-backed Gulls (*Larus marinus*) in the New York Bight. *Waterbirds* 39(1):74–86.

Watson, R. T., M. Fuller, M. Pokras, and W. G. Hunt. 2009. *Ingestion of Lead from Spent Ammunition: Implications for Wildlife and Humans*. Idaho: The Peregrine Fund.

Watts, B. 2013. Chesapeake Bay cormorants continue steep ascent. *Center for Conservation Biology Newsletter*. http://www.ccbbirds.org/2013/12/03/chesapeake-bay-cormorants-continue-steep-ascent/ (accessed March 5, 2015).

Watts, B. D., and M. A. Byrd. 2006. Status and distribution of colonial waterbirds in coastal Virginia: The 2003 breeding season. *Raven* 77:3–22.

Wayland, M., J. J. Smits, H. G. Gilchrist, T. Marchant, and J. Keating. 2003. Biomarker responses in nesting, common eiders in the Canadian arctic in relation to tissue cadmium, mercury and selenium concentrations. *Ecotoxicology* 12:225–237.

Watts, B. D., and B. R. Truitt. 2014. Spring migration of Red Knots along the Virginia Barrier Islands. *Journal of Wildlife Management* 79:288–295.

Wayland, M., and A. M. Scheuhammer. 2011. Cadmium in birds. In: *Environmental Contaminants in Wildlife: Interpreting Tissue Concentrations*, ed. W. N. Beyer and J. P. Meador, 645–668. Boca Raton, FL: CRC Press.

Wayland, M., H. G. Gilchrist, and E. Neugebauer. 2005. Concentrations of cadmium, mercury and selenium in common eider ducks in the eastern Canadian arctic: Influence of reproductive stage. *Science Total Environment* 351–352:323–332.

Wayland, M., K. A. Hobson, and J. Sirois. 2000. Environmental contaminants in colonial waterbirds from Great Slave lake, NWT: Spatial, temporal, and food-chain considerations. *Artic* 53:221–233.

Weber, D. N., and W. M. Dingel. 1997. Alterations in neurobehavioral responses in fishes exposed to lead and lead-chelating agents. *American Zoology* 37:354–362.

Weimerskirch, H. 2001. Seabird demography and its relationship with the marine environment. In: *Biology of Marine Birds*, ed. E. A. Schreiber and J. Burger, 115–135. Boca Raton: CRC Press.

Weinstein, M. P., and D. A. Kreeger. 2002. Preface, In: *Concepts and Controversies in Tidal Marsh Ecology*, ed. M. P. Weinstein and D. A. Kreeger, xv–xvi. New York: Kluwer Academic.

Weis, P., and J. S. Weis. 1977. Methylmercury teratogenesis in the killifish, *Fundulus heteroclitus*. *Teratology* 16:317–325.

Weis, J. S., and C. A. Butler. 2009. Salt marshes: A natural and unnatural history. New Brunswick, NJ: Rutgers University Press.

Weis, J. S., and P. Weis. 1998. Effects of exposure to lead on behavior of Mummichog (*Fundulus heteroclitus L.*) larvae. *Journal of Experimental Marine Biology and Ecology* 222:1–10.

Weis, J. S., L. Bergey, J. Reichmuth, and A. Candelmo. 2011. Living in a contaminated estuary: Behavioral changes and ecological consequences for five species. *Bioscience* 61:375–385.

Weis, J. S., G. Smith, T. Zhou, C. Santiago-Bass, and P. Weis. 2001. Effects of Contaminants on Behavior: Biochemical Mechanisms and Ecological Consequences. Killifish from a contaminated site are slow to capture prey and escape predators; altered neurotransmitters and thyroid may be responsible for this behavior, which may produce population changes in the fish and their major prey, the grass shrimp. *Bioscience* 51:209–217.

Weisberg, S. B., J. A. Ranasinghe, D. M. Dauer, L. C. Schaffner, R. J. Diaz, and J. B. Frithsen. 1997. An estuarine benthic index of biotic integrity (B-IBI) for Chesapeake Bay. *Estuaries* 20:149–158.

Weisberg, S. D., J. A. Ranasinghe, J. S. O'Connor, and D. A. Adams. 1998. A benthic index of biotic integrity (B-IBI) for the New York/New Jersey Harbor. Appendix C. for Sediment Quality of the NY/NJ Harbor System EPA/902-R-98-001.http://www.epa.gov/emap/remap/html/docs/nynjsedapp1.pdf (accessed April 26, 2015).

Wells, J. V., B. Robertson, K. V. Rosenberg, and D. W. Mehlman. 2010. Global versus local conservation focus of U.S. state agency endangered bird species list. *PLoS ONE* 5:e8608.

Welty, J. C. 1975. *The Life of Birds*. Philadelphia: W.B. Saunders.

Weseloh, D. V. C., and D. Moore. 2006. Abstract only: Variable impacts of cormorants on Black-crowned Night-Herons at four Ontario breeding colonies. Colonial Waterbirds of the New York/New Jersey Harbor. Fort Wadsworth, Staten Island, New York. http://nbiinin.ciesin.columbia.edu/jamaicabay/stake holder/HarborHerons_Cormorants_More_113006-120106.pdf (accessed April 18, 2015).

Weseloh, D. V., T. W. Custer, and B. M. Braune. 1989. Organochlorine contaminants in eggs of Common Terns from Canadian Great Lakes, 1981. *Environmental Pollution* 59:141–160.

Weseloh, D. V. C., D. J. Moore, C. E. Hebert, S. R. de Solla, B. M. Braune, and D. J. McGoldrick. 2011. Current concentrations and spatial and temporal trends in mercury in Great Lakes Herring gull eggs, 1974–2009. *Ecotoxicology* 20:1644–1658.

West, A. D., J. D. Goss-Custard, R. A. Stillman, R. W. Caldow, S. E. L. V. dit Durell, and S. McGrorty. 2002. Predicting the impacts of disturbance on shorebird mortality using a behaviour-based model. *Biological Conservation* 106:319–328.

Western Hemisphere Shorebird Reserve Network (WHSRN). 2015. WHSRN List of Sites. www.whsrn.org /sites/list-sites (accessed February 18, 2015).

White, D. H., and M. T. Finley. 1978. Uptake and retention of dietary cadmium in Mallard Ducks. *Environmental Research* 17:53–59.

White, D. H., M. F. Finley, and J. F. Ferrel. 1978. Histopathologic effects of dietary cadmium on kidneys and testes of Mallard Ducks. *Journal of Toxicology and Environmental Health* 4:551–558.

Whittow, G. C. 2001. Seabird reproductive physiology and energetics. In: *Biology of Marine Birds*, ed. E. A. Schreiber and J. Burger, 409–437. Boca Raton, FL: CRC Press.

Wickliffe, L. C., and P. G. R. Jodice. 2010. Seabird attendance at shrimp trawlers in nearshore waters of South Carolina. *Marine Ornithology* 38:31–39.

Wiener, J. G., and D. J. Spry. 1996. Toxicological significance of mercury in freshwater fish. In: *Environmental Contaminants in Wildlife: Interpreting Tissue Concentrations*, ed. W. N. Beyer, G. H. Heinz, A. W. Redmon-Norwood. Boca Raton, FL: SETAC, Lewis Publishing.

Wilcox, C., E. Van Sebille, and B. D. Hardesty. 2015. Threat of plastic pollution to seabirds is global, pervasive, and increasing. *Proceedings of the National Academy of Sciences of the United States of America*. Published online before print August 31, 2015, doi: 10.1073/pnas.1502108112.

Williams, R. N. 2006. *Return to the River: Restoring Salmon to the Columbia River*. New York: Elsevier.

Williams, B., D. F., Brinker, and B. D. Watts. 2007. The status of colonial nesting wading bird populations within the Chesapeake Bay and Atlantic barrier island–lagoon system. *Waterbirds* 30:82–92.

Wilson, S. G., and T. R. Fischetti. 2010. Coastline Population Trends in the United States: 1960 to 2008 Population Estimates and Projections. U.S. Bureau of Census. http://www.census.gov/prod/2010pubs /p25-1139.pdf (accessed February 6, 2015).

Wilson, B., and B. Vermillion. 2006. Habitat mosaics to meet the needs of priority Gulf coast birds. US Fish Wildlife Service, Gulf Coast Joint Venture. https://www.fws.gov/southwest/AboutUs/PDFs/Region 2AllEmpMtngJuly2014BWilson(1).pdf (accessed March 14, 2016).

Winston, T. 2014. New York City Audubon's Harbor Herons Project: 2014 Nesting Survey Report. New York City Audubon, New York. http://www.harborestuary.org/reports/harborheron/2014_HH_Interim_Survey_ Report.pdf (accessed March 14, 2016).

Withers, K. 2002. Shorebird use of coastal wetlands and barrier island habitat in the Gulf of Mexico. *Science World Journal* 2:514–536.

Wolf, S. G., M. A. Snyder, W. J. Sydeman, D. F. Doak, and D. A. Croll. 2010. Predicting population consequences of ocean climate change for an ecosystem sentinel, the seabird Cassin's auklet. *Global Change Biology* 16:1923–1935.

Wolfe, M. F., T. Atkeson, W. Bowerman, J. Burger, D. C. Evers, M. W. Murray, and E. Zillioux. 2007. Wildlife indicators. In: *Ecosystem Responses to Mercury Contamination: Indicators of Change*, ed. R. Harris, D. P. Krabbenhoft, R. Mason, M. W. Murray, R. Reash, and T. Saltman, 123–190. Boca Raton, FL: CRC Press.

Wolfe, M. F., S. Schwarzbach, and R. A. Sulaiman. 1998. Effects of mercury on wildlife: A comprehensive review. *Environmental Toxicology and Chemistry* 17:146–160.

Wolkovich, E. M., and E. E. Cleland. 2011. The phenology of plant invasions: A community ecology perspective. *Frontiers in Ecology & the Environment* 9:287–294.

Wolkovich, E. M., B. I. Cook, and T. J. Davies. 2013. Progress towards an interdisciplinary science of plant phenology: Building predictions across space, time and species diversity. *New Phytologist* 201:1156–1162.

Work, T. M., and M. R. Smith. 1996. Lead exposure in Laysan Albatross adults and chicks in Hawaii: Prevalence, risk factors, and biochemical effects 1996. *Archives Environmental Contamination and Toxicology* 24:478–482.

World Health Organization (WHO). 1990. IPCS—Methylmercury. *Environmental Health Criteria* 101:42–58.

World Health Organization (WHO). 1991. IPCS—Inorganic mercury. *Environmental Health Criteria* 118:30–50.

World Health Organization (WHO). 2013. Climate change and human health: Ecosystem goods and services for health. http://www.who.int/globalchange/ecosystems/en/ (accessed March 14, 2016).

Wren, C. D., S. Harris, and N. Harttrup.1995. Ecotoxicology of mercury and cadmium. In: *Handbook of Toxicology*, ed. D. J. Hoffman, B. A. Rattner, G. A. Burton, and J. Cairns, 392–423. Boca Raton, FL: Lewis Publishers.

Wu, L., J. Chen, K. K. Tanji, and G. S. Banuelos. 1995. Distribution and biomagnification of selenium in a restored upland grassland contaminated by selenium from agricultural drain water. *Environmental Toxicology and Chemistry* 14:733–742.

Xing, M., P. Zhao, G. Guo, Y. Guo, K. Zhang, L. Tian, Y. He, H. Chai, and W. Zhang. 2015. Inflammatory factor alterations in the gastrointestinal tract of cocks overexposed to arsenic trioxide. *Biological Trace Element Research* 167:288–299.

Yang, D. Y., Y. W. Chen, J. M. Gunn, and N. Belzile. 2008. Selenium and mercury in organisms: Interactions and mechanisms. *Environmental Reviews* 16:71–92.

Yang, S., Z. Zhang, J. He, J. Li, J. Zhang, H. Xing H., and S. Xu. 2012. Ovarian toxicity induced by dietary cadmium in hen. *Biological Trace Element Research* 148:53–60.

Yarnold, D. 2014. The challenge. It's time to act. *Audubon Magazine* 116. http://www.audubon.org/magazine/september-october-2014/its-time-act (accessed February 19, 2016).

Yeardley, R. B., J. M. Lazorchak, and S. G. Paulsen. 1998. Elemental fish tissue contamination in northeastern US lakes: Evaluation of an approach to regional assessment. *Environmental Toxicology and Chemistry* 17:1875–1884.

Zamanii-Ahmadmahmoodi, R., M. Alahverdi, and R. Mirzaei. 2014. Mercury concentrations in Common Tern *Sterna hirundo* and Slender-billed Gull *Larus genei* from the Shadegan Marshes of Iran, in the northwestern corner of the Persian Gulf. *Biological Trace Elements Research* 159:161–166.

Zarudsky, J. 1981. Forster's Tern breeding on Long Island. *Kingbird* 31:212–213.

Zhang, K., B. C. Douglas, and S. P. Leatherman. 2004. Global warming and coastal erosion. *Climatic Change* 64:41–58.

Zhang, W., L. Huang, and W. X. Wang. 2012. Biotransformation and detoxification of inorganic arsenic in a marine juvenile fish *Terapon jarbua* after waterborne and dietborne exposure. *Journal of Hazardous Materials* 221:162–169.

Zhang, Y., L. Ruan, M. Fasola, E. Boncompagni, Y. Dong, N. Dai, C. Gandini, E. Orvini, and X. Ruiz. 2006. Little egrets (*Egretta garzetta*) and trace-metal contamination in wetlands of China. *Environmental Monitoring and Assessment* 118:355–368.

Zillioux, E. J., D. B. Porcella, and J. M. Benoit. 1993. Mercury cycling and effects in freshwater wetland ecosystems. *Environmental Toxicology and Chemistry* 12:2245–2264.

Zimmerman, J., S. F. Michels, D. Smith, and S. Bennett. 2014. Horseshoe Crab spawning activity in Delaware Bay: 1999–2007. Report to the ASMFC Horseshoe Crab Technical Committee, July 15, 2014.

Index

Page numbers followed by f and t indicate figures and tables, respectively.